Conjugate Construction of Quantum Optics: From Foundations to Applications

Jeffrey Zheng (郑智捷)

Conjugate Construction of Quantum Optics: From Foundations to Applications

量子光学的共轭体系基础及其应用

Quantum Measurement Foundation on Multiple Variables of Complex Vector Function

nova
science publishers
New York

Library of Congress Cataloging-in-Publication Data

ISBN: 979-8-88697-651-9

Published by Nova Science Publishers, Inc. † New York

Dedicated to

I Ching — First Conjugate Construction
Xinjian Zheng — Grand Father
Lanze Yin & Sumin Zheng — Mother & Father
Qingshi Gao — Mentor on Parallel Algorithm & Architecture
Tasiyasu L. Kunii — Master on Meta Knowledge
Bob Beaumont — Adviser on Optimization
Shoujue Wang — Master on Multivalued Logic
Youping Li — Master on Measuring Time Order
Harold Szu — Master on Natural Intelligence
Ping Zhang — Wife

Contents

Preface

One Century Debates of Quantum Entangled Measurements

Since the 1920s, the debate on quantum entanglement has become the most important topic of the quantum foundation. In 1935, Einstein proposed the EPR paradox and concluded that quantum mechanics is incomplete; Bohr disagreed with Einstein's conclusion and actively fought back in his lifetime.

In 1964, Bell proposed the Bell inequality on probability measurements influencing a group of scientists. After a series of advanced quantum entangled global experiments from the 1970s, the experimental results successfully indicate the measures that do not meet the Bell inequality and definitely support quantum mechanics. The set of negative experiments causes popular opinions to believe that Einstein's conclusion is wrong.

In 2022, three top experts won the Nobel Prizes in Physics for their contributions to advanced quantum entanglement theory and experiments.

This book provides structural evidence on the conjugate transformation structure and conjugate complex vector system to support the Einstein conclusion. **Quantum Mechanics Is Incomplete!**

Incomplete Evidence in History

Following de Moivre, Euler works to trace the incomplete history of complex functions and geometric paradoxes. Gauss provided a rigorous proof of the fundamental theorem of algebra and named a pair of real and imaginary numbers as a complex number.

n Coefficients $\in R \leftrightarrow n$ Solutions $\in C$	Gaussian Theorem
$\{x, x^*\} \in C \leftrightarrow \{\theta, -\theta\} \in R, \theta = \angle x, -\theta = \angle x^*$	Complex Conjugate Pair

Hamilton, de Morgan, Clifford, Heaviside, Mobius, Listing, Klein, and Poincare followed Gaussian paths in both complex functions and logic/geometric paradoxes, non-Euclidean geometry; Poincare proposed upper half-plane UHP and lower half-plane LHP for complex hypergeometric functions.

$$x \in C \rightarrow \left\{ \begin{array}{ll} x & +\theta \\ x^* & -\theta \\ r & = |x| \end{array} \right\} : \left\{ \begin{array}{c} \{x^* x = x x^* = 1\} \\ \{\theta, -\theta\} \in R \end{array} \right\} \rightarrow \frac{+\theta \in UHP}{-\theta \in LHP} \rightarrow \frac{\{re^{i\theta}\}}{\{re^{-i\theta}\}}$$

Classical Quantum Logical and Mathematical Foundation

During the initial period of quantum mechanics, quantum theory creators Planck, Lorentz, Einstein, Bohr, Heisenberg, Schrodinger, Dirac, Born, Pauli, Weyl, and von Neumann all agreed that only real eigenvalues can be observed and measured. They discovered a list of algebra and differential expressions such as the Heisenberg matrix and Schrodinger/Dirac equations on pair complex conjugate parameters to resolve pair real eigenvalues as solutions.

In the 1920s, Hilbert provided a complex inner product space in a complete metric space with an orthogonal complement of eigenvalue solutions on complex fields to be the infinite-dimensional bilinear matrices in Hilbert spaces to support quantum exploration in the twentieth century as the mathematical foundation to reverse the Gaussian fundamental theorem of algebra.

$x, y \in C, \langle x, y\rangle = \overline{\langle y, x\rangle}, \parallel x \parallel = \sqrt{\langle x, x\rangle} \leftrightarrow \langle \mathbf{z}, \mathbf{w}\rangle = \sum_{n=1}^{\infty} z_n \overline{w}_n, \sum_{n=1}^{\infty} \parallel z_n \parallel < \infty$
$\{n$ Pair Variables$\} \in C \leftrightarrow \{$Hilbert Space + Lie Brackets$\} \leftrightarrow \{n$ Eigenvalues$\} \in R$

In 1995, a single quantum bit, qubit, was proposed by Schumacher and Wootters that expresses two orthogonal basis vectors and their combinations. The qubit is well accepted by early explorers, including Deutsch, Bennett, Wieser, Yao, Shor, Hruby, and Reimer, in quantum Turing machines, quantum information, quantum computing, and qubit field theory environments.

Single Qubit	$\left\{ \vert 0\rangle = \begin{pmatrix}1\\0\end{pmatrix}, \vert 1\rangle = \begin{pmatrix}0\\1\end{pmatrix} \right\}$	Two Orthogonal Basis Vectors
Multiple n Qubits	2^n Configurations	Quantum Information Phase Space
$\{n$ Pair Variables$\} \in C$	$\{+1, -1\}^n$ Solutions	$\{n$ Eigenvalues$\} \in R$

Complete Quantum Logical and Mathematical Foundation

In 1994, the author's PhD thesis proposed conjugate classification and transformation of regular plane lattices for binary images including conjugate transformation structure and conjugate algebra. Let a conjugate or complete quantum bit / a complete qubit be a cobit.

$2n$ Conjugate Vectors: $\{\Gamma_j(X)\}_{j=0}^{n}, \{\tilde{\Gamma}_k(X)\}_{k=0}^{n}$	$2n$ Feature Vectors
n Pair Indices: $\langle A, B\rangle X = \{\{\Gamma_j(X)\}_{j\in A}, \{\tilde{\Gamma}_k(X)\}_{k\in B}\}$	$2n$ Feature Indices
Trans. Equation: $F(\langle A, B\rangle X) = (\Gamma(X) \cap (\neg\Gamma_A(X))) \cup (\neg\tilde{\Gamma}_B(X))$	$\{2n$ Vars$\} \in \{0, \tilde{0}, 1, \tilde{1}\}^n$
$\{n$ Pair Measuring Vectors$\} \in C \rightarrow \{+1, +i, -1, -i\}^n$ Solutions	$\{2n$ Eigenvalues$\} \in C$

Both constructions can be compared in the total number of phase spaces.

Classical Foundation: n Qubits $\rightarrow 2^n$ Phase Space $\equiv \{+1, -1\}^n$ Eigenvalues
Conjugate Foundation: n Cobits $\rightarrow 2^{2n}$ Phase Space $\equiv \{+1, +i, -1, -i\}^n$ Eigenvalues

Three logic foundations, {Classic Logic, Quantum Logic, Conjugate Logic}, can be compared as three sets $\{\{0,1\},\{|0\rangle,|1\rangle\},\{(1,0),(0,1),(-1,0),(0,-1)\}\}$ on three metric coordinates.

$$\begin{array}{|c|}\hline \begin{matrix}\text{Classic}\\ \text{Logic}\end{matrix} : 0-1 \quad \begin{matrix}\text{Quantum}\\ \text{Logic}\end{matrix} : \begin{matrix}|1\rangle & \\ | & \\ \circ & -\ |0\rangle\end{matrix} \quad \begin{matrix}\text{Conjugate}\\ \text{Logic}\end{matrix} : \begin{matrix} & (0,1) & \\ & | & \\ (-1,0)\ - & \circ & -\ (1,0) \\ & | & \\ & (0,-1) & \end{matrix} \\ \hline\end{array}$$

The foundation of the conjugate complex vector system is composed of four quadrants of attribute regions and eight feature vectors in four conjugate pairs.

$$x \in C \rightarrow \begin{Bmatrix} y & \theta \\ \tilde{y} & \tilde{\theta} \\ r = |y| & \end{Bmatrix} \rightarrow \begin{Bmatrix} YY & Y\tilde{Y} \\ \tilde{Y}Y & \tilde{Y}\tilde{Y} \end{Bmatrix} \rightarrow \frac{\{2,\tilde{2}\} \,\|\, \{0,\tilde{0}\}}{\{3,\tilde{3}\} \,\|\, \{1,\tilde{1}\}} \rightarrow \frac{\left\{re_2^{i\theta}, re_2^{i\tilde{\theta}}\right\} \,\|\, \left\{re_0^{i\theta}, re_0^{i\tilde{\theta}}\right\}}{\left\{re_3^{i\theta}, re_3^{i\tilde{\theta}}\right\} \,\|\, \left\{re_1^{i\theta}, re_1^{i\tilde{\theta}}\right\}}$$

Three phase spaces can be compared in the n variables condition and three eigenvalue solutions.

Classical Phase Space : $\boxed{\{0,1\}/\text{bit}} \rightarrow \boxed{n \text{ bits}} \rightarrow$ 2^n States $\{0,1\}$ Linear Combination Solutions

Quantum Phase Space : $\boxed{\{|0\rangle,|1\rangle\}/\text{qubit}} \rightarrow \boxed{n \text{ qubits}} \rightarrow$ 2^n States $\{1,-1\}$ Real Number Eigensolutions

Conjugate Phase Space : $\boxed{\{0,1,\tilde{0},\tilde{1}\}/\text{cobit}} \rightarrow \boxed{n \text{ cobits}} \rightarrow$ 2^{2n} States $\{1,i,-1,-i\}$ Complex Eigensolutions

There are **Huge Gaps** between classical and conjugate foundations.

Classic Bases $2^n \ll 2^{2n}$ **Conjugate Bases**

$$(2^n)^2 = 2^n \times 2^n = 2^{2n} = 4^n$$

Therefore, there are so many ghosts and paradoxes in classical quantum mechanics and quantum information.

Book Organization

The book is composed of three parts I–III: I. Conjugate Foundation on Complex Number Plane; II. Applications on Entangled Measurements of Quantum Optics; III. Appendices.

Part I Conjugate Foundation on Complex Number Plane

In Part I, four chapters are arranged as follows: Chapter 1. Conjugate Operators of Complex Vector Functions on Complex Number Plane; Chapter 2. Measurement Operators of Conjugate Transformation Structure on Complex Plane; Chapter 3. Conjugate Complex Measurements of Clock Reference Signals; and Chapter 4. Combinatorial Lattice Clusters of Conjugate Complex Vector Functions

Fundamental Conjugate Complex Vector System

Chapter 1, proposes a parallel model on synchronous conditions to expand the classical complex vector system from two complex conjugate vectors to eight conjugate vectors to have a pair of conjugate vectors in each quadrant. Full recursive operations are described in conjugate complex vector systems. The description of the whole structure is composed of six lemmas, twenty-four theorems, and sixteen corollaries.

The foundation of the conjugate complex vector structure is established.

For a complex number $x = (p, q), r = \sqrt{p^2 + q^2}$, the fundamental theorem of algebra is a necessary and sufficient condition to correspond to the solution space on the condition of $\{$real coefficients$\} \leftrightarrow \{+\theta, -\theta\}$ and $\{x, x^*\}$ a pair of measures, complex vectors, and complex conjugate operator $*$.

Gauss's fundamental theorem of algebra is incomplete in covering the solution spaces under the coefficients on imaginary numbers to naturally contain the Mobius ring and intrinsic logic paradoxes $A = \neg A$ in the phase space.

Four conjugate operators $\{\sim, \sharp, \prime, *\}$ are defined associated with a pair of conjugate angles $\{\theta, \tilde{\theta}\}$ and the four attribute regions in a complex plane on the synchronous parallel processing and recursive organization conditions. Eight feature vectors are composed of two crosses, and the whole construction provides six dynamic transformation modes associated with the basic recursive operations.

Based on the four distribution regions (four colors of the conjugate ring), a pair of conjugate vectors (opposite rotations), and other comprehensive approaches, this extends the fundamental theorem of algebra to cover $\{$complex coefficients$\}$ in a complete solution space. Conjugate complex vector systems guarantee without logic paradoxes on the vector phase space in synchronous dynamic environments.

From {0,1} Vectors on Conjugate Transformation Structure to Conjugate Algebra

In Chapter 2, the description of the whole structure is composed of three lemmas, twenty-two theorems, and seven corollaries in two parts. The first part describes the eight complex vectors in a pair of conjugate feature classes generated by the conjugate transformation structure from $\{0, 1\}$ elements of an N length sequence.

The conjugate transformation structure processes the initial 0–1 vector of N elements to be $2n$ feature classes and transforms feature vectors into 2^{2n} configurations. To output a pair of conjugate classes, eight elements are conjugate complex number vectors.

The second part describes expanding operations to conjugate algebra with complete brackets identifying symmetric/antisymmetric measurements with the time order.

Poisson, Lagrange, Lie, and Dirac brackets support multiple real eigenvalue solutions $\{-1, +1\}$ of a dynamic system to identify symmetric/antisymmetric properties of the principal diagonal eigenvalues on the Hermite matrices, and those that ignore multiple imaginary solutions of the auxiliary diagonal eigenvalues completely.

Using conjugate algebra, conjugate brackets provide the complex eigenvalue solutions $\{+1, +i, -1, -i\}$, $\{+1, -i, -1, +i\}$ beyond Poisson, Lagrange, Lie, and Dirac brackets to identify vector operators $\{-1, \leftarrow\}$ of symmetric/antisymmetric properties under {Group, Ring, Field} domains. The time order satisfies vector and tensor operational environments under a single circle of the eigenvalue matrix condition.

Precise Clock Signals on Complex Planes

In Chapter 3, clock reference signals sufficiently provide the controllable quantitative bases of the precise measuring foundation on a complex number plane.

The description of the whole structure is composed of one lemma, ten theorems, and eight corollaries.

Conjugate measurements of clock reference signals provide the conjugate complementary measurement space, the narrow projection on the conjugate configuration, and the general projection on the complex number plane.

A narrow projection provides 2^{2m} positions or 4×2^{2m} positions in a configuration space of one or four 2D uniform distributions without a metric order, and a general projection provides $(m + 1)^2$ positions or $4(m + 1)^2$ positions in 2D double-binormal distributions corresponding to a complex number plane on one quadrant or four quadrants.

Strong Interactions on Conjugate Group Combinatorial Lattice

In Chapter 4, a conjugate group combinatorial lattice is proposed. The description of the whole structure is composed of one lemma, thirteen theorems, and twelve corollaries.

Three extension steps are required from a single vector to generate a conjugate group combinatorial lattice. Step 1. The conjugate Klein group expands one vector to eight vectors of two conjugate crosses in a conjugate cluster; Step 2. Four vectors of each cross take sixteen combinations to generate nine nontrivial vectors; Step 3. Two sets of eighteen nontrivial vectors create an integrated Abel group under complex addition to generate a conjugate group combinatorial lattice.

Normally entangled interactions appear in the eight vectors of a conjugate cluster for the conjugate Klein group in Step 1, and extended entanglements are restricted in the same eigenvalue on a cycle with a constant radius r. However, the extensions in Steps 2 and 3 generate multiple cycles under combinatorial and Abel group operators to indicate that there are strong interactions involved beyond a single-cycle condition.

Part II Applications on Entangled Measurements of Quantum Optics

In Part II, three chapters are arranged as follows: Chapter 5. Modern Progress in Entangled Measurements of Quantum Optics; Chapter 6. Entanglement Foundation of Quantum

Optics on Conjugate Phase Space; Chapter 7. Conjugate Construction in Future Explorations.

Historic Development of Complex Conjugate Systems and Entangled Measurements

In Chapter 5, the historic development of complex conjugate systems, complex geometry, multiple variable complex functions, and entangled measurements of quantum optics are reviewed. The description of the whole structure is composed of twenty-two lemmas and twenty corollaries.

The frontier quantum theory system and its corresponding various representations of the quantum entangled state are reviewed, and their typical complex geometric invariants satisfying the characteristics of global transformation are described.

Wootters have listed various measurement schemes to indicate entanglement vectors without orthogonal properties in quantitative measuring conditions.

In modern quantum optics theory and experimental devices, various reflection mirrors play a key role in implementing basic complex operators in quantum optical and photonic environments.

For example, a full-reflection mirror implements an imaginary operator i, and a semi-transparent mirror implements a conjugate operator $\sim$.

The proving procedure of the **Hardy Theorem** is a perfect case to illustrate a series of complex operators applied as the basic reasons in the derivation.

Foundation of Quantum Optics on Entangled Measurements

Chapter 6 is divided into three parts. The first part discusses the core matrix structure of quantum interaction. The second part summarizes the most important results of quantum entanglements during 2002–2022.

The third part describes the necessary and sufficient conditions of quantum entanglement to establish a self-consistent description of space and time in the quanta optical measurement system under a formal description of the measurement paradox-free. The description of the whole structure is composed of five lemmas, six assumptions, six predictions, sixteen theorems, and sixteen corollaries.

Quantum optics and quantum information under Dirac brackets provide a necessary and sufficient condition to support {Real Eigenvalue} solutions of $\{+1, -1\} \leftrightarrow \{+\theta, -\theta\}$ and $\{x, x^*\}$ a pair of measures, complex vectors, and complex conjugate operator $*$.

The general solution of the Dirac bracket and Lie algebra schemes is a reverse version of the Gaussian fundamental theorem of algebra. It is incomplete to cover the complex solution spaces under the parameters of any imaginary numbers containing the Mobius ring and intrinsic logic paradoxes $A = \neg A$ in the phase space.

From primitive distributions, classical logic is a linear structure, quantum logic is generalized to be two vectors, and conjugate logic is generalized to be four meta-vectors on balanced distributions. The numbers of state spaces involved in the n meta-variables are $\{2^n, 2^n, 2^{2n}\}$.

From Lie algebra to conjugate algebra, the classical quantum formula simplifies the eight-variable vectors into two vectors on a pair of complex conjugate representations. The complete quantum solution needs to be equipped with principal and auxiliary double diagonal complex eigenvalues, and the n qubit phase space representation expands as the n cobit (cobit: complete or conjugate qubit) phase space representation from 2^n to 2^{2n}.

Eight vectors of a conjugate cluster are located in the four attribute regions; in the same region, paired conjugated states provide measurable characteristic information carried in the representation of paired entangled states in the description of quantum information interaction.

Select a single diamond-shaped loop on the complex plane and use the spider web transformation of the square root mode to determine the density matrix of each coordinate point on the complex plane of the accompanying projected density values, forming a circular feature visualization distribution.

Both complete distribution and product distribution are superimposed to be a distribution with maximum values in eight directions $\{0, \pi/4, \pi/2, \cdots, 7\pi/4\}$ that can be decomposed into two conjugate crosses.

Through $n = 1$ cobweb transformation, the density matrix in the complex plane shows from diamond to rectangular feature distributions, and fine eigenvalues are determined by the quantization scales of real and imaginary parts of the input vector and the number of iterations of the interactions.

Future Exploration of Conjugate Complex Vector Systems

Chapter 7 explores a series of problems related to conjugate vector algebra operations from several aspects, including classical paradox, modern logic, topological geometry, visual paradox, classical Tai Chi, resolving Mobius paradox, holistic conjugate representation and measurement of coherent states, geometry of complex numbers, modern representation of complex conjugate numbers, theory of complex functions, mathematical physics, complex function equations, variational methods, two-body entanglement and time-reversal symmetry, symmetry + supersymmetry criteria, and variant constructions + knowledge model.

The 0-1 vector logic is established, leading to the unparalleled support of the multivariable complex function system and the basic module of the description system. Future challenges are predicted for further exploration. The description of the whole structure is composed of ten lemmas, twelve theorems, eleven corollaries, and three predictions.

Sample Results

Sample statements are selected.

Reflective Mirrors ≡ Basic Complex Operators In modern quantum optics theory and experimental devices, various reflection mirrors play a key role in implementing basic complex operators in quantum optical and photonic environments.

Paradox in Real Eigenvalue Solutions The general solution of the Dirac bracket and Lie algebra schemes is a reverse version of the Gaussian fundamental theorem of algebra. It is incomplete to cover the complex solution spaces under the parameters of any imaginary numbers containing the Mobius ring and intrinsic logic paradoxes $A = \neg A$ in the phase space.

Classical Quantum Formula ≡ Simplified Conjugate Structure From Lie algebra to conjugate algebra, the classical quantum formula simplifies the eight-variable vectors into two vectors on a pair of complex conjugate representations. The complete quantum solution needs to be equipped with principal and auxiliary double diagonal complex eigenvalues, and the n qubit phase space representation expands as the n cobit phase space representation from 2^n to 2^{2n}.

Variant Construction Variant construction is composed of variant logic, variant measurement, variant map, and hierarchical knowledge model, including description + process knowledge model, local + global analysis structure, and classification + retrieval mechanism. According to the relevant local characteristics of the system, variant construction forms a foundation to provide integrated holistic structural support on the multiple levels of a hierarchy on complex system analysis, design, implementation, and other engineering activities.

Predictions

Three predictions A–C are provided.

A. Conjugate Complex Vector Solutions Useful in Vortex Mechanisms Conjugate complex vector functions provide powerful synchronous functions to recursively generate dynamic effects of increasing and decreasing radii in the complex plane to provide a foundation to support future exploration and applications on complicated vortex mechanisms.

B. Coming General Computers Modern computers are restricted in Type B $n = 0, State = 2$ mode on $\{0, 1\}$ Turing machine in 1D real manifolds. The coming general computers are based on Type {A,B} $n = 1, State = 4$ mode on full $\{(1, 0), (0, 1), (-1, 0), (0, -1)\}$ eigenvalue solutions on conjugate machines to support 1D complex manifolds for complete quantum computations.

C. Extensive Applications on Supercomputation and Visualizations Resolved Mobius paradoxes will be useful to implement higher performance supercomputing and visualization targets on conjugate complex vector systems in hypersynchronous network environments for knowledge and intelligent content management for subsequent generations.

Part III Appendix

In Part III, seven subappendixes are listed for each chapter to list relevant important results in lemmas, theorems, and corollaries.

Appendix 1: Lemmas 1–6, Theorems 1–24, and Corollaries 1–16 in Chapter 1
Appendix 2: Lemmas 1–3, Theorems 1–22, and Corollaries 1–7 in Chapter 2
Appendix 3: Lemma 1, Theorems 1–10, and Corollaries 1–8 in Chapter 3
Appendix 4: Lemma 1, Theorems 1–13, and Corollaries 1–12 in Chapter 4
Appendix 5: Lemmas 1–22 and Corollaries 1–20 in Chapter 5
Appendix 6: Lemmas 1–5, Assumptions 1–6, Predictions 1–6, Theorems 1–16, and Corollaries 1–16 in Chapter 6
Appendix 7: Lemmas 1–10, Theorems 1–12, Corollaries 1–11, and Predictions 1–3 in Chapter 7

Jeffrey Zheng
Kunming, Yunnan China

November 2022

Acknowledgments

I would like to thank all respected teachers: Xinjian Zheng, Shiying Li, Lanze Yin, Sumin Zheng, Fumin Zheng, Aiyue Jin, Xiuwen Zheng, Wenqing Liu, Zemin Zheng, Qin Tu, Tianmin Zheng, Lixia Li, Zhize Yin, Huiyin Sun, Runze Yin, Shiding Yang, Yaoze Yin, Wenzhen Zhang, Kangze Yin, Zhanchun Yang, Dahui Li, Yongnian Dai, Ruifang Zhao, Hanqiao Jiang, Linling Sheng, Dexiang Zhu, Wen-an Xu, Jixi Yang, Shangjie Tang, Guojun Ren, Naijian Zhang, Jiamo Liu, Dasong Zhang, Qingshi Gao, Yunmei Dong, Xiang Zhang, Jikang Wu, Tingjun Wang, Shaobai Zhou, Baoji Yuan, Guangnan Ni, Kai Hwang, Daozheng Wei, Renji Tao, Xuyan Tu, Kingsun Fu, Tse-yun Feng, Yaohan Chu, Ruwei Dai, Yanpei Liu, Xiru Cheng, Shuqing Chen, Hoon Heng Teh, Wellington Yu, Kaiching Chu, Ifey Chang, Bengchin Ooi, Anthony Meader, Rod Worley, Bihn Pham, John Chrossley, Chris Wallace, Terry Caelli, Les Katchen, Tasiyasu L. Kunii, Clement Leung, John Mo, Laszlo Nemes, Bob Beaumont, Jerry Madakbas, Shoujue Wang, Harold Szu, Zhenya He, Deyi Li, Barry Robson, Shuwan Lu, Youping Li, Xijin Xiong, Xuemou Wu, Hugh Ching, Ni-bin Chang, Ling-an Wu, Weimou Zheng, Yongde Zhang and Chenghong Wang, and other mentors for encouragement and help in over sixty years of conjugate complex exploration.

I would like to thank my colleagues: Jianzhong Liu, Chris Zheng, Gang Liu, Yi Liu, Yaoheng Xiong, Jiming Chen, Weiming Zhou, Jiangyun Cui, Xinshi Peng, Bo Li, Kang Dai, Yun Dai, Wen Zhang, Weizong Zhu, Jiawen Tang, Jiaping Tang, Peixian Zhu, Yifang Zhang, Xiaoyun Kang, Rong Yang, Zhenchuan Shi, Zhengzhong Ping, Dapeng Tian, Wilson Wen, Benjian Lin, Rongquan Wang, Xianzhao Zeng, Xiangzhen Qiao, Xiaolan Tang, Zhiyong Liu, Yeqin Hou, Mingtao Zhou, Feibi Xia, Weizhong Shao, Xue Wang, Peizhuang Wan, Yongling Shen, Zhilie Li, Henry Wu, Xingdong Wu, Guozhu Dong, Tao Li, Bing Li, Ming Yan, Alice Guo, Tao Chen, Tong Li, Yixian Yang, Lizhen Li, Zhengfu Han, Chunhua Miao, Lian Lu, Yinfu Xie, Shu Zhang, Xiazhou Yang, Xiaoyun Pu, Bing Zhang, Yiming Xiao, Zhaotian Zhang, Siyu Zheng, Yimin Li, Weilian Wang, Fei Gao, Siwen Bi, Dennis Heim, Olga Heim, Xing Chen, Shijun Xin, Jin Zhang, Tianrong Zhang, Hepu Deng, Dawu Gu, Yuncai Wang, Runyao Duan, Changgeng Peng, Yanli Shi, Yi-nan Fang, Zangi, Xing Chu, Niansheng Tang, Zhenyu Guo, Xiubo Chen, Xiuguo Yang, Fanyao Qu, Li Wang, Licheng Wang, Ying Zhao, Lu Shan, Yueping Wang, Shiyun Tian, Wei Hao, Ping Lin, Guojie Li,Huimin Lin, Shijia Zhou, Wenqi Liu, Qin Liu, Zhigang Zhang, Desheng Li, Jin Li, Weidong Ma, Yuzhu Rao, Nanhua Shi, Lin Zhang, Jinming Hu, Weizhong Yang,Shipu Wang, Wei Zhou, Shaowen Yao, Yu Lian, Wufu Li, Yujian Pei, Chao Zheng, Veronica Lieseputra, Xiaoxue Zhong, Yanzhen Qu, Wei Wang, Yun Yang, Jianxue Yan, Hua Zhang, Zizhen Zhang, Fei Yan, Heng Yang, Yi Yang, Han Han,

Jiawen Qiu, and Kunhui Hong for their criticism, encouragement, suggestions, corrections, and discussions for the early version of the book.

I am particularly grateful to my students for the past twenty years: Tingting Chen, Liting Yao, Bingjing Cai, Wenjia Zhao, Qin Kang, Binyu Yan, Yeyu Yin, Qinping Li, Zhiqian Yu, Yao Zhou, Jie Wan, Huan Wang, Jie-ao Zhu, Qinxian Bu, Weiqiong Zhang, Zu Wan, An Wang, Yuqian Liu, Lei Du, Ruoyu Shen, Heyuan Chen, Yan Ji, Guoxiu Zhai, Pingan Zeng, Wenjia Liu, Ruoxue Wu, Lixin Wu, Zhonghao Yang, Lihua Leng, Zhihui Hou, Yuyuan Mao, Yamin Luo, Zhefei Li, Yifeng Zheng, Chao Fan, Mu Qiao, Huanxian Zheng, Xin Zhang, Liuyun Du, Tao Li, Zhongwei Zhang, Yang Zhou, Zhenhui Wang, Xinmei Li, Tan Yao, Fan Peng, Rui Wang, Qi Wang, Yujin Lu, Wanyi Zhou, Yangyang Tao, and many other students in a series of exploring projects on higher performance of vector logic, computation, and visualization for conjugate and variant schemes in frontier research for various typical applications.

I would like to thank following the organizations: Factory of Medical Devices, Kunming Medical Colleague (1971–1978); Department of Physics, Yunnan University (1978); Graduate School, University of Science and Technology of China USTC (1978–1981); Institute of Computing Technology, Chinese Science Academy (1981–1987); Institute of Systems Science, National University of Singapore (1987–1990); Department of Computer Science, Monash University (1991–1994); Department of Mathematics and Computation, Victoria University of Technology (1994–1997); Institute of Manufactural Science & Technology, CSIRO (1998–2002); School of Software, Yunnan University (2004–2022); Key Laboratory of Software Engineering of Yunnan; Key Laboratory of Quantum Information of Yunnan; International Association of Advanced Materials IAAM; IEEE, SIPE, ACM, SIAM, CACR, IACR, CGI, InTech Press, Springer Verlag Press, ResearchGate, ResearchSquare; RNA/DNA sequences Nextstrain, GISAID, NCBI GenBank, CNGBdb, {USTC, CAS ANU} quantum random number server for quantum cryptographic sequences, and other organizations for their comprehensive support for earlier exploration and activities on conjugate and variant construction over the past fifty years.

I sincerely thank the following four partial funding resources for supporting the early work of the book.

NSF of China (61362014,62041213)
The Key Project of Yunnan on Quantum Sequence Measurements (2018ZI002)
Yunnan Advanced Overseas Scholar Project
Australian Commercialising Emerging Technologies (COMET) program

Finally, I thank Pinterest ... Springer Nature Press for the copyright permission to use some interesting pictures for comparison in the book.

List of Contributors

Jeffrey Zheng
Conjugate System Research e-mail: conjugatelogic@yahoo.com

Jianzhong Liu
Conjugate System Research e-mail: liujianz6655@126.com

Youping Li
Chinese Academy of Engineering Physics e-mail: liyp@cac.cn

Qi Wang
Yunnan University e-mail: 173216604@qq.com

Yujin Lu
Yunnan University e-mail: 839639760@qq.com

Yangyang Tao
Yunnan University e-mail: innogentao@163.com

Wanyi Zhou
Yunnan University e-mail: evanchow1024@outlook.com

Fan Peng
Yunnan University e-mail: fanpengv@163.com

Rui Wang
Yunnan University e-mail: richyvong0828@gmail.com

Part I

Conjugate Foundation on Complex Number Plane

Awareness of ignorance is beginning of wisdom
– Socrates

Measure what can be measured & make measureable what cannot be measured.
– Galileo Galilei

Everything must be made as simple as possible But not simpler.
– Albert Einstein

Part I includes four chapters: Chapters 1–4.

Chapter 1. Conjugate Operators of Complex Vector Functions on Complex Number Plane The whole structure is composed of six lemmas, twenty-four theorems, and sixteen corollaries. The foundation of the conjugate complex vector structure is established.

The four distinct regions (four colors of the conjugate ring) and a pair of conjugate vectors (opposite rotations) and other comprehensive approaches extend the fundamental theorem of algebra to cover {complex coefficients} in a complete solution space. The conjugate complex vector systems guarantee without logic paradoxes on the vector phase space in synchronous dynamic environments.

Chapter 2. Measurement Operators of Conjugate Transformation Structure on Complex Plane The whole structure is composed of three lemmas, twenty-two theorems, and seven corollaries in two parts. The first part describes the eight complex vectors in a pair of conjugate feature classes generated by the conjugate transformation structure from {0, 1} elements of an N length sequence. The second part describes expanding operations to conjugate algebra with complete brackets identifying symmetric/antisymmetric measurements with the time order.

Chapter 3. Conjugate Complex Measurements of Clock Reference Signals Clock reference signals sufficiently provide the controllable quantitative bases of the precise measuring foundation on a complex number plane. The whole structure is composed of one lemma, ten theorems, and eight corollaries.

Conjugate measurements of clock reference signals provide the conjugate complementary measurement space, the narrow projection on the conjugate configuration, and the general projection on the complex number plane on one quadrant or four quadrants.

Chapter 4. Combinatorial Lattice Clusters of Conjugate Complex Vector Functions A conjugate group combinatorial lattice is proposed. The description of the whole structure is composed of one lemma, thirteen theorems, and twelve corollaries.

Normally entangled interactions appear in the eight vectors of a conjugate cluster for the conjugate Klein group and extended entanglements are restricted in the same eigenvalue on a cycle with a constant radius r. The extensions generate multiple cycles under combinatorial and Abel group operators to indicate that there are strong interactions involved beyond a single cycle condition.

Chapter 1

Conjugate Operators of Complex Vector Functions on Complex Number Plane

Parallel Time Synchronization Analysis

Jeffrey Zheng*
Conjugate System Research, Melbourne, Australia

Abstract

Gauss proved the fundamental theorem of algebra to represent n polygons with n real coefficients and n complex roots pairing complex conjugate structures in the complex plane. With the rapid development, multiple disciplines have been explored, such as relativity, quantum mechanics, negative probability, negative energy, and quantum entanglement, which have attracted popular attention. The series of paradoxes inherent in modern science require ultrabroad support from mathematical physics, logics, mathematics, and complex function theories. Expanding the coefficients of algebraic equations from real numbers to complex numbers, from serial computing methods to superparallel vector clusters, the intrinsic structure lacks the necessary support. Based on the paired conjugate angles $\{\theta, \tilde{\theta}\}, \theta + \tilde{\theta} = \pi/2$, a vector structure of the solution on the conjugate complex plane is proposed for complete complex coefficients of algebraic equations. The chapter consists of ten sections. The core content includes that any complex number forms a four-point rectangle through the Klein group and converts the eight points of two rectangles into a circle to transfer the relationship from position (rectangle) to time (circle). Using exponential representations. Attribute classification forms a four-region partition to support four vector groups of inner products $\{1, i, -1, -i\}$. Six groups of iterative changes are in the global and local variant characteristics. The modular structure is formed. The coefficient solution space of the binomial equation extends from m-D real to $4m$-D complex numbers in the Galois fields. For example, $\{x-1=0\} \rightarrow \{\{x-1, x-i, x+1, x+i\} = 0\} \equiv \{x^4-1=0\}$. Lemmas 1–6 cite the classical results, and Theorems 1–24 and Corollaries 1–16 describe the core models. All related supports are listed and summarized in Tables 7–9.

Keywords: Cyclotomic field, real coefficient solution, complex coefficient solution, paired conjugate angles, global invariant, conjugate complex plane, galois field

---

*Corresponding Author's Email: conjugatelogic@yahoo.com.

In: Conjugate Construction of Quantum Optics
Editors: Jeffrey Zheng
ISBN: 979-8-88697-651-9

Logic is foundation of certainty of all knowledge we acquire
—Leohard Euler

All mathematical science is founded on relations between physical laws & laws of numbers
—James Clerk Maxwell

Shortest path between two truths in real domain passes through complex domain
—Jacques Hadamard

1. Introduction

The complex number plane or complex plane is the first part of the complex variable functions [1]–[13] to provide foundational support for the development of modern logics, mathematics, science, engineering, and computing technology.

1.1 Cyclotomic Field, Complex Conjugate Operator, and Time Science

Tracing historical events in the expansion of the complex plane, de Moivre [16] as a friend of Coles [17], Newton, Halley, and Stirling derived the formula in 1707

$$\cos x = \frac{1}{2}\left(cos(nx) + i\sin(nx)\right)^{1/n} + \frac{1}{2}\left(cos(nx) - i\sin(nx)\right)^{1/n}$$

and developed in 1722 as the de Moivre equation

$$(\cos x + i\sin x)^n = \cos(nx) + i\sin(nx).$$

In 1749, Euler proved that the de Moivre equation holds for any real coefficient $n \in R$ based on Euler's formula [18].

In 1799, Gauss provided a rigorous proof [19] of the solution structure of the fundamental equations of algebra, an original contribution to the geometry of complex numbers. The n algebraic equations with real coefficients, expressed as n polygonal circle equations on the complex plane, provide n complex solutions and have complex conjugate pairing characteristics.

Hamilton submitted his paper to the Irish Academy of Sciences in 1835: Theory of Conjugate Functions, or Algebraic Couples; with a Preliminary and Elementary Essay on Algebra as the Science of Pure Time [20]. This research lays an algebraic foundation for complex variable functions to support frontier complex dynamic time measurements in the world.

From the level of duality on balanced logic analysis of complex dynamic systems, de Morgan laid the foundation for both mathematical logic and complex function theory [21]. A formal logic book was published in 1847 to establish duality rules for logical operations based on $\{\cap$ and, $\cup$ or, $\neg$ not$\}$ operators: de Morgan Law $(\neg : \cap \leftrightarrow \cup)$,

$$\neg(A \cup B) = \neg A \cap \neg B, \ \neg(A \cap B) = \neg A \cup \neg B$$

In 1849, de Morgan published a book titled *Trigonometry and Double Algebra*. The trigonometric part discussed the trigonometric functions of 2D plane rectangular coordinates and their various extensions affected on phase transition metrics: $\{\theta, \pi/2 - \theta\}$

complement, $\{\theta, \pi - \theta\}$ supplement, $\{\theta, \pi + \theta\}$ opponents, $\{\theta, 2\pi + \theta\}$ completion. On page 8, the angle of {0–90, 90–180, 180–270, 270–360} corresponds to the value symbol, and the 2D coordinate plane can be separated into four $\{++, +-, --, -+\}$ subplanes. Various trigonometric functions represent a function cluster of

$$\{\cos\theta, \sin\theta, \tan\theta, \cot\theta, \sec\theta, \operatorname{cosec}\theta, \operatorname{vers}\theta, \operatorname{covers}\theta\}$$

Page 14 used paired reciprocal angles $\{\theta, -\theta\}$ as eight groups of $\{\pi/2 - \theta, \pi - \theta, 3\pi/2 - \theta, 2\pi - \theta\}$, $\{\theta, \pi/2 + \theta, \pi + \theta, 3\pi/2 + \theta\}$ directions and explored each transforming mode represented by trigonometric functions. Page 15 graphically represents two sets of crosses and eight sets of relevant points on 2D Cartesian coordinates. In the subsequent double algebra part, a series of complex function representations were discussed using algebraic representations. However, there is no related description for the subsequent complex function on the trigonometric expansion effects. Using the unit circle in Chapter 8, the roots of unity were discussed as complex solutions to n-degree binomials.

In 1878, Clifford [22] published the book *Elements of Dynamic* to describe the space of vector representations with an algebraic dimension on $(m+1)m/2$ growth. Using complex and quaternion algebraic structures, complex motion trajectories were comprehensively represented as trigonometric functions, circles, ellipses, parabolas, spirals, and so on, and the common motion patterns of complex system dynamics were systematically explored.

In 1873, Heaviside [23] was inspired by Maxwell's work on electromagnetism [24] to study the theory of electronic circuits. In 1880, the differential operator was developed using the curl and divergence/convergence representation of a vector and quaternion structures. The original Maxwell equations involved twenty unknown variables and twenty differential equations to be simplified to four differential equations with four variables as the representation of modern Maxwell equations suitable for comparison with quantum physics.

Heaviside proposed a step function to simulate an electronic switch, and the unit pulse is called the Dirac δ operator. His outstanding research achievements have comprehensively changed the face of modern network communications, applied mathematics, advanced science, and innovative technologies in the world and made fundamental contributions to the development of frontal technologies such as long-distance telecommunications, quantum physics, cyber network communications, complex electronic circuit analysis, and super digital computer clusters.

Klein [25] and Poincare [26] in the 1880s, based on non-Euclidean geometry, extended the transforming group to complex coefficients and obtained a new family of complex functions: automorphic functions, which are generalizations within a compact equation on circular functions, hyperbolic functions, elliptic functions, elementary functions, and other functions. The defined upper half-plane UHP of the complex plane is called the Poincare upper half-plane, which provides basic support for modern frontier mathematical physics, elementary particles, astronomy, and cosmology.

Following Newton, Leibniz, Bernoulli, Cotes, de Moivre, Euler, Bessel, Bessel, Poisson, Legendre, Monge, Laplace, and Green [34], the complex variable function was deeply explored from the application levels of multivariable differential and integral equations on analyzing dynamic systems. Associated with Hamilton [20], de Morgan [21], Fourier [37], Dirichlet, Kronecker, Poocelet, Lobatchevsky, Bolyai, Abel [38], Jacobi [35], Galois [39], Kummer [40], Riemann [41], Cauchy [32], Laguerre, Grassmann [36], Maxwell [24],

Gibbs [44], Clifford [22], Heaviside [23], Dedekind, Sylvester, Frobenius, Jordan, Klein, Lie, Weierstrass [43], Beltrami, Christoffel, Lipschitz, Helmhotz, Hermite, Ball, Cayley, Poincare, Hill, Liapounoff, Hadamard, Noether, Weyl [47], Hilbert [48], Picard, von Neumann [49], Birkhoff [50, 51], Courant [52], Carton, Hardy [53], QL Xiong [54], LK Hua [55], SS Chern [56], CN Yang [57], CT Yau [58], XJ Xiong [59], Arnold [60], Bourbaki [61], Penrose [62], Langlands [63], Wiles, and Perelman, huge active explorations of in-depth investigation were carried out by world-class mathematicians from multivariable complex functions, Riemann geometry, large-scale differential and integral equations, variational principles, lattice theory, spiral theory, complex function matrix theory, differential geometry, gauge fields, complex geometry, classic group, spinor algebra, general complex variable function, and other broad basic and application fields. Further references can be retrieved from [1]–[15].

With the special importance of prime numbers, classical number theory problems and extended complex number structures are closely related to the cyclotomic field and the measurement of time variation.

1.2 Formal System Paradox, Compatibility, and Completeness

In the early 20th century, the frontier research of modern mathematics and logic collided violently. Russell's paradox [82, 95] proposed by Russell marked a feasible decision of a logic system based on a compatibility mechanism. With Hilbert's notable achievements in geometric logic and formal mathematical foundation, whether logical paradoxes are included in a theoretical framework is the core issue that modern logic and mathematics must be concerned with.

In exploratory studies on axiomatic model theory in 1930, Godel established incompleteness theorems for axiomatic systems [85]. These theorems ended a half-century of attempts, beginning with the work of Gottlob Frege and culminating in Principia Mathematica and Hilbert's formalism, to find a set of axioms sufficient for all mathematics.

Along with the Turing machine, recursive function theory, λ calculus, Shannon switch theory, information theory, finite automata, etc., many studies have comprehensively promoted the origin and development of modern computers, information science, and communication technology [86, 87]. Whether the formal description system itself is compatible and complete has become a necessary condition for judging the stability place of the foundation for computational theories and modern sciences.

The traditional logic system is based on binary logic, starting from Maxwell's equation and general relativity. The mathematical and physical foundations used are gradually transformed from the measurement system of real variable functions to complex variable functions, such as quantum mechanics and Hilbert space [88]–[92], and the logical system that may be properly supported has become a fundamental issue that is always difficult to determine.

For this reason, Bohr used the ancient oriental Tai-Chi model [95, 99]–[101] to simulate the laws of quantum mechanics. Complex dynamic changes require more powerful theoretical models and method support. From the fundamental level of mathematical differential and integral equations, rationally explaining the complex number characteristics unique to the Schrodinger equation has become a lifelong problem that has troubled

Schrodinger himself for a long time. The famous paradox proposed, Schrodinger's cat, which has been standing for a hundred years, is still an unsolved mystery for modern mathematical physicists.

1.3 Frontier Technological Problems and Inherent Paradoxes

In the 20th century, the rapid development of modern technology reasonably used hierarchically structured system analysis. A series of paradoxes and problems are inherent in modern technologies: wave-particle duality, antiparticles, gravitational waves, standard models, cosmic rays, global and local conservation, nonconservation, supersymmetry, dark matter, dark energy, black holes, negative probability, negative energy, quantum entanglement, quantum cryptography, and quantum computers. Appropriate support is required for basic disciplines on logic, mathematics, complex functions, mathematical physics, combinatorial group theory, quantum mechanics, quantum information, quantum gravity, quantum meta logic, meta-knowledge models, complex dynamic system analysis, nonequilibrium statistics, information science, and high-performance parallel computation.

1.4 Fundamental Difficulties and Obstacles to Overcome

From the perspective of system architecture, the foundation of previous researchers on complex variable functions and complex planes is reliable and credible for a serial computing model. From a parallel computing scheme, the limitation can be identified to be regulated to the complex conjugate operator and properties of two $\{\theta, -\theta\}$ angular measurements. Two complex conjugate states are located in different attribute regions of the complex plane. Such a composition mechanism constrains the basic analysis framework of the general complex function to be the Poincare upper half-plane model.

Complex conjugate operators play a central role in the development of wider fields such as modern physical mathematics, complex functions, complex geometry, complex analysis, and space technology and applications. From the perspective of the parallel transformation of complex variable functions, it is necessary for the coefficients of algebraic equations to be expanded from real numbers to complex numbers to additionally apply complex geometric operators.

From the viewpoint of numerical computation on a single argument θ, correlation computation is a serial algorithm architecture. Initial states can be calculated separately by adjusting the transformation modes, such as starting point, rotation, and translation. With the support of modern parallel computing clusters, what expansion is required to ensure that the complex functions of complex dynamic systems can be naturally applied? Under the conditions of synchronization time and high concurrency, it can efficiently meet the needs of multiprocessors, network computing systems, large-scale computing clusters, parallel computing, and analytical processing. Simulations and calculations are supported for simulation analysis, cyberspace communication, quantum physics, quantum information, astrophysics, quantum gravity, cosmology, etc.

Briefly, it is essential to effectively extend the asynchronous time architecture of the complex variable function on a serial algorithm to a synchronous time computing architecture of a parallel algorithm on multiple arguments and vector cluster conditions to explore multiple conjugate operators on the complex plane to extend the basic properties of clustering vector operators.

1.5 Conjugate Complex Number Plane

After more than forty years of continuous research, the formation and establishment of a conjugate transformation system can be traced back to the author's doctoral thesis: *Conjugate Transformation of Regular Plane Lattices for Binary Images* in 1994 [108, 109], in 2010 proposed Variant Logic [110], in 2019 completed Springer Nature monograph: *Variant Construction from Theoretical Foundation to Applications* [111], in 2021 Science Press published the monograph: *Variant Construction Theory and Its Applications Volume 1 Theoretical Foundation and Applications* [112].

Based on the paired conjugate complex angle $\{\theta, \tilde{\theta}\}, \theta + \tilde{\theta} = \pi/2$, this chapter creates an extended set of conjugate vector operators: $\{\sim, *, \natural, \prime\}$ corresponds to {conjugation, complex conjugate, real conjugate, negative conjugate}, which supports the complex conjugate operations on the system of complex vector functions.

Using global and local conjugate operations, the conjugate complex plane is established to support the solution of complex coefficient algebraic equations.

The chapter consists of ten sections. The core content includes eight groups of conjugated eigenvector clusters, forming a four-region partition as the four quadrants according to the attribute classification and supporting four groups of expanded inner product representations $\{1, i, -1, -i\}$ and four categories of product conversion modes. The system displays six groups of global and local visual characteristics of iterative changes and related global and regional variant trends. Focusing the attribute space on the first quadrant, the modular structure of the representation feature is formed.

Using $\{x - 1 = 0\} \in \mathrm{Gal}(1) \rightarrow \{\{x - 1, x - i, x + 1, x + i\} = 0\} \equiv \{x^4 - 1 = 0\} \in \mathrm{Gal}(4)$, which expands the coefficients of the binomial formula from real numbers to complex numbers.

Combining multiple sets of {Lemma, Theorem, Corollary}, Lemmas 1–6 cite the classical results of complex functions, and Theorems 1–24 and Corollaries 1–16 describe the core planned results. All symbols, parameters, and quantitative measures are listed and summarized in Tables 7–9 for convenience of illustration.

1.6 Chapter Content

The first section comprehensively describes the central problems of the development of modern science and technology and their relationship with the complex number plane. The second section discusses the basis of complex vectors and a pair of complex conjugate geometric representations. The third section provides the extended conjugate vector clustering operator representation under necessary and sufficient conditions. The fourth section studies the interactive state of clusters on conjugate vector operators. The fifth section presents the complex functions of power exponents. The sixth section studies the complex functions of different quadrants on power exponents. The seventh section discusses the variation in the iterative interaction radius. The eighth section describes the modular representation of the power exponent complex function. The ninth section studies the global division and interval division on the complex number plane. The tenth section summarizes the research results and possible future explorations.

2. Fundamental Issues on Complex Number Plane

Using complex geometric models to explain the results of number theory, Gauss [19] in 1831 described positive definite binary quadratic forms and their necessary and sufficient conditions for positive definiteness.

In 1881, Poincare described hyperbolic geometry in terms of the hyperboloid model, formulating transformations. In Klein and Poincare's [26] series of studies on elliptic functions and a theory of automorphic functions, the research focus is on the upper half-plane UHP of the complex plane, and imaginary numbers or ratios are restricted to take values greater than 0.

2.1 *Complex Vectors and Complex Conjugate Representation*

Any n elemental algebraic equation has n complex solutions with geometric representations. A complex conjugate operator $*$ provides an algebraic basis extending vector $x = (p, q) = p + iq$ to another vector $x^* = (p, -q) = p - iq$ as a pair of vectors on the complex plane. In Fig. 1, a single vector and pairwise complex conjugate vectors are provided.

2.2 *Complex Conjugate Pair*

Lemma 1 (Complex Conjugate Pairs of Elementary Algebraic Equations). *The elementary algebraic equations in n variables have n complex solutions, two related complex solutions, with pairwise complex conjugate properties.*

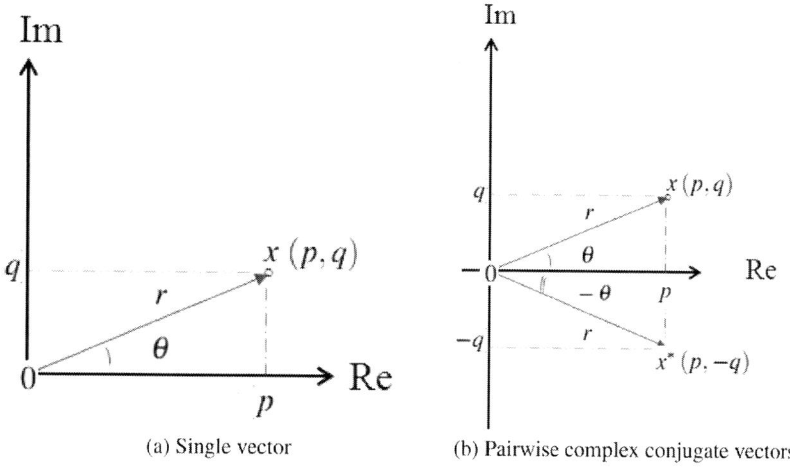

Figure 1. Complex vector representation: (a) basis vector $x = (p, q)$; (b) pairwise complex conjugate vectors $\{(p, q), (p, -q)\}$.

In the definition of the Hilbert space, the vector modulus $|x| = \sqrt{\langle x|x^*\rangle} > 0, x \neq 0$ describes the inner product of complex conjugate operators as the most important part. The scribed structure is constrained in the upper half-plane.

Lemma 2 (Length of Complex Conjugate Vector). *The modulo $|x|$ represents the inner product of pairwise complex conjugate vectors in Hilbert space, quantifying the general invariant of the vector length for the canonical projection of complex conjugate vectors.*

Lemma 3 (Noncompact Equivalent Representation). *The 2×2 real rotation matrix is an extended representation of complex ordinal pairs with more distinguishable symbols of combination states to be compared with the complex number states on sequence pairs.*

Proof. The complex number states of sequence pairs $\{(a, b), (b, a)\}$ can be represented as the $\{+, -\}$ symbol for each component, and there are eight combinations. The notation of the four elements in the 2×2 matrix proves $2^4 = 16$ combinations. In addition to swapping $\{a, b\}$ positions, 32 combinations can be formed.

Due to the extended mode of matrix representation, from the view of symbolic symmetry, $(p, q) \to \begin{pmatrix} p & -q \\ q & p \end{pmatrix}$ expresses complex pairs as a 2×2 real rotation matrix, which is a similar representation of the Hadamard matrix $(1, 1) \to \begin{pmatrix} 1 & -1 \\ 1 & 1 \end{pmatrix}$ that can simulate and calculate related values. Compared with a complex couple, the number of positive and negative signs of matrix elements cannot be described in a balanced style.

2.3 *Pairwise Complex Conjugate Operators and Upper and Lower Divisions*

Starting from any point x in the first quadrant on the complex plane CP, the geometric method was used to describe vectors. de Moivre [16], Gauss [19], Wessel [27], Argand [29], de Morgan [21], Poincare [26], and so on, focused on pairwise complex conjugate vectors $\{x, x^*\}, r = |x|$ and related pairwise arguments $\{\theta, -\theta\}, \theta + (-\theta) = 0, \theta \in R$. The angle of θ rotates counter-clockwise, the angle $-\theta$ rotates clockwise, and the complex plane is divided into $\{+, -\}$ upper and lower regions: the upper half-plane UHP and the lower half-plane LHP, respectively.

$$\boxed{x \in \mathrm{CP}} \to \begin{Bmatrix} x & +\theta \\ x^* & -\theta \end{Bmatrix} : \{x^* x = x x^* = 1\} \to \frac{+\theta \in \mathrm{UHP}}{-\theta \in \mathrm{LHP}} \to \frac{\{re^{i\theta}\}}{\{re^{-i\theta}\}}, \theta \in R$$

Pairwise complex conjugate description supports any complex number on the complex plane, local representation, and global division and lays the foundation for the fine algebraic features of complex variable functions.

3. Necessary and Sufficient Conditions for Cluster Representation of Conjugate Operators

3.1 Extended Conjugate Operators

Let $\sim$ be a conjugate operator to swap the two components of a complex vector: $x = (p, q), \theta = \angle(p, q), \tilde{x} = (q, p), \tilde{\theta} = \angle(q, p), \theta + \tilde{\theta} = \pi/2$. For two sets of vectors, the relevant complex conjugate vectors are four states $\{x, x^*\}, \{\tilde{x}, \tilde{x}^*\}$, shown in Fig. 2.

From the view of complex geometric representation, it is not sufficient to have the conjugate operator restricted in the {[00], [01]} quadrants, and it is necessary for the vector cluster to cover the four quadrants as a whole.

The conjugate Klein group plays a central role in the analysis of ultracomplex dynamic systems. The complex number structure is represented as pairs based on complex number geometry. Compared with complex conjugate operators and conjugate operators, it is sufficient to form a complete basis of interaction for the multivariate operational structure on complex functions.

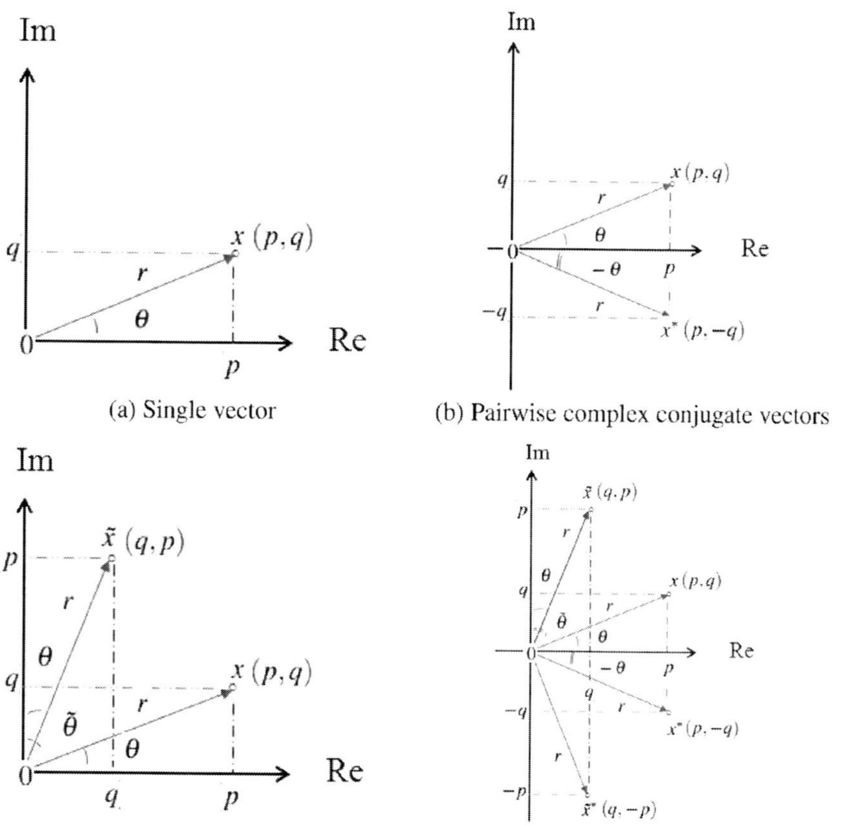

Figure 2. Extended conjugate complex vectors: (a) basic vector (x, y); (b) pairwise complex conjugate vectors $\{(x, y), (x, -y)\}$; (c) pairwise conjugate vectors $\{(x, y), (y, x)\}$; (d) conjugate + complex quaternion $\{(x, y), (x, -y), (y, x), (y, -x)\}$.

3.2 Conjugate Klein Groups

To retain the consistency of the formula, four conjugate operators are presented: $\sim$ conjugate, $*$ complex conjugate, $'$ negative conjugate, and $\sharp$ real conjugate. The eight corresponding states are expressed as

$$x = (p,q), x^{\sharp} = (-p,q), x' = (-p,-q), x^{*} = (p,-q) \quad (1)$$
$$\tilde{x} = (q,p), \tilde{x}^{\sharp} = (-q,p), \tilde{x}' = (-q,-p), \tilde{x}^{*} = (q,-p) \quad (2)$$

From the four quadrants, the conjugate state defined in this way can be uniquely restricted in the same attribute quadrant. The conjugate operator is active simply to swap imaginary and real numbers. In the first quadrant, the two components are positive real numbers. The complex plane is divided into four quadrants according to the attribute of the combinations of conjugate operators.

Under the combination of the four conjugate operators, two clusters of eight vectors are formed.

$$K(x) = X = \{x, x^{\sharp}, x', x^{*}\} \quad (3)$$
$$\tilde{K}(x) = \tilde{X} = \{\tilde{x}, \tilde{x}^{\sharp}, \tilde{x}', \tilde{x}^{*}\} \quad (4)$$
$$\{X, \tilde{X}\} := \{x, x^{\sharp}, x', x^{*}, \tilde{x}, \tilde{x}^{\sharp}, \tilde{x}', \tilde{x}^{*}\} \quad (5)$$

Lemma 4 (Klein Group). *The four states associated with $K(x)$ occupy the four vertex positions on the complex plane, forming a rectangle.*

Proof. $K(x) = \{x, x^{\sharp}, x', x^{*}\}$, the four vertex positions refer to Fig. 3(a).

The four states associated with $K(p,q)$ occupy the four vertex positions of the complex plane using the quantized coordinate projection to form a rectangle. The four states associated with $\tilde{K}(p,q)$ form another rectangle.

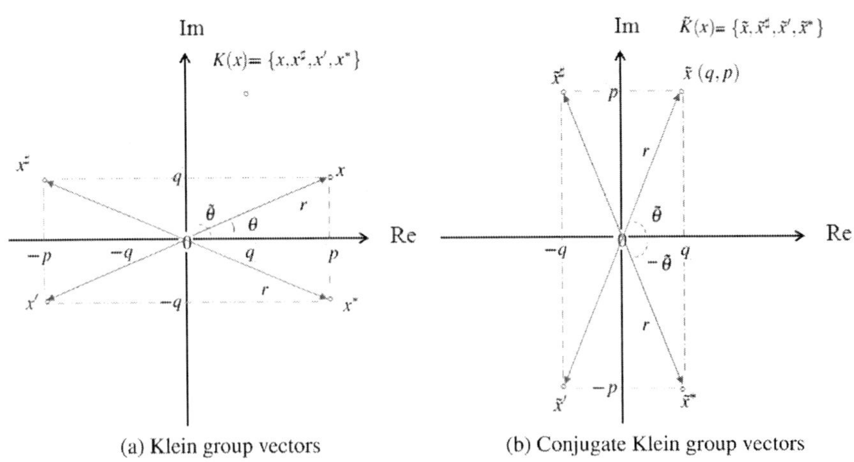

Figure 3. Conjugate klein group vectors: (a) basic quaternary vectors $K(x) = \{x, x^{\sharp}, x', x^{*}\}$; (b) conjugate quaternary vectors $\tilde{K}(x) = \{\tilde{x}, \tilde{x}^{\sharp}, \tilde{x}', \tilde{x}^{*}\}$.

Theorem 1 (Conjugate Klein Groups). *$K(x)$ and $\tilde{K}(x)$ consist of two sets of Klein groups on eight states, and the cluster can generally be represented as a single rectangle or two rectangles.*

Proof. From a single complex number $x = (p, q)$, two operators K and $\tilde{K}$ transform it from a single state expansion into eight states

$$K(x) \rightarrow \{x, x^{\sharp}, x', x^{*}\}$$

$$\tilde{K}(x) \rightarrow \{\tilde{x}, \tilde{x}^{\sharp}, \tilde{x}', \tilde{x}^{*}\}.$$

When $p = q$, only a single rectangle is distinguishable.

Two four-point rectangles of $K(x)$ and $\tilde{K}(x)$ are shown in Fig. 3.

3.3 Two Groups of Conjugate Imaginary Numbers

The eight states involved in $\{K(x), \tilde{K}(x)\}$ can be expressed in another form [25]. Using the imaginary number $i = \sqrt{-1}$ as a basic operator, the Gauss integer has four units $i^{\lambda}, \lambda \in \{0, 1, 2, 3\}$. Two conjugate clusters are denoted as $\{V(x), \tilde{V}(x)\}$ to describe the sequence formed under the iterative operation.

$$x = i^0 x = (p, q), x^{\sharp} = i^1 \tilde{x} = (-p, q) \tag{6}$$

$$x' = i^2 x = -x = (-p, -q), x^{*} = i^3 \tilde{x} = (p, -q) \tag{7}$$

$$\tilde{x} = i^0 \tilde{x} = (q, p), \tilde{x}^{\sharp} = i^1 x = (-q, p) \tag{8}$$

$$\tilde{x}' = i^2 \tilde{x} = (-q, -p), \tilde{x}^{*} = i^3 x = (q, -p) \tag{9}$$

$$i \rightarrow \{i^0, i^1, i^2, i^3\} = \{1, i, -1, -i\} : \{x\} \rightarrow \{x, \tilde{x}^{\sharp}, x', \tilde{x}^{*}\} \tag{10}$$

$$i \rightarrow \{i^0, i^1, i^2, i^3\} = \{1, i, -1, -i\} : \{\tilde{x}\} \rightarrow \{\tilde{x}, x^{\sharp}, \tilde{x}', x^{*}\} \tag{11}$$

$$V(x) = \{x, \tilde{x}^{\sharp}, x', \tilde{x}^{*}\} \tag{12}$$

$$\tilde{V}(x) = \{\tilde{x}, x^{\sharp}, \tilde{x}', x^{*}\} \tag{13}$$

It is worth emphasizing that the quaternary division formed by the imaginary iterative sequence is different from the two sets of conjugated Klein groups with two rectangles, and the two sets of imaginary quaternions are formed in two crosses on the complex plane. Four states of each cross were separated as a right triangle on the $\pi/2$ radian (rad).

3.4 Necessary and Sufficient Conditions on Conjugate Clusters

Theorem 2 (Unique Eight/Four Vectors of Conjugate Clusters beyond Gaussian Integers). *On the complex plane for any $x = (p, q), p, q \in N, \theta = \angle(p, q), 0 \leq \theta < \pi/2$, there is a unique conjugate state $\tilde{x} = (q, p), p, q \in N, \tilde{\theta} = \angle(q, p), 0 < \tilde{\theta} \leq \pi/2$ and $\tilde{\theta} = \pi/2 - \theta$. Beyond the four elements of Gaussian integers, it can simultaneously satisfy $\{K(x), \tilde{K}(x)\}, \{V(x), \tilde{V}(x)\}$ on the complex plane. Only two cases are satisfied: 1. Eight distinguishable points; 2. Four distinguishable points.*

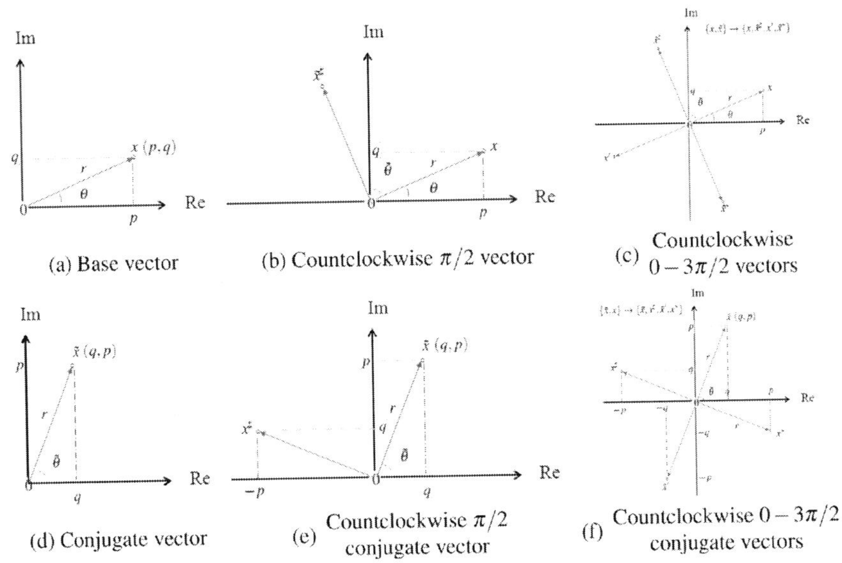

Figure 4. Two groups of imaginary rotation of four vectors: (a) base vector $x = (p, q)$; (b) pairwise counter-clockwise vectors $\{x, \tilde{x}^{\sharp}\}$; (c) four countclockwise vectors $V(x) = \{x, \tilde{x}^{\sharp}, x', \tilde{x}^{*}\}$; (d) conjugate vector $\tilde{x} = (q, p)$; (e) pairwise counter-clockwise vectors $\{\tilde{x}, x^{\sharp}\}$; (f) four countclockwise conjugate vectors $\tilde{V}(x) = \{\tilde{x}, x^{\sharp}, \tilde{x}', x^{*}\}$.

Proof. For any vector $x = (p, q), p, q > 0, p \neq q \rightarrow \{K(x), \tilde{K}(x)\} \cup \{V(x), \tilde{V}(x)\}$, it is natural to form an eight-point cluster that can be decomposed into two rectangles or two crosses; however, standard Gauss integers involve only four elements. In two cases of two values, the four-point states are reduced into a four-point cluster:

A. $p = q \rightarrow$ Four vectors are $\{(1, 1), (-1, 1), (-1, -1), (1, -1)\}$. In the [00] quadrant, the two conjugate vectors are congruently coincident at the $\pi/4$ position.
B. $\{p > 0, q = 0\}$ or $\{q > 0, p = 0\} \rightarrow \{(0, 1)(0, -1)(1, 0)(-1, 0)\}$; when one number of the complex couple is 0, it makes coordinate vectors in four directions, and the degenerate case satisfies the Gauss integer structure.

Conversely, when there are eight states that satisfy the conditions, two conjugate states in the [00] quadrant can be selected to determine the order and pair relations of the conjugate mapping.

Theorem 3 (Octary/Quaternary Vectors on Circle). *The octary vectors that satisfy Theorem 2 are located on the circle with radius* $r = \sqrt{p^2 + q^2}$.

Proof. All vectors of $\{K(x), \tilde{K}(x)\}$ or $\{V(x), \tilde{V}(x)\}$ have the same radius length.

In Fig. 5 (a)–(f), the multiple vectors are all clustered on the same circle of equal radius.

Corollary 1 (Necessary and Sufficient Conditions of Octary/Quaternary Vectors on Circle). *When* $p, q > 0, p \neq q$, *each quadrant contains two vectors, and the four quadrants contain a total of eight vectors. When* $p, q > 0, p = q$, *each quadrant contains one vector, and four quadrants contain four vectors. When* $\{p > 0, q = 0\}$ *or* $\{q > 0, p = 0\}$, *the four vectors lie on the axes.*

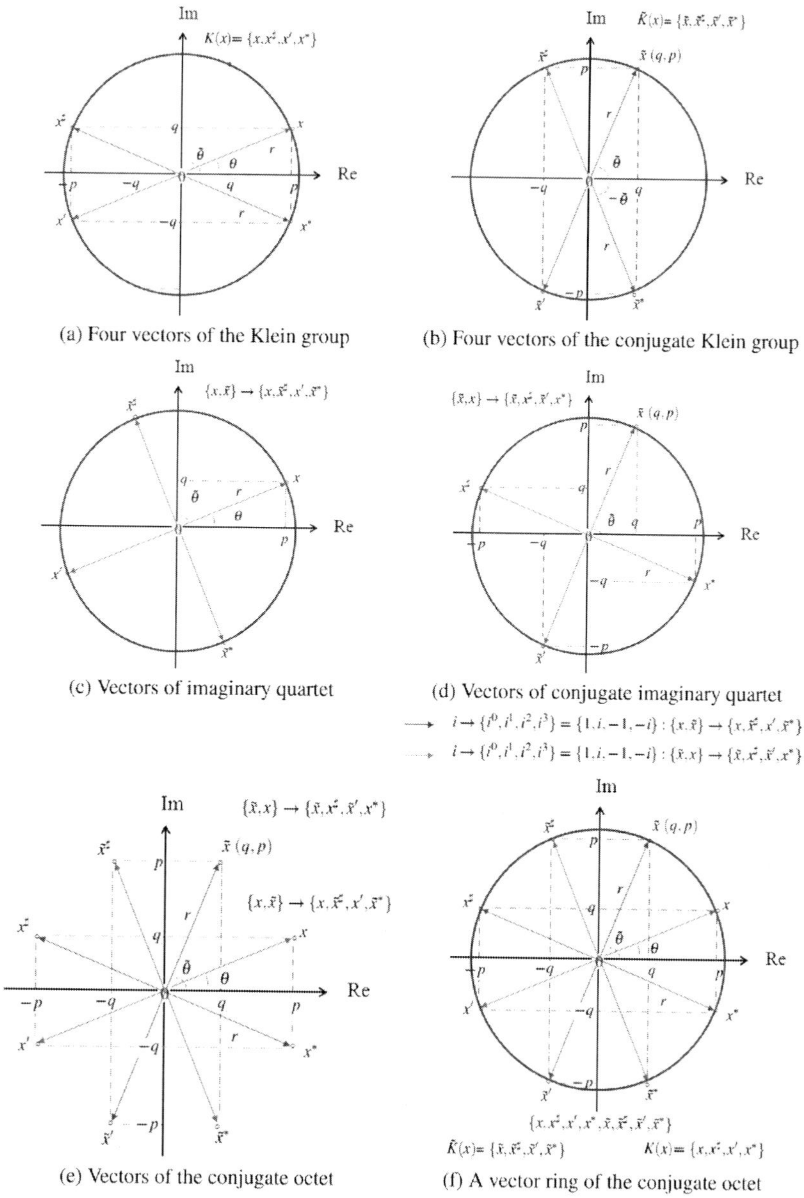

(a) Four vectors of the Klein group (b) Four vectors of the conjugate Klein group

(c) Vectors of imaginary quartet (d) Vectors of conjugate imaginary quartet

(e) Vectors of the conjugate octet (f) A vector ring of the conjugate octet

Figure 5. Vectors of conjugate klein groups and vectors of imaginary quartet group. (a) Base vectors $K(x)$. (b) Conjugate vectors $\tilde{K}(x)$. (c) Imaginary rotations $\{x, ix, i^2x, i^3x\}$. (d) Imaginary vectors $\{\tilde{x}, i\tilde{x}, i^2\tilde{x}, i^3\tilde{x}\}$. (e) Conjugate octet. (f) Conjugate octet cycle.

Proof. The selected vectors satisfy the integer feature. In the same quadrant, only two vectors of the conjugate pair fit the condition. When the two vectors become the same, only the unique midpoint satisfies the condition. When one of the parameters is zero, the vector is located on the axes.

Theorem 4 (Number of Cocircular Clusters of Octary/Quaternary Vectors). *For the* $[0, N]^2$ *quadrant, the number of cocircular clusters* $C_I(N)$ *is* $C_I(N) = (N + 1)N/2$.

Proof. In the quadrant, only integer positions can be selected. Starting from the $(0, 0)$ point, when the area is expanded by 1, the incremental sequence satisfies $1, 2, 3, \cdots, N$, so the total number of circles is

$$C_I(N) = \sum_{j=1}^{N} j = (N+1)N/2.$$

From the geometric distances on the complex plane, each cocircular cluster forms an equivalent class that can naturally distinguish various cocircular clusters as separated classes. Theorem 4 provides simultaneous processing of $O(N^2/2)$ classes as a potential foundation of concurrent-hyperscale/parallel-computing.

The numbers of $N = 1 - 12$ circles are listed as examples:

N	1	2	3	4	5	6	7	8	9	10	11	12
$C_I(N)$	1	3	6	10	15	21	28	36	45	55	66	78

Corollary 2 (Number of Clusters in Four Quadrants Equivalent to Number of Circles). *Four quadrants* $\{[00], [10], [11], [01]\}$ *satisfy the condition of Theorem 4, and all quadrants are shared on the* $C_I(N)$ *distinct circles.*

Proof. The number of differences of the attribute in each quadrant is equivalent to the number of definite vector lengths in the quadrant, and the related vectors in the same length on the four quadrants are located on one of the circles.

Corollary 3 (Constraints on Number of Clusters of Eight Vectors on the Circle). *From any vector* $x = (p, q)$, *satisfying Theorem 4, there is a certain radius* $r = \sqrt{p^2 + q^2}$ *and paired conjugate angles* $\{\theta, \tilde{\theta}\}$, *related to four/eight vectors. The number of conjugate vectors in pairs satisfies* $4 \leq |\{K(x), \tilde{K}(x)\} = \{V(x), \tilde{V}(x)\}| \leq 8$.

Proof. Because $\{K(x), \tilde{K}(x)\} = \{V(x), \tilde{V}(x)\}$, eight vectors can be determined together via the Klein group and the imaginary rotation group. Under the constraint of $\theta + \tilde{\theta} = \pi/2$, it is reduced into four vectors if and only if the four vectors lie on the coordinate axis or the two conjugate angles are equal.

Corollary 4 (Number of Vector Clusters in Four Quadrants). *When the complex plane region is* $[-N, N]^2$, *the distinct vector positions and the number of points on the complex plane are related. The total number of distinguishable vectors is* $F_I = 8N^2 + 1$.

Proof. The complex plane $[-N, N]^2$ region can be divided into four types {A,B,C,D} on four quadrants. There is Type A $\{[1, N]^2, [-1, -N] \times [1, N], [-1, -N]^2, [1, N] \times [-1, -N], \}$, and each area can be identified by positions on pairwise basic and conjugate vectors. For Type B, N elements of the diagonal parts on each block overlap, and the four axes overlap in Type C $\{[\pm 0] \times [1, N], [-1, -N] \times [\pm 0], [\pm 0] \times [-1, -N], [1, N] \times [\pm 0], \}$, and the four quadrants all overlap at the zero point as Type D $(0, 0)$. Therefore, the total number of distinguishable positions is

$$F_I(N) = (A) + (B) + (C) + (D) = 8N^2 + 4N - 4N + 1 = 8N^2 + 1.$$

Excluding the contribution of zeros, each block provides $2N^2$ distinct positions. Each point makes a contribution to a pair of conjugate angles $\{\theta, \tilde{\theta}\}$.

Corollary 5 (Number of Distinct Vectors in $[-N,N]^2$). *There is a total number of $(2N+1)^2$ sample points, and each point provides a pair of conjugate angles $\{\theta,\tilde{\theta}\}$ to act once. The total number $F_V(N)$ is*

$$F_V(N) = 2(2N+1)^2 = 8N^2 + 8N + 2.$$

There is a number of $8N+1$ amounts that is larger than Corollary 4. Three excess parts are provided by the overlapped elements in the four axes, the four diagonals, and the zero position.

Proof. Each of the four axes and the diagonal provides N covering positions, plus touching zeros. There are $8N+1$ overlapping positions.

From the view of maximum parallelization, Corollary 5 provides multiple sets of $O(8N^2)$ vectors that need to be processed simultaneously; there are sixteen times more than the concurrency described on cocircular clusters in Theorem 4.

4. Interactive Clusters

According to the distinguishable position of the complex plane, the conjugate measures of each pair are provided twice as often as the base vectors of the conjugate complex functions. Additionally, eight relevant vectors can perform interaction operations. The intrinsic rules of pairwise interactions among clusters and vectors need to be fully explored.

4.1 Base Representation

For the eight vectors determined, various pairwise vectors and the four conjugate operators are organized by the multiplication rules, and the degenerate features of the ordered complex number products are observed.

$$x = (p,q), ||x|| = r = \sqrt{p^2+q^2}, p = r\cos\theta, q = r\sin\theta \tag{14}$$

$$x^n = r^n(\cos n\theta, \sin n\theta), n \in N, \theta \in R \tag{15}$$

$$r^0 = 1, r^1 = ||x||, r^2 = ||x||^2, r^3 = ||x||^3, r^4 = ||x||^4, r^5 = ||x||^5, r^n = ||x||^n$$

$$\angle x^0 = \angle 1 = 0, \angle x^1 = \theta, \angle x^2 = 2\theta, \angle x^3 = 3\theta, \angle x^4 = 4\theta, \angle x^5 = 5\theta, \angle x^n = n\theta$$

$$\tilde{x} = (q,p), ||\tilde{x}|| = r = \sqrt{q^2+p^2}, q = r\cos\tilde{\theta}, p = r\sin\tilde{\theta} \tag{16}$$

$$\tilde{x}^n = r^n(\cos n\tilde{\theta}, \sin n\tilde{\theta}), n \in N, \tilde{\theta} \in R \tag{17}$$

$$r^0 = 1, r^1 = ||\tilde{x}||, r^2 = ||\tilde{x}||^2, r^3 = ||\tilde{x}||^3, r^4 = ||\tilde{x}||^4, r^5 = ||\tilde{x}||^5, r^n = ||\tilde{x}||^n$$

$$\angle \tilde{x}^0 = \frac{\pi}{2}, \angle \tilde{x}^1 = \tilde{\theta}, \angle \tilde{x}^2 = 2\tilde{\theta}, \angle \tilde{x}^3 = 3\tilde{\theta}, \angle \tilde{x}^4 = 4\tilde{\theta}, \angle \tilde{x}^5 = 5\tilde{\theta}, \angle \tilde{x}^n = n\tilde{\theta}$$

Four pairs of complex conjugate states $+1$

$$\{\{x, x^*\}, \{\tilde{x}, \tilde{x}^*\}, \{\tilde{x}^\sharp, \tilde{x}'\}, \{x^\sharp, x'\}\}$$

$$\{e_0^{i\theta}, e_1^{i\tilde{\theta}}\}, \{e_0^{i\tilde{\theta}}, e_1^{i\theta}\}, \{e_2^{i\theta}, e_3^{i\tilde{\theta}}\}, \{e_2^{i\tilde{\theta}}, e_3^{i\theta}\}$$

Four pairs of real conjugate states -1

$$\{\{x,,x^{\sharp}\}\{\tilde{x},\tilde{x}^{\sharp}\},\{\tilde{x}^{*},\tilde{x}'\},\{x',x^{*}\}\}$$
$$\{e_0^{i\theta},e_2^{i\tilde{\theta}}\},\{e_0^{i\tilde{\theta}},e_2^{i\theta}\},\{e_1^{i\theta},e_3^{i\tilde{\theta}}\},\{e_1^{i\tilde{\theta}},e_3^{i\theta}\}$$

Four pairs of conjugated states $+i$

$$\{\{x,\tilde{x}\},\{x^{\sharp},\tilde{x}^{\sharp}\},\{x',\tilde{x}'\},\{x^{*},\tilde{x}^{*}\}\}$$
$$\{e_0^{i\theta},e_0^{i\tilde{\theta}}\},\{e_2^{i\tilde{\theta}},e_2^{i\theta}\},\{e_3^{i\theta},e_3^{i\tilde{\theta}}\},\{e_1^{i\tilde{\theta}},e_1^{i\theta}\}$$

Four pairs of negative conjugate states $-i$

$$\{\{x,\tilde{x}'\},\{x',\tilde{x}\},\{x^{*},\tilde{x}^{\sharp}\},\{x^{\sharp},\tilde{x}^{*}\}\}$$
$$\{e_0^{i\theta},e_3^{i\tilde{\theta}}\},\{e_0^{i\tilde{\theta}},e_3^{i\theta}\},\{e_2^{i\theta},e_1^{i\tilde{\theta}}\},\{e_2^{i\tilde{\theta}},e_1^{i\theta}\}$$

Four Conversions

$$1=(0)(2)(3)(1), i=(2310), -1=(30)(21), -i=(1320)$$

Four Conjugate Conversions

$$i=(0231), -1=(21)(30), -i=(3201), 1=(1)(0)(2)(3).$$

4.2 *Interaction of Eight States*

Lemma 5 (Eight Vector Interaction). *In the* $X = K(x), \tilde{X} = \tilde{K}(x)$ *expression, which is multiplied in pairs of products is arranged to form sixty-four pairwise groups of output representations.*

Proof. Eight complex numbers are multiplied two by two to create sixty-four outputs, which retain complex numbers

$(X\tilde{X})^2$	X	$\tilde{X}$
X	XX	$X\tilde{X}$
$\tilde{X}$	$\tilde{X}X$	$\tilde{X}\tilde{X}$

$\rightarrow$

$(X\tilde{X})^2$	x	$x^{\sharp}$	x'	x^{*}	$\tilde{x}$	$\tilde{x}^{\sharp}$	$\tilde{x}'$	$\tilde{x}^{*}$
x	xx	$xx^{\sharp}$	xx'	xx^{*}	$x\tilde{x}$	$x\tilde{x}^{\sharp}$	$x\tilde{x}'$	$x\tilde{x}^{*}$
$x^{\sharp}$	$x^{\sharp}x$	$x^{\sharp}x^{\sharp}$	$x^{\sharp}x'$	$x^{\sharp}x^{*}$	$x^{\sharp}\tilde{x}$	$x^{\sharp}\tilde{x}^{\sharp}$	$x^{\sharp}\tilde{x}'$	$x^{\sharp}\tilde{x}^{*}$
x'	$x'x$	$x'x^{\sharp}$	$x'x'$	$x'x^{*}$	$x'\tilde{x}$	$x'\tilde{x}^{\sharp}$	$x'\tilde{x}'$	$x'\tilde{x}^{*}$
x^{*}	xx	$x^{*}x^{\sharp}$	$x^{*}x'$	$x^{*}x^{*}$	$x\tilde{x}$	$x^{*}\tilde{x}^{\sharp}$	$x^{*}\tilde{x}'$	$x^{*}\tilde{x}^{*}$
$\tilde{x}$	$\tilde{x}x$	$\tilde{x}x^{\sharp}$	$\tilde{x}x'$	$\tilde{x}x^{*}$	$\tilde{x}\tilde{x}$	$\tilde{x}\tilde{x}^{\sharp}$	$\tilde{x}\tilde{x}'$	$\tilde{x}\tilde{x}^{*}$
$\tilde{x}^{\sharp}$	xx	$\tilde{x}^{\sharp}x^{\sharp}$	$\tilde{x}^{\sharp}x'$	$\tilde{x}^{\sharp}x^{*}$	$\tilde{x}^{\sharp}\tilde{x}$	$\tilde{x}^{\sharp}\tilde{x}^{\sharp}$	$\tilde{x}^{\sharp}\tilde{x}'$	$\tilde{x}^{\sharp}\tilde{x}^{*}$
$\tilde{x}'$	$\tilde{x}'x$	$\tilde{x}'x^{\sharp}$	$\tilde{x}'x'$	$\tilde{x}'x^{*}$	$\tilde{x}'\tilde{x}$	$\tilde{x}'\tilde{x}^{\sharp}$	$\tilde{x}'\tilde{x}'$	$\tilde{x}'\tilde{x}^{*}$
$\tilde{x}^{*}$	$\tilde{x}^{*}x$	$\tilde{x}^{*}x^{\sharp}$	$\tilde{x}^{*}x^{*}$	$\tilde{x}^{*}x^{*}$	$\tilde{x}^{*}\tilde{x}$	$\tilde{x}^{*}\tilde{x}^{\sharp}$	$\tilde{x}^{*}\tilde{x}'$	$\tilde{x}^{*}\tilde{x}^{*}$

4.3 *Twelve Output Vectors*

Multiple products of sixty-four entries in the table yield the same result, and combined states can be reduced to the twelve cases listed in Table 1.

Table 1. Sixty-four products output twelve vectors

Output Product	Product Representation	Algebraic Formula	Cluster No.
$(p^2+q^2,0)$	$\{xx^*,x^*x,x'x^\sharp,x^\sharp x',\tilde{x}\tilde{x}^*,\tilde{x}^*\tilde{x},\tilde{x}'\tilde{x}^\sharp,\tilde{x}^\sharp\tilde{x}'\}$	$r^2(\cos^2\theta+\sin^2\theta,0)=r^2(1,0)$	8
$(p^2-q^2,2pq)$	$\{xx,x'x',\tilde{x}^\sharp\tilde{x}^*,\tilde{x}^*\tilde{x}^\sharp\}$	$r^2(\cos 2\theta,\sin 2\theta)$	4
$(2pq,p^2-q^2)$	$\{\tilde{x}x^*,x^*\tilde{x},\tilde{x}'x^\sharp,x^\sharp\tilde{x}'\}$	$r^2(\sin 2\theta,\cos 2\theta)$	4
$(0,p^2+q^2)$	$\{\tilde{x}x,x\tilde{x},\tilde{x}'x',x'\tilde{x}',\tilde{x}^\sharp\tilde{x}^*,\tilde{x}^*\tilde{x}^\sharp,x^*\tilde{x}^\sharp,\tilde{x}^\sharp x^*\}$	$r^2(0,\cos^2\theta+\sin^2\theta)=r^2(0,1)$	8
$(-p^2+q^2,2pq)$	$\{x^*x^\sharp,x^\sharp x^*,\tilde{x}\tilde{x},\tilde{x}'\tilde{x}'\}$	$r^2(-\cos 2\theta,\sin 2\theta)$	4
$(-2pq,p^2-q^2)$	$\{\tilde{x}^*x',x'\tilde{x}^*,\tilde{x}^\sharp x,x\tilde{x}^\sharp\}$	$r^2(-\sin 2\theta,\cos 2\theta)$	4
$(-p^2-q^2,0)$	$\{x'x^*,x^*x',xx^\sharp,x^\sharp x,\tilde{x}\tilde{x}^\sharp,\tilde{x}^\sharp\tilde{x},\tilde{x}'\tilde{x}^*,\tilde{x}^*\tilde{x}'\}$	$r^2(-\cos^2\theta-\sin^2\theta,0)=r^2(-1,0)$	8
$(-p^2+q^2,-2pq)$	$\{xx',x'x,\tilde{x}^*\tilde{x}^*,\tilde{x}^\sharp\tilde{x}^\sharp\}$	$r^2(-\cos 2\theta,-\sin 2\theta)$	4
$(-2pq,-p^2+q^2)$	$\{\tilde{x}x^\sharp,x^\sharp\tilde{x},\tilde{x}'x^*,x^*\tilde{x}'\}$	$r^2(-\sin 2\theta,-\cos 2\theta)$	4
$(0,-p^2-q^2)$	$\{\tilde{x}x',x'\tilde{x},\tilde{x}^*x^*,x^*\tilde{x}^*,\tilde{x}'x,x\tilde{x}',x^\sharp\tilde{x}^\sharp,\tilde{x}^\sharp x^\sharp\}$	$r^2(0,-\cos^2\theta-\sin^2\theta)=r^2(0,-1)$	8
$(p^2-q^2,-2pq)$	$\{x^*x^*,x^\sharp x^\sharp,\tilde{x}'\tilde{x},\tilde{x}\tilde{x}'\}$	$r^2(\cos 2\theta,-\sin 2\theta)$	4
$(2pq,-p^2+q^2)$	$\{\tilde{x}^*x,x\tilde{x}^*,\tilde{x}^\sharp x',x'\tilde{x}^\sharp\}$	$r^2(\sin 2\theta,-\cos 2\theta)$	4

Theorem 5 (Twelve Output Vectors). *As output results of sixty-four products satisfying Lemma 5, a total number of twelve distinguishable vectors are included.*

Proof. In Table 1, clustering product pairs are listed. Each group of projected vectors contains eight pairs of products distributed along four axes. There are a total of $4\times 8=32$ pairs of products. In addition, each quadrant contains a pair of feature vectors associated with four sets of products, involving $8\times 4=32$ pairs of products.

5. Complex Exponential Functions

To simplify the complex degeneracy laws, the complex exponential function is represented in the section.

5.1 Base Representation

The eight distinguishable vertices are all located on the ring with radius r of $r=\sqrt{p^2+q^2}$.

Two states in the same quadrant are distinguished by pairwise angle θ radians and conjugate angle $\tilde{\theta}$ radians.

Two states in the same quadrant are distinguished by a pair of angles θ and conjugate angles $\tilde{\theta}$.

$$x,\tilde{x}\in[00];\ \tilde{x}^\sharp,x^\sharp\in[10];\ x',\tilde{x}'\in[11];\ \tilde{x}^*,x^*\in[01].$$

Theorem 6 (Clustering Structure of Eight Vectors). *Two group representations have equivalent elements between* $\{K(x),\tilde{K}(x)\}$ *and* $\{V(x),\tilde{V}(x)\}$ *within different clustering partitions.*

Proof. In both representations, half of the number of the vectors is located in either $K(x)$ or $\tilde{K}(x)$. Half of the vectors belong to $V(x)$, and the other half belong to $\tilde{V}(x)$. Therefore, two clusters of $\{V(x),\tilde{V}(x)\}$ are represented in a mixed mode, half of the vectors are derived from $K(x)$, and the other half are derived from $\tilde{V}(x)$.

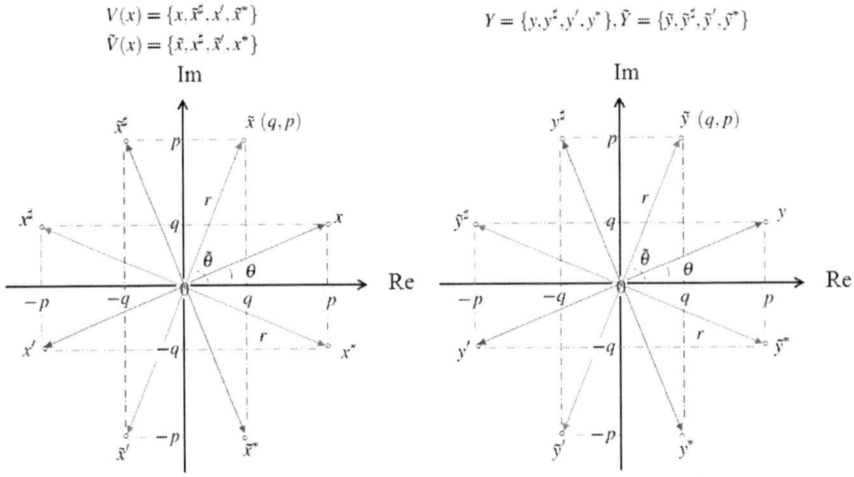

(a) Vector X (b) Vector Y

Figure 6. Original and corrected vectors: (a) base vectors $x = (p, q)$, $\{X = V(x), \tilde{X} = \tilde{V}(x)$; (b) corrected vectors $\{Y = U(y), \tilde{Y} = \tilde{U}(y)\}$.

5.2 *Corrected Representation* $\{x\} \to \{y\}$

Let y be a variable to correct for noncanonical properties in the sequence of x variables, as shown in Fig. 6. There are two crosses of quaternary structures corresponding to the pairwise base state and conjugated state. Four related states in each cross are distinguished in various quadrants.

$$y, \tilde{y} \in [00];\ y^{\sharp}, \tilde{y}^{\sharp} \in [10];\ y', \tilde{y}' \in [11];\ y^{*}, \tilde{y}^{*} \in [01] \to \{y, \tilde{y}, y^{\sharp}, \tilde{y}^{\sharp}, y', \tilde{y}', y^{*}, \tilde{y}^{*}\}$$

Theorem 7 (Eight Vectors on Corrected Representations). *Using the* $\{K(x), \tilde{K}(x)\}$ *groups, the corrected vectors can be renamed from* $\{V(x), \tilde{V}(x)\} \to \{U(y), \tilde{U}(y)\}$. *The involved vectors represent the two equivalent clusters.*

Proof. Shown in Fig. 6, starting from x and $\tilde{x}$, only two pairs of vectors in $\{[10], [01]\}$ need to be swapped in the four quadrants; then, a new coding series can be generated $\{V(x), \tilde{V}(x)\} \to \{U(y), \tilde{U}(y)\}$.

Using complex vectors, there is a set of rotation operators $i : \{1, i, -1, -i\} = \{i^0, i^1, i^2, i^3\}, i = \sqrt{-1}$ to map $\theta = \angle(p, q), \tilde{\theta} = \angle(q, p)$ into the two crosses

$$\tilde{\theta} = \frac{\pi}{2} - \theta \tag{18}$$

$$i : \theta \to \left\{\theta, \theta + \frac{\pi}{2}, \theta + \pi, \theta + \frac{3\pi}{2}\right\} \tag{19}$$

$$i : \tilde{\theta} \to \left\{\tilde{\theta}, \tilde{\theta} + \frac{\pi}{2}, \tilde{\theta} + \pi, \tilde{\theta} + \frac{3\pi}{2}\right\} \tag{20}$$

$$= \left\{\frac{\pi}{2} - \theta, \pi - \theta, \frac{3\pi}{2} - \theta, 2\pi - \theta\right\} \tag{21}$$

Table 2. Corresponding relationships on corrections

$\{Y, \tilde{Y}\}$	$\{\theta\}$	$\{\tilde{\theta}\}$	Formula	Equal	Trigonometric Rep.
y	θ		$re^{i\theta}$		$r(\cos\theta, \sin\theta)$
$\tilde{y}$	$\pi/2-\theta$	$\tilde{\theta}$	$re^{i\tilde{\theta}}$	$re^{i(\pi/2-\theta)}$	$r(\sin\theta, \cos\theta)$
$y^\sharp$	$\theta+\pi/2$		$re^{i(\theta+\pi/2)}$		$r(-\cos\theta, \sin\theta)$
$\tilde{y}^\sharp$	$\pi-\theta$	$\tilde{\theta}+\pi/2$	$re^{i(\tilde{\theta}+\pi/2)}$	$re^{i(\pi-\theta)}$	$r(-\sin\theta, \cos\theta)$
y'	$\theta+\pi$		$re^{i(\theta+\pi)}$		$r(-\cos\theta, -\sin\theta)$
$\tilde{y}'$	$3\pi/2-\theta$	$\tilde{\theta}+\pi$	$re^{i(\tilde{\theta}+\pi)}$	$re^{i(3\pi/2-\theta)}$	$r(-\sin\theta, -\cos\theta)$
y^*	$\theta+3\pi/2$		$re^{i(\theta+3\pi/2)}$		$r(\cos\theta, -\sin\theta)$
$\tilde{y}^*$	$2\pi-\theta$	$\tilde{\theta}+3\pi/2$	$re^{i(\tilde{\theta}+3\pi/2)}$	$re^{i(2\pi-\theta)}$	$r(\sin\theta, -\cos\theta)$

Related Connections:

x Series	$x\ \tilde{x}\ \tilde{x}^\sharp\ x^\sharp\ x'\ \tilde{x}'\ \tilde{x}^*\ x^*$
y Series	$y\ \tilde{y}\ y^\sharp\ \tilde{y}^\sharp\ y'\ \tilde{y}'\ y^*\ \tilde{y}^*$

$$U(y) = Y = \{y, y^\sharp, y', y^*\} \tag{22}$$
$$\tilde{U}(y) = \tilde{Y} = \{\tilde{y}, \tilde{y}^\sharp, \tilde{y}', \tilde{y}^*\}. \tag{23}$$

Corollary 6 (Eight Corrected Vectors in Order). *The corrected $\{U(y)\}$ cluster contains four vectors associated with θ as a base cross, while the $\{\tilde{U}(y)\}$ cluster has four vectors associated with $\tilde{\theta}$ as a conjugate cross.*

Proof. Referring to Table 2, from $\{y\}$ and $\{\tilde{y}\}$, the four quadrants $\{[00], [10], [11], [01]\}$ are arranged in the counter-clockwise rotation, and the angles of the consequent vectors in each cross are incremented by $\pi/2$.

5.3 *Twelve Representatives of Output Product Vectors*

For the twelve representatives generated after the binary product, it is sufficient to select twelve clusters from the sixty-four product results as the representatives.

Corollary 7 (Twelve Output Vectors After Single Product). *Satisfying Corollary 6, eight vectors take 8×8 times the number of products and then group the results as clusters. Twelve distinguishable vectors are generated.*

Proof. In Table 3, the algebraic formulas and equivalent clusters are separately listed.

The eight vectors after n times of products retain twelve representatives.

Corollary 8 (Twelve Output Vectors After *n* products). *Satisfying Corollary 6, eight vectors take multiple products and merging operations. A total of twelve distinguishable vectors are generated.*

Proof. Referring to Table 4, the algebraic formulas and equivalent arguments are separately listed.

Table 3. Correspondence between output variables and angular measures for one product

$(Y\tilde{Y})^2$	$\{2\theta\}$	$\left\{2\tilde{\theta}\right\}$	Formula	Equal.
$\{y\tilde{y}^*\} := (r,0)^2$	0		r^2e^{i0}	$r^2(1,0)$
$\{yy\} := (y)^2$	2θ		$r^2e^{i2\theta}$	
$\{\tilde{y}\tilde{y}^*\} := (\tilde{y})^2$	$\frac{\pi}{2}-2\theta$	$2\tilde{\theta}$	$r^2e^{i2\tilde{\theta}}$	$r^2e^{i(\frac{\pi}{2}-2\theta)}$
$\{\tilde{y}y\} := (0,r)^2$	$\frac{\pi}{2}$		$r^2e^{i\frac{\pi}{2}}$	$r^2(0,1)$
$\{\tilde{y}\tilde{y}\} := (\tilde{y}^\sharp)^2$	$2\theta+\frac{\pi}{2}$		$r^2e^{i(2\theta+\frac{\pi}{2})}$	
$\{yy^*\} := (\tilde{y}')^2$	$\pi-2\theta$	$2\tilde{\theta}+\frac{\pi}{2}$	$r^2e^{i(2\tilde{\theta}+\frac{\pi}{2})}$	$r^2e^{i(\pi-2\theta)}$
$\{y'\tilde{y}^*\} := (-r,0)^2$	π		$r^2e^{i\pi}$	$r^2(-1,0)$
$\{yy'\} := (y')^2$	$2\theta+\pi$		$r^2e^{i(2\theta+\pi)}$	
$\{\tilde{y}\tilde{y}^\sharp\} := (\tilde{y}')^2$	$\frac{3\pi}{2}-2\theta$	$2\tilde{\theta}+\pi$	$r^2e^{i(2\tilde{\theta}+\pi)}$	$r^2e^{i(\frac{3\pi}{2}-2\theta)}$
$\{\tilde{y}y'\} := (0,-r)^2$	$\frac{3\pi}{2}$		$r^2e^{i\frac{3\pi}{2}}$	$r^2(0,-1)$
$\{\tilde{y}^*\tilde{y}^*\} := (\tilde{y}^*)^2$	$2\theta+\frac{3\pi}{2}$		$r^2e^{i(2\theta+\frac{3\pi}{2})}$	
$\{y^*y\} := (y^*)^2$	$2\pi-2\theta$	$2\tilde{\theta}+\frac{3\pi}{2}$	$r^2e^{i(2\tilde{\theta}+\frac{3\pi}{2})}$	$r^2e^{i(2\pi-2\theta)}$

Table 4. Correspondence between variables and arguments for multiple products

$(Y\tilde{Y})^n$	$\{n\theta\}$	$n\left\{\tilde{\theta}\right\}$	Formula	Equal.
$(r,0)^n$	0		r^ne^{i0}	$r^n(1,0)$
$(y)^n$	$n\theta$		$r^ne^{in\theta}$	
$(\tilde{y})^n$	$\frac{\pi}{2}-n\theta$	$n\tilde{\theta}$	$r^ne^{in\tilde{\theta}}$	$r^ne^{i(\frac{\pi}{2}-n\theta)}$
$(0,r)^n$	$\frac{\pi}{2}$		$r^ne^{i\frac{\pi}{2}}$	$r^n(0,1)$
$(y^\sharp)^n$	$n\theta+\frac{\pi}{2}$		$r^ne^{i(n\theta+\frac{\pi}{2})}$	
$(\tilde{y}^\sharp)^n$	$\pi-n\theta$	$n\tilde{\theta}+\frac{\pi}{2}$	$r^ne^{i(n\tilde{\theta}+\frac{\pi}{2})}$	$r^ne^{i(\pi-n\theta)}$
$(-r,0)^n$	π		$r^ne^{i\pi}$	$r^n(-1,0)$
$(y')^n$	$n\theta+\pi$		$r^ne^{i(n\theta+\pi)}$	
$(\tilde{y}')^n$	$\frac{3\pi}{2}-n\theta$	$n\tilde{\theta}+\pi$	$r^ne^{i(n\tilde{\theta}+\pi)}$	$r^ne^{i(\frac{3\pi}{2}-n\theta)}$
$(0,-r)^n$	$\frac{3\pi}{2}$		$r^ne^{i\frac{3\pi}{2}}$	$r^n(0,-1)$
$(y^*)^n$	$n\theta+\frac{3\pi}{2}$		$r^ne^{i(n\theta+\frac{3\pi}{2})}$	
$(\tilde{y}^*)^n$	$2\pi-n\theta$	$n\tilde{\theta}+\frac{3\pi}{2}$	$r^ne^{i(n\tilde{\theta}+\frac{3\pi}{2})}$	$r^ne^{i(2\pi-n\theta)}$

6. Complex Exponential Functions in Different Quadrants

6.1 Four Quadrant Partition Conditions

Unlike the Knuth (quater-imaginary base) [97] and positional number systems [98] representation, this type of partition is similar to the Poincare hypergeometric distribution to divide the upper half complex plane H^+ (upper half-plane) [96] and the lower half complex plane H^- (lower half-plane) mechanism,

$$H^+ \equiv \{(a,b)|b > 0; a,b \in R\}, H^- \equiv \{(a,b)|b < 0; a,b \in R\}$$

The complex exponential function is based on attributes from relevant positions, and the four quadrants provide a natural structure for attribute features. $y = re^{i\theta}, \tilde{y} = re^{i\tilde{\theta}}, r > 0$, the complex plane can be divided into eight regions on the four quadrants $\{Q_j(y)\}_{j=0}^{3}$, $\{\tilde{Q}_j(y)\}_{j=0}^{3}$.

$$Q_0(y) \equiv \left\{re^{i\theta} | \theta \in \left[0, \frac{\pi}{2}\right)\right\} \tag{24}$$

$$\tilde{Q}_0(y) \equiv \left\{re^{i\tilde{\theta}} | \tilde{\theta} \in \left(0, \frac{\pi}{2}\right]\right\} \tag{25}$$

$$Q_2(y) \equiv \left\{re^{i\theta} | \theta \in \left[\frac{\pi}{2}, \pi\right)\right\} \tag{26}$$

$$\tilde{Q}_2(y) \equiv \left\{re^{i\tilde{\theta}} | \tilde{\theta} \in \left(\frac{\pi}{2}, \pi\right]\right\} \tag{27}$$

$$Q_3(y) \equiv \left\{re^{i\theta} | \theta \in \left[\pi, \frac{3\pi}{2}\right)\right\} \tag{28}$$

$$\tilde{Q}_3(y) \equiv \left\{re^{i\tilde{\theta}} | \tilde{\theta} \in \left(\pi, \frac{3\pi}{2}\right]\right\} \tag{29}$$

$$Q_1(y) \equiv \left\{re^{i\theta} | \theta \in \left[\frac{3\pi}{2}, 2\pi\right)\right\} \tag{30}$$

$$\tilde{Q}_1(y) \equiv \left\{re^{i\tilde{\theta}} | \tilde{\theta} \in \left(\frac{3\pi}{2}, 2\pi\right]\right\} \tag{31}$$

Corollary 9 (Common Part of Two Conjugate Values in Same Region). *Pairwise conjugate exponent representation of* $\{Q_j(y)\}_{j=0}^{3}, \{\tilde{Q}_j(y)\}_{j=0}^{3}$, *share arguments* $\theta, \tilde{\theta} \in \{(0, \frac{\pi}{2}), (\frac{\pi}{2}, \pi), (\pi, \frac{3\pi}{2}), (\frac{3\pi}{2}, 2\pi)\}$, *at four coordinates on the axis, the corresponding value ranges are different.*

Proof. Checking formula (24)–(31), the common part of the same region between the four formula representations is $\{[,), (,]\} \to (,)$.

6.2 *Invariant Properties of Attribute Sign*

Let $S(x)$ denote the sign of the x vector, for example, $x = (-p, q), S(x) = (-, +)$.

Corollary 10 (Paired Conjugate Exponents Have Same Sign). *The conjugate exponent representation of paired* $\{Q_j(y)\}_{j=0}^{3}, \{\tilde{Q}_j(y)\}_{j=0}^{3}$ *have related signs*

$$S(e_j^{i\theta}) = S(e_j^{i\tilde{\theta}}) \in \{(+,+)_{j=0}, (-,+)_{j=2}, (-,-)_{j=3}, (+,-)_{j=1}\}.$$

Proof. From the level of attributes, the four regions of the formula (24)–(31) provide the global regional invariance of the attribute property.

From the representation of multiple iterations, the intrinsic structure of the transforming system can be observed. In this section, the constraint is applied to be the unit length $r = 1, r = r^2 = r^n$ to focus attention on the conformal variations of the conjugate complex exponential function. From the series of complex action results, the law of clustering action is analyzed in an order mode.

Four Mixed Exponential Function States $\{e^{i\theta}, \tilde{e}^{i\tilde{\theta}}, e^{i\tilde{\theta}}, \tilde{e}^{i\theta}\}$ represent two distinguishable states,

$$e^{i\theta} = \tilde{e}^{i\tilde{\theta}} \tag{32}$$

$$e^{i\tilde{\theta}} = \tilde{e}^{i\theta}. \tag{33}$$

Let $\{0, 2, 3, 1\}$ mark the $\{[00], [10], [11], [01]\}$ quadrants. After calibration, $Y \to \{e_j^{i\theta}\}_{j=0}^{3}$, $\tilde{Y} \to \{e_j^{i\tilde{\theta}}\}_{j=0}^{3}$ corresponds to the complex function index expression,

$$y = \tilde{e}_0^{i\tilde{\theta}} = e_0^{i\theta} = e^{i\theta} \tag{34}$$

$$y^{\sharp} = \tilde{e}_2^{i\tilde{\theta}} = e_2^{i\theta} = e^{i(\theta+\frac{\pi}{2})} \tag{35}$$

$$y' = \tilde{e}_3^{i\tilde{\theta}} = e_3^{i\theta} = e^{i(\theta+\pi)} \tag{36}$$

$$y^{*} = \tilde{e}_1^{i\tilde{\theta}} = e_1^{i\theta} = e^{i(\theta+\frac{3\pi}{2})} \tag{37}$$

$$\tilde{y} = \tilde{e}_0^{i\theta} = e_0^{i\tilde{\theta}} = e^{i(\frac{\pi}{2}-\theta)} \tag{38}$$

$$\tilde{y}^{\sharp} = \tilde{e}_2^{i\theta} = e_2^{i\tilde{\theta}} = e^{i(\pi-\theta)} \tag{39}$$

$$\tilde{y}' = \tilde{e}_3^{i\theta} = e_3^{i\tilde{\theta}} = e^{i(\frac{3\pi}{2}-\theta)} \tag{40}$$

$$\tilde{y}^{*} = \tilde{e}_1^{i\theta} = e_1^{i\tilde{\theta}} = e^{i(2\pi-\theta)} \tag{41}$$

$$\{e_j^{i\theta}\}_{j=0}^{3} \to \{e_0^{i\theta}, e_2^{i\theta}, e_3^{i\theta}, e_1^{i\theta}\}, \{e_j^{i\tilde{\theta}}\}_{j=0}^{3} \to \{e_0^{i\tilde{\theta}}, e_2^{i\tilde{\theta}}, e_3^{i\tilde{\theta}}, e_1^{i\tilde{\theta}}\}$$

$$\{Y, \tilde{Y}\} \to \{e_0^{i\theta}, e_0^{i\tilde{\theta}}, e_2^{i\theta}, e_2^{i\tilde{\theta}}, e_3^{i\theta}, e_3^{i\tilde{\theta}}, e_1^{i\theta}, e_1^{i\tilde{\theta}}\}.$$

Theorem 8 (Four Exponential Representations). *Combined with* $\{Q_j(y)\}_{j=0}^{3}, \{\tilde{Q}_j(y)\}_{j=0}^{3}$, *the exponential representations have arguments from* θ *and* $\tilde{\theta}$, *each region contains a pair of conjugate exponents, the adjacent counter-clockwise numbering order is* $\{0, 2, 3, 1\}$, *and the adjacent angle increment is* $\pi/2$.

Proof. Checking the formula (34)–(41), there are only differences in attribute indexes between the Euler formula of the complex exponential function and the four regional formulas, and the vector attribute features can be determined from the signs of the regions.

6.3 Conjugate Clustering Operator

In contrast to the complex variable function, it is convenient to show pairwise complex conjugate representation.

$$\boxed{x \in \text{CP}} \to \begin{Bmatrix} x & \theta \\ x^* & -\theta \end{Bmatrix} : \{x^*x = xx^* = 1\} \to \frac{+\theta \in \text{UHP}}{-\theta \in \text{LHP}} \to \frac{\{re^{i\theta}\}}{\{re^{-i\theta}\}}, \theta \in R$$

From any point x in the first quadrant, conjugate complex numbers $\{x, \tilde{x}\}$ use paired conjugate angles $\{\theta, \tilde{\theta}\}, \theta + \tilde{\theta} = \pi/2$ to be mapped as $\{y, \tilde{y}\}$ after the position and time of conjugate correction.

Quaternary operators $\{\sim, *, \sharp, '\}$ are applied, and their combinations are performed in $\{Y = \{y, y^{\sharp}, y', y^{*}\}, \tilde{Y} = \{\tilde{y}, \tilde{y}^{\sharp}, \tilde{y}', \tilde{y}^{*}\}\}$. There is a natural division of the conjugate complex plane, and the two coordinate regions of adjacent quadrants can be reused. Each complex number is divided into four quadrants by the signs of their attributes. $\{[00] = \{0, \tilde{0}\} \equiv \{y, \tilde{y}\}, [10] = \{2, \tilde{2}\} \equiv \{y^{\sharp}, \tilde{y}^{\sharp}\}, [11] = \{3, \tilde{3}\} \equiv \{y', \tilde{y}'\}, [01] = \{1, \tilde{1}\} \equiv \{y^{*}, \tilde{y}^{*}\}\}$. The symbols are listed corresponding to the complex numbers in each quadrant $\{\{0, \tilde{0}\} : (+,+), \{2, \tilde{2}\} : (-,+), \{3, \tilde{3}\} : (-,-), \{1, \tilde{1}\} : (+,-)\}$. For each complex point, the common ring of the conjugate complex plane contains at least $\{1, i, -1, -i\}$ four coordinate points. There are at most twelve sites to provide necessary and sufficient conditions for the representation of conjugated complex functions.

$$\boxed{x \in \mathrm{CP}} \rightarrow \begin{Bmatrix} y\ \theta \\ \tilde{y}\ \tilde{\theta} \end{Bmatrix} \rightarrow \begin{Bmatrix} YY & Y\tilde{Y} \\ \tilde{Y}Y & \tilde{Y}\tilde{Y} \end{Bmatrix} \rightarrow \frac{\{2, \tilde{2}\} \| \{0, \tilde{0}\}}{\{3, \tilde{3}\} \| \{1, \tilde{1}\}} \rightarrow \frac{\{re_2^{i\theta}, re_2^{i\tilde{\theta}}\} \| \{re_0^{i\theta}, re_0^{i\tilde{\theta}}\}}{\{re_3^{i\theta}, re_3^{i\tilde{\theta}}\} \| \{re_1^{i\theta}, re_1^{i\tilde{\theta}}\}}$$

Based on the two generators, the conjugate complex function maps the square to be transformed into a circle, which provides a circular time variable to be arbitrarily divided. The structure is convenient to support any number of iterative operations. Unlike the classical complex function, which is divided into two partitions, UHP and LHP, conjugate complex clustering divides CP into four coordinates + four quadrants dependent on the four properties of the attributes.

6.4 Pairwise Vector Interaction of Coordinate Projection

The conjugate expressions are equally represented into four groups,

$$e^{i\theta} : e^{i\tilde{\theta}} \tag{42}$$

$$e^{i\theta} : \tilde{e}^{i\theta} \tag{43}$$

$$\tilde{e}^{i\tilde{\theta}} : \tilde{e}^{i\theta} \tag{44}$$

$$\tilde{e}^{i\tilde{\theta}} : e^{i\tilde{\theta}}. \tag{45}$$

For the convenience of analysis, there are two expressions $\{e^{i\theta}, e^{i\tilde{\theta}}\}$ selected for subsequent derivation.

There are sixteen pairs of vectors to support the $\{e_j^{i\theta}\}_{j=0}^{3} \times \{e_j^{i\tilde{\theta}}\}_{j=0}^{3}$ interaction mode. Using the conjugate exponent function, four sets of unit coordinate vectors are formed:

$1 : \{e_0^{i\theta} e_1^{i\tilde{\theta}}, e_1^{i\theta} e_0^{i\tilde{\theta}}, e_2^{i\theta} e_3^{i\tilde{\theta}}, e_3^{i\theta} e_2^{i\tilde{\theta}}\}$ vertical product
$i : \{e_0^{i\theta} e_0^{i\tilde{\theta}}, e_2^{i\theta} e_1^{i\tilde{\theta}}, e_3^{i\theta} e_3^{i\tilde{\theta}}, e_1^{i\theta} e_2^{i\tilde{\theta}}\}$ slope product
$-1 : \{e_0^{i\theta} e_2^{i\tilde{\theta}}, e_2^{i\theta} e_0^{i\tilde{\theta}}, e_1^{i\theta} e_3^{i\tilde{\theta}}, e_3^{i\theta} e_1^{i\tilde{\theta}}\}$ horizontal product
$-i : \{e_0^{i\theta} e_3^{i\tilde{\theta}}, e_3^{i\theta} e_0^{i\tilde{\theta}}, e_2^{i\theta} e_2^{i\tilde{\theta}}, e_1^{i\theta} e_1^{i\tilde{\theta}}\}$ backslope product

There are four equations.

$$e_0^{i\theta} e_1^{i\tilde{\theta}} = e_1^{i\theta} e_0^{i\tilde{\theta}} = e_2^{i\theta} e_3^{i\tilde{\theta}} = e_3^{i\theta} e_2^{i\tilde{\theta}} = 1 \tag{46}$$

$$e_0^{i\theta} e_0^{i\tilde{\theta}} = e_2^{i\theta} e_1^{i\tilde{\theta}} = e_3^{i\theta} e_3^{i\tilde{\theta}} = e_1^{i\theta} e_2^{i\tilde{\theta}} = i \tag{47}$$

$$e_0^{i\theta} e_2^{i\tilde{\theta}} = e_2^{i\theta} e_0^{i\tilde{\theta}} = e_1^{i\theta} e_3^{i\tilde{\theta}} = e_3^{i\theta} e_1^{i\tilde{\theta}} = -1 \tag{48}$$

$$e_0^{i\theta} e_3^{i\tilde{\theta}} = e_3^{i\theta} e_0^{i\tilde{\theta}} = e_2^{i\theta} e_2^{i\tilde{\theta}} = e_1^{i\theta} e_1^{i\tilde{\theta}} = -i \tag{49}$$

Theorem 9 (Four Coordinate Vectors). *Eight conjugate exponent functions are multiplied two by two to form a quaternary unit vector under the action of the formula (46)–(49):* $\{1, i, -1, -i\}$.

Proof. It is simple to substitute the corresponding conjugate exponent formula for calculation to confirm the projection value on each equation.

$$
\begin{aligned}
e_0^{i\theta} e_1^{i\tilde{\theta}} &= e^{i\theta} e^{i(2\pi-\theta)} = e_1^{i\theta} e_0^{i\tilde{\theta}} = e^{i(\frac{3\pi}{2}+\theta)} e^{i(\frac{\pi}{2}-\theta)} = e_2^{i\theta} e_3^{i\tilde{\theta}} = e^{i(\frac{\pi}{2}+\theta)} e^{i(\frac{3\pi}{2}-\theta)} \\
&= e_3^{i\theta} e_2^{i\tilde{\theta}} = e^{i(\pi+\theta)} e^{i(\pi-\theta)} = e^{i2\pi} = (1,0) = 1 \\
e_0^{i\theta} e_0^{i\tilde{\theta}} &= e^{i\theta} e^{i(\frac{\pi}{2}-\theta)} = e_2^{i\theta} e_1^{i\tilde{\theta}} = e^{i(\frac{\pi}{2}+\theta)} e^{i(2\pi-\theta)} = e_3^{i\theta} e_3^{i\tilde{\theta}} = e^{i(\pi+\theta)} e^{i(\frac{3\pi}{2}-\theta)} \\
&= e_1^{i\theta} e_2^{i\tilde{\theta}} = e^{i(\frac{3\pi}{2}+\theta)} e^{i(\pi-\theta)} = e^{i\frac{\pi}{2}} = (0,1) = i \\
e_0^{i\theta} e_2^{i\tilde{\theta}} &= e^{i\theta} e^{i(\pi-\theta)} = e_2^{i\theta} e_0^{i\tilde{\theta}} = e^{i(\frac{\pi}{2}+\theta)} e^{i(\frac{\pi}{2}-\theta)} = e_1^{i\theta} e_3^{i\tilde{\theta}} = e^{i(\frac{3\pi}{2}+\theta)} e^{i(\frac{3\pi}{2}-\theta)} \\
&= e_3^{i\theta} e_1^{i\tilde{\theta}} = e^{i(\pi+\theta)} e^{i(2\pi-\theta)} = e^{i\pi} = (-1,0) = -1 \\
e_0^{i\theta} e_3^{i\tilde{\theta}} &= e^{i\theta} e^{i(\frac{3\pi}{2}-\theta)} = e_2^{i\theta} e_2^{i\tilde{\theta}} = e^{i(\frac{\pi}{2}+\theta)} e^{i(\pi-\theta)} = e_3^{i\theta} e_0^{i\tilde{\theta}} = e^{i(\pi+\theta)} e^{i(\frac{\pi}{2}-\theta)} \\
&= e_1^{i\theta} e_1^{i\tilde{\theta}} = e^{i(\frac{3\pi}{2}+\theta)} e^{i(2\pi-\theta)} = e^{i\frac{3\pi}{2}} = (0,-1) = -i
\end{aligned}
$$

In Fig. 7(a)–(d), the projection sequence of the pair distribution shows four rotations of angle symmetry modes.

Another sixteen pairs of vectors $\{e_j^{i\tilde{\theta}}\}_{j=0}^{3} \times \{e_j^{i\theta}\}_{j=0}^{3}$ are the same as the first sixteen results to exchange two components of the product, and the whole structure can be reduced to a quaternary unit vector.

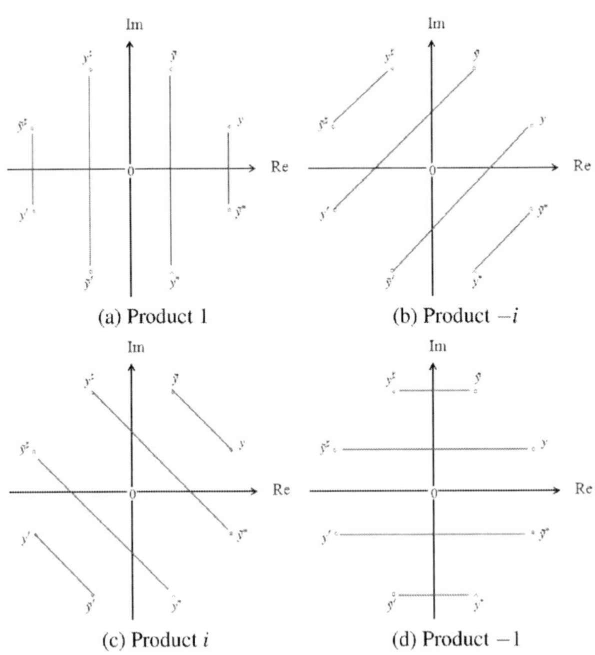

Figure 7. Four pairs of products $\{1, i, -1, -i\}$.

Corollary 11 (Products of Thirty-Two Conjugate Exponents). *Under the action of the formula (46)–(49), the eight conjugate exponent vectors involve a total of thirty-two products of sequence pairs to form a quaternion unit vector:* $\{1, i, -1, -i\}$.

Proof. In the proof process of Theorem 9, each sixteen products generate four vectors, and the result is invariant exchanging the two multiplication integrals, so the whole structure takes pairwise $4 \times 8 = 32$ products.

6.5 Other Projection for Pairwise Vector Interactions

For sixteen pairs of vectors that satisfy $\{e_j^{i\theta}\}_{j=0}^{3} \times \{e_j^{i\theta}\}_{j=0}^{3}$ action, the interactions are

$$e_0^{i\theta} e_0^{i\theta} = e^{i\theta} e^{i\theta} = e^{i2\theta} = e_0^{i2\theta} \tag{50}$$
$$e_0^{i\theta} e_2^{i\theta} = e^{i\theta} e^{i(\theta+\frac{\pi}{2})} = e^{i(2\theta+\frac{\pi}{2})} = e_2^{i2\theta} \tag{51}$$
$$e_0^{i\theta} e_3^{i\theta} = e^{i\theta} e^{i(\theta+\pi)} = e^{i(2\theta+\pi)} = e_3^{i2\theta} \tag{52}$$
$$e_0^{i\theta} e_1^{i\theta} = e^{i\theta} e^{i(\theta+\frac{3\pi}{2})} = e^{i(2\theta+\frac{3\pi}{2})} = e_1^{i2\theta} \tag{53}$$
$$e_2^{i\theta} e_2^{i\theta} = e^{i(\theta+\frac{\pi}{2})} e^{i(\theta+\frac{\pi}{2})} = e^{i(2\theta+\pi)} = e_3^{i2\theta} \tag{54}$$
$$e_2^{i\theta} e_3^{i\theta} = e^{i(\theta+\frac{\pi}{2})} e^{i(\theta+\pi)} = e^{i(2\theta+\frac{3\pi}{2})} = e_1^{i2\theta} \tag{55}$$
$$e_2^{i\theta} e_1^{i\theta} = e^{i(\theta+\frac{\pi}{2})} e^{i((\theta+\frac{3\pi}{2}))} = e^{i(2\theta+2\pi)} = e_0^{i2\theta} \tag{56}$$
$$e_2^{i\theta} e_0^{i\theta} = e^{i(\theta+\frac{\pi}{2})} e^{i\theta} = e^{i(2\theta+\frac{\pi}{2})} = e_2^{i2\theta} \tag{57}$$
$$e_3^{i\theta} e_3^{i\theta} = e^{i(\theta+\pi)} e^{i(\theta+\pi)} = e^{i(2\theta+2\pi)} = e_0^{i2\theta} \tag{58}$$
$$e_3^{i\theta} e_1^{i\theta} = e^{i(\theta+\pi)} e^{i(\theta+\frac{3\pi}{2})} = e^{i(2\theta+\frac{\pi}{2})} = e_2^{i2\theta} \tag{59}$$
$$e_3^{i\theta} e_0^{i\theta} = e^{i(\theta+\pi)} e^{i\theta} = e^{i(2\theta+\pi)} = e_3^{i2\theta} \tag{60}$$
$$e_3^{i\theta} e_2^{i\theta} = e^{i(\theta+\pi)} e^{i(\theta+\frac{\pi}{2})} = e^{i(2\theta+\frac{3\pi}{2})} = e_1^{i2\theta} \tag{61}$$
$$e_1^{i\theta} e_1^{i\theta} = e^{i(\theta+\frac{3\pi}{2})} e^{i(\theta+\frac{3\pi}{2})} = e^{i(2\theta+\pi)} = e_3^{i2\theta} \tag{62}$$
$$e_1^{i\theta} e_0^{i\theta} = e^{i(\theta+\frac{3\pi}{2})} e^{i(\theta)} = e^{i(2\theta+\frac{3\pi}{2})} = e_1^{i2\theta} \tag{63}$$
$$e_1^{i\theta} e_2^{i\theta} = e^{i(\theta+\frac{3\pi}{2})} e^{i(\theta+\frac{\pi}{2})} = e^{i(2\theta+2\pi)} = e_0^{i2\theta} \tag{64}$$
$$e_1^{i\theta} e_3^{i\theta} = e^{i(\theta+\frac{3\pi}{2})} e^{i(\theta+\pi)} = e^{i(2\theta+\frac{\pi}{2})} = e_2^{i2\theta}. \tag{65}$$

When the original angle and the output angle satisfy the $0 < \theta, 2\theta < \frac{\pi}{2}$ condition, the newly created vector in the relevant quadrant keeps the attribute properties invariant. The transfer features are divided into four types of patterns: $\{$A, B, C, D$\}$.

A: $e_0^{i\theta}$	Input	Output	B: $e_2^{i\theta}$	Input	Output	C: $e_3^{i\theta}$	Input	Output	D: $e_1^{i\theta}$	Input	Output
$e_0^{i\theta}$	0	0	$e_2^{i\theta}$	2	3	$e_3^{i\theta}$	3	0	$e_1^{i\theta}$	1	3
$e_2^{i\theta}$	2	2	$e_3^{i\theta}$	3	1	$e_1^{i\theta}$	1	2	$e_0^{i\theta}$	0	1
$e_3^{i\theta}$	3	3	$e_1^{i\theta}$	1	0	$e_0^{i\theta}$	0	3	$e_2^{i\theta}$	2	0
$e_1^{i\theta}$	1	1	$e_0^{i\theta}$	0	2	$e_2^{i\theta}$	2	1	$e_3^{i\theta}$	3	2

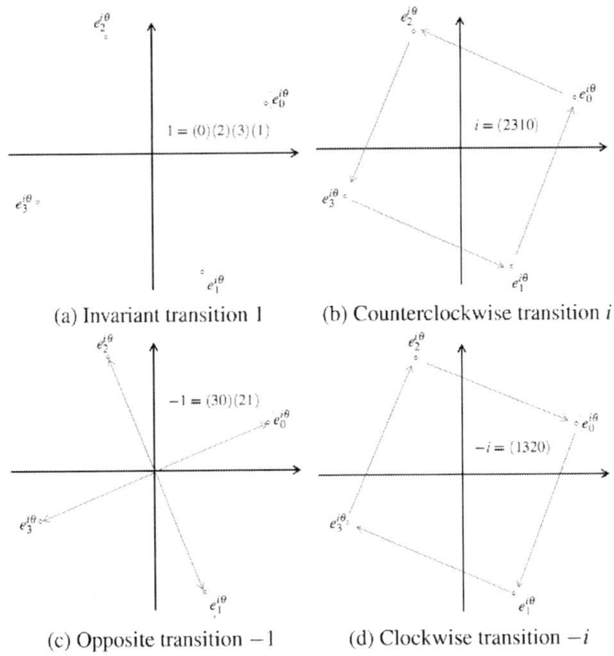

Figure 8. Four transitions of pairwise product outputs $\{1, i, -1, -i\}$: (a) A: $1 = (0)(2)(3)(1)$; (b) B: $i = (2310)$; (c) C: $-1 = (30)(21)$; (d) D: $-i = (1320)$.

Four types of transitions are shown in Fig. 8(a)–(d), and the arrows are applied to indicate the transition direction with the original vector as the starting point.

Four mapping cases are applied in related regions, where the input and output are located in the same or different regions. The four cases are A: (0)(2)(3)(1) invariant; B: (2310) counter-clockwise; C: (30)(12) cross; D: (1320) clockwise.

In the type A case, there is an element of the [00] region selected as the initial quantity, and then four elements (including itself) are selected from the four regions as the paired quantity. After the product of the initial quantity and the paired quantity, the created result remains in the same region as the paired quantity. Therefore, in the Type A case, the calculation formula maintains the region invariance $+1$.

In the Type B case, there is an element of the [10] region selected as the initial quantity, and then four elements (including itself) of four regions are selected as the paired quantity. After multiplying the initial amount and the paired amount, the resulting state represents a transition to a region that is rotated 90 degrees counter-clockwise to the paired amount. The transition can be represented by counter-clockwise $+i$.

In the Type C case, there is an element of the [11] region selected as the initial quantity, and then four elements (including itself) of four regions are selected as the paired quantity. After the product of the initial amount and the paired amount, the resulting state represents a transition to a region rotated 180 degrees from the paired amount. The transition can be represented by reflection -1.

In the Type D case, there is an element of the [01] region selected as the initial quantity, and then four elements (including itself) of four regions are selected as the paired quantity. After multiplying the initial amount and the paired amount, the resulting state represents a transition to a region that is rotated 90 degrees clockwise from the paired amount. The transition can be represented by clockwise $-i$.

Theorem 10 (Sixteen Product Vectors and Four Outputs). *In the* $\{e_j^{i\theta}\}_{j=0}^{3} \times \{e_j^{i\theta}\}_{j=0}^{3}$ *action, the four transitions of output forms are the four Types:* $\{A, B, C, D\}$.

Proof. Checking the formula (50)–(65), every four products constitute a group of transitions, and the transition rules of each group can be clearly distinguished.

For sixteen pairs of vectors that satisfy $\{e_j^{i\tilde{\theta}}\}_{j=0}^{3} \times \{e_j^{i\tilde{\theta}}\}_{j=0}^{3}$ action, the results of the interaction are

$$e_0^{i\tilde{\theta}} e_0^{i\tilde{\theta}} = e^{i(\frac{\pi}{2}-\theta)} e^{i(\frac{\pi}{2}-\theta)} = e^{i(\pi-2\theta)} = e_2^{i2\tilde{\theta}} \tag{66}$$
$$e_0^{i\tilde{\theta}} e_2^{i\tilde{\theta}} = e^{i(\frac{\pi}{2}-\theta)} e^{i(\pi-\theta)} = e^{i(\frac{3\pi}{2}-2\theta)} = e_3^{i2\tilde{\theta}} \tag{67}$$
$$e_0^{i\tilde{\theta}} e_3^{i\tilde{\theta}} = e^{i(\frac{\pi}{2}-\theta)} e^{i(\frac{3\pi}{2}-\theta)} = e^{i(2\pi-2\theta)} = e_1^{i2\tilde{\theta}} \tag{68}$$
$$e_0^{i\tilde{\theta}} e_1^{i\tilde{\theta}} = e^{i(\frac{\pi}{2}-\theta)} e^{i(2\pi-\theta)} = e^{i(\frac{\pi}{2}-2\theta)} = e_0^{i2\tilde{\theta}} \tag{69}$$
$$e_2^{i\tilde{\theta}} e_2^{i\tilde{\theta}} = e^{i(\pi-\theta)} e^{i(\pi-\theta)} = e^{i(2\pi-2\theta)} = e_1^{i2\tilde{\theta}} \tag{70}$$
$$e_2^{i\tilde{\theta}} e_3^{i\tilde{\theta}} = e^{i(\pi-\theta)} e^{i(\frac{3\pi}{2}-\theta)} = e^{i(\frac{\pi}{2}-2\theta)} = e_0^{i2\tilde{\theta}} \tag{71}$$
$$e_2^{i\tilde{\theta}} e_1^{i\tilde{\theta}} = e^{i(\pi-\theta)} e^{i(2\pi-\theta)} = e^{i(\pi-2\theta)} = e_2^{i2\tilde{\theta}} \tag{72}$$
$$e_2^{i\tilde{\theta}} e_0^{i\tilde{\theta}} = e^{i(\pi-\theta)} e^{i(\frac{\pi}{2}-\theta)} = e^{i(\frac{3\pi}{2}-2\theta)} = e_3^{i2\tilde{\theta}} \tag{73}$$
$$e_3^{i\tilde{\theta}} e_3^{i\tilde{\theta}} = e^{i(\frac{3\pi}{2}-\theta)} e^{i(\frac{3\pi}{2}-\theta)} = e^{i(\pi-2\theta)} = e_2^{i2\tilde{\theta}} \tag{74}$$
$$e_3^{i\tilde{\theta}} e_1^{i\tilde{\theta}} = e^{i(\frac{3\pi}{2}-\theta)} e^{i(2\pi-\theta)} = e^{i(\frac{3\pi}{2}-2\theta)} = e_3^{i2\tilde{\theta}} \tag{75}$$
$$e_3^{i\tilde{\theta}} e_0^{i\tilde{\theta}} = e^{i(\frac{3\pi}{2}-\theta)} e^{i(\frac{\pi}{2}-\theta)} = e^{i(2\pi-2\theta)} = e_1^{i2\tilde{\theta}} \tag{76}$$
$$e_3^{i\tilde{\theta}} e_2^{i\tilde{\theta}} = e^{i(\frac{3\pi}{2}-\theta)} e^{i(\pi-\theta)} = e^{i(\frac{\pi}{2}-2\theta)} = e_0^{i2\tilde{\theta}} \tag{77}$$
$$e_1^{i\tilde{\theta}} e_1^{i\tilde{\theta}} = e^{i(2\pi-\theta)} e^{i(2\pi-\theta)} = e^{i(2\pi-2\theta)} = e_1^{i2\tilde{\theta}} \tag{78}$$
$$e_1^{i\tilde{\theta}} e_0^{i\tilde{\theta}} = e^{i(2\pi-\theta)} e^{i(\frac{\pi}{2}-\theta)} = e^{i(\frac{\pi}{2}-2\theta)} = e_0^{i2\tilde{\theta}} \tag{79}$$
$$e_1^{i\tilde{\theta}} e_2^{i\tilde{\theta}} = e^{i(2\pi-\theta)} e^{i(\pi-\theta)} = e^{i(\pi-2\theta)} = e_2^{i2\tilde{\theta}} \tag{80}$$
$$e_1^{i\tilde{\theta}} e_3^{i\tilde{\theta}} = e^{i(2\pi-\theta)} e^{i(\frac{3\pi}{2}-\theta)} = e^{i(\frac{3\pi}{2}-2\theta)} = e_3^{i2\tilde{\theta}}. \tag{81}$$

When the original angle and output angle satisfy $0 < \tilde{\theta}, 2\tilde{\theta} < \pi/2$, the newly created vector in the relevant region follows the attribute property changes. The transitions are identified into four types: {E, F, G, H}.

E: $e_0^{i\tilde{\theta}}$	Input	Output
$e_0^{i\tilde{\theta}}$	0	2
$e_2^{i\tilde{\theta}}$	2	3
$e_3^{i\tilde{\theta}}$	3	1
$e_1^{i\tilde{\theta}}$	1	0

F: $e_2^{i\tilde{\theta}}$	Input	Output
$e_2^{i\tilde{\theta}}$	2	1
$e_3^{i\tilde{\theta}}$	3	0
$e_1^{i\tilde{\theta}}$	1	2
$e_0^{i\tilde{\theta}}$	0	3

G: $e_3^{i\tilde{\theta}}$	Input	Output
$e_3^{i\tilde{\theta}}$	3	2
$e_1^{i\tilde{\theta}}$	1	3
$e_0^{i\tilde{\theta}}$	0	1
$e_2^{i\tilde{\theta}}$	2	0

H: $e_1^{i\tilde{\theta}}$	Input	Output
$e_1^{i\tilde{\theta}}$	1	1
$e_0^{i\tilde{\theta}}$	0	0
$e_2^{i\tilde{\theta}}$	2	2
$e_3^{i\tilde{\theta}}$	3	3

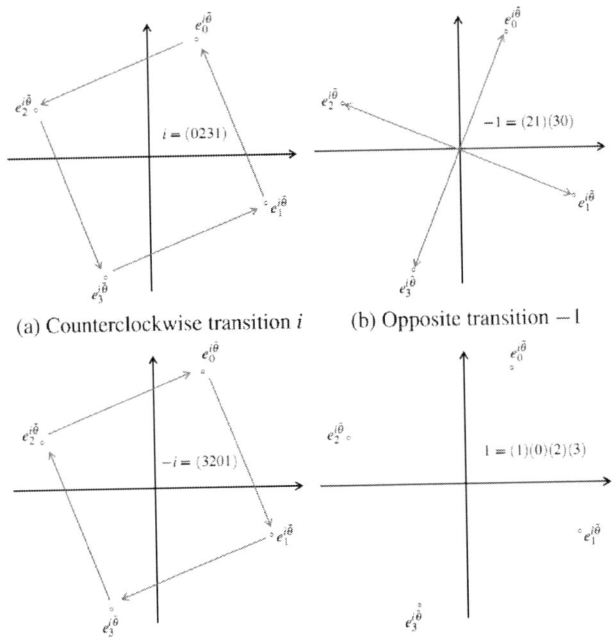

(a) Counterclockwise transition i (b) Opposite transition -1

(c) Clockwise transition $-i$ (d) Invariant transition $+1$.

Figure 9. Four conjugate product outputs $\{i, -1, -i, 1\}$: (a) E:$i = (0231)$; (b) F:$-1 = (21)(30)$; (c) G:$-i = (5301)$; (d) H:$1 = (1)(0)(2)(3)$.

The four types of transitions are shown in Fig. 9(a)–(d), and the arrows are used to describe the mapping direction with the original vector as the starting point.

For four mapping groups on related regions, pairwise conjugate input and conjugate output are located in the same or different regions. The four types of cases are E: (0231) counter-clockwise; F: (21)(30) cross; G: (3201) clockwise; and H: (1)(0)(2)(3) unchanged.

In the Type E case, a conjugated element of the [00] region is selected as the initial quantity, and then four regions corresponding to four conjugated elements (including itself) are selected as the paired quantity. After multiplying the initial amount and the paired amount, the resulting state represents a transition to a region that is rotated 90 degrees counter-clockwise to the paired amount. The transition is represented by counter-clockwise $+i$.

In the Type F case, a conjugated element of the [10] region is selected as the initial quantity, and then four regions corresponding to four conjugated elements (including itself) are selected as the paired quantity. After the product of the initial amount and the paired amount, the resulting state represents a transition to a region rotated 180 degrees from the paired amount. The transition is represented by negative -1.

In the Type G case, a conjugated element of a [11] region is selected as the initial quantity, and then four regions corresponding to four conjugated elements (including itself) are selected as the paired quantity. After multiplying the initial amount and the paired amount, the resulting state represents a transition to a region that is rotated 90 degrees clockwise from the paired element. The transition is represented by clockwise $-i$.

In the Type H case, a conjugated element of the [01] region is selected as the initial quantity, and then four regions corresponding to four conjugated elements (including itself) are selected as the paired quantity. After the product of the initial quantity and the paired

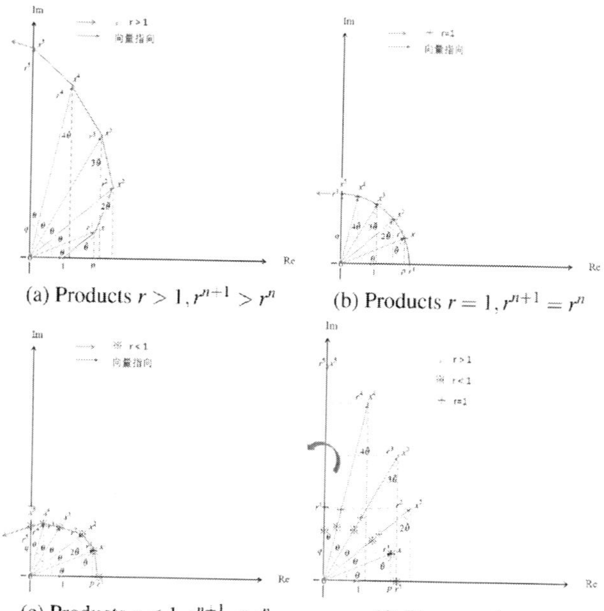

(a) Products $r > 1, r^{n+1} > r^n$ (b) Products $r = 1, r^{n+1} = r^n$

(c) Products $r < 1, r^{n+1} < r^n$ (d) Three products

Figure 10. Four groups of pair products counter-clockwise: (a) $r > 1$; (b) $r = 1$; (c) $r < 1$; (d) $r : \{> 1, = 1, < 1\}$.

quantity, the formed state representation remains in the same region as the paired quantity. The transition maintains the region invariance $+1$.

Theorem 11 (Sixteen Conjugate Product Vectors and Four Outputs). *In the* $\{e_j^{i\tilde{\theta}}\}_{j=0}^{3} \times \{e_j^{i\tilde{\theta}}\}_{j=0}^{3}$ *action, the four groups of output forms are:* $\{E, F, G, H\}$.

Proof. Checking the formula (66)–(81), every four products constitute a group of transitions, and each transition law can be clearly distinguished.

Corollary 12 (Four Transition Groups). *From the perspective of the transition attribute mode, the eight groups of pairs are clustered into four groups under the product transition:* $\{A, H\}$ *is equivalent to* $+1$*;* $\{B, E\}$ *counter-clockwise* $+i$*;* $\{C, F\}$ *negative conjugate* -1*;* $\{D, G\}$ *clockwise* $-i$.

6.6 Twelve Product Outputs

From the view of output classification, sixty-four sets of multiplication points are divided into three groups: $\{\{Y\tilde{Y}, \tilde{Y}Y\}, \{YY\}, \{\tilde{Y}\tilde{Y}\}\}$ corresponding to the product of $\{32, 16, 16\}$ groups.

Thirty-two sets of products between two sets of conjugate ring elements generate four coordinate vectors to output $\{1, i, -1, -i\}$

$$\left\{ \begin{matrix} & Y\tilde{Y} \\ \tilde{Y}Y & \end{matrix} \right\} : \left\{ \begin{matrix} y\tilde{y}^* = \tilde{y}^* y = 1 \\ y\tilde{y} = \tilde{y}y = i \\ y\tilde{y}^\sharp = \tilde{y}^\sharp y = -1 \\ y\tilde{y}' = \tilde{y}' y = -i \end{matrix} \right\} : \left\{ \quad \right\}$$

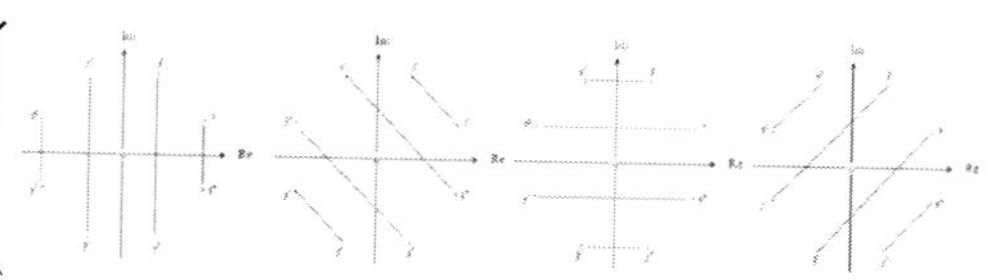

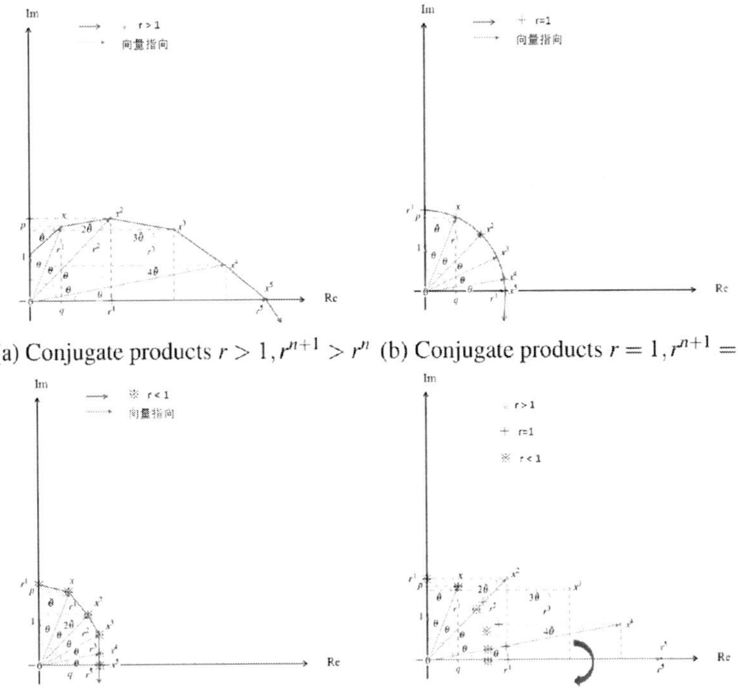

(a) Conjugate products $r > 1, r^{n+1} > r^n$ (b) Conjugate products $r = 1, r^{n+1} = r^n$

(c) Conjugate products $r < 1, r^{n+1} < r^n$ (d) Three conjugate products

Figure 11. Four groups of pair conjugate products clockwise: (a) $r > 1$; (b) $r = 1$; (c) $r < 1$; (d) $r : \{> 1, = 1, < 1\}$.

The cross quadruple of the original state is associated with the four output orders of the sixteen products: $\{1, i, -1, -i\}$

$\{ Y^2 = YY \} :$

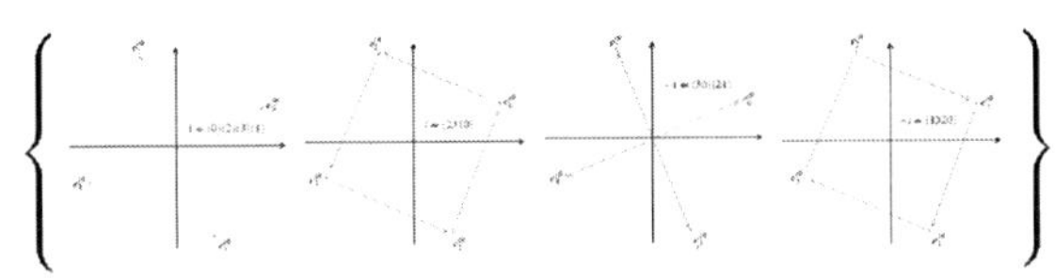

The cross quadruple of the conjugate states is associated with the four output orders of the sixteen products: $\{i, -1, -i, 1\}$

$\{ \tilde{Y}^2 = \tilde{Y}\tilde{Y} \} :$

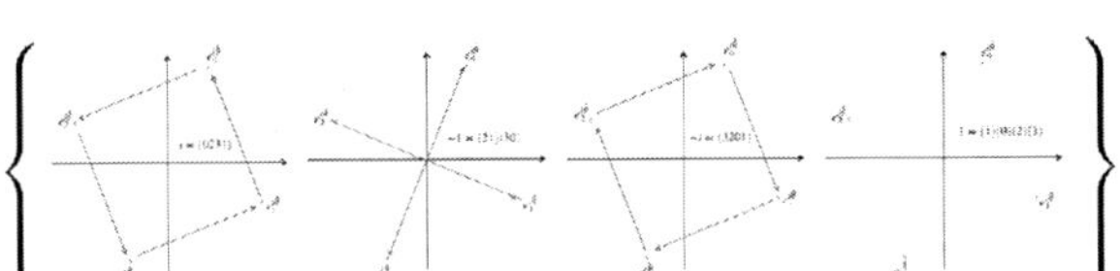

Two sets of conjugated cross structures are multiplied to form three types of output: thirty-two cross products form the four coordinate projections, sixteen cross products form four groups of new variable outputs, and sixteen conjugate cross products form four groups of new conjugate variable outputs. The sixteen cross products are projected as twelve vectors of distinguishable features.

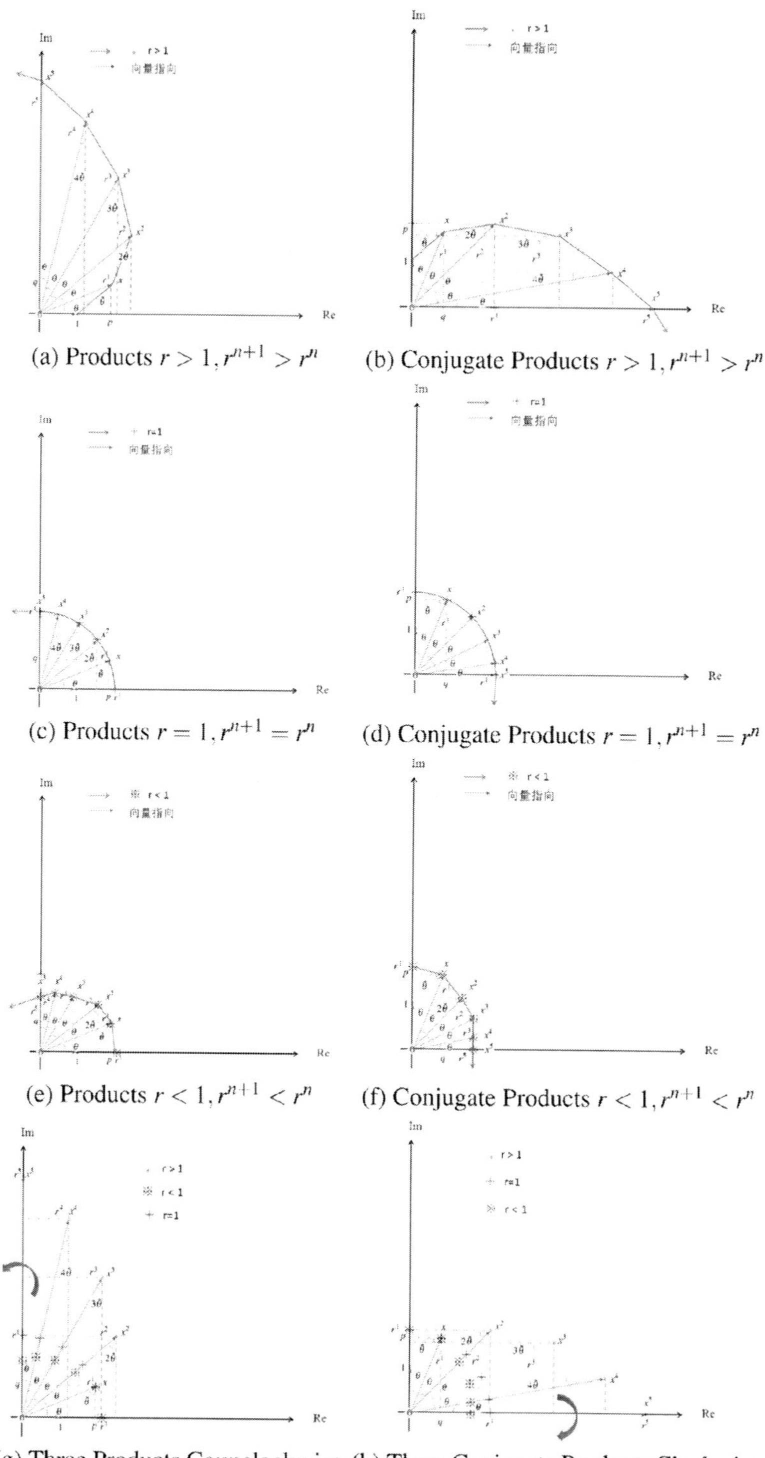

(a) Products $r > 1, r^{n+1} > r^n$ (b) Conjugate Products $r > 1, r^{n+1} > r^n$

(c) Products $r = 1, r^{n+1} = r^n$ (d) Conjugate Products $r = 1, r^{n+1} = r^n$

(e) Products $r < 1, r^{n+1} < r^n$ (f) Conjugate Products $r < 1, r^{n+1} < r^n$

(g) Three Products Counclockwise (h) Three Conjugate Products Clockwise

Figure 12. Eight product output maps: (a) (b) $r > 1$; (c) (d) $r = 1$; (e) (f) $r < 1$; (g) (h) $r : \{> 1, = 1, < 1\}$.

7. Iterative Interaction of Radius Change Mode

7.1 Base Change

Under the action of multiplication, the vector length r will be accompanied by n multiplications, and the argument may change associated with n times. The iterative vector length follows three different trends: $\{r^n < r^{n+1}\}, \{r^n = r^{n+1}\}, \{r^n > r^{n+1}\}$.

Conjugate Quadruple	$\{re_0^{i\theta}, re_2^{i\theta}, re_3^{i\theta}, re_1^{i\theta}\}, \{re_0^{i\tilde{\theta}}, re_2^{i\tilde{\theta}}, re_3^{i\tilde{\theta}},, re_1^{i\tilde{\theta}}\}$
Conjugate Octal	$re_0^{i\theta}, re_0^{i\tilde{\theta}}, re_2^{i\theta}, re_2^{i\tilde{\theta}}, re_3^{i\theta}, re_3^{i\tilde{\theta}}, re_1^{i\theta}, re_1^{i\tilde{\theta}}$
Squared Conjugate Quadruple	$\{r^2e_0^{i2\theta}, r^2e_2^{i2\theta}, r^2e_3^{i2\theta}, r^2e_1^{i2\theta}\}, \{r^2e_0^{i2\tilde{\theta}}, r^2e_2^{i2\tilde{\theta}}, r^2e_3^{i2\tilde{\theta}}, r^2e_1^{i2\tilde{\theta}}\}$
Squared Conjugate Octal	$r^2e_0^{i2\theta}, r^2e_0^{i2\tilde{\theta}}, r^2e_2^{i2\theta}, r^2e_2^{i2\tilde{\theta}}, r^2e_3^{i2\theta}, r^2e_3^{i2\tilde{\theta}}, r^2e_1^{i2\theta}, r^2e_1^{i2\tilde{\theta}}$
n Power Conjugate Quaternion	$\{r^ne_0^{in\theta}, r^ne_2^{in\theta}, r^ne_3^{in\theta}, r^ne_1^{in\theta}\}, \{r^ne_0^{in\tilde{\theta}}, r^ne_2^{in\tilde{\theta}}, r^ne_3^{in\tilde{\theta}}, r^ne_1^{in\tilde{\theta}}\}$
n Power Conjugate Octal	$r^ne_0^{in\theta}, r^ne_0^{in\tilde{\theta}}, r^ne_2^{in\theta}, r^ne_2^{in\tilde{\theta}}, r^ne_3^{in\theta}, r^ne_3^{in\tilde{\theta}}, r^ne_1^{in\theta}, r^ne_1^{in\tilde{\theta}}$
$s\theta < \frac{\pi}{2}$ Multiple Angle	$r^1, r^s, r^{2s}, r^{(t-1)s}, r^{ts}, s = \left\lfloor \frac{\pi}{2\theta} \right\rfloor, \theta > 0$
Isovalue Cycle Condition	$\{r_j\}_{j=1}^m, r_j \neq r_k, j \neq k, r_j^n = r_j^{n+1}, r_1^n = r_1^{n+1}, r_2^n = r_2^{n+1}, r_m^n = r_m^{n+1}.$

7.2 Typical Change

In Fig. 10–12, a single set and three sets of simultaneous iteration results formed by three sets of conditions are shown.

In the four diagrams of Fig. 10(a)–(d), Fig. (a) shows that the vector starts from the horizontal coordinate counter-clockwise, the vector length and argument increase synchronously with the increase of the number of iterations, and then crosses the vertical coordinate; Fig. (b) shows that the vector starts from the horizontal coordinate, the length of the vector remains constant as the number of iterations increases, and the argument increases synchronously, and then crosses the vertical coordinate; Fig. (c) shows that the vector starts from the horizontal coordinate, as the number of iterations increases, the vector length shortens, while the argument grows steadily across the vertical coordinate; Fig. (d) shows a sum of the effects of the three cases (a)–(c).

In the four diagrams of Fig. 11(a)–(d), Fig. (a) shows that the vector starts from the vertical coordinate clockwise, and the vector length and argument increase synchronously with the increase in the number of iterations, crossing the horizontal coordinates. Figure (b) shows that the vector starts from the vertical coordinate, the length of the vector remains constant as the number of iterations increases, and the argument increases synchronously, crossing the horizontal coordinate. Figure (c) shows that the vector starts from the vertical coordinate. As the number of iterations increases, the vector length shortens, while the argument grows steadily, crossing the horizontal coordinate. Figure (d) shows the combination of three cases (a)–(c).

Of the eight diagrams in Fig. 12(a)–(h), three patterns of change are shown in the first three pairs of diagrams: (a)(b) growth and (c)(d) equal (e)(f) shrinkage; the case of the (g)(h) triplet combination is shown in the fourth set of diagrams.

Theorem 12 (Three Vector Length Growths + Six Change Modes). *Use a pair of conjugate exponents* $\{e_0^{i\theta}, e_0^{i\tilde{\theta}}\}$ *iterates, from the starting point of the horizontal coordinate and the vertical coordinate, respectively forming three groups of counter-clockwise and clockwise. As the number of iterations increases, the argument increases synchronously, but the length of the vector is increased, unchanged, and shrinks. Three mechanisms showing six growth patterns are identified.*

Proof. The increasing and shrinking modes clearly form different envelopes, and when the vector length is a constant, the reciprocal rotation directions can be distinguished.

7.3 Step-by-Step Iteration

The cross-quadruple variables can be performed as the four classes of iteration outputs of n products: $\{>, =, <, \text{superposition}\}$

$\{ Y^n = Y \cdots Y \}:$

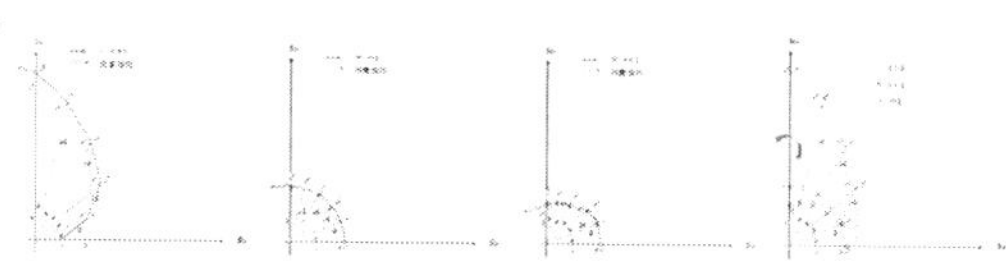

The conjugate cross-quadruple variables can be performed as four classes of iterative outputs of n products: $\{>, =, <, \text{superposition}\}$

$\{ \tilde{Y}^n = \tilde{Y} \cdots \tilde{Y} \}:$

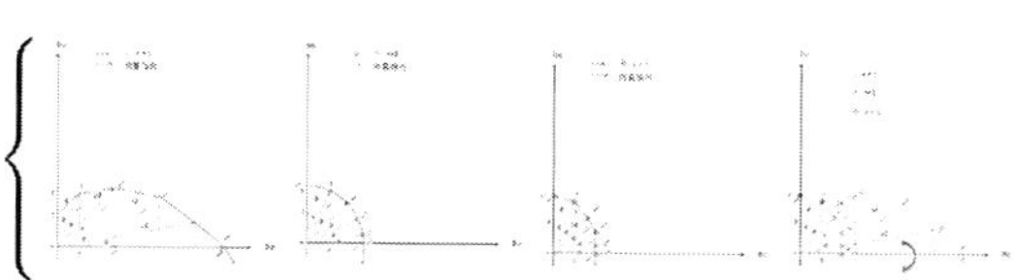

In the cases of various iterations, it is easy to observe the effect of simultaneous changes in vector lengths and arguments in the increasing or decreasing iterations. Compared with the equal-length vector mode, only the angles change regularly. Usually, a single generator is needed, and after repeated iterations, a definite circle can be generated.

7.4 Global Variation

In Fig. 13–18, a single set and multiple sets of iteration results formed by three sets of conditions are shown.

Twelve diagrams are given in Fig. 13(a)–(l).

Figures (a)–(c) depict the enlarged spiral. In Fig. (a), the initial vector is from the four points on the four coordinates counter-clockwise with the increase in the number of iterations, the vector length and the argument grow synchronously, and the rotation of the entire cycle feels like four sets of spirals from small to large. In Fig. (b), the vector starts from the

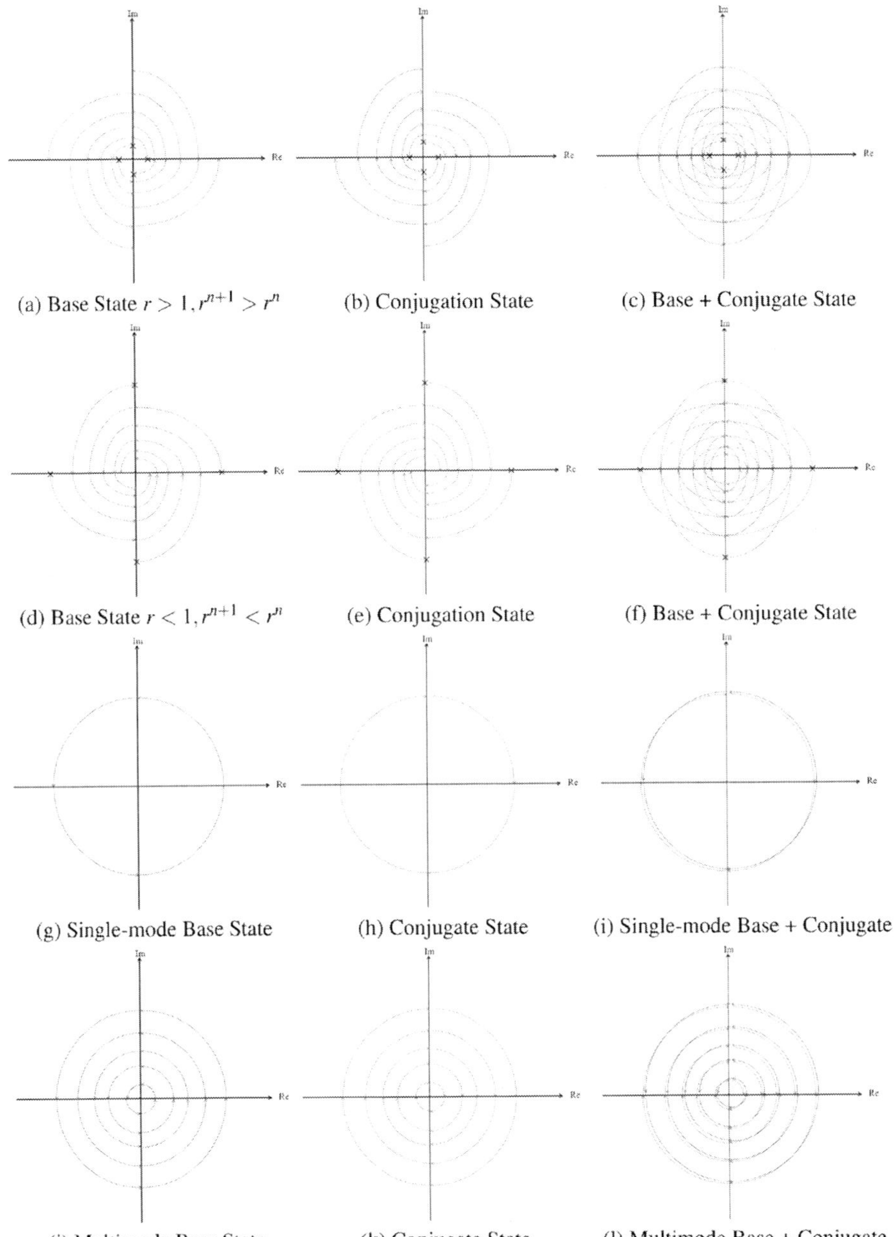

(a) Base State $r > 1, r^{n+1} > r^n$ (b) Conjugation State (c) Base + Conjugate State

(d) Base State $r < 1, r^{n+1} < r^n$ (e) Conjugation State (f) Base + Conjugate State

(g) Single-mode Base State (h) Conjugate State (i) Single-mode Base + Conjugate

(j) Multimode Base State (k) Conjugate State (l) Multimode Base + Conjugate

Figure 13. Two-color three groups of whole mapping independent + superposition output: {(a)–(c)} $r > 1$; {(d)–(e)} $r < 1$; {(g)–(i)} single mode $\{r_j\}_{j=1}^{m}, r_j \neq r_k, j \neq k, r_j^{n+1} = r_j^n$; {(j)–(l)} multiple modes $r : \{> 1, = 1, < 1\}$.

starting point position on the four clockwise coordinates, the length of the vector increases synchronously with the increase in the number of iterations, the angle of argument increases synchronously, and the visual perception of completing the entire cycle is four groups of spirals from small to large. Figure (c) shows the superposition distribution of (a)+(b), and the entire distribution is composed of multiple nested ellipses of different sizes.

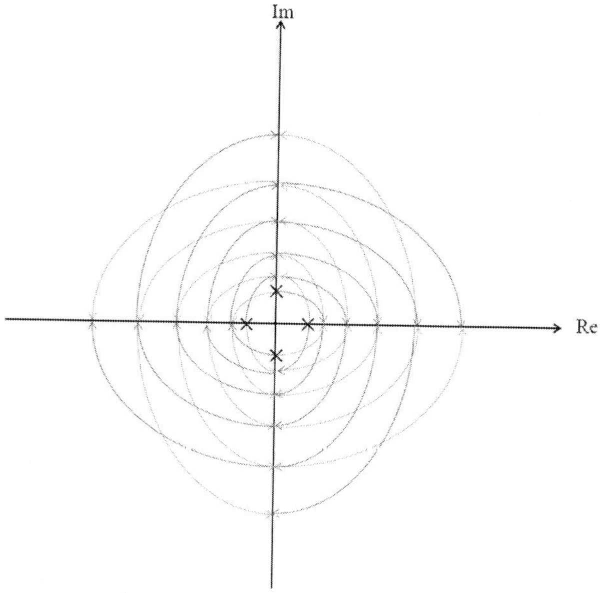

(a) Base + Conjugate States $r > 1, r^{n+1} > r^n$

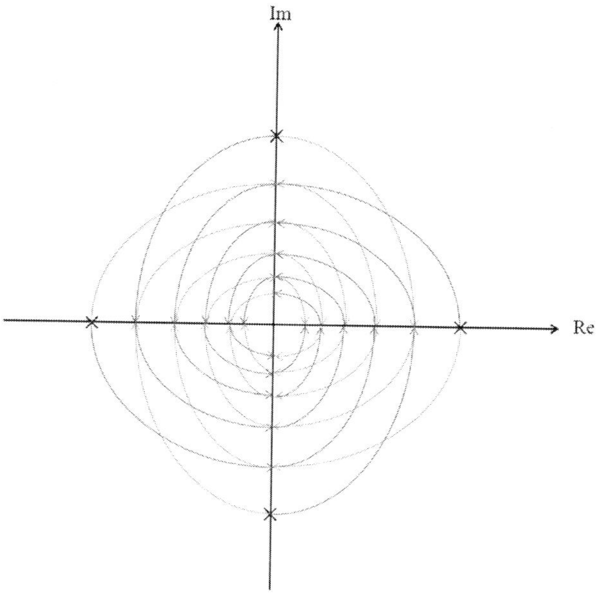

(b) Base + Conjugate States $r < 1, r^{n+1} < r^n$

Figure 14. Two-color two groups of whole mapping independent + superposition output: (a) $r > 1$; (b) $r < 1$.

Figures (d)–(f) show the reduced spiral. In Fig. (d), the initial vector starts from the four points on the four coordinates counter-clockwise and with the increase in the number of iterations. The vector length changes, the argument increases synchronously, and the rotation of the entire cycle feels like four sets of spirals from large to small. Figure (e) shows that the initial vector starts from the four points on the four coordinates clockwise. As the number of iterations increases, the length of the vector shrinks and decreases. With the

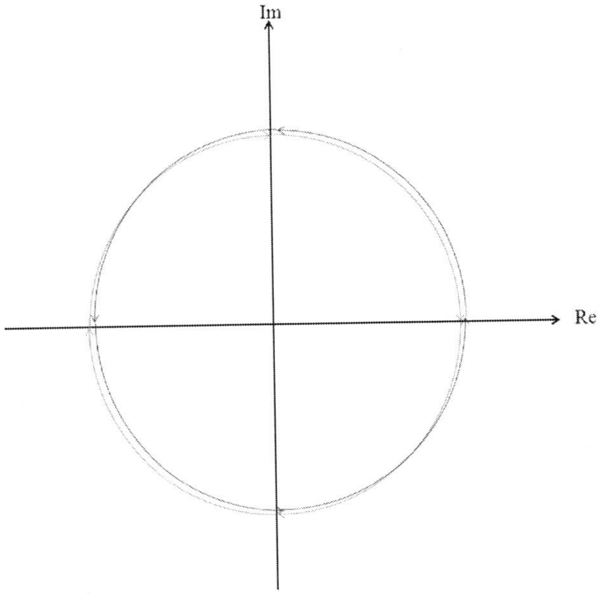

(a) Single Mode Base + Conjugate State

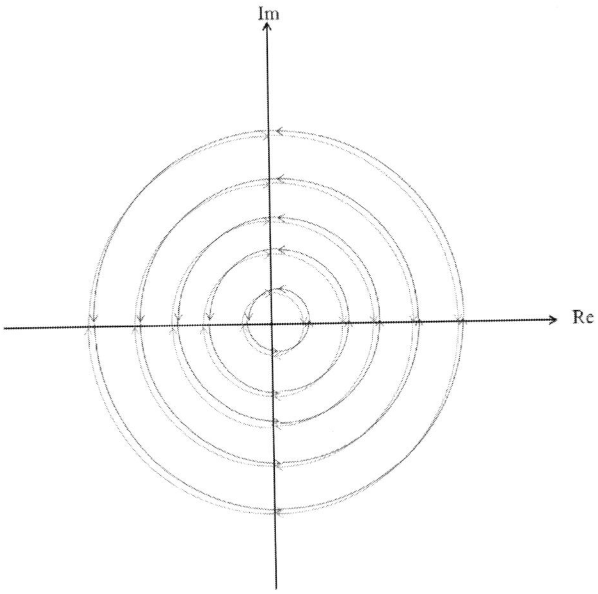

(b) Multimode Base + Conjugate State

Figure 15. Two-color two sets of whole mapping independent + superposition output: (a) single mode; $\{r_j\}_{j=1}^m, r_j \neq r_k, j \neq k, r_j^{n+1} = r_j^n$; (b) multiple modes.

synchronous increase of the argument, the visual perception of the entire cycle of rotation is from large to small. The group of spirals in Fig. (f) shows the superposition of (d)+(e), and the entire distribution is composed of multiple nested ellipses of different proportions.

Figures (g)–(i) depict a single circle. In Fig. (g), the initial vector starts from the four point positions on each coordinate counter-clockwise, and the entire cycle rotation resembles a complete circle after completion. Figure (h) shows that the initial vector starts from

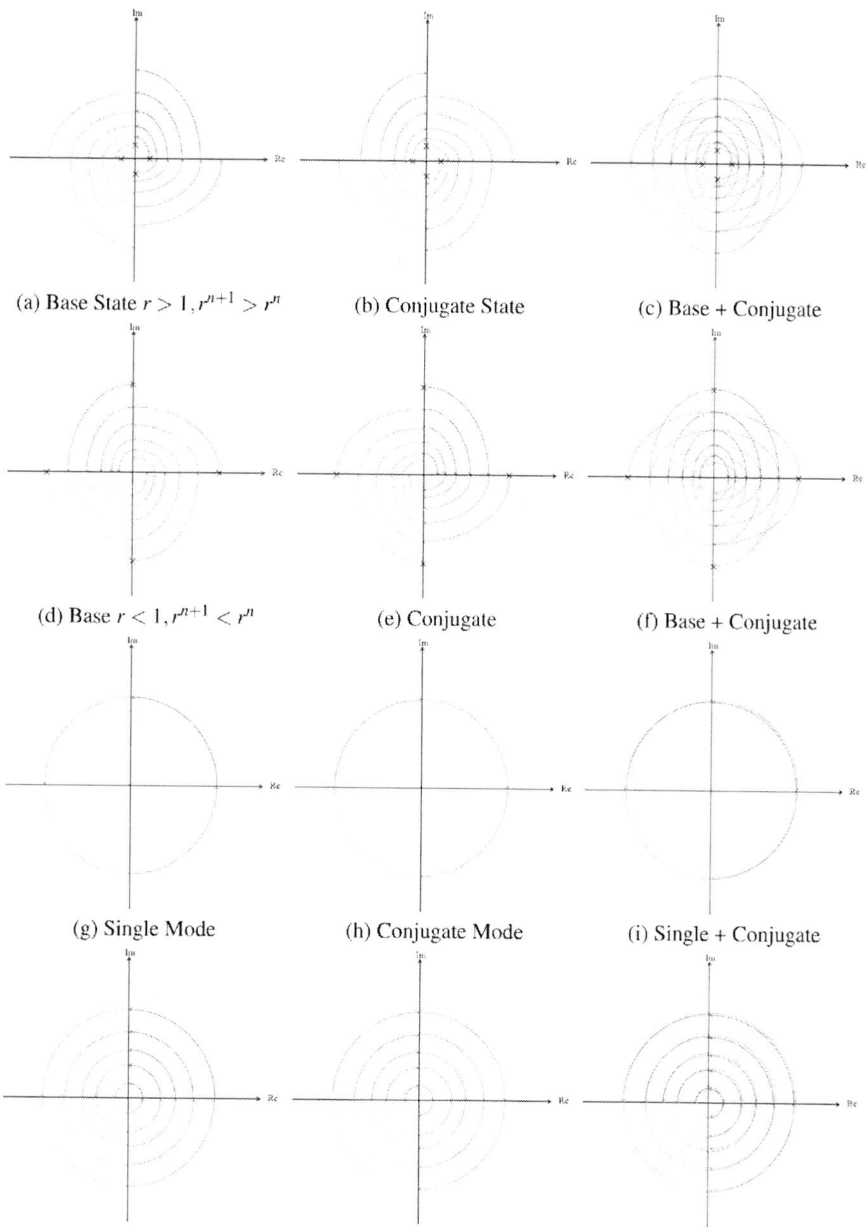

(a) Base State $r > 1, r^{n+1} > r^n$ (b) Conjugate State (c) Base + Conjugate

(d) Base $r < 1, r^{n+1} < r^n$ (e) Conjugate (f) Base + Conjugate

(g) Single Mode (h) Conjugate Mode (i) Single + Conjugate

(g) Multiple Modes (h) Conjugate Modes (i) Multimodes + Conjugate

Figure 16. Three groups on eight color outputs $\{(a)–(c)\}$ $r > 1$; $\{(d)–(e)\}$ $\{r_j\}_{j=1}^{m}, r_j \neq r_k, j \neq k, r_j^{n+1} = r_j^n$; $\{(g)–(i)\}$ $r < 1$; $r : \{> 1, = 1, < 1\}$.

the point position on a single coordinate. Starting from the clock, along with the synchronous increase of the argument, it completes a circle of the whole ring. Figure (i) shows the superimposed distribution of the two circles (g) + (h), and the overall distributions are two color rings.

Figures (j)–(l) depict multiple circles. In Fig. (j), the initial vector starts from multiple points on multiple coordinates counter-clockwise. After completion, the whole cycle

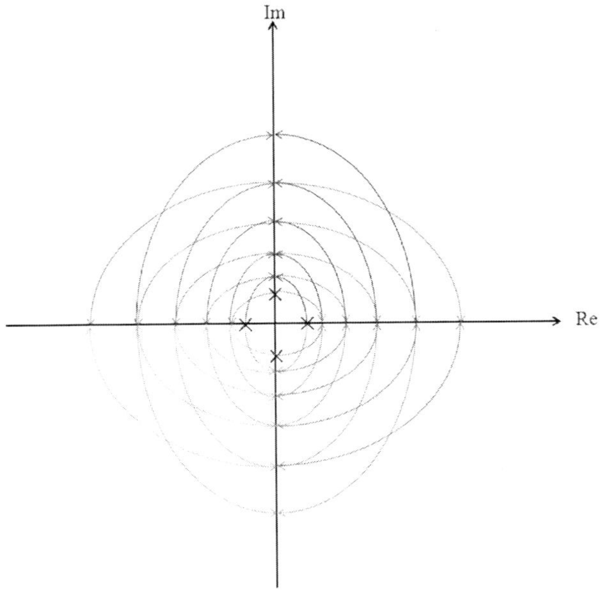

(a) Base + Conjugate $r > 1, r^{n+1} > r^n$

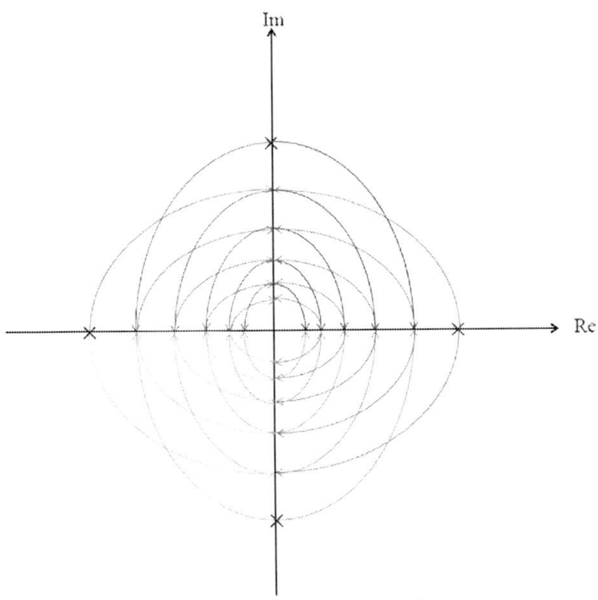

(d) Base + Conjugate $r < 1, r^{n+1} < r^n$

Figure 17. Two groups on eight color outputs: (a) $r > 1$; (b) $r < 1$.

rotation feels in multiple complete circles; Fig. (k) shows that the initial vector starts from multiple points on multiple coordinates. The position starts clockwise, along with the synchronous increase of the argument, to complete multiple circles of the whole rings; Fig. (l) shows the superimposed distribution of (j)+(k) multiple circles, and the overall distribution is composed of multilayer two-color rings.

In the two figures of Fig. 14(a)–(b), Fig. (a) is an enlarged diagram of Fig. 13(c), and the whole distribution is a number of ellipses of different sizes on the nested composition.

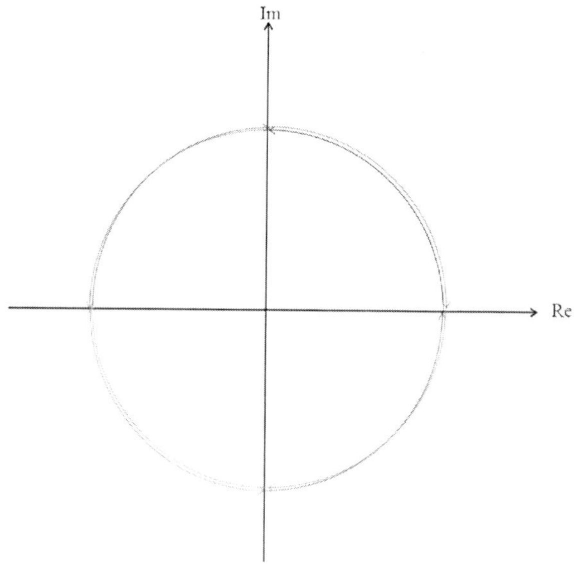

(a) Single mode + conjugate

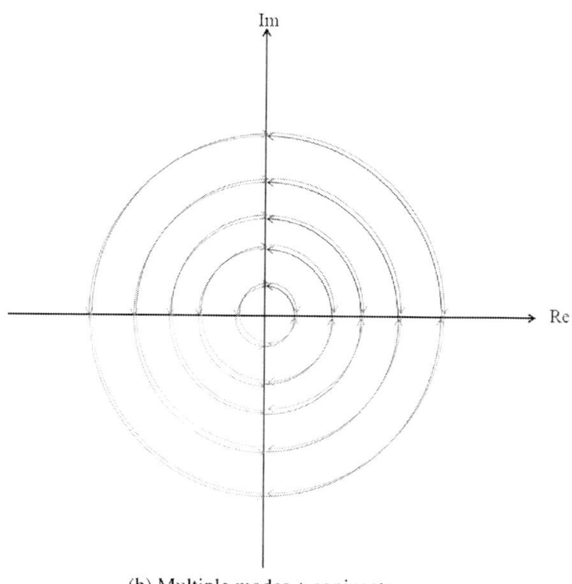

(b) Multiple modes + conjugate

Figure 18. Two groups on eight color outputs: (a) single mode $\{r_j\}_{j=1}^{m}, r_j \neq r_k, j \neq k, r_j^{n+1} = r_j^{n}$; (b) multiple modes.

Figure (b) shows the enlarged illustration of Fig. 13(f); the entire distribution is composed of multiple nested ellipses of different sizes.

In the two figures of Fig. 15(a)–(b), Fig. (a) shows an enlarged diagram of Fig. 13(i), and the whole distribution is two colored circles. Figure (b) shows an enlarged view of Fig. 13(l); the overall distribution consists of multiple two-color rings.

In the twelve diagrams of Fig. 16(a)–(l), the basic distribution is similar to that of Fig. 13(a)–(l), but the four regions use different colors to separate the differences between counter-clockwise and clockwise, and eight groups of different colors can be distinguished.

In the two diagrams of Fig. 17(a)–(b), Fig. (a) shows the enlarged diagram of Fig. 14(c), and the whole distribution shows eight colors to form multiple nesting ellipses of different sizes. Figure (b) shows the enlarged diagram of Fig. 14(f). The entire distribution is composed of eight colors to form multiple ellipses of different nested sizes.

In the two figures of Fig. 18(a)–(b), Fig. (a) shows an enlarged view of Fig. 14(i), which is distributed as an eight-color circle. Figure (b) shows an enlarged view of Fig. 14(l); the whole distribution consists of multiple eight-color rings.

Theorem 13 (Grow and Shrink Mode under Multiple Rotations). *Using four pairs of conjugate exponents* $\{\{r^k e_j^{ik\theta}\}_{j=0}^3, \{r^k e_j^{ik\tilde{\theta}}\}_{j=0}^3\}_{k=0}^n$ *to perform n iterations, eight groups of initial values are obtained from the corresponding coordinates as the starting points, and four groups of counter-clockwise and clockwise rotations are formed. With the increase in the number of iterations, the arguments grow synchronously, and the length of the vector is increased by* $r^k < r^{k+1}$*, and contracted by* $r^k > r^{k+1}$*. Two mechanisms of action, the starting point and the end point, are selected. The change patterns of the two trends are opposite, and the overall distribution characteristics are similar.*

Proof. Observe the diagram of Fig. 13(c) & (f) and Fig. 14. The intersections in the figure indicate the starting points, the direction of change is reversed, and both distributions are similar.

Theorem 14 (Single-Cycle and Multi-Cycle Modes). *Using a pair of conjugate exponents* $\{\{r_j^k e_0^{ik\theta}\}_{j=1}^m, \{r_j^k e_0^{ik\tilde{\theta}}\}_{j=1}^m\}_{k=0}^n$ *to perform n iterations, paired initial values from the corresponding coordinate points, two groups of counter-clockwise and clockwise are formed, respectively. The number of iterations increases, the argument increases synchronously, and the j vector radius* $r_j^k = r_j^{k+1}$ *iteration remains unchanged. Single- or multiple-radius bicyclic structures are formed.*

Proof. Observing Fig. 13(i)(l) and the illustration of Fig. 15, since the vector length does not change, under the condition of a given radius, only pairs of opposite generators are required to form the corresponding structure.

Corollary 13 (Eight Coloring Modes). *For the diagrams that satisfy Theorem 13–14, two different colors can be selected for each region, and the entire attribute distribution characteristics are distinguished by eight colors.*

Proof. Looking at the diagrams of Fig. 16(a)–(i) and Fig. 17–18, an eight-color representation is sufficient.

7.5 *Whole Iteration Diagram*

Three sets of iterative outputs with increasing vector length for all *n* products $r^{j+1} > r^j, 1 \leq j \leq n$.

$$\left\{\begin{bmatrix} Y^j \\ \tilde{Y}^j \\ Y^j + \tilde{Y}^j \\ r^{j+1} > r^j \end{bmatrix}^n_{j=1} : \right.$$

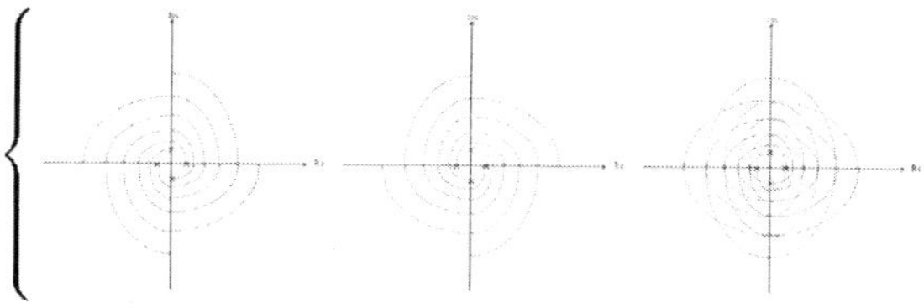

Whole n times products, three sets of iterative output results with decreasing vector length: $r^{j+1} < r^j, 1 \leq j \leq n$.

$$\left\{\begin{bmatrix} Y^j \\ \tilde{Y}^j \\ Y^j + \tilde{Y}^j \\ r^{j+1} < r^j \end{bmatrix}^n_{j=1} : \right.$$

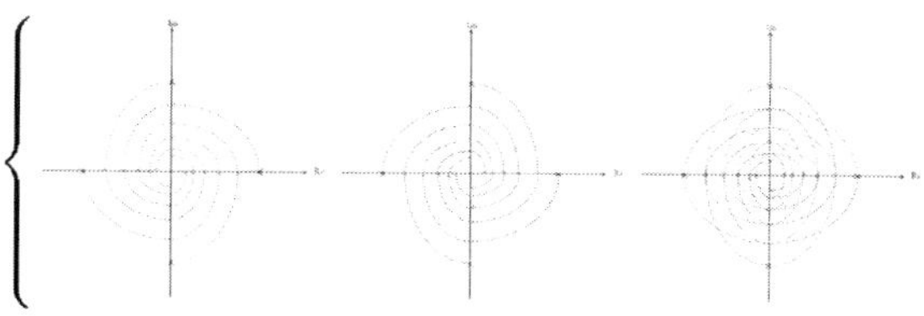

In n times of the product, the vector length is constant to support the single cycle of three-group iteration on the output diagram: $r^{j+1} = r^j, 1 \leq j \leq n$.

$$\left\{\begin{bmatrix} Y^j \\ \tilde{Y}^j \\ Y^j + \tilde{Y}^j \\ r^{j+1} = r^j \end{bmatrix}^n_{j=1} : \right.$$

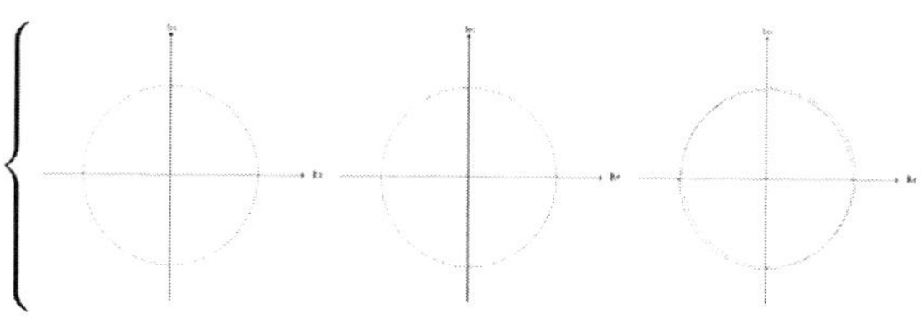

In the n products, the multicircle three-group iterations of output diagrams have constant vector lengths: $r_k^{j+1} = r_k^j, 1 \leq j \leq n, 1 \leq k \leq s$.

$$\left\{\begin{bmatrix} Y^j \\ \tilde{Y}^j \\ Y^j + \tilde{Y}^j \\ r_k^{j+1} = r_k^j \end{bmatrix}^n_{j=1} : \right.$$

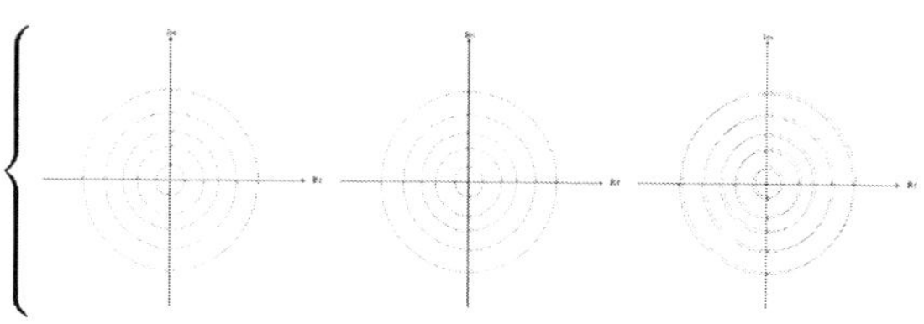

After multiple iterations from four groups, the resulting output images have various local and global features. Either growing or shrinking, the trajectory formed by a single generator is a spiral.

8. Modular Representation of Conjugate Exponential Function

8.1 Argument Modular Representation

Due to the multiclass conjugate exponential functions, the reduction creates a periodic result.

Lemma 6 (Periodic Characteristic of Global Modulo Function). *For any complex function representation, the paired conjugate arguments $\{\theta, \tilde{\theta}\}$ rotate and transform the values,* *satisfying* $\theta \bmod 2\pi \in [0, 2\pi), 0 \leq \theta < 2\pi$; $\tilde{\theta} \bmod \widetilde{2\pi} \in (0, 2\pi], 0 < \tilde{\theta} \leq 2\pi$ *on periodic characteristics.*

Theorem 15 (Quarter Attribute Partition). *The complex plane is divided into eight regions* $\{Q_j(y)\}_{j=0}^{3}, \{\tilde{Q}_j(y)\}_{j=0}^{3}$, *and four attribute partitions are formed.*

Proof. For the j region, there is a conjugate attribute feature on the region $S(Q_j(y)) = S(\tilde{Q}_j(y))$.

According to the rotational symmetry of the four attribute regions, it is important to focus attention on the $j = [00]$ region $S(Q_0(y)) = S(\tilde{Q}_0(y)) = (+, +)$. Using the expression of modular operation in this region, the conjugate exponent of the modular structure with paired positive parameters is expressed on the inherent algebraic structure characteristics.

8.2 Base Representation

From the first term of the complex plane to be limited on the [00] region, the modulo operator will retain a pair of conjugate angles in $0 \leq \theta, n\theta < \pi/2; 0 < \tilde{\theta}, n\tilde{\theta} \leq \frac{\pi}{2}, \tilde{\theta} = \pi/2 - \theta, n\tilde{\theta} = \pi/2 - n\theta$.

Theorem 16 (Modulo Representation in First Quadrant). *In the first quadrant, the paired conjugate arguments* $\{\theta, \tilde{\theta}\}$ *rotate the transformation value; they satisfy* $\theta(\bmod\ \pi/2) \in [0, \pi/2), 0 \leq \theta < \pi/2$; $\tilde{\theta}(\bmod\ \tilde{\pi}/2) \in (0, \pi/2], 0 < \tilde{\theta} \leq \pi/2$ *on periodic characteristics.*

Proof. From the argument representation, the two sets of conjugate arguments represent monotonically increasing/monotonically decreasing representations of $0 - \pi/2$ periodic effects.

Theorem 17 (Interaction on Modular Parameters). *The algebraic equation transforms any* m *element partition condition on the solvable ring to form a regular polygon with* m *vertices through the cyclotomic fields. When* m *is a prime number, the polygon is expanded by the conjugate rotation group to form a total number of* $4m$ *vertex polygons, arguments* $\{n\theta, n\tilde{\theta}\}$ *constraint in* $n\theta(\bmod\,\pi/2) \in [0, \pi/2)$; $n\tilde{\theta}(\bmod\,\tilde{\pi}/2) \in (0, \pi/2]$.

Proof. For the m solution of the prime number in the cyclotomic fields, it is necessary to take conjugate expansion of reflection and rotation $\pi/2$ to form a total number of $4m$ polygons. There are pairwise conjugate m features in the first quadrant.

$$e^{in\theta} \equiv e^{in\theta(\bmod\,\pi/2)}, n\theta \in [0, \pi/2) \tag{82}$$

$$e^{in\tilde{\theta}} \equiv e^{in\tilde{\theta}(\bmod\,\tilde{\pi}/2)}, n\tilde{\theta} \in (0, \pi/2] \tag{83}$$

For the number of vertices increased in the solution space of the first quadrant, the modulo operation on the partition characteristics of the m element ensures the positive values of the two arguments. A series of analytic number theory formulas can be used for in-depth exploration.

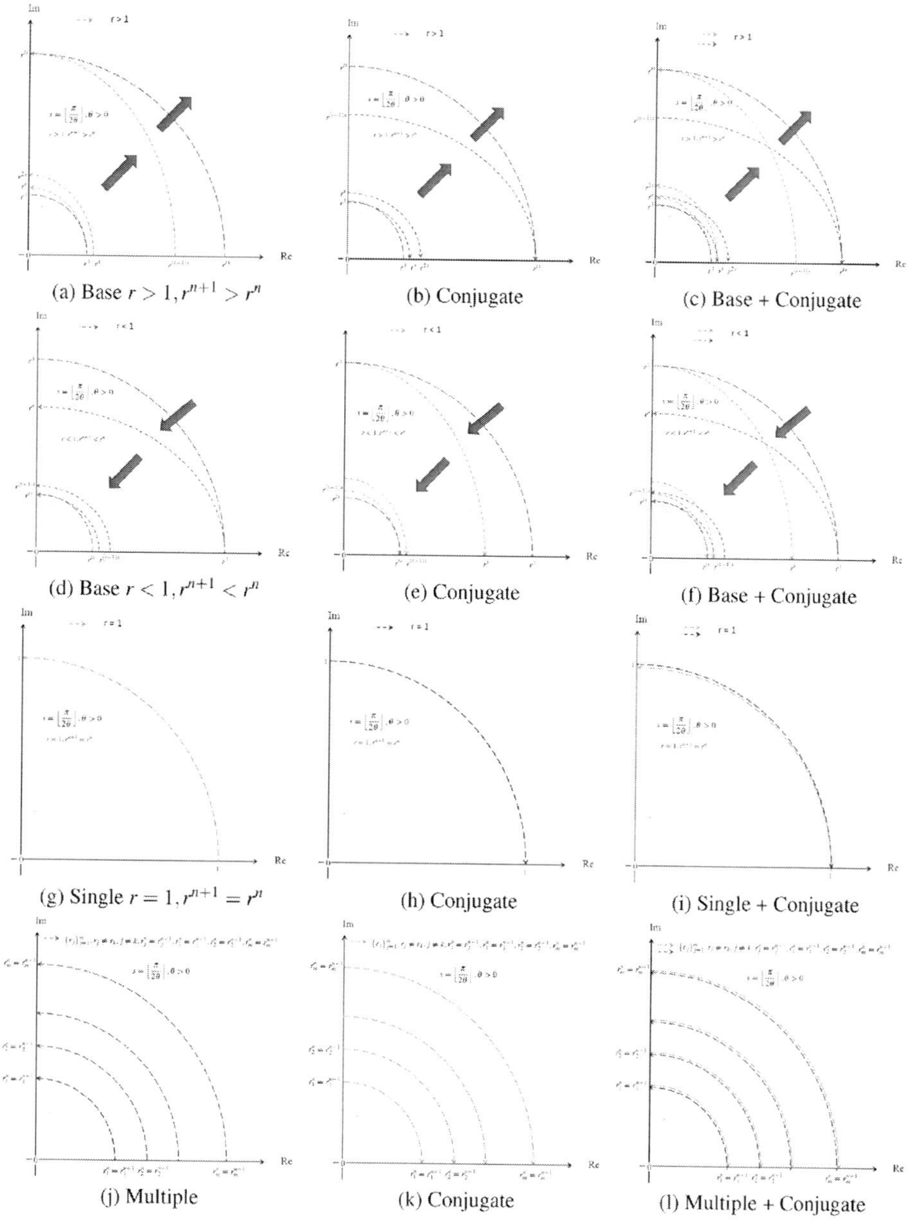

(a) Base $r > 1, r^{n+1} > r^n$ (b) Conjugate (c) Base + Conjugate

(d) Base $r < 1, r^{n+1} < r^n$ (e) Conjugate (f) Base + Conjugate

(g) Single $r = 1, r^{n+1} = r^n$ (h) Conjugate (i) Single + Conjugate

(j) Multiple (k) Conjugate (l) Multiple + Conjugate

Figure 19. Four maps of global projections $\{(a)–(c)\}$ $r > 1$; $\{(d)–(e)\} r < 1$; $\{(g)–(i)\}$ single mode $\{r_j\}_{j=1}^{m}, r_j \neq r_k, j \neq k, r_j^{n+1} = r_j^n$; $\{(j)–(l)\}$ multiple modes.

8.3 Three Radius Changes of Modulo Representation in First Quadrant

In Fig. 19–21, a single set and multiple sets of iteration results under three conditions are shown.

Twelve diagrams are presented in Fig. 19(a)–(l).

Figures (a)–(c) depict the enlarged curl segments. In Fig. (a), the initial vector starts from the point on the horizontal coordinate counter-clockwise; with the increase in the

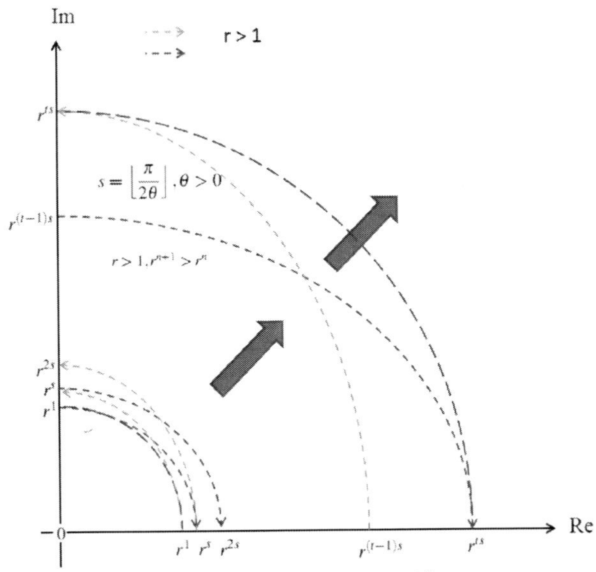

(a) Base + Conjugate $r > 1, r^{n+1} > r^n$

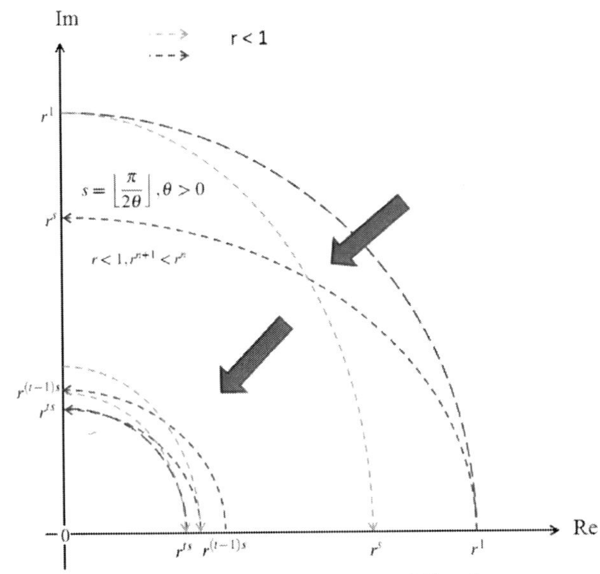

(b) Base + Conjugate $r < 1, r^{n+1} < r^n$

Figure 20. Two maps of global projections (a) $r > 1$; (b) $r < 1$.

number of iterations, the length of the vector increases synchronously, and the period of the argument changes. Several groups of spirals from small to large are rotated successively. In Fig. (b), schematically, the initial vector starts from the point on the vertical coordinate clockwise, with the increase in the number of iterations, the length of the vector increases synchronously, and the period of the argument changes. Several groups of spirals from small to large are successively rotated. Figure (c) provides a schematic (a) + (b) superposition distribution. The whole distribution is multiple groups of symmetrical spiral segments with direction marks from small to large.

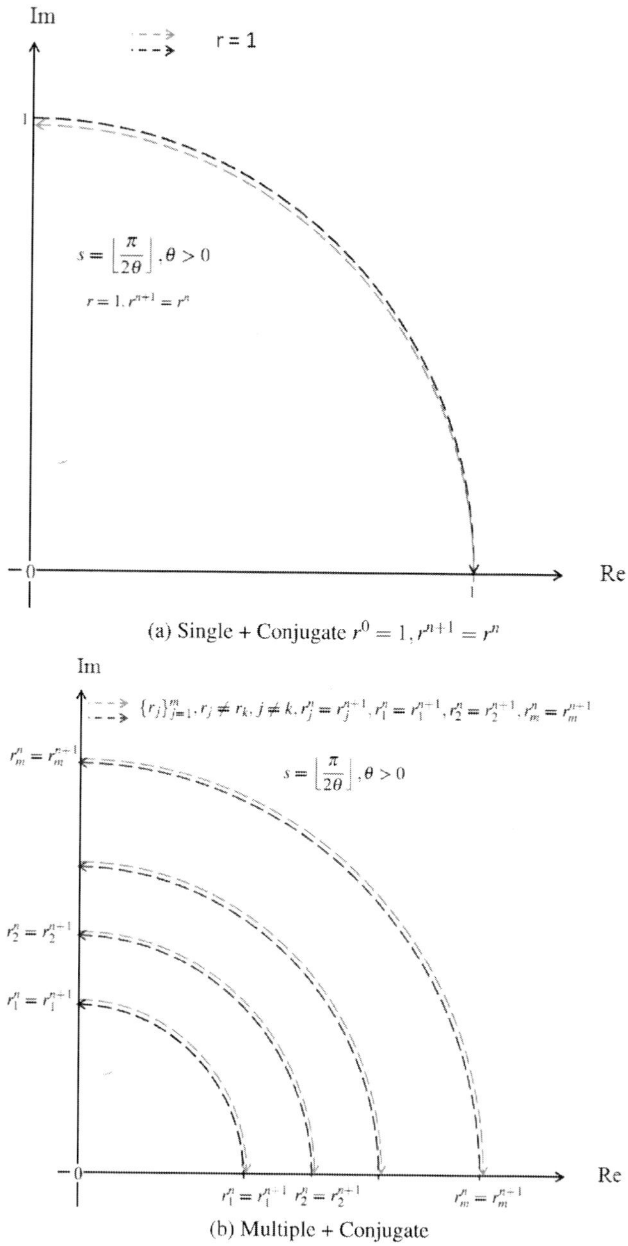

Figure 21. Two maps of global projections (a) single mode $\{r_j\}_{j=1}^m, r_j \neq r_k, j \neq k, r_j^{n+1} = r_j^n$; (b) multiple modes.

Figures (d)–(f) depict the reduced curl segments. In Fig. (d), the initial vector starts from the point on the horizontal coordinate counter-clockwise, and with the increase in the number of iterations, the vector length shrinks synchronously, and the argument period changes. Several groups of spirals from large to small are completed successively. In Fig. (e), the initial vector starts from the point on the vertical coordinate clockwise, the length of the vector shrinks and decreases as the number of iterations increases, the argument period changes, and the group of spirals rotates from large to small. In Fig. (f), representing (d) + (e) superposition, the whole distribution is multiple groups of symmetrical spiral segments with direction marks from large to small.

Figures (g)–(i) depict quarter arcs. In Fig. (g), the initial vector starts from the radial position of the point on the horizontal coordinate counter-clockwise to form a quarter arc. Figure (h) shows that the initial vector starts from the radial position of the point on the vertical coordinate clockwise, with the simultaneous increase of the argument, forming a quarter arc. Figure (i) shows the superposition of (g) + (h) on two quarter arcs changing in different directions, and the whole representation is two-color quarter arcs.

Figures (j)–(l) depict groups of quarter arcs. In Fig. (j), the initial vectors start from multiple radius positions on the horizontal coordinate counter-clockwise, and the cycles rotate into multiple groups of quarter arcs. Figure (k) shows that the initial vector starts from multiple positions of the vertical coordinate. Starting from the radius of the initial points clockwise, the periodic rotation creates multiple groups of quarter arcs. Figure (l) shows the superposition of (j)+(k) on multiple groups of quarter arcs, and the whole map is composed of multiple arcs.

In the two maps of Fig. 20(a)–(b), Fig. (a) shows the enlarged map of Fig. 19(c). Several sets of curls from small to large. Figure (b) shows an enlarged illustration of Fig. 19(f); the whole distribution is composed of several groups of curls from large to small.

In the two maps of Fig. 21(a)–(b), Fig. (a) shows an enlarged map of Fig. 19(i), and the whole distribution shows two-color quarter arcs. Figure (b) shows an enlarged map of Fig. 19(l); the whole distribution consists of multiple bicolor quarter arcs.

Theorem 18 (Growing and Shrinking of Multiple Rotations in First Quadrant). *Using a pair of conjugate exponents* $\{\{r^k e_0^{ik\theta}\}, \{r^k e_0^{ik\tilde{\theta}}\}\}_{k=0}^n$ *to perform n iterations, the paired initial values start from the corresponding coordinate points and form four counter-clockwise and clockwise groups. The number of k time iterations increases, the argument period changes, the length of the vector increases* $r^k < r^{k+1}$*, and contraction* $r^k > r^{k+1}$ *occurs on two actions. To select the starting point and the end point, the two change patterns are opposite, and the whole characteristics are similar to the two-color curl structure.*

Proof. Observing the maps of Fig. 19(c) & (f); and Fig. 20. The intersection in the figure indicates the initial point, the direction of change is reversed, and the whole distribution is similar.

Theorem 19 (Rotational Mode of Single Arc and Multi-Arc in First Quadrant). *Using a pair of conjugate exponents* $\{\{r_j^k e_0^{ik\theta}\}_{j=1}^m, \{r_j^k e_0^{ik\tilde{\theta}}\}_{j=1}^m\}_{k=0}^n$ *to perform n iterations, the initial value corresponds to the coordinate point, and two groups of counter-clockwise and clockwise are formed. As the number of iterations increases, the argument increases synchronously, and the jth vector radius* $r_j^k = r_j^{k+1}$ *iteratively remains unchanged, forming a single or multiple two-color quarter arc structures.*

Proof. Observing Fig. 19(i)–(l) and the map of Fig. 21, since the vector length does not change, under the condition of a given radius, the generators are paired with each other, forming a two-color quarter arc structure.

8.4 Schematic Projection

For n products in the first quadrant, three sets of iterative output maps present with increasing vector length: $r^{j+1} > r^j, 1 \leq j \leq n$.

$$\left\{\begin{matrix} Y_0^j \\ \tilde{Y}_0^j \\ Y_0^j + \tilde{Y}_0^j \\ r^{j+1} > r^j \end{matrix}\right\}_{j=1}^{n} :$$

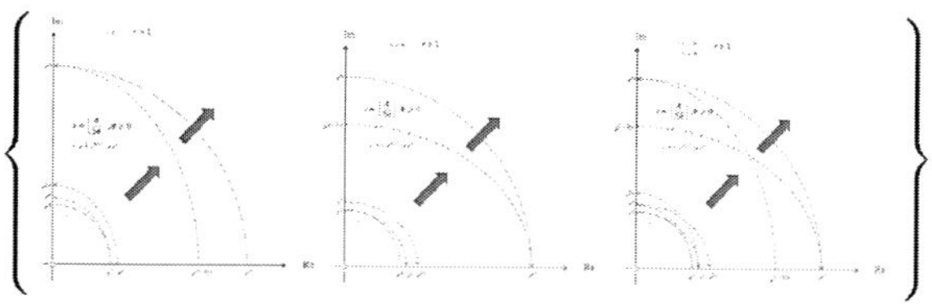

For n products in the first quadrant, three sets of iterative output maps present with decreasing vector length: $r^{j+1} < r^j, 1 \leq j \leq n$.

$$\left\{\begin{matrix} Y_0^j \\ \tilde{Y}_0^j \\ Y_0^j + \tilde{Y}_0^j \\ r^{j+1} < r^j \end{matrix}\right\}_{j=1}^{n} :$$

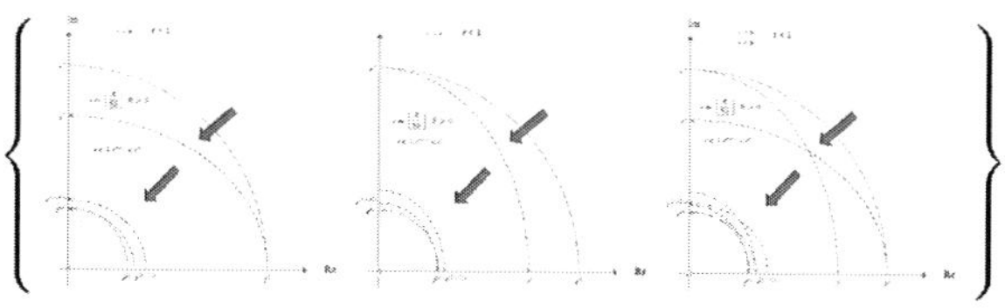

For n products in the first quadrant, the vector length is constant on the single cycle three groups of the iteration output maps: $r^{j+1} = r^j, 1 \leq j \leq n$.

$$\left\{\begin{matrix} Y_0^j \\ \tilde{Y}_0^j \\ Y_0^j + \tilde{Y}_0^j \\ r^{j+1} = r^j \end{matrix}\right\}_{j=1}^{n} :$$

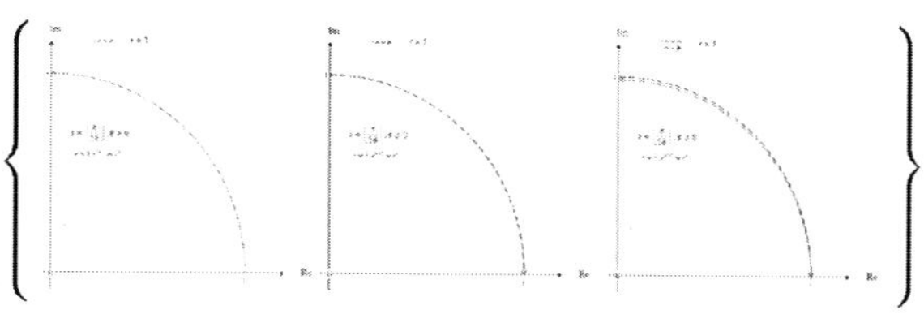

For n products in the first quadrant, multicycle three groups of iteration output maps present multiple constant vector lengths: $r_k^{j+1} = r_k^j, 1 \leq j \leq n, 1 \leq k \leq s$.

$$\left\{\begin{matrix} Y_0^j \\ \tilde{Y}_0^j \\ Y_0^j + \tilde{Y}_0^j \\ r_k^{j+1} = r_k^j \end{matrix}\right\}_{j=1}^{n} :$$

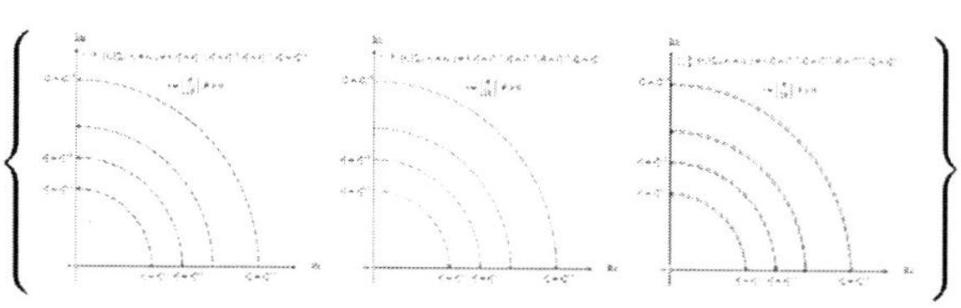

9. Whole and Interval Partitions

9.1 Four Intervals and Number of Partitions in the First Quadrant

From the view of two-dimensional decomposition, each m circular polynomial satisfies the condition of conjugate decomposition, let $\{\{F(m), F(\tilde{m})\}, \{P(m), P(\tilde{m}\}, P(m) \cup P(\tilde{m}), P(m) \cap P(\tilde{m}, P(m) + P(\tilde{m}\}$ be the number of whole divisions, the number of individual projections, the number of distinguishable projections, and the number of common projections for a

Table 5. Number of Whole Divisions and Number of Individual Projections in First Quadrant $m = 1 - 16$

m	1	2	3	4	5	6	7	8	9	10	11	12	13	14	15	16	Formula
$F(m)$	4	4	12	4	20	12	28	8	36	20	44	12	52	28	60	16	$F(m)$
$P(m) = P(\tilde{m})$	1	1	3	1	5	3	7	2	9	5	11	3	13	7	15	4	$F(m)/4$
$P(m) \cup P(\tilde{m})$	2	2	4	2	6	4	8	3	10	6	12	4	14	8	16	5	$P(m) + 1$
$P(m) \cap P(\tilde{m})$	0	0	2	0	4	2	6	1	8	4	10	2	12	6	14	3	$P(m) - 1$
$P(m) + P(\tilde{m})$	2	2	6	4	10	6	14	4	18	10	22	6	26	14	30	8	$2P(m)$

single quadrant.

$$F(m) = \begin{cases} 4m, \ m \equiv 1(\bmod 2) \\ m, \ m \equiv 0(\bmod 4) \\ 2m, \ m \equiv 2(\bmod 4) \end{cases} \tag{84}$$

$$P(m) = P(\tilde{m}) = F(m)/4 \tag{85}$$

$$P(m) \cup P(\tilde{m}) = P(m) + 1 \tag{86}$$

$$P(m) \cap P(\tilde{m}) = P(m) - 1 \tag{87}$$

$$P(m) + P(\tilde{m}) = 2P(m) \tag{88}$$

The number of whole divisions and the effective number of individual projections in each quadrant grow nonlinearly. The detailed quantitative relationship is described in the following paragraphs. A small number of quantitative relationships as examples are given in Table 5.

Theorem 20 (Number of Whole Divisions and Individual Projections on First Quadrant). *From parameters determined by formula (84)–(88) $\{F(m), P(m), P(\tilde{m}), P(m) \cup P(\tilde{m}), P(m) \cap P(\tilde{m}), P(m) + P(\tilde{m}\}$, they provide the number of whole divisions and the number of individual projections for the selected quadrant.*

Proof. Because m is an odd number, the initial point is on the unit vector $(1, 0)$, and the remaining $m - 1$ elements form in pairs of $(m - 1)/2$ complex conjugate numbers. A total of $4m$ discriminative positions are included through the conjugate extension over the four quadrants. When $m \equiv 0(\bmod 4)$, each distinguishable position has conjugate symmetry without further extension. When $m \equiv 2(\bmod 4)$ the number of correlations divided by two is an odd number, multiplied by four, and the complete conjugate structure of $2m$ is obtained. After determining $F(m)$, each quadrant is divided into $F(m)/4$ elements as $\{P(m), P(\tilde{m})\}$. There are two numbers: $\{P(m) \cup P(\tilde{m}), P(m) \cap P(\tilde{m})\}$ to add or subtract 1 to each quadrant element, respectively. The sum of two numbers of conjugate elements $P(m) + P(\tilde{m})$ is twice the number of a single quadrant division.

Corollary 14 (Equivalent Category and Even Number Partition). *For any $\{P(m), P(\tilde{m})\}$ in the first quadrant provides the same type of equivalent distinction structure, and even partition numbers appear less frequently.*

Proof. The specific quadrant has been divided into the same number and is shown in the conjugate extension of the same characteristics. From the view of statistical frequency, Table 5 lists $m = 1 - 16$ values $\{P(m), P(\tilde{m})\}$: 1(3),3(2),5(2),7(2), {9,11,13,15}(1); {2,4}(1), with the ratio 7/8 being odd and only the ratio 1/8 being even expression.

9.2 *Relative Coprime Condition*

The multiplicative group and quantitative measures are applied from the view of placeholder count. From the view of number-theoretical analysis, the representation of modular operations and the resulting complex modes are useful tools to solve the numbers from corresponding quantization matrices.

The whole and quadrant partitions determine the relationship projected to the first quadrant. Under the auxiliary observation of suitable paired conjugate arguments $\{\theta, \tilde{\theta}\}$ on the condition of modulo operation, the trajectory complexity is closely related to the pairwise conjugate rational order on the complex plane.

9.3 *Placeholder Counter Operator*

Let $\{T(n,s,m), T(n,s,\tilde{m})\}$ be the placeholder number as a conjugate coprime discriminator, where $m > 0$ is the number of interval divisions, $0 \leq n < m$ in the selection interval. Denote $[n,m] < m$ as the greatest common multiple between n and m and s positions $0 \leq s \leq m$ modulo $\{m, \tilde{m}\}$ to be occupied or not.

$$T(n,s,m) = \begin{cases} 1, & s = tn \bmod m; s,t \in [0,m) \\ 0, & s \neq tn \bmod m; s,t \in [0,m) \end{cases} \tag{89}$$

$$T(n,s,\tilde{m}) = \begin{cases} 1, & s = tn \bmod \tilde{m}; s,t \in (0,m] \\ 0, & s \neq tn \bmod \tilde{m}; s,t \in (0,m]. \end{cases} \tag{90}$$

Theorem 21 (Common Part Equivalent). *For all common parts, whose values lie in* $0 < n < m$, *the placeholder count values* $T(n,s,m) = T(n,s,\tilde{m})$ *are equal.*

Proof. The definitions of the two functions are the same, and the value ranges of the head and tail are clearly distinguished, so the values of the common part are the same.

9.4 *Summation Representation*

For a given n, m, placeholder points $\{T(n,m), T(n,\tilde{m}), T(m), T(\tilde{m})\}$ special location $T(0,m) = 1, T(m,m) = 0$; $T(0,\tilde{m}) = 0, T(m,\tilde{m}) = 1$ common part $0 < n < m$

$$T(n,m) = \sum_{s=0}^{m} T(s,n,m) = \frac{m}{[n,m]} \tag{91}$$

$$T(m) = \sum_{n=0}^{m} T(n,m) \tag{92}$$

$$T(n,\tilde{m}) = \sum_{s=0}^{m} T(s,n,\tilde{m}) = \frac{m}{[n,m]} \tag{93}$$

$$T(\tilde{m}) = \sum_{n=0}^{m} T(n,\tilde{m}) \tag{94}$$

The number of $\{T(m), T(\tilde{m})\}$ is related to all instances of $\{T(n,s,m), T(n,s,\tilde{m})\}$ summation.

Theorem 22 (Total Number of Placeholders). *For all values in the common part of* $0 < n < m$, *the placeholder count* $T(n,s,m) = T(n,s,\tilde{m})$ *value is equal. For all cases, the parameters satisfy* $T(m) = T(\tilde{m})$. *When* m *is a prime number,* $T(m) = T(\tilde{m}) = m^2 - m + 1$.

Proof. Due to the conjugate symmetry of the placeholder count, the equation is true in general. When m is a prime number, the core part of the matrix provides $(m-1)^2$ placeholders, and the edge part provides m placeholders. The summation of both contributions is

$$T(m) = T(\tilde{m}) = m + (m-1)^2 = m^2 - m + 1$$

Let $\Delta(n,m)$, $\Delta(n,\tilde{m})$ be the defect value, and the boundary value is $\Delta(0,m) = \Delta(m,m) = 0$, $\Delta(0,\tilde{m}) = \Delta(m,\tilde{m}) = 0$ on pair vectors $\{T(s,n,m), T(s,n,\tilde{m})\}_{s=0}^{m}$, $0 < n < m$ in general.

$$\sum_{s=0}^{m} T(s,n,m) = \frac{m}{[n,m]} \tag{95}$$

$$\Delta(n,m) = m - \sum_{s=0}^{m} T(s,n,m) = m - \sum_{s=0}^{m} T(s,n,\tilde{m}) = \Delta(n,\tilde{m}) \tag{96}$$

$$\Delta(n,m) = \Delta(n,\tilde{m}) = m\left(1 - \frac{1}{[n,m]}\right) \tag{97}$$

$$\Delta(m) = \Delta(\tilde{m}) = \sum_{n=0}^{m} \Delta(n,m) = \sum_{n=0}^{m} \Delta(n,\tilde{m}) \tag{98}$$

When $\{n,m\}$ has a relative prime $[n,m] = 1$, $\Delta(n,m) = 0$ property. Through the formula (98) in general, each defect value can be determined in turn, and the corresponding quantized value can be collected. The summation of the placeholders and the defects is the global conjugate invariant on the scale m, denoted as $H(m), H(\tilde{m})$

$$H(m) = H(\tilde{m}) = T(m) + \Delta(m) = T(\tilde{m}) + \Delta(\tilde{m}) = m^2 - m + 1 \tag{99}$$

Theorem 23 (Global Conjugate Invariant). *The global conjugate invariant* $H(m) = H(\tilde{m}) = m^2 - m + 1$ *is a monotonically increasing sequence, and* $\Delta(m) = \Delta(\tilde{m})$. *When* m *is prime* $\Delta(m) = \Delta(\tilde{m}) = 0$. $\{T(m) = T(\tilde{m})\}$ *is the wave growth sequence, and* $\{\Delta(m) = \Delta(\tilde{m})\}$ *is the segmental wave of the pulse sequence,* $\{T(m) - \Delta(m), T(\tilde{m}) - \Delta(\tilde{m})\}$ *is a fluctuating growth sequence.*

Proof. In the general case, the core matrix provides $(m-1)^2$ placeholders, and the edge part provides m placeholders. The summation of the global conjugate invariants is $H(m) =$

Table 6. Small number of whole conjugate invariant growth

m :	1	2	3	4	5	6	7	8	9	10	11	12	13	14	15	16	17
$T(m) = T(\tilde{m})$:	1	3	7	11	21	21	43	43	61	63	111	77	157	129	147	171	263
$\Delta(m) = \Delta(\tilde{m})$:	0	0	0	2	0	10	0	14	12	28	0	56	0	54	64	70	0
$T(m) - \Delta(m) =$:	1	3	7	9	21	11	43	19	49	35	111	21	157	75	83	101	263
$H(m) = H(\tilde{m})$:	1	3	7	13	21	31	43	57	73	91	111	133	157	183	211	241	263
$\sum_{j=1}^{m} T(j)$:	1	4	11	22	43	64	107	150	211	274	385	462	619	748	895	1066	1329
$\sum_{j=1}^{m} \Delta(j)$:	0	0	0	2	2	12	12	26	38	66	66	122	122	176	240	310	310
$\sum_{j=1}^{m}(T(j) - \Delta(j)) =$:	1	4	11	20	41	52	95	124	173	208	319	340	497	572	655	756	1019
$\sum_{j=1}^{m} H(j)$:	1	4	11	24	45	76	119	176	249	340	451	584	741	924	1135	1367	1636

$H(\tilde{m}) = m^2 - m + 1$. Under the condition of nonprime $\Delta(m) = \Delta(\tilde{m}) > 0$, the sequence provides a whole that varies by each position with increasing composite number.

For small numbers of quantization parameters, Table 6 is a reference. It is observed from the sequence that with the gradual increase of m, $\{H(m), H(\tilde{m})\}$ rises in a quadratic growth pattern. With the numbering changes in the situation between prime numbers and composite numbers, the $\{T(m), \Delta(m), T(m) - \Delta(m)\}$ sequence shows obvious fluctuation effects.

Corollary 15 (Conjugate Invariant on Galois Field of Algebraic Biperiodic Function). *For m conjugate variables, it is natural to express the m order complex variables in the functional algebraic structure $G(m)$. The whole structure is intrinsically associated with the P-adic number theory of the modular structure and the Galois field of the algebraic representation.*

Proof. The conjugate curl representation of the mini/max region from the level of fine representation of finite modular functions provides global and local invariant series to support the representation of discrete algebraic features and periodic functions. The specific Galois field algebraic representation $G(m)$ provides refined analysis and description features, which can be further explored.

Theorem 24 (Extended Galois Field Algebraic Equivalent Representation). *The solution space of the m polynomial algebraic equation and the m conjugate complex functions are equivalent to increasing the solution space from $\{x^m - 1\} \in G(m)$ to $\{x^m + 1, x^m + i, x^m - 1, x^m - i\}$, expanded in the $x^{4m} - 1 \in G(4m)$ structure.*

Proof. The equation $x^{4m} - 1 = (x^m + 1)(x^m + i)(x^m - 1)(x^m - i)\} \in G(4m)$ holds in general. When m is a prime number, the corresponding four groups of m element regular polygons correspond to the four positions where the initial point is $\{1, i, -1, -i\}$, and the relevant solution space is composed of a whole to form a $4m$ polygon.

When m is composite, the corresponding number is determined by Theorem 23.

In either case, the partition structure of m can be studied based on the first quadrant to form a refined representation.

For example, for $m = 1$, there is a single solution $x - 1 = 0$, and the quaternary solution space can be extended to the corresponding equation: $\{x-1, x-i, x+1, x+i\} = 0$. Therefore, the extended solution space is contained in $x^4 - 1 = 0$.

Extending $\{x^4-1, x^4-i, x^4+1, x^4+i\} = 0$, the solution space is contained in $x^{16} - 1 = 0$. The quarter partitions are precisely expressed in the first quadrant.

Corollary 16 (Invariant Property of 4m Dimensional Solution Space). *In the solution structure, the* $4m$ *variable algebraic equation* $\{x^{4m}-1\} \equiv \{x^{m}+1, x^{m}+i, x^{m}-1, x^{m}-i\}$ *retains the whole cluster unchanged.*

Proof. From the view of the whole partition, different equations are created from the four initial points of the solution space on the feature cluster. There are only translation phase differences among the solution space, so the whole sequence of the cluster collection remains unchanged.

9.5 Various Placeholder Distributions

In the following examples, s denotes the horizontal label, n denotes the vertical label, and the matrix shows $\{T(n,s,m), T(n,s,\tilde{m})\}$ itself and related placeholder values for a series of cumulative summations.

$T(s,n,1),s:$	0	1	$\sum$
0	1	0	1
1	0	0	0
$\sum$	1	0	1

+

$T(s,n,\tilde{1}),s:$	0	1	$\sum$
0	0	0	0
1	0	1	1
$\sum$	0	1	1

=

$T(s,n,1)+T(s,n,\tilde{1}),s:$	0	1	$\sum$
0	1	0	1
1	0	1	1
$\sum$	1	1	2

$T(s,n,2),s:$	0	1	2	$\sum$
0	1	0	0	1
1	1	1	0	2
2	0	0	0	0
$\sum$	2	1	0	3

+

$T(s,n,\tilde{2}),s:$	0	1	2	$\sum$
0	0	0	0	0
1	0	1	1	2
2	0	0	1	1
$\sum$	0	1	2	3

=

$T(s,n,2)+T(s,n,\tilde{2}),s:$	0	1	2	$\sum$
0	1	0	0	1
1	1	2	1	4
2	0	0	1	1
$\sum$	2	2	2	6

$T(s,n,3),s:$	0	1	2	3	$\sum$
0	1	0	0	0	1
1	1	1	1	0	3
2	1	1	1	0	3
3	0	0	0	0	0
$\sum$	3	2	2	0	7

+

$T(s,n,\tilde{3}),s:$	0	1	2	3	$\sum$
0	0	0	0	0	0
1	0	1	1	1	3
2	0	1	1	1	3
3	0	0	0	1	1
$\sum$	0	2	2	3	7

=

$T(s,n,3)+T(s,n,\tilde{3}),s:$	0	1	2	3	$\sum$
0	1	0	0	0	1
1	1	2	2	1	6
2	1	2	2	1	6
3	0	0	0	1	1
$\sum$	3	4	4	3	14

$T(s,n,4),s:$	0	1	2	3	4	$\sum$
0	1	0	0	0	0	1
1	1	1	1	1	0	4
2	1	0	1	0	0	2
3	1	1	1	1	0	4
4	0	0	0	0	0	0
$\sum$	4	2	3	2	0	11

+

$T(s,n,\tilde{4}),s:$	0	1	2	3	4	$\sum$
0	0	0	0	0	0	0
1	0	1	1	1	1	4
2	0	0	1	0	1	2
3	0	1	1	1	1	4
4	0	0	0	0	1	1
$\sum$	0	2	3	2	4	11

=

$T(s,n,4)+T(s,n,\tilde{4}),s:$	0	1	2	3	4	$\sum$
0	1	0	0	0	0	1
1	1	2	2	2	1	8
2	1	0	2	0	1	4
3	1	2	2	2	1	8
4	0	0	0	0	1	1
$\sum$	4	4	6	4	4	22

$T(s,n,5),s:$	0	1	2	3	4	5	$\sum$
0	1	0	0	0	0	0	1
1	1	1	1	1	1	0	5
2	1	1	1	1	1	0	5
3	1	1	1	1	1	0	5
4	1	1	1	1	1	0	5
5	0	0	0	0	0	0	0
$\sum$	5	4	4	4	4	0	21

+

$T(s,n,\tilde{5}),s:$	0	1	2	3	4	5	$\sum$
0	0	0	0	0	0	0	0
1	0	1	1	1	1	1	5
2	0	1	1	1	1	1	5
3	0	1	1	1	1	1	5
4	0	1	1	1	1	1	5
5	0	0	0	0	0	1	1
$\sum$	0	4	4	4	4	5	21

=

$T(s,n,5)+T(s,n,\tilde{5}),s:$	0	1	2	3	4	5	$\sum$
0	1	0	0	0	0	0	1
1	1	2	2	2	2	1	10
2	1	2	2	2	2	1	10
3	1	2	2	2	2	1	10
4	1	2	2	2	2	1	10
5	0	0	0	0	0	1	1
$\sum$	5	8	8	8	8	5	42

$T(s,n,6),s:$	0	1	2	3	4	5	6	$\sum$
0	1	0	0	0	0	0	0	1
1	1	1	1	1	1	1	0	6
2	1	0	1	0	1	0	0	3
3	1	0	0	1	0	0	0	2
4	1	0	1	0	1	0	0	3
5	1	1	1	1	1	1	0	6
6	0	0	0	0	0	0	0	0
$\sum$	6	2	4	3	4	2	0	21

+

$T(s,n,\tilde{6}),s:$	0	1	2	3	4	5	6	$\sum$
0	0	0	0	0	0	0	0	0
1	0	1	1	1	1	1	1	6
2	0	0	1	0	1	0	1	3
3	0	0	0	1	0	0	1	2
4	0	0	1	0	1	0	1	3
5	0	1	1	1	1	1	1	6
6	0	0	0	0	0	0	1	1
$\sum$	0	2	4	3	4	2	6	21

=

$T(s,n,6)+T(s,n,\tilde{6}),s:$	0	1	2	3	4	5	6	$\sum$
0	1	0	0	0	0	0	0	1
1	1	2	2	2	2	2	1	12
2	1	0	2	0	2	0	1	6
3	1	0	0	2	0	0	1	4
4	1	0	2	0	2	0	1	6
5	1	2	2	2	2	2	1	12
6	0	0	0	0	0	0	1	1
$\sum$	6	4	8	6	8	4	6	42

$T(s,n,7),s:$	0	1	2	3	4	5	6	7	$\sum$
0	1	0	0	0	0	0	0	0	1
1	1	1	1	1	1	1	1	0	7
2	1	1	1	1	1	1	1	0	7
3	1	1	1	1	1	1	1	0	7
4	1	1	1	1	1	1	1	0	7
5	1	1	1	1	1	1	1	0	7
6	1	1	1	1	1	1	1	0	7
7	0	0	0	0	0	0	0	0	0
$\sum$	7	6	6	6	6	6	6	0	43

+

$T(s,n,\tilde{7}),s:$	0	1	2	3	4	5	6	7	$\sum$
0	0	0	0	0	0	0	0	0	0
1	0	1	1	1	1	1	1	1	7
2	0	1	1	1	1	1	1	1	7
3	0	1	1	1	1	1	1	1	7
4	0	1	1	1	1	1	1	1	7
5	0	1	1	1	1	1	1	1	7
6	0	1	1	1	1	1	1	1	7
7	0	0	0	0	0	0	0	1	1
$\sum$	0	6	6	6	6	6	6	7	43

=

$T(s,n,7)+T(s,n,\tilde{7}),s:$	0	1	2	3	4	5	6	7	$\sum$
0	1	0	0	0	0	0	0	0	0
1	1	2	2	2	2	2	2	1	14
2	1	2	2	2	2	2	2	1	14
3	1	2	2	2	2	2	2	1	14
4	1	2	2	2	2	2	2	1	14
5	1	2	2	2	2	2	2	1	14
6	1	2	2	2	2	2	2	1	14
7	0	0	0	0	0	0	0	1	1
$\sum$	7	12	12	12	12	12	12	7	86

$T(s,n,8),s:$	0	1	2	3	4	5	6	7	8	$\sum$
0	1	0	0	0	0	0	0	0	0	1
1	1	1	1	1	1	1	1	1	0	8
2	1	0	1	0	1	0	1	0	0	4
3	1	1	1	1	1	1	1	1	0	8
4	1	0	0	0	1	0	0	0	0	2
5	1	1	1	1	1	1	1	1	0	8
6	1	0	1	0	1	0	1	0	0	4
7	1	1	1	1	1	1	1	1	0	8
8	0	0	0	0	0	0	0	0	0	0
$\sum$	8	4	6	4	7	4	6	4	0	43

+

$T(s,n,\tilde{8}),s:$	0	1	2	3	4	5	6	7	8	$\sum$
0	0	0	0	0	0	0	0	0	0	0
1	0	1	1	1	1	1	1	1	1	8
2	0	0	1	0	1	0	1	0	1	4
3	0	1	1	1	1	1	1	1	1	8
4	0	0	0	0	1	0	0	0	1	2
5	0	1	1	1	1	1	1	1	1	8
6	0	0	1	0	1	0	1	0	1	4
7	0	1	1	1	1	1	1	1	1	8
8	0	0	0	0	0	0	0	0	1	1
$\sum$	0	4	6	4	7	4	6	4	8	43

=

$T(s,n,8)+T(s,n,\tilde{8}),s:$	0	1	2	3	4	5	6	7	8	$\sum$
0	1	0	0	0	0	0	0	0	0	1
1	1	2	2	2	2	2	2	2	1	16
2	1	0	2	0	2	0	2	0	1	8
3	1	2	2	2	2	2	2	2	1	16
4	1	0	0	0	2	0	0	0	1	4
5	1	2	2	2	2	2	2	2	1	16
6	1	0	2	0	2	0	2	0	1	8
7	1	2	2	2	2	2	2	2	1	16
8	0	0	0	0	0	0	0	0	1	1
$\sum$	8	8	12	8	14	8	12	8	8	86

$T(s,n,9),s:$	0	1	2	3	4	5	6	7	8	9	$\sum$
0	1	0	0	0	0	0	0	0	0	0	1
1	1	1	1	1	1	1	1	1	1	0	9
2	1	1	1	1	1	1	1	1	1	0	9
3	1	0	0	1	0	0	1	0	0	0	3
4	1	1	1	1	1	1	1	1	1	0	9
5	1	1	1	1	1	1	1	1	1	0	9
6	1	0	0	1	0	0	1	0	0	0	3
7	1	1	1	1	1	1	1	1	1	0	9
8	1	1	1	1	1	1	1	1	1	0	9
9	0	0	0	0	0	0	0	0	0	0	0
$\sum$	9	6	6	8	6	6	8	6	6	0	61

+

$T(s,n,\tilde{9}),s:$	0	1	2	3	4	5	6	7	8	9	$\sum$
0	0	0	0	0	0	0	0	0	0	0	0
1	0	1	1	1	1	1	1	1	1	1	9
2	0	1	1	1	1	1	1	1	1	1	9
3	0	0	0	1	0	0	1	0	0	1	3
4	0	1	1	1	1	1	1	1	1	1	9
5	0	1	1	1	1	1	1	1	1	1	9
6	0	0	0	1	0	0	1	0	0	1	3
7	0	1	1	1	1	1	1	1	1	1	9
8	0	1	1	1	1	1	1	1	1	1	9
9	0	0	0	0	0	0	0	0	0	1	1
$\sum$	0	6	6	8	6	6	8	6	6	9	61

=

$T(s,n,9)+T(s,n,\tilde{9}),s:$	0	1	2	3	4	5	6	7	8	9	$\sum$
0	1	0	0	0	0	0	0	0	0	0	1
1	1	2	2	2	2	2	2	2	2	1	18
2	1	2	2	2	2	2	2	2	2	1	18
3	1	0	0	2	0	0	2	0	0	1	6
4	1	2	2	2	2	2	2	2	2	1	18
5	1	2	2	2	2	2	2	2	2	1	18
6	1	0	0	2	0	0	2	0	0	1	6
7	1	2	2	2	2	2	2	2	2	1	18
8	1	2	2	2	2	2	2	2	2	1	18
9	0	0	0	0	0	0	0	0	0	1	1
$\sum$	9	12	12	16	12	12	16	12	12	9	122

$T(s,n,10),s:$	0	1	2	3	4	5	6	7	8	9	10	$\sum$
0	1	0	0	0	0	0	0	0	0	0	0	1
1	1	1	1	1	1	1	1	1	1	1	0	10
2	1	0	1	0	1	0	1	0	1	0	0	5
3	1	1	1	1	1	1	1	1	1	1	0	10
4	1	0	1	0	1	0	1	0	1	0	0	5
5	1	0	0	0	0	1	0	0	0	0	0	2
6	1	0	1	0	1	0	1	0	1	0	0	5
7	1	1	1	1	1	1	1	1	1	1	0	10
8	1	0	1	0	1	0	1	0	1	0	0	5
9	1	1	1	1	1	1	1	1	1	1	0	10
10	0	0	0	0	0	0	0	0	0	0	0	0
$\sum$	10	4	8	4	8	5	8	4	8	4	0	63

+

$T(s,n,\tilde{10}),s:$	0	1	2	3	4	5	6	7	8	9	10	$\sum$
0	0	0	0	0	0	0	0	0	0	0	0	0
1	0	1	1	1	1	1	1	1	1	1	1	10
2	0	0	1	0	1	0	1	0	1	0	1	5
3	0	1	1	1	1	1	1	1	1	1	1	10
4	0	0	1	0	1	0	1	0	1	0	1	5
5	0	0	0	0	0	1	0	0	0	0	1	2
6	0	0	1	0	1	0	1	0	1	0	1	5
7	0	1	1	1	1	1	1	1	1	1	1	10
8	0	0	1	0	1	0	1	0	1	0	1	5
9	0	1	1	1	1	1	1	1	1	1	1	10
10	0	0	0	0	0	0	0	0	0	0	1	1
$\sum$	0	4	8	4	8	5	8	4	8	4	10	63

=

$T(s,n,10)+T(s,n,\tilde{10}),s:$	0	1	2	3	4	5	6	7	8	9	10	$\sum$
0	1	0	0	0	0	0	0	0	0	0	0	1
1	1	2	2	2	2	2	2	2	2	2	1	20
2	1	0	2	0	2	0	2	0	2	0	1	10
3	1	2	2	2	2	2	2	2	2	2	1	20
4	1	0	2	0	2	0	2	0	2	0	1	10
5	1	0	0	0	0	2	0	0	0	0	1	4
6	1	0	2	0	2	0	2	0	2	0	1	10
7	1	2	2	2	2	2	2	2	2	2	1	20
8	1	0	2	0	2	0	2	0	2	0	1	10
9	1	2	2	2	2	2	2	2	2	2	1	20
10	0	0	0	0	0	0	0	0	0	0	1	1
$\sum$	10	8	16	8	16	10	16	8	16	8	10	126

10. Summary

The conjugate complex organization on the complex plane is composed of three parts: the core concept description, formula expression, and their quantitative characteristics, and relevant symbols and expressions are collected in Tables 7–9. The summary is concluded from three levels.

10.1 Conjugate Complex Plane Representation

From the conceptual level, Table 7 collects the basic differences, representation, and connections between the conjugate complex plane and the complex plane.

From any point x in the first quadrant of the complex plane, using the geometric vector description, the paired complex conjugate variables $\{x, x^*\}, r = |x|$ are accompanied by the paired arguments $\{x, x^*\}\{\theta, -\theta\}, \theta \in R, \theta+(-\theta) = 0$. The θ angle rotates counter-clockwise, the $-\theta$ angle rotates clockwise, and the complex plane is divided into $\{H^+, H^-\}$ upper and lower regions: the upper half-plane and the lower half-plane. The value of the argument is a real number.

$$\boxed{x \in \text{CP}} \rightarrow \left\{\begin{matrix} x & \theta \\ x^* & -\theta \end{matrix}\right\} : \{x^*x = xx^* = 1\} \rightarrow \frac{H^+(\theta)\ \text{UHP}}{H^-(\theta)\ \text{LHP}} \rightarrow \frac{\{re^{i\theta}\}}{\{re^{-i\theta}\}}, \theta \in R$$

From any point x in the first quadrant, conjugate complex numbers $\{x, \tilde{x}\}$ are paired conjugate angles $\{\theta, \tilde{\theta}\}, \theta + \tilde{\theta} = \pi/2$ to be corrected as $\{y, \tilde{y}\}$ after conjugate position/time correction. Quaternary operators $\{^\sim, ^*, ^\sharp, '\}$ and their combinations $\{Y = \{y, y^\sharp, y', y^*\}, \tilde{Y} = \{\tilde{y}, \tilde{y}^\sharp, \tilde{y}', \tilde{y}^*\}\}$ provide the natural division of the conjugate complex plane, the two coordinate regions of adjacent quadrants, which can be reused together. Each complex number can be expressed in four quadrants by attribute characteristics as indexes $\{[00] = \{0, \tilde{0}\} = \{Q_0, \tilde{Q}_0\} \equiv \{y, \tilde{y}\}, [10] = \{2, \tilde{2}\} = \{Q_2, \tilde{Q}_2\} \equiv \{\tilde{y}^\sharp, \tilde{y}^\sharp\}, [11] = \{3, \tilde{3}\} = \{Q_3, \tilde{Q}_3\} \equiv \{y', \tilde{y}'\}, [01] = \{1, \tilde{1}\} = \{Q_1, \tilde{Q}_1\} \equiv \{y^*, \tilde{y}^*\}\}$. The indexes corresponding to the complex numbers in each region are $\{\{0, \tilde{0}\} : (+,+), \{2, \tilde{2}\} : (-,+), \{3, \tilde{3}\} : (-,-), \{1, \tilde{1}\} : (+,-)\}$. Each complex cluster on the common ring contains at least four coordinate points $\{1, i, -1, -i\}$ on the conjugate complex plane, occupying a maximum of twelve points, to represent a conjugate complex function on necessary and sufficient conditions.

$$\boxed{x \in \text{CP}} \rightarrow \left\{\begin{matrix} y & \theta \\ \tilde{y} & \tilde{\theta} \end{matrix}\right\} \rightarrow \left\{\begin{matrix} YY & Y\tilde{Y} \\ \tilde{Y}Y & \tilde{Y}\tilde{Y} \end{matrix}\right\} \rightarrow \frac{\{2, \tilde{2}\}|\{0, \tilde{0}\}}{\{3, \tilde{3}\}|\{1, \tilde{1}\}} \rightarrow \frac{\left\{re_2^{i\theta}, re_2^{i\tilde{\theta}}\right\}\left|\right|\left\{re_0^{i\theta}, re_0^{i\tilde{\theta}}\right\}}{\left\{re_3^{i\theta}, re_3^{i\tilde{\theta}}\right\}\left|\right|\left\{re_1^{i\theta}, re_1^{i\tilde{\theta}}\right\}}$$

10.2 Iterative Interaction Features of Conjugate Complex Functions

10.2.1 Various Iterative Product Effects and Potential Applications

Using the complex exponential function from the level of iterative interaction, Table 8 collects the whole iterative interaction effect of two and multiple products to show a series of results following three groups of conditions.

Table 7. Conjugate complex function model and formula on conjugate complex plane

Model	Feature	Formulas	Quantitative
Complex Plane CP	Complex Space	$CP = \{\forall(a,b) \mid a,b \in R\}$	Description Space
Complex Vector x Complex Conjugate x^* Vector Length r Exponential Function	Geometric Representation Complex Conjugate Measure Complex Conjugate Pair	$x = (p,q), \theta = \angle x$ $x^* = (p,-q), -\theta = \angle x^*$ $r = \sqrt{p^2+q^2} = \sqrt{xx^*}$ $x = re^{i\theta}, x^* = re^{-i\theta}, \theta \in R$	Counter-clockwise Clockwise Inner Product Square Root Reciprocal Rotation
Upper and Lower Half Planes $\{H^+,H^-\}$	Paired Half Space	$CP = \frac{H^+(+\theta)}{H^-(-\theta)} = \frac{re^{i\theta}}{re^{-i\theta}}, \theta \in R$	CP : Pair Half Planes
UHP H^+	+ Imaginary	$H^+ = \{\forall(a,b) \mid b > 0, a,b \in R\}$	Counter-clockwise
LHP H^-	- Imaginary	$H^- = \{\forall(a,b) \mid b < 0, a,b \in R\}$	Clockwise
Complex Compactness	Parameter Coverage	$\Vert\{(p,q)\}\Vert < \left\Vert\left\{\begin{pmatrix} p & -q \\ q & p \end{pmatrix}\right\}\right\Vert$	Matrix Has More Number than Complex Pair
Conjugate Vector $\tilde{x}$	Conjugate State	$\tilde{x} = (q,p), \tilde{\theta} = \pi/2 - \theta$	Conjugate Position
Conjugate Operators $\{\sim, \sharp, \prime, *\}$	Extended Operators	$\sim$ Conjugate $\sharp$ Real Conjugate $\prime$ Negative Conjugate $*$ Complex Conjugate	Basic Operators
Klein Group $X = K(x) =$ $\{x, x^\sharp, x', x^*\}$	Quaternary Position	$X = K(x) =$ $\left\{\begin{matrix} x = (p,q), x^\sharp = (-p,q), \\ x' = (-p,-q), x^* = (p,-q) \end{matrix}\right\}$	Quaternary Rectangle
Conjugate Klein Group $\tilde{X} = \tilde{K}(x) =$ $\{\tilde{x}, \tilde{x}^\sharp, \tilde{x}', \tilde{x}^*\}$	Quaternary Position	$\tilde{X} = \tilde{K}(x) =$ $\left\{\begin{matrix} \tilde{x} = (q,p), \tilde{x}^\sharp = (-q,p), \\ \tilde{x}' = (-q,-p), \tilde{x}^* = (q,-p) \end{matrix}\right\}$	Conjugate Rectangle
Attribute Sign $S(x)$	Attribute Symbol	$x^\sharp = (-p,q), S(x^\sharp) = (-,+)$	Attribute Category
Double Crosses $\{V(x), \tilde{V}(x)\}$	Basic Representation	$i : x \rightarrow \{x, \tilde{x}^\sharp, x', \tilde{x}^*\} = V(x)$ $i : \tilde{x} \rightarrow \{\tilde{x}, x^\sharp, \tilde{x}', x^*\} = \tilde{V}(x)$	$\{i^0, i^1, i^2, i^3\} =$ $\{1, i, -1, -i\}$
Double Cross Group $\left\{U(y), \tilde{U}(y)\right\}$	Correction	$x \rightarrow V(x) = U(y) = \{y, y^\sharp, y', y^*\} = Y$ $\tilde{x} \rightarrow \tilde{V}(x) = \tilde{U}(y) = \{\tilde{y}, \tilde{y}^\sharp, \tilde{y}', \tilde{y}^*\} = \tilde{Y}$	Convert States
Rectangle to Cycle	Space-Time Swap	$\{K(x), \tilde{K}(x)\} \leftrightarrow \{U(y), \tilde{U}(y)\}$	Position/Time Exchange
Conjugate Angles $\{\theta, \tilde{\theta}\}$	Conjugate Base Angle Four Basic Angles Polyphase Angles Equivalent Modes	$\tilde{\theta} = \frac{\pi}{2} - \theta$ $i : \theta \rightarrow \{\theta, \theta + \frac{\pi}{2}, \theta + \pi, \theta + \frac{3\pi}{2}\}$ $i : \tilde{\theta} \rightarrow \{\tilde{\theta}, \tilde{\theta} + \frac{\pi}{2}, \tilde{\theta} + \pi, \tilde{\theta} + \frac{3\pi}{2}\}$ $= \{\frac{\pi}{2} - \theta, \pi - \theta, \frac{3\pi}{2} - \theta, 2\pi - \theta\}$	Conjugate Polyphases
Attribute Indexes $Y = \{y, y^\sharp, y', y^*\}$ $\tilde{Y} = \{\tilde{y}, \tilde{y}^\sharp, \tilde{y}', \tilde{y}^*\}$	Attribute Expression	$Y = \left\{\begin{matrix} y = re_0^{i\theta}, y^\sharp = re_2^{i\theta} \\ y' = re_3^{i\theta}, y^* = re_1^{i\theta} \end{matrix}\right\}$ $\tilde{Y} = \left\{\begin{matrix} \tilde{y} = re_0^{i\tilde{\theta}}, \tilde{y}^\sharp = re_2^{i\tilde{\theta}} \\ \tilde{y}' = re_3^{i\tilde{\theta}}, \tilde{y}^* = re_1^{i\tilde{\theta}} \end{matrix}\right\}$	Conjugate Indexes Octet Clusters
Conjugate Quadrant QCP:$\{Q_j, \tilde{Q}_j\}_{j=0}^3$	Parameter Space	$CP = \frac{H^+}{H^-} = \frac{(-,+) \vert (+,+)}{(-,-) \vert (+,-)} : \frac{\{2,\tilde{2}\} \Vert \{0,\tilde{0}\}}{\{3,\tilde{3}\} \Vert \{1,\tilde{1}\}} :$ $\frac{\{Q_2,\tilde{Q}_2\} \vert \{Q_0,\tilde{Q}_0\}}{\{Q_3,\tilde{Q}_3\} \vert \{Q_1,\tilde{Q}_1\}} : \frac{\{re_2^{i\theta}, re_2^{i\tilde{\theta}}\} \Vert \{re_0^{i\theta}, re_0^{i\tilde{\theta}}\}}{\{re_3^{i\theta}, re_3^{i\tilde{\theta}}\} \Vert \{re_1^{i\theta}, re_1^{i\tilde{\theta}}\}}$	A Complex Plane Paired Half Planes Four Quadrants Eight Attribute Regions

Table 8. Conjugate complex function of iterative interaction feature

Model	Feature	Formula	Quantitative
Global Modulo Operator $\{n\theta, n\tilde{\theta}\}$	2π Period	$\begin{cases} n\theta(\bmod 2\pi) \in [0, 2\pi) \\ n\tilde{\theta}(\bmod \widetilde{2\pi}) \in (0, 2\pi] \end{cases}$	Whole Projection on Full Complex Plane
64 Product Outputs	Three Clusters	$\{Y, \tilde{Y}\} \times \{Y, \tilde{Y}\} = \begin{Bmatrix} YY & Y\tilde{Y} \\ \tilde{Y}Y & \tilde{Y}\tilde{Y} \end{Bmatrix}$ $= \{\{Y\tilde{Y}, \tilde{Y}Y\}, \{YY\}, \{\tilde{Y}\tilde{Y}\}\}$	Interactive Projection Clustering Invariants
Coordinate Projection $\{Y\tilde{Y}, \tilde{Y}Y\}$	4 Crosses 32 Products	$\begin{Bmatrix} Y\tilde{Y} \\ \tilde{Y}Y \end{Bmatrix} \to \begin{cases} y\tilde{y}^* = \tilde{y}^* y = 1 \\ y\tilde{y} = \tilde{y}y = i \\ y\tilde{y}^\# = \tilde{y}^\# y = -1 \\ y\tilde{y}' = \tilde{y}' y = -i \end{cases}$	$\{r^2, ir^2, -r^2, -ir^2\}$ Coordinate Projection $r = 1 \to \{1, i, -1, -i\}$
Cross Products $\{YY\}$	2 Crosses 16 Products	$\{YY\} \to \{1, i, -1, -i\}$ Order	New Cross $\{r^2 e_j^{i2\theta}\}_{j=0}^3$
Conjugate Crosses $\{\tilde{Y}\tilde{Y}\}$	2 Crosses 16 Products	$\{\tilde{Y}\tilde{Y}\} \to \{i, -1, -i, 1\}$ Order	New Conjugate Cross $\{r^2 e_j^{i2\tilde{\theta}}\}_{j=0}^3$
Multiple Iterations $\{Y^k, \tilde{Y}^k\}_{k=1}^n$	6 Trends	$\{Y^k, \tilde{Y}^k\} \to \begin{cases} r^{k+1} > r^k \\ r^{k+1} < r^k \\ r^{k+1} = r^k \end{cases}$	Y^k : Counter-clockwise $\tilde{Y}^k$: Clockwise
Increase Mode $r^{k+1} > r^k$ Small to Large	Counter-clockwise Clockwise Merged	$\begin{cases} \{r^k e_j^{ik\theta}\}_{j=0}^3 \\ \{r^k e_j^{ik\tilde{\theta}}\}_{j=0}^3 \\ \{r^k e_j^{ik\theta} \cup r^k e_j^{ik\tilde{\theta}}\}_{j=0}^3 \end{cases}$	Multiple Spirals Countclockwise Clockwise Multiple Ellipses
Reduce Mode $r^{k+1} < r^k$ Big to Small	Counter-clockwise Clockwise Merged	$\begin{cases} \{r^k e_j^{ik\theta}\}_{j=0}^3 \\ \{r^k e_j^{ik\tilde{\theta}}\}_{j=0}^3 \\ \{r^k e_j^{ik\theta} \cup r^k e_j^{ik\tilde{\theta}}\}_{j=0}^3 \end{cases}$	Multiple Spirals Countclockwise Clockwise Multiple Ellipses
Single Cycle $C = r^{k+1} = r^k$ Constant Radius	Counter-clockwise Clockwise Merged	$\begin{cases} \{Ce_j^{ik\theta}\}_{j=0}^3 \\ \{Ce_j^{ik\tilde{\theta}}\}_{j=0}^3 \\ \{Ce_j^{ik\theta} \cup Ce_j^{ik\tilde{\theta}}\}_{j=0}^3 \end{cases}$	Polyphase Circle Counter-clockwise Clockwise Compound
Multi-Cycles $C_l = r_l^{k+1} = r_l^k$ Multiple Constant Radii	Counter-clockwise Clockwise Merged	$\begin{cases} \{\{C_l e_j^{ik\theta}\}_{j=0}^3\}_{l=1}^m \\ \{\{C_l e_j^{ik\tilde{\theta}}\}_{j=0}^3\}_{l=1}^m \\ \{\{C_l e_j^{ik\theta} \cup C_l e_j^{ik\tilde{\theta}}\}_{j=0}^3\}_{l=1}^m \end{cases}$	Multiple Polyphase Circles Counter-clockwise Clockwise Compound
Modulo Operator First Quadrant $\{n\theta, n\tilde{\theta}\}$	$\pi/2$ Period	$\begin{cases} n\theta(\bmod \pi/2) \in [0, \pi/2) \\ n\tilde{\theta}(\bmod \tilde{\pi}/2) \in (0, \pi/2] \end{cases}$	First Quadrant Projection
Multiple Iterations $\{y^k, \tilde{y}^k\}_{k=1}^n$	6 Trends	$\{y^k, \tilde{y}^k\} \to \begin{cases} r^{k+1} > r^k \\ r^{k+1} < r^k \\ r^{k+1} = r^k \end{cases}$	y^k : Counter-clockwise $\tilde{y}^k$: Clockwise
Increase Mode $r^{k+1} > r^k$ Small to Large	Countclockwise Clockwise Merged	$\begin{cases} r^k e_0^{ik\theta} \\ r^k e_0^{ik\tilde{\theta}} \\ r^k e_0^{ik\theta} \cup r^k e_0^{ik\tilde{\theta}} \end{cases}$	Multiple Curls Counter-clockwise Clockwise Multiple Closed Curls
Reduce Mode $r^{k+1} < r^k$ Big to Small	Countclockwise Clockwise Merged	$\begin{cases} r^k e_0^{ik\theta} \\ r^k e_0^{ik\tilde{\theta}} \\ r^k e_0^{ik\theta} \cup r^k e_0^{ik\tilde{\theta}} \end{cases}$	Multiple Curls Counter-clockwise Clockwise Multiple Closed Curls
Single Arc Mode $C = r^{k+1} = r^k$ Constant Radius	Countclockwise Clockwise Merged	$\begin{cases} Ce_0^{ik\theta} \\ Ce_0^{ik\tilde{\theta}} \\ Ce_0^{ik\theta} \cup Ce_0^{ik\tilde{\theta}} \end{cases}$	Countclockwise Arc Clockwise Arc Compound Arcs
Multi-Arc Mode $C_l = r_l^{k+1} = r_l^k$ Multiple Constant Radii	Countclockwise Clockwise Merged	$\begin{cases} \{C_l e_0^{ik\theta}\}_{l=1}^m \\ \{C_l e_0^{ik\tilde{\theta}}\}_{l=1}^m \\ \{C_l e_0^{ik\theta} \cup C_l e_0^{ik\tilde{\theta}}\}_{l=1}^m \end{cases}$	Multiple Arcs Counter-clockwise Clockwise Multiple Compound Arcs

Table 9. Conjugate complex function on quantitative features

Model	Feature	Formula	Quantitative
Number of Cocycle Clusters $C_I(N)$	$[-N, N]^2$	$C_I(N) = (N+1)N/2$	Total Number of Cocycle Clusters
Number of Vector Clusters $F_I(N)$	$[-N, N]^2$	$F_I(N) = 8N^2 + 1$	Vector Clustering
Number of Vectors $F_V(N)$	$[-N, N]^2$	$F_V(N) = 8N^2 + 8N + 2$	Counting Vectors
Algebraic Equations on Real Coefficients	m Variable Real Equation	$\exists m$ Solution $\in Gal(m)$	Galois Field Solution
Galois Field Complex Solutions	$Gal(m)$ Specific and Complete Solutions	$Gal(F(m)) \leq Gal(4m)$	Quadruple Expansion
Number of Full Divisions $F(m)$	Three Categories	$F(m) = \begin{cases} 4m, m \equiv 1(\text{mod } 2) \\ m, m \equiv 0(\text{mod } 4) \\ 2m, m \equiv 2(\text{mod } 4) \end{cases}$	Full Partition
Single Interval Number $P(m) = P(\tilde{m})$	Single Quadrant	$P(m) = F(m)/4$	One-quadrant Division
	Distinguishable	$P(m) \cup P(\tilde{m}) = P(m) + 1$	Distinct Number
	Common	$P(m) \cap P(\tilde{m}) = P(m) - 1$	Same Number
	Sum	$P(m) + P(\tilde{m}) = 2P(m)$	Sum Number
Placeholder $T(m) = T(\tilde{m})$	Prime Number	$T(m) \leq m^2 - m + 1$	Maximum Number
Line Defect $\Delta(n,m) = \Delta(n,\tilde{m})$	Statistics by Line	$\Delta(n,m) = m(1 - 1/[n,m])$	Determine Coefactors
Number of Defects $\Delta(m) = \Delta(\tilde{m})$	0 When Prime	$\Delta(m) = \sum_{n=0}^{m} \Delta(n,m)$	Composite Numbers More Complex
Total Number $H(m) = H(\tilde{m})$	Total Number of Seats	$H(m) = T(m) + \Delta(m) = m^2 - m + 1$	Global Invariant

On whether it is a binary or multivariate product result, from a given argument, the eight input elements are multiplied with each other on sixty-four products, and the output results are merged into twelve vectors: four coordinate vectors, eight double-angle or multiple-angle vectors that maintain the conjugate relationship of the two crosses.

The full iteration mode produces three types of observables: Type (A) small to large; Type (B) large to small; and Type (C) constant radius. In the determination of the maximum and minimum intervals, if the generation order of Types (A) and (B) cannot be distinguished, then the iterative output results show similar effects. Typical visual effects are counter-clockwise rotating spirals, clockwise rotating spirals, multiple ellipses, etc., showing the multiple complexities of complex dynamic characteristics.

Spirals are everywhere in nature, from the universe beyond the Milky Way to hurricanes, from the vines of creeping plants to pineapples, the rinds of pine cones, the swirling patterns on nautilus shells and the radial arrangement of sunflower seeds, the structure of human cells to delicate flower tower of cauliflower. These observable ultra/hyper/macro/micro structures are closely related to cosmic spirals, natural spirals, growth curves, plant spirals, and more beautiful curves. The detailed description of the principle model and method is still an unknown mystery today, and it is worth exploring in depth.

Type (C) provides the iteration mode and is closely related to $\{SO(m), O(m), GL(m), SU(m)\}$ transformation groups to satisfy quantum mechanics, Galois theory, standard model, and other gauge metric transformations. The transformation of the cyclotomic complex function supports the core part of modern mathematical physics. The multicircle structure expands the visual output of the classification from the level of description and analysis on each feature vector.

From the view of measuring parameters, the parameters in the first quadrant are a pair of conjugate variables with synchronous large-to-small or small-to-large characteristics. The typical outputs are maintained in the quarter-cone range as the modulo function correlated. In addition to keeping the information of the entire iteration intact, it can provide compact and efficient features for computational and structural observation.

In three visualization maps, binary product, multivariate product, and multivariate product, in the first quadrant, the relevant results are briefly displayed.

10.2.2 Binary Product

The thirty-two products between the two conjugate crosses generate four coordinate vectors, corresponding to $\{1, i, -1, -i\}$

$$\begin{Bmatrix} Y\tilde{Y} \\ \tilde{Y}Y \end{Bmatrix} : \begin{Bmatrix} y\tilde{y}^* = \tilde{y}^*y = 1 \\ y\tilde{y} = \tilde{y}y = i \\ y\tilde{y}^\sharp = \tilde{y}^\sharp y = -1 \\ y\tilde{y}' = \tilde{y}'y = -i \end{Bmatrix} :$$

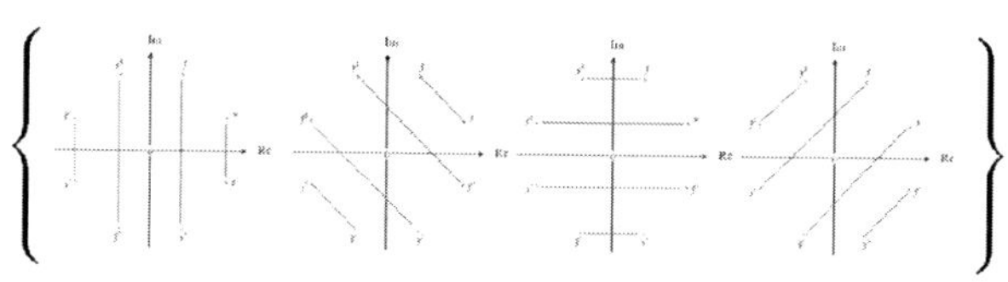

In the cross associated with the original variable, the four outputs are merged from the sixteen products: $\{1, i, -1, -i\}$

$$\{Y^2 = YY\} :$$

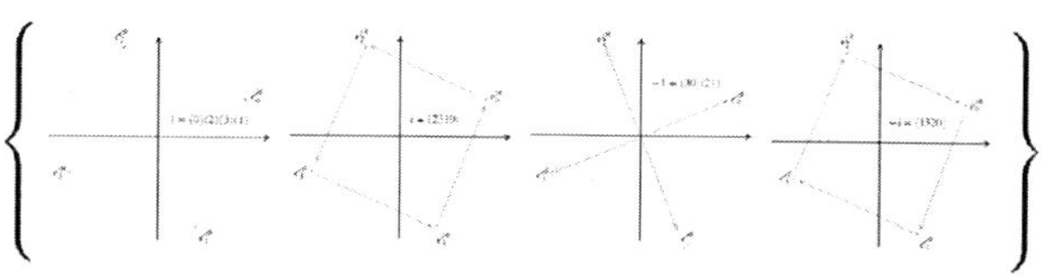

In the cross associated with the conjugate variable, the four outputs are merged from the sixteen products: $\{i, -1, -i, 1\}$

$$\{\tilde{Y}^2 = \tilde{Y}\tilde{Y}\} :$$

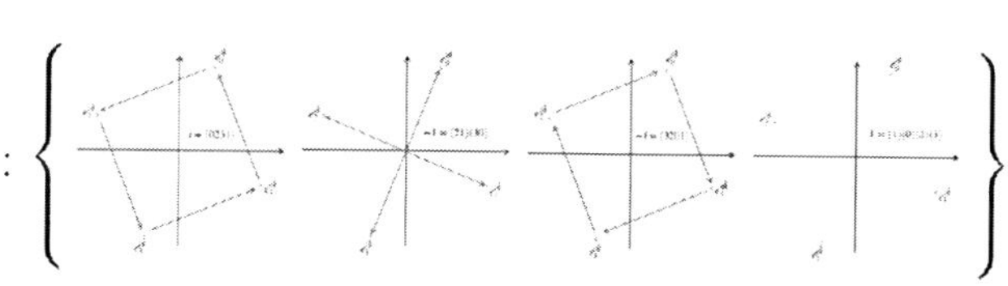

10.2.3 Multiple Products

In the quadruple associated with the original variable, the four iterative outputs are provided in the n products: $\{>, =, <,$ superposition$\}$

$$\{\, Y^n = Y \cdots Y \,\} :$$

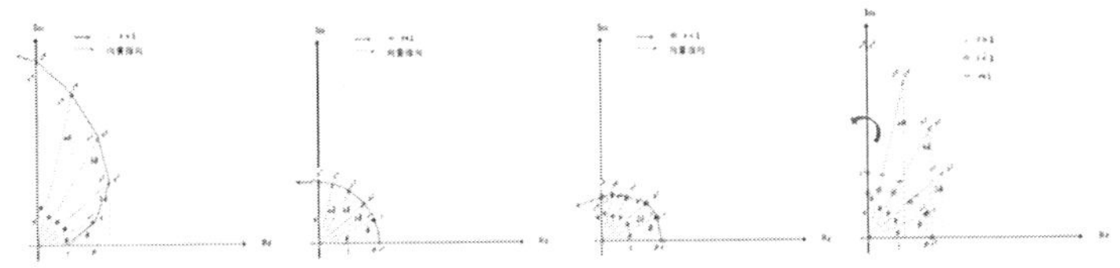

In the quadruple associated with the conjugate variable, the four iterative outputs are provided in the n products: $\{>, =, <,$ superposition$\}$

$$\{\, \tilde{Y}^n = \tilde{Y} \cdots \tilde{Y} \,\} :$$

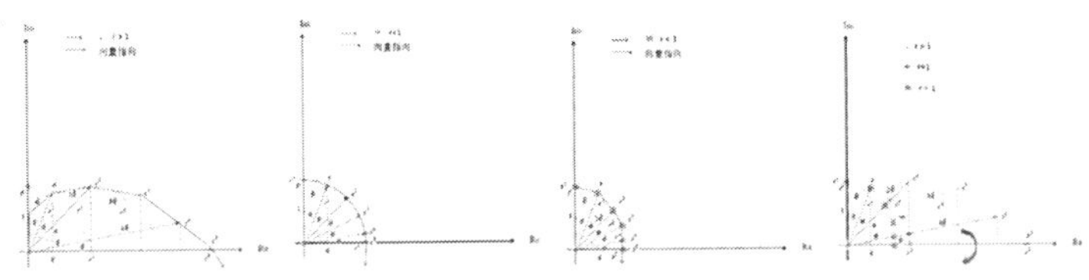

In all n products, three iterative outputs are shown with increasing vector length:

$$\left\{ \begin{matrix} Y^j \\ \tilde{Y}^j \\ Y^j + \tilde{Y}^j \\ r^{j+1} > r^j \end{matrix} \right\}_{j=1}^{n} :$$

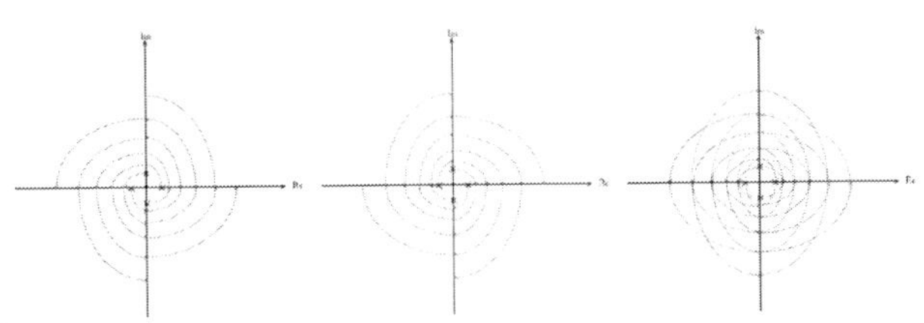

In all n products, three iterative outputs are shown with decreasing vector length:

$$\left\{ \begin{matrix} Y^j \\ \tilde{Y}^j \\ Y^j + \tilde{Y}^j \\ r^{j+1} < r^j \end{matrix} \right\}_{j=1}^{n} :$$

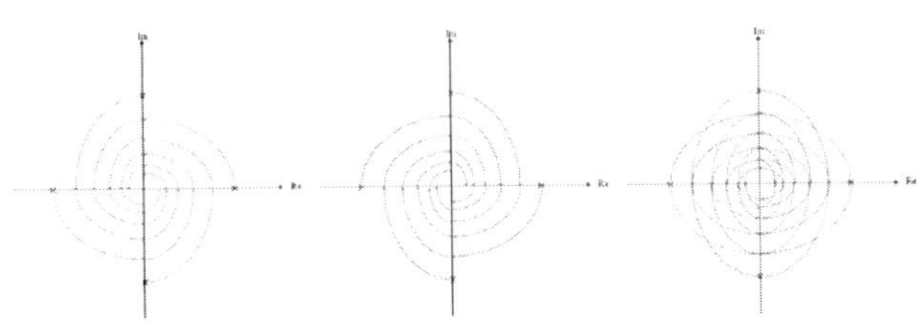

In the full n products, the vector length is constant on a single cycle of three group iteration output maps: $r^{j+1} = r^j, 1 \leq j \leq n$

$$\left\{ \begin{matrix} Y^j \\ \tilde{Y}^j \\ Y^j + \tilde{Y}^j \\ r^{j+1} = r^j \end{matrix} \right\}_{j=1}^{n} :$$

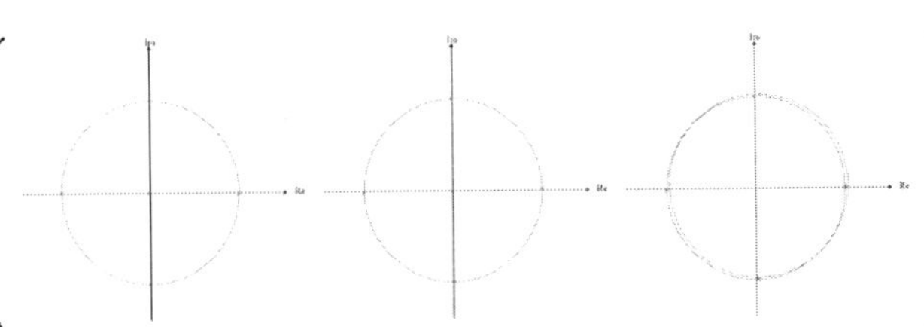

In the full n products, the multicircles are shown in three group iteration output maps with a set of constant vector lengths: $r_k^{j+1} = r_k^j, 1 \leq j \leq n, 1 \leq k \leq s$

$$\left\{\begin{bmatrix} Y^j \\ \tilde{Y}^j \\ Y^j + \tilde{Y}^j \\ r_k^{j+1} = r_k^j \end{bmatrix}_{j=1}^{n}\right\} :$$

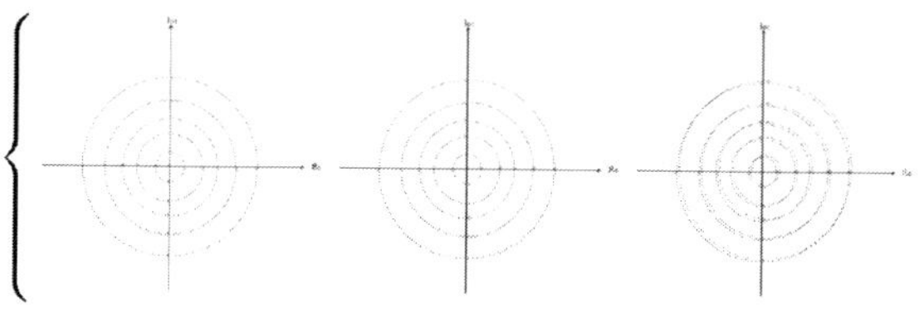

10.2.4 Multivariate Product in First Quadrant

In the n product of the first quadrant, three iterative output maps are shown with increasing vector length: $r^{j+1} > r^j, 1 \leq j \leq n$

$$\left\{\begin{bmatrix} Y_0^j \\ \tilde{Y}_0^j \\ Y_0^j + \tilde{Y}_0^j \\ r^{j+1} > r^j \end{bmatrix}_{j=1}^{n}\right\} :$$

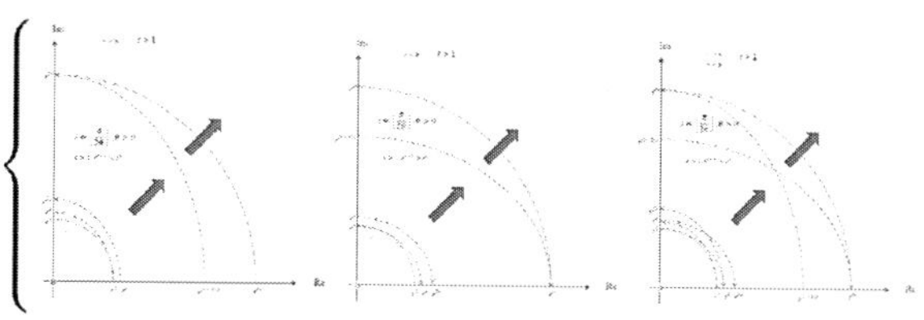

In the n products of the first quadrant, three iterative output maps are shown with decreasing vector length: $r^{j+1} < r^j, 1 \leq j \leq n$

$$\left\{\begin{bmatrix} Y_0^j \\ \tilde{Y}_0^j \\ Y_0^j + \tilde{Y}_0^j \\ r^{j+1} < r^j \end{bmatrix}_{j=1}^{n}\right\} :$$

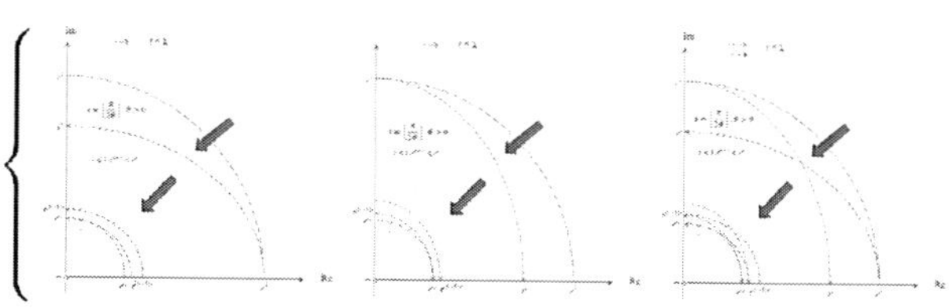

In the n products of the first quadrant, the vector length is constant, and a single arc is shown in three group iteration output maps: $r^{j+1} = r^j, 1 \leq j \leq n$

$$\left\{\begin{bmatrix} Y_0^j \\ \tilde{Y}_0^j \\ Y_0^j + \tilde{Y}_0^j \\ r^{j+1} = r^j \end{bmatrix}_{j=1}^{n}\right\} :$$

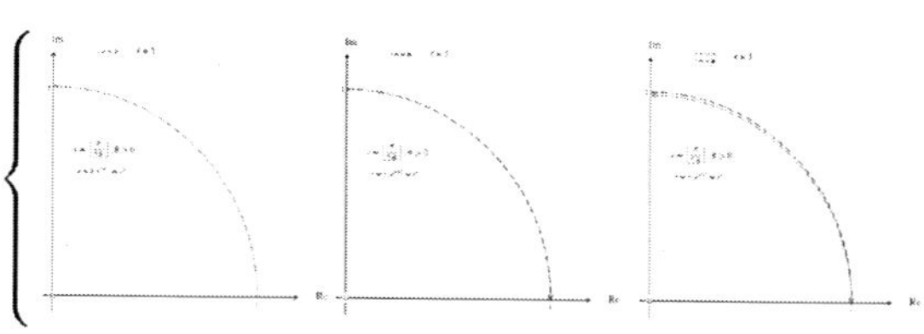

In the n products of the first quadrant, multi-arcs are shown in three group iteration output maps with a set of constant vector lengths: $r_k^{j+1} = r_k^j, 1 \leq j \leq n, 1 \leq k \leq s$

$$\left\{\begin{bmatrix} Y_0^j \\ \tilde{Y}_0^j \\ Y_0^j + \tilde{Y}_0^j \\ r_k^{j+1} = r_k^j \end{bmatrix}_{j=1}^{n}\right\} :$$

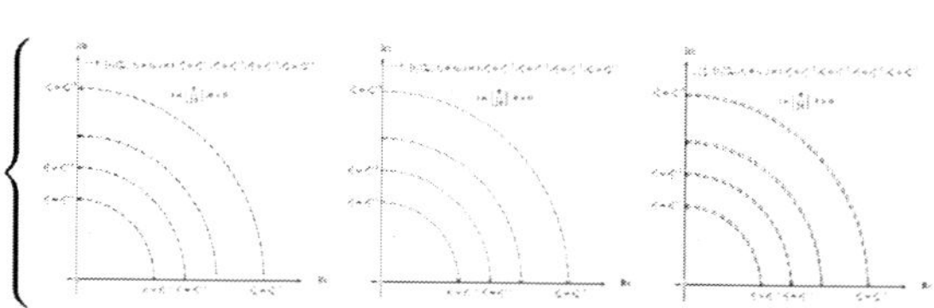

10.3 Conjugate Complex Function of Quantization Features

Full quantitative features are assembled in Table 9. For the $[-N, N]^2$ region, the number of independent cycles increases by the order of $C_I(N) = (N+1)N/2 = O(N^2/2)$. The number of feature vectors is $F_V(N) = 2(2N+1)^2$. The number of vector clusters shrinks to $F_I(N) = 8N^2 + 1$, and the number of cycles is approximately 1/16 of the number of vector clusters.

The number of various features shows that if the scale of the region increases, then the complexity of the structure is increased and the pattern of simplicity changes, which provides the basic support for high-performance simulation and computation for modern ultralarge-scale network clusters in parallel and distributed cloud computing environments.

The number of divisions $F(m)$ shows the full structural complexity. When m is an odd number, the feature space is expanded four times; when m is an even number but not a multiple of four, the feature space is expanded twice; and when m is a multiple of four, the feature space remains unchanged.

The intrinsic m solutions of the binomial m algebraic equation with real coefficients are contained in $Gal(m)$, extended to complex coefficients, and the solution space needs $Gal(4m)$ support.

$$\{x^m - 1 = 0\} \rightarrow \{\{x^m - 1, x^m - i, x^m + 1, x^m + i\} = 0\} \equiv \{x^{4m} - 1 = 0\} \in Gal(4m)$$

This type of division is extended by the number of intervals in the first quadrant, which is $F(m)/4$ for a single quadrant, and the numbers of pairwise quantizations for distinguishable, common, and summation are calculated in $\{\cup, \cap, +\}$ on quantity change after operation. It is a necessary expansion to use a single quadrant for system analysis.

For the placeholder $T(m)$, the number of defects $\Delta(m)$, and the total number $H(m) = m^2 - m + 1$ with the theoretical characteristics of number theory, there are series formulas and prime numbers, composite numbers, and calculation mode associations such as the greatest common multiple factor. This type of representation is closely related to the P-adic algebraic structure, cooperating with conjugate exponential functions. From the example table of placeholder distribution, the composite number contains multiple prime factors to provide more complex characteristics. Compared with any prime number distribution, the number of defects is 0, and the placeholder number is equal to the total number.

10.4 Differences between Conjugate Complex and Complex Conjugate Operators

From the architectural analysis of the full-space structure and the hierarchic levels of macroscopic complex functions,

$$\{\text{Conjugate Operators} + \text{Four Quadrants}\} \equiv \{\text{Complex Conjugate Operator} + \text{UHP/LHP}\}$$

There is no essential difference between the two constructions, and the entire structure is equivalent!

From the basic levels of refined calculations, in the 2D rectangular coordinate of trigonometric functions represented by de Morgan [21] balanced logic, the pairwise arguments $\{\theta, -\theta\}$ are recognized, and four attribute properties are identified.

In this chapter, four attribute features and eight correlated states in 2D Cartesian coordinates are extended in the conjugate vector clusters on the complex plane. The Clifford [22] number $(m + 1)m/2$ based on vector analysis is the same number of cocircular clusters in the chapter. For specific frontier applications, it is worth exploring in depth.

From the levels of cluster analysis, to solve the complex coefficient solution of the n-th degree binomial in the Galois fields, this chapter provides the support of necessary and sufficient conditions. Various conjugate operators have opened the comprehensive use of various harmonic functions, modular functions, algebraic number theory, differential geometry, multivariable complex functions, and complex geometry to solve application problems.

Due to the extensiveness and diversity of applications in complex function research, these historic research results have not attracted much attention from later generations and expanded applications.

At present, the development of science and technology in the world and cutting-edge research are faced with many computing bottlenecks in parallel distributed cloud computing environments in the communication of global cyberspace and the higher performance of ultralarge-scale systems. These parallel characteristics of optimal vector schemes need to be restudied.

It is expected that the vector expansion of the conjugate operators on the conjugated complex plane can systematically surpass the complex conjugate operator on the complex plane, effectively promoting the levels of the foundation and applications of the complex functions in parallel environments and the in-depth development of advanced science and technology in the future.

References

[1] Complex number. https://en.wikipedia.org/wiki/Complex_number.
Eric W. Weisstein. Complex number. mathworld.wolfram.com.

[2] Morris Kline. *A History of Mathematical Thought*, Vol. 1–3, Oxford University Press, 1972.

[3] Lars Ahlfors. *Complex Analysis*, 3rd edition, McGraw-Hill, 1979.

[4] Tom Apostol. *Mathematical Analysis*. Addison-Wesley, 1981.

[5] John B. Conway. *Functions of One Complex Variable I*, Springer, 1986. ISBN 978-0-387-90328-6.

[6] Kapil D. Joshi. *Foundations of Discrete Mathematics*, John Wiley & Sons, 1989.

[7] H. D. Ebbinghaus, H. Hermes, F. Hirzebruch, M. Koecher, K. Mainzer, J. Neukirch, A. Prestel and R. Remmert. *Numbers* (hardcover ed.). Springer, 1991. — An advanced perspective on the historical development of the concept of number.

[8] Tristan Needham. *Visual Complex Analysis*, Oxford University Press Inc., 1997.

[9] E. D. Solomentsev. Complex number. In *Encyclopedia of Mathematics*, Hazewinkel, M. (Ed.), EMS Press, 2001 [1994].

[10] Victor J. Katz. *A History of Mathematics, Brief Version*, Addison-Wesley, 2004.

[11] John Derbyshire. *Unknown Quantity: A Real and Imaginary History of Algebra*, Joseph Henry Press, 2006.

[12] W. H. Press, S. A. Teukolsky, W. T. Vetterling and B. P. Flannery, eds. Section 5.5 complex arithmetic. In *Numerical Recipes: The Art of Scientific Computing*, 3rd edition, Cambridge University Press, 2007.

[13] Paul J. Nahin. *An Imaginary Tale: The Story of $\sqrt{-1}$*, Princeton University Press, 2007.

[14] Leo Corry. *A Brief History of Numbers*, Oxford University Press, 2015.

[15] Lizhen Ji, Athanese Papadopoulos and Sumio Yamada, eds. *From Riemann to Differential Geometry and Relativity*, Springer, 2017.

[16] Abraham De Moivre. https://en.wikipedia.org/wiki/Abraham_de_Moivire.
A. de Moivre. A method of extracting roots of an infinite equation. *Philos. Trans. R. Soc.* 20(240), 1698, 190–193. http://doi.org/10.1098/rstl.1698.0034.
A. de Moivre. *Miscellanea Analytica de Seriebus et Quadraturis (in Latin)*, J. Tonson and J. Watts, 1730.

[17] Roger Cotes. https://en.wikipedia.org/wiki/Roger_Coles.
Roger Cotes. Logometria. *Philos. Trans. R. Soc.* 29(338), 1714, 5–45.
Roger Cotes with Robert Smith, eds., *Harmonia Mensurarum* …, Cambridge, 1722, chapter: “Logometria”.
Harmonia Mensurarum a more complete account of Cotes's involvement with principia, followed by an even more thorough discussion of his mathematical work. https://www.mathpages.com/home/kmath192/kmath192.htm.

[18] Leonard Euler. *Introductio in Analysin Infinitorum [Introduction to the Analysis of the Infinite] (in Latin)*, Vol. 1, Marc Michel Bosquet & Co, 1748.
Leonhard Euler. Recherches sur les racines imaginaires des équations. *Histoire de l'Académie Royale des Sciences et des Belles-Lettres de Berlin* 5, 1751, 222–288.
English translation: Leonhard Euler. Investigations on the imaginary roots of equations. *Histoire de l'Académie Royale des Sciences et des Belles-Lettres de Berlin* 5, 1751, 222–288.

[19] Carl Friedrich Gauss. https://en.wikipedia.org/wiki/Carl_Friedrich_Gauss. Carl Friedrich Gauss. Demonstratio nova theorematis omnem functionem algebraicam rationalem integram unius variabilis in factores reales primi vel secundi gradus resolvi posse, 1799, Helmstedt: C. G. Fleckeisen (tr. New proof of the theorem that every integral rational algebraic function of one variable can be resolved into real factors of the first or second degree).
C. F. Gauss. Theoria residuorum biquadraticorum. Commentatio secunda [Theory of biquadratic residues. Second memoir.]. *Commentationes Societatis Regiae Scientiarum Gottingensis Recentiores (in Latin)* 7, 1831, 89–148. 高斯 算术探索 潘丞彪 张明尧 译 哈尔滨工业大学出版社 2011.

[20] William Rowan Hamilton. https://en.wikipedia.org/wiki/William_Rowan_Hamilton.
William Rowan Hamilton. Theory of conjugate functions or algebraic couples: With a preliminary and elementary essay on algebra as the science of pure time. *Trans. R. Ir. Acad.* 17(part 1), 1837, 293–422. https://www.maths.tcd.ie/pub/HistMath/People/Hamilton/PureTime/PureTime.pdf. 1868 – Über die Hypothesen, welche der Geometrie zugrunde liegen. Abh. Kgl. Ges. Wiss., Göttingen 1868. Translation EMIS, pdfOn the hypotheses which lie at the foundation of geometry, translated by W. K.Clifford, Nature 8 1873 183 – reprinted in Clifford's Collected Mathematical Papers, London 1882 (MacMillan); New York 1968 (Chelsea).
Wm. Hamilton. On a new species of imaginary quantities connected with a theory of quaternions. *Proc. R. Ir. Acad.* 2, 1844, 424–434.

[21] Augustus De Morgan. https://en.wikipedia.org/wiki/Augustus_De_Morgan.
Augustus De Morgan. *Trigonometry and Double Algebra*, Taylor, Walton & Malbery, 1849.
Augustus De Morgan. *Syllabus of a Proposed System of Logic*. University of California Libraries, Walton and Maberly, 1860.

[22] William Clifford. *Elements of Dynamic: An Introduction to the Study of Motion and Rest in Solid and Fluid Bodies I, II, III*, Macmillan & Co., 1878.
William Clifford. Applications of Grassmann's extensive algebra. *Am. J. Math.* 1(4), 1878, 350–358. http://doi.org/10.2307/2369379.JSTOR2369379.

[23] Oliver Heaviside. https://en.wikipedia.org/wiki/Oliver_Heaviside.

[24] James Clerk Maxwell. *An Elementary Treatise on Electricity*, Clarendon Press, 1881.

[25] Felix Klein. https://en.wikipedia.org/wiki/Felix_Klein.
克莱因瓶 https://baike.baidu.com/item/克莱因瓶/1187845.
F. Klein. Riemannschen Flächen, Springer, reprint, 1986. 菲利克斯 克莱因著 舒湘芹 等译 高观点下的初等数学, 复旦大学出版社 2008.
F 克莱因 著, 齐民友 译, 数学在19世纪的发展 第一卷 高等教育出版社 2010 p275 Gauss 整数.

[26] Hanri Poincare. https://en.wikipedia.org/wiki/Hanri_Poincare.
Poincare half plane model. https://en.wikipedia.org/wiki/Poincar%C3%A9_half-plane_model.

[27] Caspar Wessel. Om Directionens analytiske Betegning, et Forsog, anvendt fornemmelig til plane og sphæriske Polygoners Oplosning [On the analytic representation of direction, an effort applied in particular to the determination of plane and spherical polygons]. *Nye Samling af det Kongelige Danske Videnskabernes Selskabs Skrifter [New Collection of the Writings of the Royal Danish Science Society] (in Danish)* 5, 1799, 469–518.

[28] John Wallis. *A treatise of algebra, both historical and practical ...*, printed by John Playford, for Richard Davis, 1685, 264–273.

[29] Argand. *Essai sur une manière de représenter les quantités imaginaires dans les constructions géométriques [Essay on a Way to Represent Complex Quantities by Geometric Constructions] (in French)*, Madame Veuve Blanc, 1806.
Argand. Reflexions sur la nouvelle théorie des imaginaires, suives d'une application à la demonstration d'un theorème d'analise [Reflections on the new theory of complex numbers, followed by an application to the proof of a theorem of analysis]. *Annales de mathématiques pures et appliquées (in French)* 5, 1814, 197–209.

[30] Buée. Mémoire sur les quantités imaginaires [Memoir on imaginary quantities]. *Philos. Trans. R. Soc.* (in French) 96, 1806, 23–88. http://doi.org/10.1098/rstl.1806.0003. S2CID 110394048.

[31] John Warren. *A Treatise on the Geometrical Representation of the Square Roots of Negative Quantities*, Cambridge University Press, 1828.

[32] Augustin-Louis Cauchy. *Cours d'Analyse de l'École Royale Polytechnique, 1ère partie: Analyse Algébrique*, Éditions Jacques Gabay (published 1992), 1821. ISBN 978-2-87647-053-8 (tr. Course on Analysis of the Royal Polytechnic Academy, Part 1: Algebraic Analysis).

[33] F. Bessel. Untersuchung des Theils der planetarischen Störungen, Berlin Abhandlungen, article 14, 1824.

[34] G. Green. *An Essay on the Application of Mathematical Analysis to the Theories of Electricity and Magnetism*, T. Wheelhouse, 1828.

[35] C. G. J. Jacobi. *Fundamenta Nova Theoriae Functionum Ellipticarum (in Latin)*, Königsberg, 1829. ISBN 978-1-108-05200-9, Reprinted by Cambridge University Press 2012.

[36] Hermann Grassmann. *A New Branch of Mathematics* (Translated by Lloyd C. Kannenberg), Open Court, 9–297, 1844 [1994].

[37] Joseph Fourier. *The Analytical Theory of Heat*, Cambridge University Press (reissued by Cambridge University Press, 2009.

[38] Niels Henrik Abel. https://en.wikipedia.org/wiki/Niels_Henrik_Abel.

[39] Evariste Galois. https://en.wikipedia.org/wiki/%C3%89variste_Galois.
Évariste Galois. Démonstration d'un théorème sur les fractions continues périodiques. *Annales de Mathématiques* XIX, 1828, 294.
Jean-Pierre Tignol. *Galois' Theory of Algebraic Equations*, World Scientific, 2001. ISBN 978-981-02-4541-2. Historical development of Galois theory.

[40] Ernst Kummer. https://en.wikipedia.org/wiki/Ernst_Kummer.
Bryan Birch. Cyclotomic fields and Kummer extensions. In *Algebraic Number Theory*, Chap. III, edited by J. W. S. Cassels and A. Frohlich, Academic Press, 1973, 85–93.
Ernst Eduard Kummer. *Collected Papers. Volume 1: Contributions to Number Theory* (André Weil, ed.), Springer-Verlag, 1975. ISBN 978-0-387-06835-0, MR 0465760[2] Ernst Eduard Kummer. *Collected Papers. Volume II: Function Theory, Geometry and Miscellaneous* (André Weil, ed.), Springer-Verlag, 1975.
"Kummer extension" (https://www.encyclopediaofmath.org/index.php?title=Kummer_extension), *Encyclopedia of Mathematics*, EMS Press, 2001 [1994].

[41] Bernhard Riemann. https://en.wikipedia.org/wiki/Bernhard_Riemann.
Bernhard Riemann. On the hypotheses which lie at the bases of geometry (Habilitationsschrift) (translated by W. K. Clifford. – via School of Mathematics), Trinity College Dublin, 1867 [1854].
B. Riemann. *Collected Works*, Dover, reprint, 1953.
B. Riemann. *Foundations for a General Theory of Functions of a Complex Variable. Collected papers*

of Bernhard Riemann (translated by R. C. Baker, C. Christensen and H. Orde), Kendrick Press, 2004, 1–41.

[42] Hermann Hankel. Vorlesungen über die complexen Zahlen und ihre Functionen [*Lectures About the Complex Numbers and Their Functions*] (in German), Vol. 1, Leopold Voss, 71. From p. 71: Wir werden den Factor (cos φ + i sin φ) haüfig den Richtungscoefficienten nennen. (We will often call the factor (cos φ + i sin φ) the "coefficient of direction".), 1867.

[43] Karl Weierstrass. Neuer Beweis des Satzes, dass jede ganze rationale Function einer Veränderlichen dargestellt werden kann als ein Product aus linearen Functionen derselben Veränderlichen. *Sitzungsberichte der königlich preussischen Akademie der Wissenschaften zu Berlin*, 1891, 1085–1101 (tr. New proof of the theorem that every integral rational function of one variable can be represented as a product of linear functions of the same variable).

[44] J. W. Gibbs. *Elementary Principles in Statistical Mechanics, Developed with Especial Reference to the Rational Foundation of Thermodynamics*, Dover Publications, 1960 [1902].
E. B. Wilson. *Vector Analysis, A Text-Book for the Use of Students of Mathematics and Physics, Founded Upon the Lectures of J. Willard Gibbs*, Yale University Press, 1929 [1901].

[45] Charles Hermite. https://en.wikipedia.org/wiki/Charles_Hermite.

[46] Jacques Hadamard. https://en.wikipedia.org/wiki/Jacques_Hadamard.

[47] Hermann Weyl. https://en.wikipedia.org/wiki/Hermann_Weyl.
Hermann Weyl. *Raum Zeit Materie*. Springer-Verlag, 1923. (Translated by H. L. Brose). *Space Time Matter*, Dover, 1950.
H. Weyl. *Algebraic Theory of Numbers*, Princeton University Press, 1959. J. W. S. Cassels (ed.) A. Fröhlich (ed.), *Algebraic Number Theory*, Acad. Press, 1967.

[48] David Hilbert. https://en.wikipedia.org/wiki/David_Hilbert.
http://www.emis.de/classics/Riemann/. Also in William B. Ewald, ed. *From Kant to Hilbert: A Source Book in the Foundations of Mathematics*, Vol. 2. Oxford Universtiy Press, 1996, 65–261.

[49] John von Neumann. https://en.wikipedia.org/wiki/John_von_Neumann.
J. von Neumann. Allgemeine Eigenwerttheorie Hermitescher Funktionaloperatoren. *Math. Ann.* 102, 1930, 49–131.

[50] G. Birkhoff. *Lattice Theory*, The American Mathematical Society, 1948.

[51] G. Birkhoff. Von Neumann and lattice theory, momories of the American mathematical society, no. 10, 1955.

[52] Richard Courant. https://en.wikipedia.org/wiki/Richard_Courant.

[53] G. H. Hardy. https://en.wikipedia.org/wiki/G._H._Hardy.
G. H. Hardy and E. M. Wright. *An Introduction to the Theory of Numbers*, 4th edition, OUP, 2000 [1938], 189.

[54] Xiong Qinglai. https://en.wikipedia.org/wiki/Xiong_Qinglai.

[55] Hua Luogeng. https://en.wikipedia.org/wiki/Hua_Luogeng.
华罗庚 数论导论 科学出版社 1957.
Additive Theory of Prime Numbers (Translations of Mathematical Monographs: Vol. 13). Amer Mathematical Society, 1966.
Loo-keng Hua. *Starting with the Unit Circle: Background to Higher Analysis*, Springer-Verlag, 1981.
Loo-keng Hua. *Selected Papers*, Springer Verlag, 1983.
Introduction to Number Theory, Springer, 1987.
华罗庚 (2013). 华罗庚文集 - 多复变函数论卷 *II*, 科学出版社 2013.

[56] Shiing-Shen Chern. https://en.wikipedia.org/wiki/Shiing-Shen_Chern.
张奠宙 王善平, 陈省身文集,华东师范大学出版社 2002.

[57] 杨振宁 Yang Chen-Ning. https://en.wikipedia.org/wiki/Yang_Chen-Ning.
杨振宁,葛墨林,何杨辉编, 拓扑与物理,江西科学技术出版社 2021.

[58] Shing-Tung Yau. https://en.wikipedia.org/wiki/Shing-Tung_Yau.

[59] 熊锡金 (1988) 泛复变函数及其在数学与物理中的应用, 东北师范大学出版社 1988.

[60] V. I. Arnold. *Mathematical Methods of Classical Mechanics*, Springer-Verlag, 1978.

[61] Nicolas Bourbaki. *Foundations of Mathematics Logic: Set Theory. Elements of the History of Mathematics*. Springer, 1998.
[62] Roger Penrose. https://en.wikipedia.org/wiki/Roger_Penrose.
R. Penrose. *The Emperor's Mind: Concerning Computers, Minds and the Law of Physics*, Penguin, 1989.
R. Penrose. *The Road to Reality - A Complete Guide to the Laws of the Universe*, Vintage Book, 2005.
[63] Robert Langlands. https://en.wikipedia.org/wiki/Robert_Langlands.
[64] David Marker. Introduction to the model theory of fields. In *Model Theory of Fields. Lecture Notes in Logic, Vol. 5*, edited by D. Marker, M. Messmer and A. Pillay, Springer-Verlag, 1996, 1-37.
[65] Michael Monastyrsky. *Riemann, Topology and Physics*, Birkhäuser, 1999.
[66] John Derbyshire, *Prime Obsession: Bernhard Riemann and the Greatest Unsolved Problem in Mathematics*, John Henry Press, 2003.
[67] S. Lang. *Algebraic Number Theory*, Addison-Wesley, 1970.
S. Lang. *Cyclotomic Fields*, Springer, 1978.
Serge Lang. *Algebra, Graduate Texts in Mathematics, Vol. 211* (Revised third ed.), Springer-Verlag, 2002.
[68] E. Artin. *Theory of Algebraic Numbers*, Gottingen, 1975.
[69] A. Weil. *Basic Number Theory*, Springer, 1967.
[70] K. Iwasawa. *Lectures on -Adic -Functions*, Springer, 1972.
K. Iwasawa. *Lectures on p-Adic Functions*, Princeton University Press, 1972.
K. Iwasawa. On some modules in the theory of cyclotomic fields. *J. Math. Soc. Japan* 16, 1964, 42–82.
[71] K. Uchida. Class numbers of imaginary abelian number fields III. *Tôhoku Math. J.* 23, 1971, 573–580.
[72] John Stillwell. *Introduction to Theory of Algebraic Integers by Richard Dedekind.* Cambridge Mathematical Library, Cambridge University Press, Great Britain, 1996.
[73] L. C. Washington. *Cyclotomic Fields*, Springer, 1982. MR0718674. J. Coates, The work of Mazur and Wiles on cyclotomic fields. *Sem. Bourbaki*, 33, 1980/81, 575. Lecture Notes in Math., 901, 1981, 220–242.
[74] 邹青 从库默到朗兰兹猜想的历史,哈尔滨工业大学出版社 2014.
[75] G. M. Goluzin. Geometric theory of functions of a complex variable. *Amer. Math. Soc.* 1969, Vol. 26, 676 pp.
[76] R. Remmert. *Classical Topics in Complex Function Theory.* Springer, 1998.
[77] 吉姆 艾尔–哈利利著 戴凡惟译, 悖论:破解科学史上最复杂的9大谜团,中国青年出版社 2015.
[78] 庞元正 李建华, 系统论 控制论 信息论 经典文献选编, 求实出版社 1989.
[79] 编辑委员会, 逻辑学词典,吉林人民出版社 1983.
[80] 斯图尔特 夏皮罗 著 郝兆宽 杨睿之 译,数学哲学,复旦大学出版社 2009.
[81] 罗伊 索伦森著 贾红雨译, 悖论简史 – 哲学和心灵的迷宫,北京大学出版社 2007.
[82] 逻辑悖论. https://baike.baidu.com/item/逻辑悖论/126875.
[83] 莫比乌斯带. https://baike.baidu.com/item/莫比乌斯带/4457881.
[84] S. Barr. *Experiments in Topology*, Dover Publications Inc., 1964.
[85] S. C. Kleene. Godel's impression on students of logic in the 1930s. In Gödel Remembered, Weingartner and Schmetterer, (Eds.) 1987, 49–64.
[86] T. Siegfried. *The Bit and the Pendulum: From Quantum Computing to M Theory - The New Physics of Information*, John Wiley & Sons, 2000.
[87] C. Seife. *Decoding the Universe: How the Science of Information is Explaining Everything in the Cosmos, from our Brains to Black Holes*, Penguin, 2006.
[88] 吉姆 巴戈特 著;徐彬 于秀秀译, 量子通史 - 量子物理史上的40个重大时刻,中信出版社 2021.
J. Baggott. *The Quantum Story in 40 Moments*, Oxford University Press, 2011.
[89] 许天周 李炳照, 线性正则变换及其应用,科学出版社 2013.
[90] 马天, 从数学观点看物理世界 - 基本粒子与统一场理论, 科学出版社 2015.
[91] G.'t Hooft. *The Cellular Automaton Interpretation of Quantum Mechanics*, Springer, 2016.

[92] 艾小文, 数论问题 - 指数和的均值分析与研究,电子工业出版社 2021.

[93] L. Z. Xu, M. S. Jiang and Z. Q. Zhu. *Combinatorial Mathematics of Computation*, Shanghai Science & Technology Press, 1983 (in Chinese).

[94] J. H. van Lint and R. M. Wilson. *A Course in Combinatorics*, 2nd edition, Cambridge University Press, 2001.

[95] M. C. Escher. *The Graphic Work*, Taco, 1989.

[96] Upper half-plane. https://en.wikipedia.org/wiki/Upper_half-plane.

[97] D. E. Knuth. An imaginary number system. *Communications of the ACM* 3(4), 1960, 245–247. https://doi.org/10.1145/367177.367233.

[98] D. E. Knuth. *Positional Number Systems, The Art of Computer Programming, Vol. 2*, Addison-Wesley, 1996, 162.

[99] [明] 来知德, 来易经图解, 中央编译出版社 2010.

[100] 施维, 周易八卦图解,四川出版集团巴蜀书社 2003.

[101] 雷泽, 十分钟周易, 中山大学出版社 1993.

[102] L. X. Wang. *An Elementary Treatise on Combinations*, Harbin Institute of Technology Press, 2012 (in Chinese).

[103] R. J. Simpson. Convex lattice polygons on mimimum area. *Bull. Austral. Math. Soc.* 42, 1990, 353–367.

[104] A. Day. Doubling constructions in lattice theory. *Can. J. Math.* 44(2), 1992, 252–269.

[105] M. Stern. *Semimodular Lattices Theory and Applications*, Cambridge University Press, 1999.

[106] 许光午,王小云, 格的计算和密码学应用, 中国科学:数学 2020年 第50卷第10期 1417–1436.

[107] W. K. Wootters. Entenglement of formation of an arbitrary state of two qubits. *Phys. Rev. Lett.* 80, 1998, 2245.

[108] Z. J. Zheng and A. Maeder. The conjugate classification of the kernel form of the hexagonal grid. In *Modern Geometric Computing for Visualization*, T.L. Kunii and Y. Shinagawa (Eds.), Springer-Verlag, 1992, 73–89.

[109] Z. J. Zheng. *Conjugate Transformation of Regular Plan Lattices for Binary Images*, PhD Thesis, Monash University, 1994.

[110] Jeffrey Zheng and Chris Zheng. A framework toexpress variant and invariant spaces for binary logic. *Front. Electr. Electron. Eng. China* 5(2), 2010, 163–172.

[111] Jeffrey Z. J. Zheng. *Variant Construction from Theoretical Foundation to Applications*, Springer Nature, 2019.

[112] 郑智捷, 变值体系理论及其应用 第一卷 理论基础及其应用,科学出版社 2021. Jeffrey Z. J. Zheng. *Vairant Construction Theory and Applications, Vol. 1. Theoretical Foundation and Applications*, Science Press, 2021.

Chapter 2

Measurement Operators of Conjugate Transformation Structure on Complex Plane

Conjugate Algebra on Pair 0–1 Feature Vectors and Symmetric/Antisymmetric Brackets

Jeffrey Zheng*
Conjugate System Research, Melbourne, Australia

Abstract

Poisson, Lagrange, Lie, and Dirac brackets play a key role in the foundation of modern physics and mathematics with wider applications in multiple advanced sciences and technologies. This chapter proposes a conjugate transformation structure and its measurement operators on a hierarchy of multiple levels to support intermediate transforming structures from pairs of logic states as microensembles to feature vector transformations as global measurements. Using logic equations and pairs of partitions on phase spaces, conjugate 0–1 vectors provide hypercomplex number systems. Multiple operators can be created and linked with Poisson, Lagrange, Lie, and Dirac brackets. The main constructions of conjugate transformation structures are described, and complex conjugate operators are discussed under a pair of symmetric and antisymmetric parameters with $O(2^{2n} \times 2^N)$, $1 \leq n \leq 2^m$ structural complexity. Using new operators, the Poisson, Lagrange, Lie, Dirac, and conjugate brackets describe symmetric/antisymmetric properties of the conjugate algebra. Conjugate algebra processes output vectors of the conjugate transformation structure to link conjugate complex vector functions to the conjugate transformation structure. In addition, the intrinsic time order of complex operators has been discussed.

Keywords: Conjugate logic hierarchy, conjugate transformation, micro ensemble, phase space, global measurement, hypercomplex number system, measurement, symmetry, antisymmetry, bracket operators, time order

---

*Corresponding Author's Email: conjugatelogic@yahoo.com.

In: Conjugate Construction of Quantum Optics
Editors: Jeffrey Zheng
ISBN: 979-8-88697-651-9

Only thing permanent is change
—Immanuel Kant

Theory yields lot But it hardly brings us any closer to secret of old one
In any case I am convinced that he does not throw dice
—Albert Einstein

Quarks came in number of varieties Only three explain all hundreds of particles
different kinds of quarks called u-type, d-type, s-type
—Richard Feynman

1. Introduction

Quantum mechanics is one of the most important theories established in the 20th century; however, after developments for 100 years, there is no consistent logic framework to support quantum mechanics from microstructures to global dynamics still associated with in a series of mysteries such as multiple paradoxes: Schrodinger cat, quantum entanglement, EPR, Bell inequality, double slits, ghost images, and quantum interference [7,8].

In modern theoretical frameworks, there are many existing works from discrete logic frameworks to continuous constructions such as complex analysis/complex operators [1, 2], multiple complex function theory [3], probability theory/measurement theory [4, 5], multiple statistical analysis [6], classical dynamics [9, 10], Hamiltonian dynamics [11, 12], Hilbert space/group theory/statistical mechanics/quantum mechanics [13], classical vector logic [15], quantum logic [14], information theory [18], cellular automata and complex theory [40, 42], conjugate transformation [48], and variant construction [49, 50].

In the subsequent sections, essential structures are briefly constructed and described: classical logic, complex number system, Hamiltonian dynamics, quantum logic, quantum information, cellular automaton interpretation of quantum mechanics, conjugate transformation, and proposed structures.

1.1 Classic Logic System and Complex Number Representation

From a modern mathematical viewpoint, a real number system is constructed from Euclid geometry under a classical logic system via set theory and measure theory. Based on a pair of real numbers and imaginary numbers *i* in two dimensions, it has been a natural way to extend complex number systems in advanced algebraic frameworks for many years.

From a foundational viewpoint, higher geometric construction via real numbers to complex numbers is a typical top-down approach that makes lower levels of refined structures extremely difficult to describe in detail. Intrinsic internal complexities may produce complex stochastic properties in further analysis.

For example, the foundation of quantum mechanics is the Hilbert space, which is based on complex number systems to be a top-down construction. There is no direct linkage between modern logic systems and complex algebraic systems in their detailed levels of basic construction. A series of problems in quantum foundations cannot apply classical logic in consistent supports. It is easier to bring various uncertain paradoxes and mysteries into systems.

1.2 Measure Theory, Probability Theory, and Statistical Mechanics

Based on classical logic, real function theory, and measure theory, modern probability theory and statistical mechanics are constructed from set theory and real number theory. Multiple variables of statistical analysis are applied to the matrix and complex analysis methodologies and toolkits.

From a foundation viewpoint, systematic operations are strongly relevant to complex geometry in corresponding measurements. These systems contain similar restrictions on detailed levels between classical logic and complex number systems.

1.3 1D Cellular Automata and Discrete Dynamics

Based on switching logic algebra [15], Wolfram established 1D cellular automata by 0–1 logic functions in 1982. This construction provides modern complexity sciences with useful methodologies and simulation tools with powerful visualization and dynamic simulation environments.

The basic components of a cellular automaton contain a cell in m bits that correspond to 2^m states in a state space, and there are 2^{2^m} logic functions in its function space. Variation patterns are based on a 0–1 vector with N bits in a total number of 2^N configurations as a configuration space. Selecting a function in repeat recursion, the whole system appears with time variation properties. Combining function space and configuration space, a united variation space has a structural complexity corresponding to $O(2^{2^m} \times 2^N)$ [41].

It is interesting to note that this discrete dynamic system is constructed by the bottom-up principle to extend efficiently on discrete probability and multiple variable statistics. It provides a solid logical foundation for discrete dynamics.

1.4 Bracket Dynamics

From the least action viewpoint, key distinctions between the Lagrange and Hamilton formulas are shown in significantly symmetric differences. From basic levels of measurements, Bracket operators were organized as a pair of symmetric/antisymmetric structures. From a foundation viewpoint, this type of pair of symmetric/antisymmetric structures is extremely important.

From a logic viewpoint, what types of balanced dynamics can satisfy both global and local logic systems with a hierarchical construction to support local and global transformations from microensembles to global universals? It is a challenging and extremely difficult problem in dynamic systems.

Considering bracket dynamics to play an essential role in classical dynamics and quantum dynamics, is it feasible to be an initial point using a logic framework to support its foundation? Further explorations are needed.

1.5 Quantum Logic, Quantum Information, and Cellular Automaton Interpretation of Quantum Mechanics

1.5.1 Quantum Logic

After Bohr and Heisenberg proposed the uncertainty principle, uncertainty was active as a new meta-parameter introduced into the logic system. Due to quantum logic being associated with superior properties different from classical logic, three-valued logic and multiple-valued logic have become popular candidates as potential logic foundations for quantum logic in the direction [14].

Since either three-valued or multiple-valued logic contains various paradoxes with intrinsic incomplete logic, it is extremely difficult for such logic systems to compare with modern switching theory [15] in foundation levels of classical logic.

1.5.2 Quantum Information

Quantum computing is an advanced approach based on quantum information using a matrix to generate quantum logic gates [17] to simulate quantum computing and quantum interferences.

From a quantum information viewpoint, this type of model and methodology is different from classical information theory [18]. Along this direction, advanced quantum technologies such as quantum communication and quantum encryptions have been developed.

1.5.3 Cellular Automaton Interpretation of Quantum Mechanics

Applying the results of 1D cellular automata [41], cellular automaton interpretation of quantum mechanics was proposed by G. ’t Hooft (Nobel Prize Winner in Physics) in 2016 in a monograph published by Springer [19]. Different from other interpretations of quantum mechanics, such as Copenhagen, multiverse, and guide waves, this interpretation focuses attention on basic 0–1 logic as a foundation ([19], page 33) to establish proper computational structures supporting a complete framework of quantum mechanics.

In the book, a larger number of examples are listed to show that in the foundation of both classical 0–1 logic, 1D cellular automata, and quantum mechanics, there are common parts in the bottom levels. Although advanced cellular automata cannot provide a hierarchical construction to support different applications, their excellent capacities in detailed simulation powers on the lowest levels are extremely important. From a measuring requirement, there are $10^{20} \leq 2^{70}$ gaps ([19], page 98) to describe the large energy gaps between the Planck scale 10^{19} GeV and the nuclear scale 1 GeV.

To satisfy various scaling requirements, it is necessary for the foundation structure to have a hierarchy in multiple levels to support extra models from microensembles to global universals to meet various special requirements.

From an architectural viewpoint, it is a basic and hardest challenge in the cellular automaton interpretation of quantum mechanics to apply the lowest levels of states establishing multiple levels of structures to support dynamic systems.

1.6 Conjugate Transformation Structure CTS

Applying balanced organization and pairs of partitions on phase spaces, Jeffrey Z.J. Zheng proposed conjugate classification, conjugate transformation [45, 46], and reversible logic equations [47] in the 1990s. In his PhD thesis [48], Zheng applied four regular plane lattices for binary images in analysis and processing and established a conjugate transformation structure (in Section 5.4). In this direction, the newest progress is two monographs published by Springer [49] in 2019 and Science Press [50] in 2021 proposing a 0–1 vector logic system from logic, measurement, and visualization as key components for variant construction.

This chapter establishes a proper logic foundation from 0–1 logic initially through multiple levels of conjugate constructions, and consequently a matrix algebraic system to support discrete dynamics.

1.7 Measurement Operators in Conjugate Transformation Structure—Chapter's Structure

Based on the introduction in Section 1, in Section 2, basic logic constructions are described from variables and states to form conjugate states and function spaces. In Section 3, a sequence of state vectors is composed of configuration spaces. In Section 4, conjugate transformation is established by pairs of two sets of feature vectors to be a reversible logic expression. In Section 5, the conjugate transformation structure (CTS) is described. In Section 6, measurement operators of CTS were proposed to form an eight element group in an algebraic construction. In Section 7, complex-conjugate operators are compared in classical dynamics and CTS. In Section 8, the conjugate algebra processing output vector of the conjugate transformation structure links conjugate complex vector functions to the conjugate transformation structure. In Section 9, the intrinsic time order of complex operators is discussed.

Phase spaces of discrete dynamics and their measurements are restricted with bracket operators. In this chapter, discrete vectors and transformation rules are provided to link conjugate transformation structures and conjugate complex vector functions via various operators.

2. Transforming Structure

Conjugate logic construction is composed of hierarchical levels on 0–1 vector structures. Using $m+1$ logic variables to make a kernel form, multiple invariants are proposed to partition pairs of vector states into various clusters to construct multiple levels of phase spaces.

Applying a pair of balanced foreground and background feature clusters, conjugate classification organizes multiple invariants to partition its phase spaces into $2n$ classes. For a given configuration, it is possible to define a feature class to partition the relevant configuration into $2n$ representatives to express this possible phase space consistently. Under transformation, $2n$ classes play the most important role. Using this type of meta-feature expression, it is feasible to establish an elementary equation of the conjugate transformation structure.

2.1 Kernel Form

Let K be a kernel form; it has a certain geometric-logic shape with $m+1$ vector logic variables: $K = \{x_0, x_1, \ldots, x_k, \ldots, x_m\}, x_k \in \{0,1\}, 0 \le k \le m$.

Let x be a corresponding position of the kernel form K, $\overrightarrow{x}$ be a vector with $m+1$ elements and $\overrightarrow{x} = (x, x_1, \ldots, x_k, \ldots, x_m)$, $K(\overrightarrow{x})$ be the kernel form of the $m+1$ elements.

For multiple variables, a kernel form has a special geometric structure, and multiple geometric and topologic invariants, such as translation, reflection, and rotation, can be identified.

2.2 Phase Spaces and Conjugate Phase Spaces

Let Ω be a phase space—a state space of a kernel form $K(\overrightarrow{x})$. For a given $K(\overrightarrow{x})$, there are distinguished $m+1$ logic variables. For a given x, there are two values (0,1). Using this value as a reference, it is feasible to organize other m variables to form a pair of two sets with conjugate vector states: $S(\overrightarrow{x})$ and its conjugation $\tilde{S}(\overrightarrow{x})$, $x_k \in \{0,1\}$, $1 \le k \le m$.

$$S(\overrightarrow{x}) = K(x = 1, x_1, \ldots, x_k, \ldots, x_m) \tag{1}$$
$$\tilde{S}(\overrightarrow{x}) = K(x = 0, x_1, \ldots, x_k, \ldots, x_m) \tag{2}$$

Let Θ and $\tilde{\Theta}$ express a pair of conjugate phase spaces associated with two set operators: $\{\wedge, \vee\}$,

$$\Theta = \{S(\overrightarrow{x})|x = 1, x_k \in \{0,1\}, 1 \le k \le m\} \tag{3}$$
$$\tilde{\Theta} = \{S(\overrightarrow{x})|x = 0, x_k \in \{0,1\}, 1 \le k \le m\} \tag{4}$$
$$\Omega = \Theta \vee \tilde{\Theta} \tag{5}$$
$$\emptyset = \Theta \wedge \tilde{\Theta} \tag{6}$$

Let $|R|$ be the number of states in the set R.

Lemma 1. *For a given $K(\overrightarrow{x})$ and its phase space Ω, $|\Theta| = |\tilde{\Theta}| = 2^m$,*

$$|\Omega| = |\Theta| + |\tilde{\Theta}| = 2^{m+1}$$

Two conjugate phase spaces partition the phase space in balance, and two conjugate phase spaces contain conjugate clusters as one-to-one pairs to organize all possible states as a balanced hierarchical structure.

2.3 2n Feature Clusters of Conjugate Classification

Due to Θ and $\tilde{\Theta}$ with a pair of two vector state sets, each set contains 2^m states. Using a specific partition, it is feasible to define relevant clusters as n groups. This classification is a conjugate classification if each group has a conjugate group and all groups are composed of all the states.

Let $\Gamma_j(\tilde{\Gamma}_j)$ be the j-th cluster of $\Theta(\tilde{\Theta})$, for any state $S(\overrightarrow{x}) \in \Gamma_j$, iff, there is one corresponding state $\tilde{S}(\overrightarrow{x}) \in \tilde{\Gamma}_j, y = \neg x = 0, y_k = \neg x_k; x_k, y_k \in \{0, 1\}, 0 < k \leq m$, and vice versa.

Using a conjugate pair of 1–1 mappings, equations of $|\Gamma_j| = |\tilde{\Gamma}_j| \geq 1$ are true on the following equations.

$$\Theta = \bigvee_{j=1}^{n} \Gamma_j \tag{7}$$

$$\tilde{\Theta} = \bigvee_{j=1}^{n} \tilde{\Gamma}_j \tag{8}$$

$$\emptyset = \Gamma_k \wedge \Gamma_j = \tilde{\Gamma}_k \wedge \Gamma_j = \Gamma_k \wedge \tilde{\Gamma}_j = \tilde{\Gamma}_k \wedge \tilde{\Gamma}_j, 1 \leq j \neq k \leq n \tag{9}$$

Theorem 1. *For a given conjugate classification, 2n feature classes $\{\{\Gamma_j\}_{j=1}^{n}, \{\tilde{\Gamma}_j\}_{j=1}^{n}\}$ partition a pair of two conjugate phase spaces Θ and $\tilde{\Theta}$, respectively.*

3. Configuration

Let X be a configuration that is composed of a sample set with N elements, and each element is a state.

$$X = (X_1, \ldots, X_l, \ldots, X_N), X_l \in \Omega.1 \leq l \leq N \tag{10}$$

$$X_l = \begin{cases} S(\overrightarrow{x^l}), X_l \in \Theta, \\ \tilde{S}(\overrightarrow{x^l}), X_l \in \tilde{\Theta}; \end{cases} \tag{11}$$

$$\overrightarrow{x^l} = (x^l, x_1^l, \ldots, x_k^l, \ldots, x_m^l), x^l, x_k^l \in \{0, 1\}, 1 \leq k \leq m \tag{12}$$

From a representative viewpoint, X may be equivalent to a $(m+1) \times N$ 0–1 sequence to be composed of N states, and each X_l state contains $m+1$ bits.

For example, when consequential states have overlapped parts in m bits, X could be represented as a 1D cellular automata with a total of N cells and each cell with $m+1$ bits.

3.1 Two Sets of Feature Vectors

For convenient analysis, suppose two states can be identified, they are in two distinguished groups. Under this condition, the configuration vector can be mapped into $2n$ feature vectors.

Using defined $2n$ classes, their feature values are:

$$\Gamma_j(X_l) = \begin{cases} 1, X_l \in \Gamma_j; \\ 0, X_l \notin \Gamma_j \end{cases} \tag{13}$$

$$\tilde{\Gamma}_j(X_l) = \begin{cases} 0, X_l \in \tilde{\Gamma}_j; \\ 1, X_l \notin \tilde{\Gamma}_j \end{cases} \quad (14)$$

Under this paired mapping rule, all $X_l, 1 \leq l \leq N$ correspond to $2n$ 0–1 feature vectors.

$$\Gamma_j(X) = (\Gamma_j(X_1), \ldots, \Gamma_j(X_l), \ldots, \Gamma_j(X_N)), 1 \leq l \leq N, 1 \leq j \leq n \quad (15)$$

$$\tilde{\Gamma}_j(X) = (\tilde{\Gamma}_j(X_1), \ldots, \tilde{\Gamma}_j(X_l), \ldots, \tilde{\Gamma}_j(X_N)), 1 \leq l \leq N, 1 \leq j \leq n; \quad (16)$$

3.2 Two Constant Vectors

In addition to these feature vectors, there are two constant vectors.

$$\Gamma_0(X_l) = 0, 1 \leq l \leq N; \quad (17)$$

$$\tilde{\Gamma}_0(X_l) = 1, 1 \leq l \leq N \quad (18)$$

Let two constant vectors be $\vec{0}$ and $\vec{1}$,

$$\vec{0} = \Gamma_0(X) = (\Gamma_0(X_1) \cdots \Gamma_0(X_l) \cdots \Gamma_0(X_N)) = (0 \cdots 0 \cdots 0), 1 \leq l \leq N; \quad (19)$$

$$\vec{1} = \tilde{\Gamma}_0(X) = (\tilde{\Gamma}_0(X_1) \cdots \tilde{\Gamma}_0(X_l) \cdots \tilde{\Gamma}_0(X_N)) = (1 \cdots 1 \cdots 1), 1 \leq l \leq N \quad (20)$$

Two constant vectors in configuration select Ø feature groups.

3.3 Feature Counting

In a 1D cellular automata, the l-th position of feature vector $\Gamma_j(X)$ has a 1 value if $X^l \in \Gamma_j\}$, and the l-th position of conjugate feature vector $\tilde{\Gamma}_j(X)$ has 0 values if $X^l \in \tilde{\Gamma}_j$. Both feature vectors are 0–1 vectors with N bits to be a feature projection from the configuration X to be composed of two sets of $2n$ feature vectors.

Counting the number of 1 elements in $\Gamma_j(X)$ and the number of 0 elements in $\tilde{\Gamma}_j(X)$, two coefficients a_j, b_j can be calculated.

$$a_j(X) = \sum_{l=1}^{N} (\Gamma_j(X_l) == 1) \quad (21)$$

$$b_j(X) = \sum_{l=1}^{N} (\tilde{\Gamma}_j(X_l) == 0) \quad (22)$$

In general, the feature vector length N satisfies the following equation for $2n$ coefficients.

$$N = \sum_{j=1}^{n} (a_j(X) + b_j(X)) \quad (23)$$

4. Basic Logic Operations on Feature Vectors

4.1 Two Canonical Feature Vectors

Using two logic operators $\{\cap, \cup\}$, combinations of whole feature vectors are represented as two canonical feature vectors

$$\Gamma(X) = \bigcup_{j=0}^{n} \Gamma_j(X); \tag{24}$$

$$\tilde{\Gamma}(X) = \bigcap_{j=0}^{n} \tilde{\Gamma}_j(X) \tag{25}$$

Theorem 2. *For a given configuration X, two canonical feature vectors satisfy $\Gamma(X) = \tilde{\Gamma}(X)$.*

Proof. For a state in a feature group, its feature value is 1, and for a state in a conjugate feature group, its special value is 0. When two types of feature vectors are merged together, only positions in feature groups keep values 1 under OR operations, and conjugate positions in conjugate groups keep values 0 under AND operations that make two integrated feature vectors equally valued.

4.2 Special Logic Operators

Using two logic operators $\{\neg, \tilde{}\}$, the following equations are satisfied.

Theorem 3. *For two feature vectors $\{\Gamma_j(X), \tilde{\Gamma}_j(X)\}$, conjugate operator $\tilde{}$ reverse feature vector and original configuration.*

$$\tilde{\Gamma}_j(X) = \neg\Gamma_j(\neg X); \tag{26}$$

$$\neg\tilde{\Gamma}_j(\tilde{X}) = \Gamma_j(\neg X); \tag{27}$$

$$\Gamma_j(X) = \neg\tilde{\Gamma}_j(\neg X) = \tilde{\tilde{\Gamma}}_j(X); \tag{28}$$

$$\neg\Gamma_j(X) = \tilde{\Gamma}_j(\neg X) \tag{29}$$

Proof. For the first equation, the l-th 0 element of $\tilde{\Gamma}_j(X)$ describes the j-th group of $X_l = \tilde{S}(\overrightarrow{y^l}) \in \tilde{\Gamma}_j$, conjugate operator $\neg X_l = \tilde{\tilde{S}}(\overrightarrow{y^l}) = (1, x_1^l, \ldots, x_k^l, \ldots, x_m^l) = S(\overrightarrow{y^l}), x_k^l = \neg y_k^l, \tilde{S}(\overrightarrow{y^l}) \in \tilde{\Gamma}_j \to \neg X_l \in \Gamma_j$. One 1-element of $\Gamma_j(\neg X)$ corresponds to one 0-element of $\tilde{\Gamma}_j(X)$ undertaken $\neg$ operations. For other equations, similar operations can be performed in the proof procedure.

4.3 Irreducible Expression

When two feature vectors are on a logical expression, if the expression cannot be reduced as a constant vector or simplified as a single feature vector, then this logical expression is an irreducible expression; otherwise, it is a reducible expression.

Theorem 4. *For two distinguished feature vectors under* $\{\neg, \cap, \cup, \tilde{}\}$ *operators, there are only six groups of irreducible expressions.*

$$\neg\Gamma_j(X) \cap \neg\Gamma_k(X) = \tilde{\Gamma}_j(\neg X) \cap \tilde{\Gamma}_k(\neg X); \tag{30}$$

$$\Gamma_j(X) \cup \Gamma_k(X) = \neg\tilde{\Gamma}_j(\neg X) \cup \neg\tilde{\Gamma}_k(\neg X); \tag{31}$$

$$\neg\tilde{\Gamma}_j(X) \cup \neg\tilde{\Gamma}_k(X) = \Gamma_j(\neg X) \cup \Gamma_k(\neg X); \tag{32}$$

$$\tilde{\Gamma}_j(X) \cap \tilde{\Gamma}_k(X) = \neg\Gamma_j(\neg X) \cap \neg\Gamma_k(\neg X); \tag{33}$$

$$\neg\Gamma_j(X) \cap \tilde{\Gamma}_k(X) = \neg\Gamma_j(X) \cap \neg\Gamma_k(\neg X); \tag{34}$$

$$\Gamma_j(X) \cup \neg\tilde{\Gamma}_k(X) = \Gamma_j(X) \cup \Gamma_k(\neg X) \tag{35}$$

5. Conjugate Transformation

For a convenient description of a given configuration, an elementary equation of conjugate transformation can be established.

5.1 Pair of Parameter Sets

Let $I = \{1, \ldots, n\}$ be an index set of feature vectors, and let $A, B \subseteq I$ be two subsets of the index set.

$$\langle A, B\rangle = \{\{\Gamma_j\}_{j\in A}, \{\tilde{\Gamma}_k\}_{k\in B}\} \tag{36}$$

In $\langle A, B\rangle$, the first parameter is the index set of the phase space Θ, and the second parameter is the index set of the conjugate phase space $\tilde{\Theta}$.

e.g., if $A = \{2, 4\}, B = \{3, 4\}$, then $\langle A, B\rangle = \{\Gamma_2, \Gamma_4, \tilde{\Gamma}_3, \tilde{\Gamma}_4\}$.

5.2 Pair Sets of Feature Vectors

For any configuration X,$\langle A, B\rangle X$ represents a pair set of feature vectors.

$$\langle A, B\rangle X = \{\{\Gamma_j(X)\}_{j\in A}, \{\tilde{\Gamma}_k(X)\}_{k\in B}\} \tag{37}$$

To express two sets of selected feature vectors, let $\Gamma_A(X)$ be a merged set of the index set A under OR operations and $\tilde{\Gamma}_B(X)$ be a joined set of the index set B of feature vectors under AND operations.

$$\Gamma_A(X) = \bigcup_{j\in A} \Gamma_j(X) \tag{38}$$

$$\tilde{\Gamma}_B(X) = \bigcap_{k\in B} \tilde{\Gamma}_k(X) \tag{39}$$

$$\langle A, B\rangle X = \{\Gamma_A(X), \tilde{\Gamma}_B(X)\} \tag{40}$$

$\langle A, B\rangle X$ is composed of two feature vectors.

While $A = B = \emptyset$, let $\Gamma_{\emptyset}(X) = \Gamma_0(X) = \overrightarrow{0}$ and $\tilde{\Gamma}_{\emptyset}(X) = \tilde{\Gamma}_0(X) = \overrightarrow{1}$ be two constant vectors.

5.3 Elementary Equation of Conjugate Transformation

Let $\natural$ be a reversing operator; it reverses selected parts of the configuration to satisfy the following equations.

Let $F(\langle A, B\rangle X)$ be an elementary equation of conjugate transformation

$$F(\langle A, B\rangle X) = \Gamma(X)\natural\langle A, B\rangle X \tag{41}$$

$$= (\Gamma(X) \cap (\neg\Gamma_A(X))) \cup (\neg\tilde{\Gamma}_B(X)) \tag{42}$$

Operator $\natural$ has the following meaning: selected components $\langle A, B\rangle X$, the operator changes selected sets of feature vectors and retains other unselected sets of feature vectors that are invariant.

Theorem 5. *The elementary equation is a vector logic equation to reverse the selected sets of feature vectors and keep other unselected sets of feature vectors invariant.*

Theorem 6. *Under the $\neg$ operator, the elementary equation is a self-conjugate equation to exchange index sets of feature vectors and to be reversed on the configuration vector.*

$$\neg F(\langle A, B\rangle X) = F(\langle B, A\rangle \neg X) \tag{43}$$

Proof.

$$\begin{aligned}
\neg F(\langle A, B\rangle X) &= \neg((\Gamma(X) \cap (\neg\Gamma_A(X))) \cup (\neg\tilde{\Gamma}_B(X))) \\
&\quad \text{de Morgan Rule} \\
&= \neg(\Gamma(X) \cap (\neg\Gamma_A(X))) \cap \tilde{\Gamma}_B(X) \\
&\quad \text{Distribution Rule} \\
&= (\neg\Gamma(X) \cup \Gamma_A(X)) \cap \tilde{\Gamma}_B(X) \\
&\quad \because \Gamma_A(X) \cap \tilde{\Gamma}_B(X) = \Gamma_A(X) \\
&= (\neg\Gamma(X) \cap \tilde{\Gamma}_B(X)) \cup (\Gamma_A(X) \cap \tilde{\Gamma}_B(X)) \\
&\quad \because \tilde{\Gamma}_B(X) = \neg\Gamma_B(\neg X), \Gamma_A(X) = \neg\tilde{\Gamma}_A(\neg X) \\
&= (\neg\Gamma(X) \cap \tilde{\Gamma}_B(X)) \cup \Gamma_A(X) \\
&\quad \because \neg\Gamma(X) = \tilde{\Gamma}(\neg X) = \Gamma(\neg X) \\
&= (\Gamma(\neg X) \cap \neg\Gamma_B(\neg X)) \cup \neg\tilde{\Gamma}_A(\neg X) \\
&= F(\langle B, A\rangle \neg X)
\end{aligned}$$

Theorem 7. *For a given $\langle A, B\rangle$, if $A = \emptyset$ or $B = \emptyset$, then $F(\langle A, B\rangle X)$ for components $\langle A, B\rangle X$ is an equation of dilation or erosion.*

Proof.

$$\begin{aligned}
F(\langle\emptyset, B\rangle X) &= (\Gamma(X) \cap (\neg\Gamma_0(X))) \cup (\neg\tilde{\Gamma}_B(X)) \\
&= \Gamma(X) \cup (\neg\tilde{\Gamma}_B(X))(\text{Dilation}) \\
F(\langle A, \emptyset\rangle X) &= (\Gamma(X) \cap (\neg\Gamma_A(X))) \cup (\neg\tilde{\Gamma}_0(X)) \\
&= \Gamma(X) \cap (\neg\Gamma_A(X))(\text{Erosion})
\end{aligned}$$

Theorem 8. *The elementary equation of conjugate transformation generates four extreme vectors* $\{\Gamma(X), \vec{0}, \vec{1}, \neg\Gamma(X)\}$*; four extreme vectors are* $\langle A, B\rangle = \{\langle\emptyset,\emptyset\rangle, \langle\emptyset, I\rangle, \langle I,\emptyset\rangle, \langle I, I\rangle\}$.

Proof.

$$\begin{aligned}
F(\langle\emptyset,\emptyset\rangle X) &= (\Gamma(X) \cap (\neg\Gamma_0(X))) \cup (\neg\tilde{\Gamma}_0(X)) \\
&= \Gamma(X) \rightarrow \text{ An Invariant Vector} \\
F(\langle\emptyset, I\rangle X) &= (\Gamma(X) \cap (\neg\Gamma_0(X))) \cup (\neg\tilde{\Gamma}_I(X)) \\
&= \Gamma(X) \cup \neg\tilde{\Gamma}_I(X) \\
&= \Gamma(X) \cup \neg\tilde{\Gamma}(X) \\
&= \Gamma(X) \cup \neg\Gamma(X) \\
&= \vec{1} \rightarrow \text{ A Constant 1 Vector} \\
F(\langle I, \emptyset\rangle X) &= (\Gamma(X) \cap (\neg\Gamma_I(X))) \cup (\neg\tilde{\Gamma}_0(X)) \\
&= \Gamma(X) \cap (\neg\Gamma_I(X)) \\
&= \Gamma(X) \cap (\neg\Gamma(X)) \\
&= \vec{0} \rightarrow \text{ A Constant 0 Vector} \\
F(\langle I, I\rangle X) &= (\Gamma(X) \cap (\neg\Gamma_I(X))) \cup (\neg\tilde{\Gamma}_I(X)) \\
&= (\Gamma(X) \cap (\neg\Gamma(X))) \cup (\neg\tilde{\Gamma}(X)) \\
&= \neg\tilde{\Gamma}(X) \\
&= \neg\Gamma(X) \rightarrow \text{ A Reversed Vector}
\end{aligned}$$

Theorem 9. *For a configuration* X *and its two sets of feature vectors* $\{\Gamma_j(X)\}_{j=0}^{n}, \{\tilde{\Gamma}_k(X)\}_{k=0}^{n}$, *the elementary equation of conjugate transformation can generate* 2^{2n} *vector functions.*

Proof. Each selection of $\langle A, B\rangle$ determines a vector logic function. Two sets are composed of all possible combinations of $A, B \subseteq I = \{1, \ldots, n\}$, each index set has 2^n selections, and a total of 2^{2n} vector logic functions can be generated.

6. Conjugate Transformation Structure (CTS)

6.1 Transformation Structure

Feature vectors are 0–1 vectors. Using defined $2 \times (n+1)$ logic vectors, it is feasible to define a new transformation structure. Let Π be a set of feature vectors.

$$\Pi(X) = \{\Gamma(X), \tilde{\Gamma}(X), \{\Gamma_j(X)\}_{j=0}^{n}, \{\tilde{\Gamma}_k(X)\}_{k=0}^{n}\} \tag{44}$$

Let the conjugate transformation structure be CTS.

$$CTS(X) = \{\Pi(X), \neg, \cap, \cup, \sim\} \tag{45}$$

Each feature vector is a logic vector variable, and two sets of conjugate feature vectors make each $CTS(X)$ a vector logic algebra—conjugate vector algebra. It is helpful for comparison to explore the global properties of the structure using vector logic algebra [15, 16] to count the total number of logic functions in a vector logic system.

6.2 Transformation Structures

Let $\{Y_k\}_{k=1}^{2n}$ be $2n$ Boolean vectors. Using these vectors under three logic operators, it is possible to establish Boolean vector algebra $BVA(Y)$.

$$BVA(Y) = \{\{Y_k\}_{k=1}^{2n}, \neg, \cap, \cup\} \tag{46}$$

Boolean expressions of BVA can be formed as a canonical sum-of-products form (disjunctive normal form DNF) or a canonical product-of-sums form (conjunctive normal form CNF).

$$\bigcup_{J \in A} \left(\bigcap_{i=1}^{2n} Y_i^{J_i} \right) \text{(DNF)} \tag{47}$$

$$\bigcap_{J \notin A} \left(\bigcup_{i=1}^{2n} Y_i^{\neg J_i} \right) \text{(CNF)} \tag{48}$$

where A expresses a subset of $\{0, 1, 2, \ldots, 2^{2n} - 1\}$.

$$J = \sum_{i=1}^{2n} J_i \times 2^{i-1}, J_i \in \{0, 1\} \tag{49}$$

$$Y_i^{J_i} = \begin{cases} Y_i, J_i = 1; \\ \neg Y_i, J_i = 0. \end{cases} \tag{50}$$

Lemma 2. *For an $BVA(Y)$ with $2n$ distinct vector variables, there are 2^{2n} states and $2^{2^{2n}}$ logic functions.*

Proof. Using a selected or unselected role to handle each vector as a bit, it is natural for $2n$ bit variables in a Boolean algebra system to contain 2^{2n} states and $2^{2^{2n}}$ functions. Either DNF or CNF provides the same number.

Theorem 10. *For an $CTS(X)$ on $1 \le n \le 2^m$, there are 2^{2n} functions.*

Proof. Using an DNF form, a simplifying process can be described as follows.

$$\left(\bigcap_{i=1}^{n} \Gamma_i(X)^{J_i} \right) \cap \left(\bigcup_{k=1}^{n} \tilde{\Gamma}_k(X)^{J_k} \right)$$

In the first part, each item can be reduced.

Type	Simplified	Condition
1	$\neg\Gamma(X)$	$\forall i, J_i = 0$
2	one of $\{\Gamma_1(X), \ldots, \Gamma_n(X)\}$	$\exists! \, i, J_i = 0$
3	$\vec{0}$	$\exists i \neq k, J_i = J_k = 0$

In the second part, each item can also be reduced.

Type	Simplified	Condition
1	$\neg\tilde{\Gamma}(X)$	$\forall k, J_k = 0$
2	One of $\{\neg\tilde{\Gamma}_1(X), \ldots, \neg\tilde{\Gamma}_n(X)\}$	$\exists! \, k, J_k = 1$
3	$\vec{0}$	$\exists k \neq i, J_k = J_i = 1$

Since $\neg\Gamma(X) \cap \neg\tilde{\Gamma}_j(X) = \neg\tilde{\Gamma}_j(X), \neg\Gamma_i(X) \cap \neg\tilde{\Gamma}_k(X) = \vec{0}$.
Only items in DNF are

$$\{\Gamma_i(X)\}_{i=1}^n, \{\neg\tilde{\Gamma}_k(X)\}_{k=1}^n$$

Using two feature vector sets, there are 2^{2n} combinations as functions.
A CNF form can be simplified in the same way.

It is important to note that $2n$ vector variables in BVA have $2^{2^{2n}}$ logic functions; however, $2n$ feature vectors in CTS have at most $2^{2^{m+1}}$ functions. From a numeric viewpoint, it is feasible to compare two numbers of functions between BVS and CTS on their largest ones.

Theorem 11. *For $2n$ vector variables in BVA and $2n$ feature vectors on the $m+1$ kernel form in CTS, if $n \leq \lceil \frac{m+1}{2} \rceil$, then the function number of CTS is systematically much larger than the function number of BVA.*

Proof. Compared with the worst condition on $2^{2^{2n}}$ and $2^{2^{m+1}}$, if $2n \leq m+1$, then CTS contains larger numbers than BVA.

Corollary 1. *In general, CTS provides a hierarchy of flexible function representatives in wider regions from $2^2, \ldots, 2^{2n}$ to $2^{2^{m+1}}$ instead of $2^{2^{2n}}$ on BVA.*

In other words, when selected clusters are less than half of the state length, the phase spaces of CTS have much simpler properties that are significantly superior to BVA, even in the worst condition.

Since each configuration has a CTS structure, combining possible configurations and feature functional spaces of CTS can be integrated.

Let Σ be a number of combining spaces on CTS.

Theorem 12. *For a configuration with N elements under a kernel form with $m+1$ bits in $2n$ clusters, the total number of combinations $\Sigma(N,m,n)$ is determined by*

$$\Sigma(N,m,n) = 2^N \times 2^{2n}, 1 \leq n \leq 2^m \tag{51}$$

Proof. For each configuration, the distinguished number is 2^N, and the number of functions on $2n$ clusters is 2^{2n}, as shown in Corollary 1.

Corollary 2. *Combining various configurations and feature functions, their controllable spaces have numeric regions from* $\Sigma(N,m,1) = 2^{N+2}$ *to* $\Sigma(N,m,2^m) = 2^N \times 2^{2^{m+1}}$.

Proof. Selecting $n = 1$ and 2^m, two numbers for minimal and maximal ones are determined.

6.3 Three Set Operators

Let $\{', \wedge, \vee\}$ be the {complement, union, intersection} operators of index sets. For an index set I, A, B are subsets and A', B' are complement subsets, $A, B \subset I$ and $A', B' \subset I$ satisfy the following equation.

$$\emptyset' = I \tag{52}$$
$$I' = \emptyset \tag{53}$$
$$A \vee A' = I \tag{54}$$
$$A \wedge A' = \emptyset \tag{55}$$

Under conjugate condition,

$$B \vee B' = \emptyset \tag{56}$$
$$B \wedge B' = I \tag{57}$$

Applying this expression, a canonical vector $\Gamma(X)$ or $\tilde{\Gamma}(X)$ can be described as two parts linked with a $\cup$ or $\cap$ operator and three set operators $\{', \wedge, \vee\}$.

$$\Gamma(X) = \bigcup_{j=0}^{n} \Gamma_j(X) = \bigcup_{j=1}^{n} \Gamma_j(X) = \Gamma_{A\vee A'}(X) = \Gamma_A(X) \cup \Gamma_{A'}(X) \tag{58}$$

$$\tilde{\Gamma}(X) = \bigcap_{k=0}^{n} \tilde{\Gamma}_k(X) = \bigcap_{k=1}^{n} \tilde{\Gamma}_k(X) = \tilde{\Gamma}_{B\wedge B'}(X) = \tilde{\Gamma}_B(X) \cap \tilde{\Gamma}_{B'}(X) \tag{59}$$

Theorem 13. *For a* $\langle A, B\rangle X \in CTS(X)$, *the elementary equation of CTS has four equivalent expressions.*

$$F(\langle A, B\rangle X) = (\Gamma(X) \cap \neg\Gamma_A(X)) \cup \neg\tilde{\Gamma}_B(X) \tag{60}$$
$$= \Gamma_{A'}(X)) \cup \neg\tilde{\Gamma}_B(X) \tag{61}$$
$$= \neg\Gamma_A(X)) \cap \tilde{\Gamma}_{B'}(X) \tag{62}$$
$$= (\Gamma(X) \cup \neg\tilde{\Gamma}_B(X)) \cap \neg\Gamma_A(X) \tag{63}$$

Proof. From the first to the second, we have

$$\begin{aligned} F(\langle A, B\rangle X) &= (\Gamma(X) \cap \neg\Gamma_A(X)) \cup \neg\tilde{\Gamma}_B(X) \\ &\because \Gamma(X) = (\Gamma_A(X) \cup \Gamma_{A'}(X)) \\ &= ((\Gamma_A(X) \cup \Gamma_{A'}(X)) \cap \neg\Gamma_A(X)) \cup \neg\tilde{\Gamma}_B(X) \\ &= (\Gamma_A(X) \cap \neg\Gamma_A(X)) \cup (\Gamma_{A'}(X) \cap \neg\Gamma_A(X)) \cup \neg\tilde{\Gamma}_B(X) \end{aligned}$$

$$\begin{aligned}
\because \vec{0} &= (\Gamma_A(X) \cap \neg\Gamma_A(X)) \\
&= (\vec{0}) \cup (\Gamma_{A'}(X) \cap \neg\Gamma_A(X)) \cup \neg\tilde{\Gamma}_B(X) \\
&= (\Gamma_{A'}(X) \cap \neg\Gamma_A(X)) \cup \neg\tilde{\Gamma}_B(X) \\
&= \Gamma_{A'}(X) \cup \neg\tilde{\Gamma}_B(X)
\end{aligned}$$

The other two expressions can be deduced in a similar way.

From the four equivalent expressions, symmetric properties can be observed in the formula.

6.4 Four Conjugate-Logic Operators

Applying refined set operators, four conjugate-logic operators $\{\neg, \cap, \cup, \sim\}$ {NOT, AND, OR, Conjugate} are used to explore the symmetric properties of the elementary equation.

For the $\neg$ operator, the elementary equation can be described in two forms.

Theorem 14. *For an elementary equation, its $\neg$ description has two forms.*

$$\begin{aligned}
\neg F(\langle A, B\rangle X) &= F(\langle A', B'\rangle X) \\
&= F(\langle B, A\rangle \tilde{X})
\end{aligned} \tag{64}$$

Proof. Checking the new form of the equation, we have

$$\begin{aligned}
\neg F(\langle A, B\rangle X) &= \neg((\Gamma(X) \cap \neg\Gamma_A(X)) \cup \neg\tilde{\Gamma}_B(X)) \\
&= \neg(\Gamma(X) \cap \neg\Gamma_A(X)) \cap \tilde{\Gamma}_B(X) \\
&\quad \because \Gamma(X) = \tilde{\Gamma}(X) \\
&= \neg(\tilde{\Gamma}(X) \cap \neg\Gamma_A(X)) \cap \tilde{\Gamma}_B(X) \\
&\quad \because \tilde{\Gamma}(X) = \tilde{\Gamma}_B(X) \cap \tilde{\Gamma}_{B'}(X) \\
&= \neg(\tilde{\Gamma}_B(X) \cap \tilde{\Gamma}_{B'}(X) \cap \neg\Gamma_A(X)) \cap \tilde{\Gamma}_B(X) \\
&= (\neg\tilde{\Gamma}_B(X) \cup \neg\tilde{\Gamma}_{B'}(X) \cup \Gamma_A(X)) \cap \tilde{\Gamma}_B(X) \\
&= \neg\tilde{\Gamma}_B(X) \cap \tilde{\Gamma}_B \cup \neg\tilde{\Gamma}_{B'}(X) \cap \tilde{\Gamma}_B \cup \Gamma_A(X) \cap \tilde{\Gamma}_B \\
&= \vec{0} \cup \neg\tilde{\Gamma}_{B'}(X) \cup \Gamma_A(X) \\
&= \neg\tilde{\Gamma}_{B'}(X) \cup \Gamma_A(X) \\
&= \Gamma_A(X) \cup \neg\tilde{\Gamma}_{B'}(X) \\
&= \Gamma_{(A')'}(X) \cup \neg\tilde{\Gamma}_{B'}(X) \\
&= F(\langle A', B'\rangle X)
\end{aligned}$$

For processing more complex operations in *CTS*, let $Y_1 = F(\langle A_1, B_1\rangle X)$ and $Y_2 = F(\langle A_2, B_2\rangle X)$ be two vectors in $CTS(X)$.

Theorem 15. *For any $Y_1, Y_2 \in CTS(X)$ under logic operators $\{\cap, \cup\}$, two combined vectors are in $CTS(X)$.*

$$Y_1 \cap Y_2 = F(\langle A_1 \vee A_2, B_1 \wedge B_2\rangle X) \tag{65}$$

$$Y_1 \cup Y_2 = F(\langle A_1 \wedge A_2, B_1 \vee B_2\rangle X) \tag{66}$$

Proof.

$$
\begin{aligned}
Y_1 \cap Y_2 &= ((\Gamma(X) \cap \neg\Gamma_{A_1}(X)) \cup \neg\tilde{\Gamma}_{B_1}(X)) \cap ((\Gamma(X) \cap \neg\Gamma_{A_2}(X)) \cup \neg\tilde{\Gamma}_{B_2}(X)) \\
&= (\Gamma(X) \cap \neg\Gamma_{A_1}(X) \cap \neg\Gamma_{A_2}(X)) \cup (\neg\tilde{\Gamma}_{B_1}(X)) \cap \neg\tilde{\Gamma}_{B_2}(X)) \\
&= (\Gamma(X) \cap \neg(\Gamma_{A_1}(X) \cup \Gamma_{A_2}(X)) \cup \neg(\tilde{\Gamma}_{B_1}(X) \cup \tilde{\Gamma}_{B_2}(X)) \\
&= (\Gamma(X) \cap \neg(\Gamma_{A_1 \vee A_2}(X)) \cup \neg(\tilde{\Gamma}_{B_1 \wedge B_2}(X)) \\
&= F(\langle A_1 \vee A_2, B_1 \wedge B_2 \rangle X) \\
Y_1 \cup Y_2 &= ((\Gamma(X) \cap \neg\Gamma_{A_1}(X)) \cup \neg\tilde{\Gamma}_{B_1}(X)) \cup ((\Gamma(X) \cap \neg\Gamma_{A_2}(X)) \cup \neg\tilde{\Gamma}_{B_2}(X)) \\
&= (\Gamma(X) \cap (\neg\Gamma_{A_1}(X) \cup \neg\Gamma_{A_2}(X)) \cup (\neg\tilde{\Gamma}_{B_1}(X)) \cup \neg\tilde{\Gamma}_{B_2}(X)) \\
&= (\Gamma(X) \cap \neg(\Gamma_{A_1}(X) \cap \Gamma_{A_2}(X)) \cup \neg(\tilde{\Gamma}_{B_1}(X) \cap \tilde{\Gamma}_{B_2}(X)) \\
&= (\Gamma(X) \cap \neg(\Gamma_{A_1 \wedge A_2}(X)) \cup \neg(\tilde{\Gamma}_{B_1 \vee B_2}(X)) \\
&= F(\langle A_1 \wedge A_2, B_1 \vee B_2 \rangle X)
\end{aligned}
$$

For any feature vector, the conjugate operator $\sim$ exchanges a pair of two feature vectors and two logic operators.

$$
\sim: \begin{cases} \Gamma_j(X) \leftrightarrow \tilde{\Gamma}_j(X) \\ \cap \leftrightarrow \cup \end{cases} \tag{67}
$$

Theorem 16. *For an elementary equation, the conjugate operator exchanges two index sets of feature vectors.*

$$
\tilde{F}(\langle A, B \rangle X) = F(\langle B, A \rangle X) \tag{68}
$$

Proof.

$$
\begin{aligned}
\tilde{F}(\langle A, B \rangle X) &= \sim ((\Gamma(X) \cap \neg\Gamma_A(X)) \cup \neg\tilde{\Gamma}_B(X)) \\
&= (\tilde{\Gamma}(X) \cup \neg\tilde{\Gamma}_A(X)) \cap \neg\Gamma_B(X) \\
&= (\Gamma(X) \cap \neg\Gamma_B(X)) \cup \neg\tilde{\Gamma}_A(X) \\
&= F(\langle B, A \rangle X)
\end{aligned}
$$

Lemma 3. *For an elementary equation, the conjugate operator and $\neg$ operator generate different results.*

$$
\tilde{F}(\langle A, B \rangle X) \neq \neg F(\langle A, B \rangle X) \tag{69}
$$

6.5 Conjugate Twist Operators

Let ζ be a conjugate twist operator or a twistor, and let E be a unit operator.

$$
\begin{aligned}
\zeta \langle A, B \rangle &= \langle B', A \rangle && (70) \\
\zeta : \langle A, B \rangle &\Rightarrow \{\langle A, B \rangle, \langle B', A \rangle, \langle A', B' \rangle, \langle B, A' \rangle\} && (71) \\
\zeta(F(\langle A, B \rangle X)) &= F(\zeta \langle A, B \rangle X) = F(\langle B', A \rangle X) && (72) \\
&= (\Gamma(X) \cap \neg\Gamma_{B'}(X)) \cup \neg\tilde{\Gamma}_A(X) && (73)
\end{aligned}
$$

Theorem 17. *For an elementary equation, a NOT operation can be performed by two twistors.*

$$\zeta^2 F(\langle A, B\rangle X) = F(\zeta^2 \langle A, B\rangle X) = F(\langle A', B'\rangle X) = \neg F(\langle A, B\rangle X) \tag{74}$$

Proof.

$$\begin{aligned}\zeta^2 F(\langle A, B\rangle X) &= \zeta(\zeta(F(\langle A, B\rangle X))) \\ &= \zeta(F(\zeta\langle A, B\rangle X)) \\ &= \zeta F(\langle B', A\rangle X) \\ &= F(\zeta\langle B', A\rangle X) \\ &= F(\langle A', B'\rangle X) \\ &= \neg F(\langle A, B\rangle X)\end{aligned}$$

Corollary 3. *For an elementary equation, four twistors are performed to be a unit operator; it is back to itself.*

$$\zeta^4 F(\langle A, B\rangle X) = E(F(\langle A, B\rangle X)) = F(\langle A, B\rangle X) \tag{75}$$

6.6 Twist and Conjugate Operators

Corollary 4. *Applying two operators $\{\sim, \zeta\}$ repeats, an elementary equation has eight expressions.*

$$\{\sim, \zeta\} : \langle A, B\rangle \Rightarrow \left\{ \begin{array}{c} \langle A, B\rangle, \langle B', A\rangle, \langle A', B'\rangle, \langle B, A'\rangle \\ \langle B, A\rangle, \langle A, B'\rangle, \langle B', A'\rangle, \langle A', B\rangle \end{array} \right\} \tag{76}$$

From various corollaries on $\{\sim, \zeta\}$ operators, the conjugate operator and twistor play a more essential role than a NOT operator in *CTS*. From the viewpoint of extending logic transformations, twistor is a new logic operator, and no corresponding operators exist in classical logic systems.

Applying Corollary 4, transforming expressions are in eight forms; for a given configuration, selected feature sets can be determined by $\langle A, B\rangle X$.

7. Measurement Structure

For a given $\langle A, B\rangle X$, let H be a measurement operator in a complex measurement mode as follows.

$$\begin{aligned} H(\langle A, B\rangle X) &= (a_A, b_B) = (a_A + i \cdot b_B) \\ &= \begin{pmatrix} \delta_1^A \cdot a_1, \delta_1^B \cdot b_1 \\ \cdots \\ \delta_j^A \cdot a_j, \delta_j^B \cdot b_j \\ \cdots \\ \delta_n^A \cdot a_n, \delta_n^B \cdot b_n \end{pmatrix} = \begin{pmatrix} \delta_1^A \cdot a_1 + i \cdot \delta_1^B \cdot b_1 \\ \cdots \\ \delta_j^A \cdot a_j + i \cdot \delta_j^B \cdot b_j \\ \cdots \\ \delta_n^A \cdot a_n + i \cdot \delta_n^B \cdot b_n \end{pmatrix} \end{aligned} \tag{77}$$

where $\delta_j^A = \begin{cases} 1, j \in A; \\ 0, j \notin A; \end{cases}$ $\delta_j^B = \begin{cases} 1, j \in B; \\ 0, j \notin B; \end{cases}$ $a_j, b_j \in \mathscr{R}, j \in \{1, \ldots, n\}, i = \sqrt{-1}$ be an imaginary number, $\mathscr{R}$ be a real number field.

Each δ_j^A and δ_j^B are extracted nontrivial measurements to satisfy relevant pairs of combinations in feature spaces.

7.1 Various Equations

Using $H(\langle A, B\rangle X)$ expressions, four special measurements can be produced.

$$H(\langle\emptyset, \emptyset\rangle X) = (a_0, b_0) = (a_0 + i \cdot b_0) \tag{78}$$
$$H(\langle I, \emptyset\rangle X) = (a_I, b_0) = (a_I + i \cdot b_0) \tag{79}$$
$$H(\langle\emptyset, I\rangle X) = (a_0, b_I) = (a_0 + i \cdot b_I) \tag{80}$$
$$H(\langle I, I\rangle X) = (a_I, b_I) = (a_I + i \cdot b_I) \tag{81}$$

From a measurement viewpoint, selecting a measurement and its complement measurement have opposite values.

$$H(\langle\emptyset, \emptyset\rangle X) + H(\langle I, I\rangle X) = H(\langle I, \emptyset\rangle X) + H(\langle\emptyset, I\rangle X) = 0$$
$$\Rightarrow a_0 = -a_I, b_0 = -b_I$$

For a selection $H(\langle A, B\rangle X)$,

$$H(\langle A, B\rangle X) + H(\langle A', B'\rangle X) = 0$$
$$\Rightarrow a_A = -a_{A'}, b_B = -b_{B'}$$

7.2 Hypercomplex Numbers and Octonion in CTS

Using such corresponding relationships, a list of formulas can be established and shown in Table 1.

Using the eight base matrices, two representations of Octonion groups can be identified, as shown in Table 2.

7.3 Eight Corresponding Complex Number Formulas

Using complex operators $\{i, *, (r \leftrightarrow i)\}$ (imaginary number, complex conjugation, and imaginary and real exchange) and conjugate operators $\{\zeta, \sim\}$ (twist and conjugate), eight equations are established as follows.

$$\begin{aligned} H(\langle A, B\rangle X) &= a_A + i \cdot b_B = i^4 \cdot (a_A + i \cdot b_B) = i^4 \cdot H(\langle A, B\rangle X) \\ &= H(\zeta^4\langle A, B\rangle X) \end{aligned} \tag{82}$$

$$\begin{aligned} H(\langle B', A\rangle X) &= b_{B'} + i \cdot a_A = -b_B + i \cdot a_A = i \cdot (a_A + i \cdot b_B) = i \cdot H(\langle A, B\rangle X) \\ &= H(\zeta\langle A, B\rangle X) \end{aligned} \tag{83}$$

Table 1. Eight expressions in complex equation, matrix, and conjugate transformation structure

Complex Op.	Complex Exp.	Complex Formula	Conjugate Op.	Matrix Base	Conjugate Formula
1	$1 \cdot H(.)$	$H(\langle A, B\rangle X)$	$E = \zeta^0 = \zeta^4$	$e_0 = \begin{pmatrix} 1 & 0 \\ 0 & 1 \end{pmatrix}$	$H(\langle A, B\rangle X)$
i	$i \cdot H(.)$	$i \cdot H(\langle A, B\rangle X)$	ζ	$e_1 = \begin{pmatrix} 0 & 1 \\ -1 & 0 \end{pmatrix}$	$H(\langle B', A\rangle X)$
$i^2 = -1$	$i^2 \cdot H(.)$	$i^2 \cdot H(\langle A, B\rangle X)$	$\zeta^2 = \neg E$	$e_2 = \begin{pmatrix} -1 & 0 \\ 0 & -1 \end{pmatrix}$	$H(\langle A', B'\rangle X)$
$i^3 = -i$	$i^3 \cdot H(.)$	$i^3 \cdot H(\langle A, B\rangle X)$	$\zeta^3 = \neg\zeta$	$e_3 = \begin{pmatrix} 0 & -1 \\ 1 & 0 \end{pmatrix}$	$H(\langle B, A'\rangle X)$
$(\leftrightarrow)$	$(i^3 \cdot H(.))^*$	$(i^3 \cdot H(\langle A, B\rangle X))^*$	$\sim$	$e_4 = \begin{pmatrix} 0 & 1 \\ 1 & 0 \end{pmatrix}$	$H(\langle B, A\rangle X)$
$i \cdot (\leftrightarrow)$	$i \cdot (i^3 \cdot H(.))^*$	$i \cdot (i^3 \cdot H(\langle A, B\rangle X))^*$	$\zeta \sim$	$e_5 = \begin{pmatrix} -1 & 0 \\ 0 & 1 \end{pmatrix}$	$H(\langle A', B\rangle X)$
$i^2 \cdot (\leftrightarrow)$	$i^2 \cdot (i^3 \cdot H(.))^*$	$i^2 \cdot (i^3 \cdot H(\langle A, B\rangle X))^*$	$\zeta^2 \sim$	$e_6 = \begin{pmatrix} 0 & -1 \\ -1 & 0 \end{pmatrix}$	$H(\langle B', A'\rangle X)$
$i^3 \cdot (\leftrightarrow)$	$i^3 \cdot (i^3 \cdot H(.))^*$	$i^3 \cdot (i^3 \cdot H(\langle A, B\rangle X))^*$	$\zeta^3 \sim$	$e_7 = \begin{pmatrix} 1 & 0 \\ 0 & -1 \end{pmatrix}$	$H(\langle A, B'\rangle X)$

where abbreviations are used: Op—Operator; Complex Exp—Complex Expression; Complex Comp—Complex Computation; Conj Op—Conjugate Operator; Conj Exp—Conjugate Expression; Conj Comp—Conjugate Computation.

Table 2. Octonion groups in two representations (a) octonion bases (b) ± four bases

(a):

$e_i e_j$	e_0	e_1	e_2	e_3	e_4	e_5	e_6	e_7
e_0	e_0	e_1	e_2	e_3	e_4	e_5	e_6	e_7
e_1	e_1	e_2	e_3	e_0	e_7	e_4	e_5	e_6
e_2	e_2	e_3	e_0	e_1	e_6	e_7	e_4	e_5
e_3	e_3	e_0	e_1	e_2	e_5	e_6	e_7	e_4
e_4	e_4	e_5	e_6	e_7	e_0	e_1	e_2	e_3
e_5	e_5	e_6	e_7	e_4	e_3	e_0	e_1	e_2
e_6	e_6	e_7	e_4	e_5	e_2	e_3	e_0	e_1
e_7	e_7	e_4	e_5	e_6	e_1	e_2	e_3	e_0

(b):

$e_i e_j$	e_0	e_1	$-e_0$	$-e_1$	e_4	e_5	$-e_4$	$-e_5$
e_0	e_0	e_1	$-e_0$	$-e_1$	e_4	e_5	$-e_4$	$-e_5$
e_1	e_1	$-e_0$	$-e_1$	e_0	$-e_5$	e_4	e_5	$-e_4$
$-e_0$	$-e_0$	$-e_1$	e_0	e_1	$-e_4$	$-e_5$	e_4	e_5
$-e_1$	$-e_1$	e_0	e_1	$-e_0$	e_5	$-e_4$	$-e_5$	e_4
e_4	e_4	e_5	$-e_4$	$-e_5$	e_0	e_1	$-e_0$	$-e_1$
e_5	e_5	$-e_4$	$-e_5$	e_4	$-e_1$	e_0	e_1	$-e_0$
$-e_4$	$-e_4$	$-e_5$	e_4	e_5	$-e_0$	$-e_1$	e_0	e_1
$-e_5$	$-e_5$	e_4	e_5	$-e_4$	e_1	$-e_0$	$-e_1$	e_0

$$H(\langle A', B'\rangle X) = a_{A'} + i \cdot b_{B'} = -a_A - i \cdot b_B = i^2 \cdot (a_A + i \cdot b_B) = i^2 \cdot H(\langle A, B\rangle X) \\ = H(\zeta^2 \langle A, B\rangle X) \tag{84}$$

$$H(\langle B, A'\rangle X) = b_B + i \cdot a_{A'} = b_B - i \cdot a_A = i^3 \cdot (a_A + i \cdot b_B) = i^3 \cdot H(\langle A, B\rangle X) \\ = H(\zeta^3 \langle A, B\rangle X) \tag{85}$$

$$H(\langle B, A\rangle X) = b_B + i \cdot a_A = (i^3 \cdot H(\langle A, B\rangle X))^* \\ = H(\sim \langle A, B\rangle X) \tag{86}$$

$$H(\langle A', B\rangle X) = a_{A'} + i \cdot b_B = -a_A + i \cdot b_B = i \cdot H(\langle B, A\rangle X) \\ = H(\zeta \langle B, A\rangle X) \tag{87}$$

Table 3. Eight operations in both the complex and conjugate formula in comparison

Op	Complex Exp	Complex Comp	Conj Op	Conj Exp	Conj Comp	Notes
1	$H(\langle A,B\rangle X)$	$H(\langle A,B\rangle X)$	E	$H(\langle A,B\rangle X)$	$H(\langle A,B\rangle X)$	Original
i	$H^{i}(\langle A,B\rangle X)$	$i\cdot H(\langle A,B\rangle X)$	ζ	$H(\langle B',A\rangle X)$	$H(\zeta\langle A,B\rangle X)$	Twist
i^2	$H^{i^2}(\langle A,B\rangle X)$	$i^2\cdot H(\langle A,B\rangle X)$	$\zeta^2=\neg$	$H(\langle A',B'\rangle X)$	$H(\zeta^2\langle A,B\rangle X)$	Negative
i^3	$H^{i^3}(\langle A,B\rangle X)$	$i^3\cdot H(\langle A,B\rangle X)$	$\zeta^3=\neg\zeta$	$H(\langle B,A'\rangle X)$	$H(\zeta^3\langle A,B\rangle X)$	Negative-Twist
$\leftrightarrow$	$H^{\leftrightarrow}(\langle A,B\rangle X)$	$(i^3\cdot H(\langle A,B\rangle X))^*$	$\sim$	$H(\langle B,A\rangle X)$	$\tilde{H}(\langle A,B\rangle X)$	Conjugate
r	$H^{r}(\langle A,B\rangle X)$	$i\cdot(i^3\cdot H(\langle A,B\rangle X))^*$	$\zeta\sim$	$H(\langle A',B\rangle X)$	$\tilde{H}(\zeta\langle A,B\rangle X)$	Real-Conjugate
$i^2\leftrightarrow$	$H^{i^2\leftrightarrow}(\langle A,B\rangle X)$	$(i\cdot H(\langle A,B\rangle X))^*$	$\zeta^2\sim$	$H(\langle B',A'\rangle X)$	$\tilde{H}(\zeta^2\langle A,B\rangle X)$	Negative-Conjugate
$*$	$H^{*}(\langle A,B\rangle X)$	$i\cdot(i\cdot H(\langle A,B\rangle X))^*$	$\zeta^3\sim$	$H(\langle A,B'\rangle X)$	$\tilde{H}(\zeta^3\langle A,B\rangle X)$	Complex-Conjugate

Where abbreviations are used: Op—Operator; Complex Exp—Complex Expression; Complex Comp—Complex Computation; Conj Op—Conjugate Operator; Conj Exp—Conjugate Expression; Conj Comp—Conjugate Computation.

$$\begin{aligned} H(\langle B',A'\rangle X) &= b_{B'}+i\cdot a_{A'}=-b_B-i\cdot a_A=i^2\cdot(b_B+i\cdot a_A)=i^2\cdot H(\langle B,A\rangle X)\\ &= H(\zeta^2\langle B,A\rangle X) \end{aligned} \tag{88}$$

$$\begin{aligned} H(\langle A,B'\rangle X) &= a_A+i\cdot b_{B'}=a_A-i\cdot b_B=i^3\cdot H(\langle B,A\rangle X)\\ &= H(\zeta^3\langle B,A\rangle X) \end{aligned} \tag{89}$$

Theorem 18. *For a selected $H(\langle A,B\rangle X)$, eight complex operators and eight conjugate measurement operators have 1–1 correspondence in equivalent relationships.*

Proof. Check Equations (83)–(89), which have 1–1 equivalent properties.

Under this correspondence, eight operators satisfy both the complex and conjugate formulas in the comparison listed in Table 3.

Theorem 19 (Numbers of Configurations in Conjugate Phase Space). *Conjugate transformation structure $2n$ group output variable cluster dimension $2^{2(n+1)}$, the corresponding complex vector function state occupies $(-2^n,2^n]^2$ configuration space.*

Proof. $2n$ set of variables in a single attribute area has 2^{2n} configurations on $[0,2^n)^2$, the complete space of configurations contains four attribute areas and complex variable function positions 1–1 correspond, $4\times 2^{2n}=2^{2(n+1)}$.

8. Conjugate Algebra Processing Output Vectors of Conjugate Transformation Structure

Combined with the results of the conjugate complex vector functions in Chapter 1, based on the conjugate orthogonality of the $2n$ group feature clusters of the conjugate transformation structure, the focus is on the symmetric and antisymmetric interaction under the representation characteristics of specific group interactions. Select a pair of conjugate variables from the output vectors and use the eight eigenvectors corresponding to the associated $\{\sim,\sharp,\prime,*\}$ conjugate operators to analyze the conjugate algebraic structure.

$$y = (p, q) = (a_j, b_j) = H(\langle A_j, B_j \rangle X) \tag{90}$$
$$y^{\sharp} = (-q, p) = (b'_j, a_j) = H(\langle B'_j, A_j \rangle X) \tag{91}$$
$$y' = (-p, -q) = (a'_j, b'_j) = H(\langle A'_j, B'_j \rangle X) \tag{92}$$
$$y^{*} = (q, -p) = (b_j, a'_j) = H(\langle B_j, A'_j \rangle X) \tag{93}$$
$$\tilde{y} = (q, p) = (b_j, a_j) = H(\langle B_j, A_j \rangle X) \tag{94}$$
$$\tilde{y}^{\sharp} = (-p, q) = (a'_j, b_j) = H(\langle A'_j, B_j \rangle X) \tag{95}$$
$$\tilde{y}' = (-q, -p) = (b'_j, a'_j) = H(\langle B'_j, A'_j \rangle X) \tag{96}$$
$$\tilde{y}^{*} = (p, -q) = (a_j, b'_j) = H(\langle A_j, B'_j \rangle X) \tag{97}$$

Corollary 5 (1-1 Correspondences of Eight Eigenvectors). *Select a pair of conjugate vectors from the 2n group of variables and expand to four attribute regions to become a single-channel conjugate structure vector of output representation. The structure and the conjugate complex vector function are corrected and transformed into octuples of the complex cluster in a 1-1 correspondence.*

Proof. The total number of vectors is $8 = 2 \times 2^2 = 2^3$. Checking the equations (90)–(97), the eight sets of complex numbers in the two systems correspond to 1-1.

Through the conjugate transformation structure, the corresponding eight eigenstates are expressed as the corrected four-attribute interval vectors through the conjugate complex vector function:

$$Y = \{y, y^{\sharp}, y', y^{*}\} \tag{98}$$
$$\tilde{Y} = \{\tilde{y}, \tilde{y}^{\sharp}, \tilde{y}', \tilde{y}^{*}\} \tag{99}$$

Using the inverse vector $\{Y^{-1}, \tilde{Y}^{-1}\}$ state representation, the quaternary conjugate vector cluster $\{Y, Y^{-1}, \tilde{Y}, \tilde{Y}^{-1}\}$ interacts to construct the Galois field algebraic structure.

Clusters of pairwise conjugate eigenvectors in the vector multiplicative space form the complex basis interactions through the following products. Use the complex function exponent to express

$$x = (p, q) \tag{100}$$
$$r = \sqrt{p^2 + q^2} \tag{101}$$
$$x = y = re_0^{i\theta} \tag{102}$$
$$\tilde{x} = \tilde{y} = re_0^{i\tilde{\theta}} \tag{103}$$

The corresponding feature vector set is

$$Y = \{re_0^{i\theta}, re_2^{i\theta}, re_3^{i\theta}, re_1^{i\theta}\} \tag{104}$$
$$\tilde{Y} = \{re_0^{i\tilde{\theta}}, re_2^{i\theta}, re_3^{i\tilde{\theta}}, re_1^{i\theta}\} \tag{105}$$
$$Y^{-1} = \overrightarrow{\tilde{Y}} = \{re_1^{i\tilde{\theta}}, re_3^{i\tilde{\theta}}, re_2^{i\tilde{\theta}}, re_0^{i\tilde{\theta}}\} \tag{106}$$
$$\tilde{Y}^{-1} = \overrightarrow{Y} = \{re_1^{i\theta}, re_3^{i\theta}, re_2^{i\theta}, re_0^{i\theta}\} \tag{107}$$

For the product matrix, the two types of exchange features $\{[x,y]_d,\{x,y\}_d\}$ operate along the main diagonal; $\{[x,y]^u,\{x,y\}^u\}$ operates along the auxiliary diagonal, $[x,y] = xy - yx, \{x,y\} = xy + yx$.

Let S, A be the symmetric/antisymmetric projection operator, and the attribute parameters $\{0, R_s, R_a, I_s, I_a\}$ are marked, respectively: 0 matrix, R_s real symmetry, R_a real antisymmetry, I_s imaginary symmetry, I_a imaginary antisymmetry, using the | operator, the left is the principal diagonal, and the left is the auxiliary diagonal feature. The attribute feature matrix is $S(\,) = \begin{pmatrix} S([\,]_d) & S([\,]^u) \\ S(\{\,\}_d) & S(\{\,\}^u) \end{pmatrix}$.

Theorem 20 (Pairwise Conjugate [,], {, } Brackets). *Two types of interaction operators $\{[x,y]_d,\{x,y\}_d\},\{[x,y]^u,\{x,y\}^u\}$ provide complementary symmetric and antisymmetric information for the selected product matrix cluster.*

Proof. From a local measurement, whether along the diagonal or auxiliary diagonal, the paired variables associated with each position in the product matrix are different. From the matrix geometric transformation, the diagonal/auxiliary diagonal arrangement is equivalent to a 90-degree rotation on the entire matrix, and the eigenvalue of the transformed matrix is multiplied by the original eigenvalue and $\{i, -i\}$. From the order of operator action, whether it is the local symmetry characteristic or the symmetric/antisymmetric quantization characteristic after the overall integration, it is intrinsically related to the arrangement order of the row/column generating vector of the measured matrix.

The modern quantum formula is based on the paired argument description $\{\theta, -\theta\}$, and the corresponding vector state is

$$Y_c = \{y, \tilde{y}^*\} = \{re_0^{i\theta}, re_1^{i\tilde{\theta}}\} = \{re^{i\theta}, re^{-i\theta}\}$$

$$Y_c^2 = \begin{array}{|c|cc|}\hline \{y,\tilde{y}^*\}^2 & y & \tilde{y}^* \\ \hline y & yy & y\tilde{y}^* \\ \tilde{y}^* & \tilde{y}^*y & \tilde{y}^*\tilde{y}^* \\ \hline \end{array} = \begin{array}{|c|cc|}\hline \{y,\tilde{y}^*\}^2 & y & \tilde{y}^* \\ \hline y & r^2e^{i2\theta} & r^2 \\ \tilde{y}^* & r^2 & r^2e_1^{i2\tilde{\theta}} \\ \hline \end{array} = r^2\begin{pmatrix} e^{i2\theta} & 1 \\ 1 & e^{-i2\theta} \end{pmatrix}$$

$$[Y_c, Y_c]_d = r^2\begin{pmatrix} e^{i2\theta} - e^{i2\theta} & 1-1 \\ 1-1 & e^{-i2\theta} - e^{-i2\theta} \end{pmatrix} = r^2\begin{pmatrix} 0 & 0 \\ 0 & 0 \end{pmatrix}$$

$$S([Y_c, Y_c]_d) = 0 : 0(1)$$

$$\{Y_c, Y_c\}_d = r^2\begin{pmatrix} e^{i2\theta} + e^{i2\theta} & 1+1 \\ 1+1 & e^{-i2\theta} + e^{-i2\theta} \end{pmatrix} = 2r^2\begin{pmatrix} e^{i2\theta} & 1 \\ 1 & e^{-i2\theta} \end{pmatrix}$$

$$S(\{Y_c, Y_c\}_d) = R_sI_s|R_sI_a : S(3)A(1)$$

$$[Y_c, Y_c]^u = r^2\begin{pmatrix} e^{i2\theta} - e^{-i2\theta} & 1-1 \\ 1-1 & e^{-i2\theta} - e^{i2\theta} \end{pmatrix} = 2r^2\begin{pmatrix} i\sin 2\theta & 0 \\ 0 & -i\sin 2\theta \end{pmatrix}$$

$$S([Y_c, Y_c]^u) = I_s|I_a : S(1)A(1)$$

$$\{Y_c, Y_c\}^u = r^2\begin{pmatrix} e^{i2\theta} + e^{-i2\theta} & 1+1 \\ 1+1 & e^{-i2\theta} + e^{i2\theta} \end{pmatrix} = 2r^2\begin{pmatrix} \cos 2\theta & 1 \\ 1 & \cos 2\theta \end{pmatrix}$$

$$\{Y_c, Y_c\}^u) = R_s|R_s : S(2)$$

$$S(Y_c, Y_c) = \begin{pmatrix} 0 & I_s|I_a \\ R_sI_s|R_sI_a & R_s|R_s \end{pmatrix}$$

Symmetric property statistics: $0(1), \{R_s(2), I_s(2)\}|\{R_s(2), I_a(2)\} : 0(1)S(6)A(2)$

These four groups of operators act on the result, and $[,]^u, \{,\}^u$ provide feature divisions.

Exchange the order of action, diagonalize the unit measure, and obtain the classical 2×2 matrix representation,

$$Y_c \times \overrightarrow{Y}_c = \begin{array}{|c|cc|} \hline \{y, \tilde{y}^*\} \times \{\tilde{y}^*, y\} & \tilde{y}^* & y \\ \hline y & r^2 & r^2 e^{i2\theta} \\ \tilde{y}^* & r^2 e^{i2\tilde{\theta}}_1 & r^2 \\ \hline \end{array} = r^2 \begin{pmatrix} 1 & e^{i2\theta} \\ e^{-i2\theta} & 1 \end{pmatrix}$$

$$[Y_c, \overrightarrow{Y}_c]_d = 2r^2 \begin{pmatrix} 0 & i\sin 2\theta \\ -i\sin 2\theta & 0 \end{pmatrix}, S([Y_c, \overrightarrow{Y}_c]_d) = I_a|I_s$$

$$\{Y_c, \overrightarrow{Y}_c\}_d = 2r^2 \begin{pmatrix} 1\,0 \\ 0\,1 \end{pmatrix}, S(\{Y_c, \overrightarrow{Y}_c\}_d) = R_s|R_s$$

$$[Y_c, \overrightarrow{Y}_c]^u = r^2 \begin{pmatrix} 0\,0 \\ 0\,0 \end{pmatrix}, S([Y_c, \overrightarrow{Y}_c]_u) = 0$$

$$\{Y_c, \overrightarrow{Y}_c\}^u = 2r^2 \begin{pmatrix} 1 & \cos 2\theta \\ \cos 2\theta & 1 \end{pmatrix}, S(\{Y_c, \overrightarrow{Y}_c\}_u) = R_s|R_s$$

$$S(Y_c, \overrightarrow{Y}_c) = \begin{pmatrix} I_a|I_s & 0 \\ R_s|R_s & R_s|R_s \end{pmatrix}$$

Symmetric property statistics: $0(1), \{R_s(2), I_a(1)\}|\{R_s(2), I_a(1)\} : 0(1)S(4)A(2)$

The difference in the symmetry between the various matrices can be quantitatively observed from the attribute statistics parameters.

The subsequent four groups of operators act on the result, $[,]_d, \{,\}_d$, providing the feature division function.

Observed from different output results, the main diagonal and auxiliary diagonal operators have complementary effects, which cannot be simply ignored.

Corollary 6 (Pauli Matrix Solutions). *The exchange operator associated with the eigenvalues of the Pauli matrix, implemented by the* $[,]_d, \{,\}_d, 2 \times 2$ *eigenmatrix.*

Proof. The Pauli matrix has the diagonal associated eigenvalues real numbers and the corresponding antidiagonal eigenvalues imaginary numbers.

The combined inverse vector is expressed as $\{Y_c, Y_c^{-1}\} = \{y, \tilde{y}^*, y^{-1}, (\tilde{y}^*)^{-1}\} = \{y, \tilde{y}^*, \tilde{y}^*, y\}$, to obtain the classical 4×4 matrix representation,

$$\{Y_c, Y_c^{-1}\}^2 = \begin{array}{|c|cccc|} \hline \{y, \tilde{y}^*\}^2 & y & \tilde{y}^* & \tilde{y}^* & y \\ \hline y & yy & y\tilde{y}^* & y\tilde{y}^* & yy \\ \tilde{y}^* & \tilde{y}^*y & \tilde{y}^*\tilde{y}^* & \tilde{y}^*\tilde{y}^* & \tilde{y}^*y \\ \tilde{y}^* & \tilde{y}^*y & \tilde{y}^*\tilde{y}^* & \tilde{y}^*\tilde{y}^* & \tilde{y}^*y \\ y & yy & y\tilde{y}^* & y\tilde{y}^* & yy \\ \hline \end{array} = r^2 \begin{pmatrix} e^{i2\theta} & 1 & 1 & e^{i2\theta} \\ 1 & e^{-i2\theta} & e^{-i2\theta} & 1 \\ 1 & e^{-i2\theta} & e^{-i2\theta} & 1 \\ e^{i2\theta} & 1 & 1 & e^{i2\theta} \end{pmatrix}$$

$$[\{Y_c, Y_c^{-1}\}, \{Y_c, Y_c^{-1}\}]_d = r^2 \begin{pmatrix} 0\,0\,0\,0 \\ 0\,0\,0\,0 \\ 0\,0\,0\,0 \\ 0\,0\,0\,0 \end{pmatrix}, S([\{Y_c, Y_c^{-1}\}, \{Y_c, Y_c^{-1}\}]_d) = 0$$

$$\{\{Y_c, Y_c^{-1}\}, \{Y_c, Y_c^{-1}\}\}_d = r^2 \begin{pmatrix} e^{i2\theta} & 1 & 1 & e^{i2\theta} \\ 1 & e^{-i2\theta} & e^{-i2\theta} & 1 \\ 1 & e^{-i2\theta} & e^{-i2\theta} & 1 \\ e^{i2\theta} & 1 & 1 & e^{i2\theta} \end{pmatrix}$$

$$S(\{\{Y_c, Y_c^{-1}\}, \{Y_c, Y_c^{-1}\}\}_d) = R_s I_s | R_s I_s$$

$$[\{Y_c, Y_c^{-1}\}, \{Y_c, Y_c^{-1}\}]^u = r^2 \begin{pmatrix} 0\,0\,0\,0 \\ 0\,0\,0\,0 \\ 0\,0\,0\,0 \\ 0\,0\,0\,0 \end{pmatrix}, S([\{Y_c, Y_c^{-1}\}, \{Y_c, Y_c^{-1}\}]^u) = 0$$

$$\{\{Y_c, Y_c^{-1}\}, \{Y_c, Y_c^{-1}\}\}^u = r^2 \begin{pmatrix} e^{i2\theta} & 1 & 1 & e^{i2\theta} \\ 1 & e^{-i2\theta} & e^{-i2\theta} & 1 \\ 1 & e^{-i2\theta} & e^{-i2\theta} & 1 \\ e^{i2\theta} & 1 & 1 & e^{i2\theta} \end{pmatrix}$$

$$S(\{\{Y_c, Y_c^{-1}\}, \{Y_c, Y_c^{-1}\}\}_d) = R_s I_s | R_s I_s$$

$$S(\{Y_c, Y_c^{-1}\}, \{Y_c, Y_c^{-1}\}) = \begin{pmatrix} 0 & 0 \\ R_s | R_s & R_s | R_s \end{pmatrix}$$

Symmetric property statistics: $0(2), \{R_s(2), I_s(2)\} | \{R_s(2), I_s(2)\} : 0(2) S(8)$

$$\{Y_c, Y_c^{-1}\} \times \{Y_c^{-1}, Y_c\} = \begin{array}{|c|cccc|} \hline & \tilde{y}^* & y & y & \tilde{y}^* \\ \hline y & y\tilde{y}^* & yy & yy & y\tilde{y}^* \\ \tilde{y}^* & \tilde{y}^*\tilde{y}^* & \tilde{y}^*y & \tilde{y}^*y & \tilde{y}^*\tilde{y}^* \\ \tilde{y}^* & \tilde{y}^*\tilde{y}^* & \tilde{y}^*y & \tilde{y}^*y & \tilde{y}^*\tilde{y}^* \\ y & y\tilde{y}^* & yy & yy & y\tilde{y}^* \\ \hline \end{array} = r^2 \begin{pmatrix} 1 & e^{i2\theta} & e^{i2\theta} & 1 \\ e^{-i2\theta} & 1 & 1 & e^{-i2\theta} \\ e^{-i2\theta} & 1 & 1 & e^{-i2\theta} \\ 1 & e^{i2\theta} & e^{i2\theta} & 1 \end{pmatrix}$$

$$[\{Y_c, Y_c^{-1}\}, \{Y_c^{-1}, Y_c\}]_d = r^2 \begin{pmatrix} 0 & i\sin 2\theta & i\sin 2\theta & 0 \\ -i\sin 2\theta & 0 & 0 & -i\sin 2\theta \\ -i\sin 2\theta & 0 & 0 & -i\sin 2\theta \\ 0 & i\sin 2\theta & i\sin 2\theta & 0 \end{pmatrix}$$

$$S([\{Y_c, Y_c^{-1}\}, \{Y_c^{-1}, Y_c\}]_d) = I_a | I_a$$

$$\{\{Y_c, Y_c^{-1}\}, \{Y_c^{-1}, Y_c\}\}_d = r^2 \begin{pmatrix} 1 & \cos 2\theta & \cos 2\theta & 1 \\ \cos 2\theta & 1 & 1 & \cos 2\theta \\ \cos 2\theta & 1 & 1 & \cos 2\theta \\ 1 & \cos 2\theta & \cos 2\theta & 1 \end{pmatrix}$$

$$S(\{\{Y_c, Y_c^{-1}\}, \{Y_c^{-1}, Y_c\}\}_d) = R_s | R_s$$

$$[\{Y_c, Y_c^{-1}\}, \{Y_c^{-1}, Y_c\}]^u = r^2 \begin{pmatrix} 0 & i\sin 2\theta & i\sin 2\theta & 0 \\ -i\sin 2\theta & 0 & 0 & -i\sin 2\theta \\ -i\sin 2\theta & 0 & 0 & -i\sin 2\theta \\ 0 & i\sin 2\theta & i\sin 2\theta & 0 \end{pmatrix}$$

$$S([\{Y_c, Y_c^{-1}\}, \{Y_c^{-1}, Y_c\}]^u) = I_a | I_a$$

$$\{\{Y_c, Y_c^{-1}\}, \{Y_c^{-1}, Y_c\}\}^u = r^2 \begin{pmatrix} 1 & \cos 2\theta & \cos 2\theta & 1 \\ \cos 2\theta & 1 & 1 & \cos 2\theta \\ \cos 2\theta & 1 & 1 & \cos 2\theta \\ 1 & \cos 2\theta & \cos 2\theta & 1 \end{pmatrix}$$

$$S(\{\{Y_c, Y_c^{-1}\}, \{Y_c^{-1}, Y_c\}\}^u) = R_s|R_s$$

$$S(\{Y_c, Y_c^{-1}\}, \{Y_c^{-1}, Y_c\}) = \begin{pmatrix} I_a|I_a & I_a|I_a \\ R_s|R_s & R_s|R_s \end{pmatrix}$$

Symmetric property statistics: $\{R_s(2), I_a(2)\}|\{R_s(2), I_a(2)\} : S(4)A(4)$

For the primary and secondary diagonally symmetric structures, the two sets of operators give the same result.

$S(\{Y_c, Y_c^{-1}\}, \{Y_c, Y_c^{-1}\})$ compared with $S(\{Y_c, Y_c^{-1}\}, \{Y_c^{-1}, Y_c\})$ contains stronger symmetry.

Corollary 7 (Dirac Matrix Solutions). *The exchange operator associated with the eigenvalues of the Dirac matrix, implemented by the* $[,]_d, \{,\}_d$, 4×4 *eigenmatrix.*

Proof. Dirac matrices have real diagonal associated eigenvalues, and the corresponding antidiagonal eigenvalues are imaginary.

Theorem 21 (4 x 4 Matrix Kernel). *If the vector* Y *is an exchange operator associated with a variety of extended vector operators* $\{-1, \leftarrow\}$ *matrix eigenvalues, the* 4×4 *eigenmatrix can be used as the kernel,* $\{-1, \leftarrow\}$ $\{[,]_d, \{,\}_d\}, \{[,]^u, \{,\}^u\}$ *matrix representation implementation.*

Proof. Y is a quaternary vector, and large matrices can be represented by the 4×4 block mode under the expansion operators, which is suitable for more complex expansions.

For larger matrices, the symmetry properties of each submatrix can be counted first, and then statistical processing is performed. The basic matrix unit is the dimension of 4×4.

Adding other attributes to expand the formula yields the pairwise $8 \times 8 = 64$ products.

$\{Y, \tilde{Y}\}^2 =$

$\{Y, \tilde{Y}\}^2$	Y	$\tilde{Y}$
Y	YY	$Y\tilde{Y}$
$\tilde{Y}$	$\tilde{Y}Y$	$\tilde{Y}\tilde{Y}$

$$YY = \begin{pmatrix} e_0^{i2\theta} & e_2^{i2\theta} & e_3^{i2\theta} & e_1^{i2\theta} \\ e_2^{i2\theta} & e_3^{i2\theta} & e_1^{i2\theta} & e_0^{i2\theta} \\ e_3^{i2\theta} & e_1^{i2\theta} & e_0^{i2\theta} & e_2^{i2\theta} \\ e_1^{i2\theta} & e_0^{i2\theta} & e_2^{i2\theta} & e_3^{i2\theta} \end{pmatrix}$$

$$= e^{i2\theta} \begin{pmatrix} 1 & i & -1 & -i \\ i & -1 & -i & 1 \\ -1 & -i & 1 & i \\ -i & 1 & i & -1 \end{pmatrix}$$

$$\tilde{Y}\tilde{Y} = \begin{pmatrix} e_0^{i2\tilde{\theta}} & e_2^{i2\tilde{\theta}} & e_3^{i2\tilde{\theta}} & e_1^{i2\tilde{\theta}} \\ e_2^{i2\tilde{\theta}} & e_3^{i2\tilde{\theta}} & e_1^{i2\tilde{\theta}} & e_0^{i2\tilde{\theta}} \\ e_3^{i2\tilde{\theta}} & e_1^{i2\tilde{\theta}} & e_0^{i2\tilde{\theta}} & e_2^{i2\tilde{\theta}} \\ e_1^{i2\tilde{\theta}} & e_0^{i2\tilde{\theta}} & e_2^{i2\tilde{\theta}} & e_3^{i2\tilde{\theta}} \end{pmatrix} = e^{i2\tilde{\theta}} \begin{pmatrix} 1 & i & -1 & -i \\ i & -1 & -i & 1 \\ -1 & -i & 1 & i \\ -i & 1 & i & -1 \end{pmatrix}$$

$$Y\tilde{Y} = \tilde{Y}Y = \begin{pmatrix} i & -1 & -i & 1 \\ -1 & -i & 1 & i \\ -i & 1 & i & -1 \\ 1 & i & -1 & -i \end{pmatrix}$$

$$\{Y,\tilde{Y}\}^2=\begin{pmatrix} \begin{matrix} e_0^{i2\theta} & e_2^{i2\theta} & e_3^{i2\theta} & e_1^{i2\theta} \\ e_2^{i2\theta} & e_3^{i2\theta} & e_1^{i2\theta} & e_0^{i2\theta} \\ e_3^{i2\theta} & e_1^{i2\theta} & e_0^{i2\theta} & e_2^{i2\theta} \\ e_1^{i2\theta} & e_0^{i2\theta} & e_2^{i2\theta} & e_3^{i2\theta} \end{matrix} & \begin{matrix} i & -1 & -i & 1 \\ -1 & -i & 1 & i \\ -i & 1 & i & -1 \\ 1 & i & -1 & -i \end{matrix} \\ \begin{matrix} i & -1 & -i & 1 \\ -1 & -i & 1 & i \\ -i & 1 & i & -1 \\ 1 & i & -1 & -i \end{matrix} & \begin{matrix} e_0^{i2\tilde{\theta}} & e_2^{i2\tilde{\theta}} & e_3^{i2\tilde{\theta}} & e_1^{i2\tilde{\theta}} \\ e_2^{i2\tilde{\theta}} & e_3^{i2\tilde{\theta}} & e_1^{i2\tilde{\theta}} & e_0^{i2\tilde{\theta}} \\ e_3^{i2\tilde{\theta}} & e_1^{i2\tilde{\theta}} & e_0^{i2\tilde{\theta}} & e_2^{i2\tilde{\theta}} \\ e_1^{i2\tilde{\theta}} & e_0^{i2\tilde{\theta}} & e_2^{i2\tilde{\theta}} & e_3^{i2\tilde{\theta}} \end{matrix} \end{pmatrix}$$

$$=\begin{pmatrix} e^{i2\theta}\begin{pmatrix} 1 & i & -1 & -i \\ i & -1 & -i & 1 \\ -1 & -i & 1 & i \\ -i & 1 & i & -1 \end{pmatrix} & \begin{pmatrix} i & -1 & -i & 1 \\ -1 & -i & 1 & i \\ -i & 1 & i & -1 \\ 1 & i & -1 & -i \end{pmatrix} \\ \begin{pmatrix} i & -1 & -i & 1 \\ -1 & -i & 1 & i \\ -i & 1 & i & -1 \\ 1 & i & -1 & -i \end{pmatrix} & e^{i2\tilde{\theta}}\begin{pmatrix} 1 & i & -1 & -i \\ i & -1 & -i & 1 \\ -1 & -i & 1 & i \\ -i & 1 & i & -1 \end{pmatrix} \end{pmatrix}$$

$$[\{Y,\tilde{Y}\},\{Y,\tilde{Y}\}]_d=r^2\begin{pmatrix} 0 & 0 \\ 0 & 0 \end{pmatrix}, S([\{Y,\tilde{Y}\},\{Y,\tilde{Y}\}]_d)=0(4)$$

$$\{\{Y,\tilde{Y}\},\{Y,\tilde{Y}\}\}_d=2r^2\begin{pmatrix} e^{i2\theta}\begin{pmatrix} 1 & i & -1 & -i \\ i & -1 & -i & 1 \\ -1 & -i & 1 & i \\ -i & 1 & i & -1 \end{pmatrix} & \begin{pmatrix} i & -1 & -i & 1 \\ -1 & -i & 1 & i \\ -i & 1 & i & -1 \\ 1 & i & -1 & -i \end{pmatrix} \\ \begin{pmatrix} i & -1 & -i & 1 \\ -1 & -i & 1 & i \\ -i & 1 & i & -1 \\ 1 & i & -1 & -i \end{pmatrix} & e^{i2\tilde{\theta}}\begin{pmatrix} 1 & i & -1 & -i \\ i & -1 & -i & 1 \\ -1 & -i & 1 & i \\ -i & 1 & i & -1 \end{pmatrix} \end{pmatrix}$$

$$S(\{\{Y,\tilde{Y}\},\{Y,\tilde{Y}\}\}_d)=\begin{pmatrix} R_sI_s|R_aI_s & R_sI_s|R_sI_a \\ R_sI_s|R_sI_a & R_sI_s|R_aI_s \end{pmatrix}=\{R_s(4),I_s(4)\}|\{R_s(2),R_a(2),I_s(2),I_a(2)\}$$

$$[\{Y,\tilde{Y}\},\{Y,\tilde{Y}\}]^u=r^2\begin{pmatrix} (e^{i2\theta}-e^{i2\tilde{\theta}})\begin{pmatrix} 1 & i & -1 & -i \\ i & -1 & -i & 1 \\ -1 & -i & 1 & i \\ -i & 1 & i & -1 \end{pmatrix} & 0 \\ 0 & (e^{i2\tilde{\theta}}-e^{i2\theta})\begin{pmatrix} 1 & i & -1 & -i \\ i & -1 & -i & 1 \\ -1 & -i & 1 & i \\ -i & 1 & i & -1 \end{pmatrix} \end{pmatrix}$$

$$S([\{Y,\tilde{Y}\},\{Y,\tilde{Y}\}]^u)=\begin{pmatrix} R_sI_s|R_aI_s & 0 \\ 0 & R_sI_s|R_aI_s \end{pmatrix}=0(2),\{R_s(2),I_s(2)\}|\{R_a(2),I_s(2)\}$$

$$\{\{Y,\tilde{Y}\},\{Y,\tilde{Y}\}\}^u = r^2\begin{pmatrix} (e^{i2\theta}+e^{i2\tilde{\theta}})\begin{pmatrix} 1 & i & -1 & -i \\ i & -1 & -i & 1 \\ -1 & -i & 1 & i \\ -i & 1 & i & -1 \end{pmatrix} & 2\begin{pmatrix} i & -1 & -i & 1 \\ -1 & -i & 1 & i \\ -i & 1 & i & -1 \\ 1 & i & -1 & -i \end{pmatrix} \\ 2\begin{pmatrix} i & -1 & -i & 1 \\ -1 & -i & 1 & i \\ -i & 1 & i & -1 \\ 1 & i & -1 & -i \end{pmatrix} & (e^{i2\tilde{\theta}}+e^{i2\theta})\begin{pmatrix} 1 & i & -1 & -i \\ i & -1 & -i & 1 \\ -1 & -i & 1 & i \\ -i & 1 & i & -1 \end{pmatrix} \end{pmatrix}$$

$$S(\{\{Y,\tilde{Y}\},\{Y,\tilde{Y}\}\}^u) = \begin{pmatrix} R_sI_s|R_aI_s & R_sI_s|R_sI_a \\ R_sI_s|R_sI_a & R_sI_s|R_aI_s \end{pmatrix} = \{R_s(4),I_s(4)\}|\{R_s(2),R_a(2),I_s(2),I_a(2)\}$$

$$S(\{\{Y,\tilde{Y}\},\{Y,\tilde{Y}\}\}) = \begin{pmatrix} 0(4) & 0(2),\{R_s(2),I_s(2)\}|\{R_a(2),I_s(2)\} \\ \{R_s(4),I_s(4)\}|\{R_s(2),R_a(2),I_s(2),I_a(2)\} & \{R_s(4),I_s(4)\}|\{R_s(2),R_a(2),I_s(2),I_a(2)\} \end{pmatrix}$$
$$= 0(6),\{R_s(10),I_s(10)\}|\{R_s(4),R_a(6),I_s(6),I_a(4)\}0(6)S(30)A(10)$$

The global symmetry characteristics of the transformation matrix can be used to compare and analyze the inherent attribute characteristics between different projection modes by quantifying the characteristics of the symmetry attributes.

Expanding the associated block matrix, the relevant modules are

$$YY^{-1} = \begin{pmatrix} 1 & -i & -1 & i \\ i & 1 & -i & -1 \\ -1 & i & 1 & -i \\ -i & -1 & i & 1 \end{pmatrix}, YY = e^{i2\theta}\begin{pmatrix} 1 & i & -1 & -i \\ i & -1 & -i & 1 \\ -1 & -i & 1 & i \\ -i & 1 & i & -1 \end{pmatrix},$$

$$Y\tilde{Y}^{-1} = e^{i2\theta}\begin{pmatrix} -i & -1 & i & 1 \\ 1 & -i & -1 & i \\ i & 1 & -i & -1 \\ -1 & i & 1 & -i \end{pmatrix}, Y\tilde{Y} = \begin{pmatrix} i & -1 & -i & 1 \\ -1 & -i & 1 & i \\ -i & 1 & i & -1 \\ 1 & i & -1 & -i \end{pmatrix}$$

$$Y^{-1}Y^{-1} = e^{i2\tilde{\theta}}\begin{pmatrix} -1 & i & 1 & -i \\ i & 1 & -i & -1 \\ 1 & -i & -1 & i \\ -i & -1 & i & 1 \end{pmatrix}, Y^{-1}Y = \begin{pmatrix} 1 & i & -1 & -i \\ -i & 1 & i & -1 \\ -1 & -i & 1 & i \\ i & -1 & -i & 1 \end{pmatrix},$$

$$Y^{-1}\tilde{Y}^{-1} = \begin{pmatrix} -i & -1 & i & 1 \\ -1 & i & 1 & -i \\ i & 1 & -i & -1 \\ 1 & -i & -1 & i \end{pmatrix}, Y^{-1}\tilde{Y}e^{i2\tilde{\theta}} = \begin{pmatrix} -i & 1 & i & -1 \\ -1 & -i & 1 & i \\ i & -1 & -i & 1 \\ 1 & i & -1 & -i \end{pmatrix}$$

$$\tilde{Y}Y^{-1} = e^{i2\tilde{\theta}}\begin{pmatrix} -i & -1 & i & 1 \\ 1 & -i & -1 & i \\ i & 1 & -i & -1 \\ -1 & i & 1 & -i \end{pmatrix}, \tilde{Y}Y = \begin{pmatrix} i & -1 & -i & 1 \\ -1 & -i & 1 & i \\ -i & 1 & i & -1 \\ 1 & i & -1 & -i \end{pmatrix},$$

$$\tilde{Y}\tilde{Y}^{-1} = \begin{pmatrix} 1 & -i & -1 & i \\ i & 1 & -i & -1 \\ -1 & i & 1 & -i \\ -i & -1 & i & 1 \end{pmatrix}, \tilde{Y}\tilde{Y} = e^{i2\tilde{\theta}}\begin{pmatrix} 1 & i & -1 & -i \\ i & -1 & -i & 1 \\ -1 & -i & 1 & i \\ -i & 1 & i & -1 \end{pmatrix}$$

$$\tilde{Y}^{-1}Y^{-1}=\begin{pmatrix}-i&-1&i&1\\-1&i&1&-i\\i&1&-i&-1\\1&-i&-1&i\end{pmatrix},\ \tilde{Y}^{-1}Y=e^{i2\theta}\begin{pmatrix}-i&1&i&-1\\-1&-i&1&i\\i&-1&-i&1\\1&i&-1&-i\end{pmatrix},$$

$$\tilde{Y}^{-1}\tilde{Y}^{-1}=e^{i2\theta}\begin{pmatrix}-1&i&1&-i\\i&1&-i&-1\\1&-i&-1&i\\-i&-1&i&1\end{pmatrix},\ \tilde{Y}^{-1}\tilde{Y}=\begin{pmatrix}1&i&-1&-i\\-i&1&i&-1\\-1&-i&1&i\\i&-1&-i&1\end{pmatrix}$$

Using the generated matrix module, the $16\times16=256$ pair product is expressed as:

$\{Y,Y^{-1},\tilde{Y},\tilde{Y}^{-1}\}^2=$

$\{Y,Y^{-1},\tilde{Y},\tilde{Y}^{-1}\}^2$	Y	Y^{-1}	$\tilde{Y}$	$\tilde{Y}^{-1}$
Y	YY	YY^{-1}	$Y\tilde{Y}$	$Y\tilde{Y}^{-1}$
Y^{-1}	$Y^{-1}Y$	$Y^{-1}Y^{-1}$	$Y^{-1}\tilde{Y}$	$Y^{-1}\tilde{Y}^{-1}$
$\tilde{Y}$	$\tilde{Y}Y$	$\tilde{Y}Y^{-1}$	$\tilde{Y}\tilde{Y}$	$\tilde{Y}\tilde{Y}^{-1}$
$\tilde{Y}^{-1}$	$\tilde{Y}^{-1}Y$	$\tilde{Y}^{-1}Y^{-1}$	$\tilde{Y}^{-1}\tilde{Y}$	$\tilde{Y}^{-1}\tilde{Y}^{-1}$

$$=\begin{pmatrix}
e^{i2\theta}\begin{pmatrix}1&i&-1&-i\\i&-1&-i&1\\-1&-i&1&i\\-i&1&i&-1\end{pmatrix} & \begin{pmatrix}1&-i&-1&i\\i&1&-i&-1\\-1&i&1&-i\\-i&-1&i&1\end{pmatrix} & \begin{pmatrix}i&-1&-i&1\\-1&-i&1&i\\-i&1&i&-1\\1&i&-1&-i\end{pmatrix} & e^{i2\theta}\begin{pmatrix}-i&-1&i&1\\1&-i&-1&i\\i&1&-i&-1\\-1&i&1&-i\end{pmatrix}\\
\begin{pmatrix}1&i&-1&-i\\-i&1&i&-1\\-1&-i&1&i\\i&-1&-i&1\end{pmatrix} & e^{i2\tilde{\theta}}\begin{pmatrix}-1&i&1&-i\\i&1&-i&-1\\1&-i&-1&i\\-i&-1&i&1\end{pmatrix} & e^{i2\tilde{\theta}}\begin{pmatrix}-i&1&i&-1\\-1&-i&1&i\\i&-1&-i&1\\1&i&-1&-i\end{pmatrix} & \begin{pmatrix}-i&-1&i&1\\-1&i&1&-i\\i&1&-i&-1\\1&-i&-1&i\end{pmatrix}\\
\begin{pmatrix}i&-1&-i&1\\-1&-i&1&i\\-i&1&i&-1\\1&i&-1&-i\end{pmatrix} & e^{i2\tilde{\theta}}\begin{pmatrix}-i&-1&i&1\\1&-i&-1&i\\i&1&-i&-1\\-1&i&1&-i\end{pmatrix} & e^{i2\tilde{\theta}}\begin{pmatrix}1&i&-1&-i\\i&-1&-i&1\\-1&-i&1&i\\-i&1&i&-1\end{pmatrix} & \begin{pmatrix}1&-i&-1&i\\i&1&-i&-1\\-1&i&1&-i\\-i&-1&i&1\end{pmatrix}\\
e^{i2\theta}\begin{pmatrix}-i&1&i&-1\\-1&-i&1&i\\i&-1&-i&1\\1&i&-1&-i\end{pmatrix} & \begin{pmatrix}-i&-1&i&1\\-1&i&1&-i\\i&1&-i&-1\\1&-i&-1&i\end{pmatrix} & \begin{pmatrix}1&i&-1&-i\\-i&1&i&-1\\-1&-i&1&i\\i&-1&-i&1\end{pmatrix} & e^{i2\theta}\begin{pmatrix}-1&i&1&-i\\i&1&-i&-1\\1&-i&-1&i\\-i&-1&i&1\end{pmatrix}
\end{pmatrix}$$

$[\{Y,Y^{-1},\tilde{Y},\tilde{Y}^{-1}\},\{Y,Y^{-1},\tilde{Y},\tilde{Y}^{-1}\}]_d=$

$$2r^2\begin{pmatrix}
0 & \begin{pmatrix}0&-i&0&i\\i&0&-i&0\\0&i&0&-i\\-i&0&i&0\end{pmatrix} & 0 & e^{i2\theta}\begin{pmatrix}0&-1&0&1\\1&0&-1&0\\0&1&0&-1\\-1&0&1&0\end{pmatrix}\\
\begin{pmatrix}0&i&0&-i\\-i&0&i&0\\0&-i&0&i\\i&0&-i&0\end{pmatrix} & 0 & e^{i2\tilde{\theta}}\begin{pmatrix}0&1&0&-1\\-1&0&1&0\\0&-1&0&1\\1&0&-1&0\end{pmatrix} & 0\\
0 & e^{i2\tilde{\theta}}\begin{pmatrix}0&-1&0&1\\1&0&-1&0\\0&1&0&-1\\-1&0&1&0\end{pmatrix} & 0 & \begin{pmatrix}0&-i&0&i\\i&0&-i&0\\0&i&0&-i\\-i&0&i&\end{pmatrix}\\
e^{i2\theta}\begin{pmatrix}0&1&0&-1\\-1&0&1&0\\0&-1&0&1\\1&0&-1&0\end{pmatrix} & 0 & \begin{pmatrix}0&i&0&-i\\-i&0&i&0\\0&-i&0&i\\i&0&-i&0\end{pmatrix} & 0
\end{pmatrix}$$

$$S([\{Y, Y^{-1}, \tilde{Y}, \tilde{Y}^{-1}\}, \{Y, Y^{-1}, \tilde{Y}, \tilde{Y}^{-1}\}]_d) = \begin{pmatrix} 0 & I_a|I_s & 0 & R_a|R_s \\ I_a|I_s & 0 & R_a|R_s & 0 \\ 0 & R_a|R_s & 0 & I_a|I_s \\ R_a|R_s & 0 & I_a|I_s & 0 \end{pmatrix}$$
$$= 0(8), \{R_a(4), I_a(4)\}|\{R_s(4), I_s(4)\} : 0(8)S(8)A(8)$$

$$\{\{Y, Y^{-1}, \tilde{Y}, \tilde{Y}^{-1}\}, \{Y, Y^{-1}, \tilde{Y}, \tilde{Y}^{-1}\}\}_d =$$
$$= 2r^2 \begin{pmatrix} e^{i2\theta}\begin{pmatrix} 1 & i & -1 & -i \\ i & -1 & -i & 1 \\ -1 & -i & 1 & i \\ -i & 1 & i & -1 \end{pmatrix} & \begin{pmatrix} 1 & 0 & -1 & 0 \\ 0 & 1 & 0 & -1 \\ -1 & 0 & 1 & 0 \\ 0 & -1 & 0 & 1 \end{pmatrix} & \begin{pmatrix} i & -1 & -i & 1 \\ -1 & -i & 1 & i \\ -i & 1 & i & -1 \\ 1 & i & -1 & -i \end{pmatrix} & e^{i2\theta}\begin{pmatrix} -i & 0 & i & 0 \\ 0 & -i & 0 & i \\ i & 0 & -i & 0 \\ 0 & i & 0 & -i \end{pmatrix} \\ \begin{pmatrix} 1 & 0 & -1 & 0 \\ 0 & 1 & 0 & -1 \\ -1 & 0 & 1 & 0 \\ 0 & -1 & 0 & 1 \end{pmatrix} & e^{i2\tilde{\theta}}\begin{pmatrix} -1 & i & 1 & -i \\ i & 1 & -i & -1 \\ 1 & -i & -1 & i \\ -i & -1 & i & 1 \end{pmatrix} & e^{i2\tilde{\theta}}\begin{pmatrix} 0 & 1 & 0 & -1 \\ -1 & 0 & 1 & 0 \\ 0 & -1 & 0 & 1 \\ 1 & 0 & -1 & 0 \end{pmatrix} & \begin{pmatrix} -i & -1 & i & 1 \\ -1 & i & 1 & -i \\ i & 1 & -i & -1 \\ 1 & -i & -1 & i \end{pmatrix} \\ \begin{pmatrix} i & -1 & -i & 1 \\ -1 & -i & 1 & i \\ -i & 1 & i & -1 \\ 1 & i & -1 & -i \end{pmatrix} & e^{i2\tilde{\theta}}\begin{pmatrix} 0 & -1 & 0 & 1 \\ 1 & 0 & -1 & 0 \\ 0 & 1 & 0 & -1 \\ -1 & 0 & 1 & 0 \end{pmatrix} & e^{i2\tilde{\theta}}\begin{pmatrix} 1 & i & -1 & -i \\ i & -1 & -i & 1 \\ -1 & -i & 1 & i \\ -i & 1 & i & -1 \end{pmatrix} & \begin{pmatrix} 1 & 0 & -1 & 0 \\ 0 & 1 & 0 & -1 \\ -1 & 0 & 1 & 0 \\ 0 & -1 & 0 & 1 \end{pmatrix} \\ e^{i2\theta}\begin{pmatrix} -i & 0 & i & 0 \\ 0 & -i & 0 & i \\ i & 0 & -i & 0 \\ 0 & i & 0 & -i \end{pmatrix} & \begin{pmatrix} -i & -1 & i & 1 \\ -1 & i & 1 & -i \\ i & 1 & -i & -1 \\ 1 & -i & -1 & i \end{pmatrix} & \begin{pmatrix} 1 & 0 & -1 & 0 \\ 0 & 1 & 0 & -1 \\ -1 & 0 & 1 & 0 \\ 0 & -1 & 0 & 1 \end{pmatrix} & e^{i2\theta}\begin{pmatrix} -1 & i & 1 & -i \\ i & 1 & -i & -1 \\ 1 & -i & -1 & i \\ -i & -1 & i & 1 \end{pmatrix} \end{pmatrix}$$

$$S(\{\{Y, Y^{-1}, \tilde{Y}, \tilde{Y}^{-1}\}, \{Y, Y^{-1}, \tilde{Y}, \tilde{Y}^{-1}\}\}_d)$$
$$= \begin{pmatrix} R_sI_s|R_aI_s & R_s|R_s & R_sI_s|R_sI_a & I_s|I_s \\ R_s|R_s & R_sI_s|R_aI_s & R_a|R_s & R_sI_s|R_sI_a \\ R_sI_s|R_sI_a & R_a|R_s & R_sI_s|R_aI_s & R_s|R_s \\ I_s|I_s & R_sI_s|R_sI_a & R_s|R_s & R_sI_s|R_aI_s \end{pmatrix}$$
$$= \{R_s(12), R_a(2), I_s(2)\}|\{R_s(10), R_a(4), I_s(6), I_a(4)\} : S(30)A(10)$$

$$[\{Y, Y^{-1}, \tilde{Y}, \tilde{Y}^{-1}\}, \{Y, Y^{-1}, \tilde{Y}, \tilde{Y}^{-1}\}]^u =$$
$$2r^2 \begin{pmatrix} e^{i2\theta}\begin{pmatrix} 1 & 0 & -1 & 0 \\ 0 & -1 & 0 & 1 \\ -1 & 0 & 1 & 0 \\ 0 & 1 & 0 & -1 \end{pmatrix} & 0 & \begin{pmatrix} i & 0 & -i & 0 \\ 0 & -i & 0 & i \\ -i & 0 & i & 0 \\ 0 & i & 0 & -i \end{pmatrix} & 0 \\ 0 & e^{i2\tilde{\theta}}\begin{pmatrix} -1 & 0 & 1 & 0 \\ 0 & 1 & 0 & -1 \\ 1 & 0 & -1 & 0 \\ 0 & -1 & 0 & 1 \end{pmatrix} & 0 & \begin{pmatrix} -i & 0 & i & 0 \\ 0 & i & 0 & -i \\ i & 0 & -i & 0 \\ 0 & -i & 0 & i \end{pmatrix} \\ \begin{pmatrix} i & 0 & -i & 0 \\ 0 & -i & 0 & i \\ -i & 0 & i & 0 \\ 0 & i & 0 & -i \end{pmatrix} & 0 & e^{i2\tilde{\theta}}\begin{pmatrix} 1 & 0 & -1 & 0 \\ 0 & -1 & 0 & 1 \\ -1 & 0 & 1 & 0 \\ 0 & 1 & 0 & -1 \end{pmatrix} & 0 \\ 0 & \begin{pmatrix} -i & 0 & i & 0 \\ 0 & i & 0 & -i \\ i & 0 & -i & 0 \\ 0 & -i & 0 & i \end{pmatrix} & 0 & e^{i2\theta}\begin{pmatrix} -1 & 0 & 1 & 0 \\ 0 & 1 & 0 & -1 \\ 1 & 0 & -1 & 0 \\ 0 & -1 & 0 & 1 \end{pmatrix} \end{pmatrix}$$

$$S([\{Y, Y^{-1}, \tilde{Y}, \tilde{Y}^{-1}\}, \{Y, Y^{-1}, \tilde{Y}, \tilde{Y}^{-1}\}]^u) = \begin{pmatrix} R_s|R_a & 0 & I_s|I_a & 0 \\ 0 & R_s|R_a & 0 & I_s|I_a \\ I_s|I_a & 0 & R_s|R_a & 0 \\ 0 & I_s|I_a & 0 & R_s|R_a \end{pmatrix}$$
$$= 0(8), \{R_s(4), I_s(4)\}|\{R_a(4), I_a(4)\}$$

$$\{\{Y, Y^{-1}, \tilde{Y}, \tilde{Y}^{-1}\}, \{Y, Y^{-1}, \tilde{Y}, \tilde{Y}^{-1}\}\}^u =$$

$$2r^2\begin{pmatrix}
e^{i2\theta}\begin{pmatrix}0 & i & 0 & -i\\ i & 0 & -i & 0\\ 0 & -i & 0 & i\\ -i & 0 & i & 0\end{pmatrix} & \begin{pmatrix}1 & -i & -1 & i\\ i & 1 & -i & -1\\ -1 & i & 1 & -i\\ -i & -1 & i & 1\end{pmatrix} & \begin{pmatrix}0 & -1 & 0 & 1\\ -1 & 0 & 1 & 0\\ 0 & 1 & 0 & -1\\ 1 & 0 & -1 & 0\end{pmatrix} & e^{i2\theta}\begin{pmatrix}-i & -1 & i & 1\\ 1 & -i & -1 & i\\ i & 1 & -i & -1\\ -1 & i & 1 & -i\end{pmatrix}\\
\begin{pmatrix}1 & i & -1 & -i\\ -i & 1 & i & -1\\ -1 & -i & 1 & i\\ i & -1 & -i & 1\end{pmatrix} & e^{i2\tilde{\theta}}\begin{pmatrix}0 & i & 0 & -i\\ i & 0 & -i & 0\\ 0 & -i & 0 & i\\ -i & 0 & i & 0\end{pmatrix} & e^{i2\tilde{\theta}}\begin{pmatrix}-i & 1 & i & -1\\ -1 & -i & 1 & i\\ i & -1 & -i & 1\\ 1 & i & -1 & -i\end{pmatrix} & \begin{pmatrix}0 & -1 & 0 & 1\\ -1 & 0 & 1 & 0\\ 0 & 1 & 0 & -1\\ 1 & 0 & -1 & 0\end{pmatrix}\\
\begin{pmatrix}0 & -1 & 0 & 1\\ -1 & 0 & 1 & 0\\ 0 & 1 & 0 & -1\\ 1 & 0 & -1 & 0\end{pmatrix} & e^{i2\tilde{\theta}}\begin{pmatrix}-i & -1 & i & 1\\ 1 & -i & -1 & i\\ i & 1 & -i & -1\\ -1 & i & 1 & -i\end{pmatrix} & e^{i2\tilde{\theta}}\begin{pmatrix}0 & i & 0 & -i\\ i & 0 & -i & 0\\ 0 & -i & 0 & i\\ -i & 0 & i & 0\end{pmatrix} & \begin{pmatrix}1 & -i & -1 & i\\ i & 1 & -i & -1\\ -1 & i & 1 & -i\\ -i & -1 & i & 1\end{pmatrix}\\
e^{i2\theta}\begin{pmatrix}-i & 1 & i & -1\\ -1 & -i & 1 & i\\ i & -1 & -i & 1\\ 1 & i & -1 & -i\end{pmatrix} & \begin{pmatrix}0 & -1 & 0 & 1\\ -1 & 0 & 1 & 0\\ 0 & 1 & 0 & -1\\ 1 & 0 & -1 & 0\end{pmatrix} & \begin{pmatrix}1 & i & -1 & -i\\ -i & 1 & i & -1\\ -1 & -i & 1 & i\\ i & -1 & -i & 1\end{pmatrix} & e^{i2\theta}\begin{pmatrix}0 & i & 0 & -i\\ i & 0 & -i & 0\\ 0 & -i & 0 & i\\ -i & 0 & i & 0\end{pmatrix}
\end{pmatrix}$$

$$S(\{\{Y, Y^{-1}, \tilde{Y}, \tilde{Y}^{-1}\}, \{Y, Y^{-1}, \tilde{Y}, \tilde{Y}^{-1}\}\}^u) = \begin{pmatrix}
I_s|I_s & R_sI_a|R_sI_s & R_s|R_s & R_aI_s|R_sI_s\\
R_sI_a|R_sI_s & I_s|I_s & R_aI_s|R_sI_s & R_s|R_s\\
R_s|R_s & R_aI_s|R_sI_s & I_s|I_s & R_sI_a|R_sI_s\\
R_aI_s|R_sI_s & R_s|R_s & R_sI_a|R_sI_s & I_s|I_s
\end{pmatrix}$$
$$= \{R_s(8), R_a(4), I_s(8), I_a(4)\}|\{R_s(12), I_s(12)\}$$

$$S(\{Y, Y^{-1}, \tilde{Y}, \tilde{Y}^{-1}\}, \{Y, Y^{-1}, \tilde{Y}, \tilde{Y}^{-1}\}) =$$
$$\begin{pmatrix}
0(8), \{R_a(4), I_a(4)\}|\{R_s(4), I_s(4)\} & 0(8), \{R_s(4), I_s(4)\}|\{R_a(4), I_a(4)\}\\
\{R_s(12), R_a(2), I_s(2)\}|\{R_s(10), R_a(4), I_s(6), I_a(4)\} & \{R_s(8), R_a(4), I_s(8), I_a(4)\}|\{R_s(12), I_s(12)\}
\end{pmatrix}$$
$$= 0(16), \{R_s(24), R_a(10), I_s(14), I_a(8)\}|\{R_s(26), R_a(8), I_s(22), I_a(8)\}$$

The arrangement order of the variables is changed to obtain another matrix arrangement,

$$\{Y, Y^{-1}, \tilde{Y}, \tilde{Y}^{-1}\} \times \{Y^{-1}, Y, \tilde{Y}^{-1}, \tilde{Y}\} =$$

	Y^{-1}	Y	$\tilde{Y}^{-1}$	$\tilde{Y}$
Y	YY^{-1}	YY	$Y\tilde{Y}^{-1}$	$Y\tilde{Y}$
Y^{-1}	$Y^{-1}Y^{-1}$	$Y^{-1}Y$	$Y^{-1}\tilde{Y}^{-1}$	$Y^{-1}\tilde{Y}$
$\tilde{Y}$	$\tilde{Y}Y^{-1}$	$\tilde{Y}Y$	$\tilde{Y}\tilde{Y}^{-1}$	$\tilde{Y}\tilde{Y}$
$\tilde{Y}^{-1}$	$\tilde{Y}^{-1}Y^{-1}$	$\tilde{Y}^{-1}Y$	$\tilde{Y}^{-1}\tilde{Y}^{-1}$	$\tilde{Y}^{-1}\tilde{Y}$

$$= r^2\begin{pmatrix}
\begin{pmatrix}1 & -i & -1 & i\\ i & 1 & -i & -1\\ -1 & i & 1 & -i\\ -i & -1 & i & 1\end{pmatrix} & e^{i2\theta}\begin{pmatrix}1 & i & -1 & -i\\ i & -1 & -i & 1\\ -1 & -i & 1 & i\\ -i & 1 & i & -1\end{pmatrix} & e^{i2\theta}\begin{pmatrix}-i & -1 & i & 1\\ 1 & -i & -1 & i\\ i & 1 & -i & -1\\ -1 & i & 1 & -i\end{pmatrix} & \begin{pmatrix}i & -1 & -i & 1\\ -1 & -i & 1 & i\\ -i & 1 & i & -1\\ 1 & i & -1 & -i\end{pmatrix}\\
e^{i2\tilde{\theta}}\begin{pmatrix}-1 & i & 1 & -i\\ i & 1 & -i & -1\\ 1 & -i & -1 & i\\ -i & -1 & i & 1\end{pmatrix} & \begin{pmatrix}1 & i & -1 & -i\\ -i & 1 & i & -1\\ -1 & -i & 1 & i\\ i & -1 & -i & 1\end{pmatrix} & \begin{pmatrix}-i & -1 & i & 1\\ -1 & i & 1 & -i\\ i & 1 & -i & -1\\ 1 & -i & -1 & i\end{pmatrix} & e^{i2\tilde{\theta}}\begin{pmatrix}-i & 1 & i & -1\\ -1 & -i & 1 & i\\ i & -1 & -i & 1\\ 1 & i & -1 & -i\end{pmatrix}\\
e^{i2\tilde{\theta}}\begin{pmatrix}-i & -1 & i & 1\\ 1 & -i & -1 & i\\ i & 1 & -i & -1\\ -1 & i & 1 & -i\end{pmatrix} & \begin{pmatrix}i & -1 & -i & 1\\ -1 & -i & 1 & i\\ -i & 1 & i & -1\\ 1 & i & -1 & -i\end{pmatrix} & \begin{pmatrix}1 & -i & -1 & i\\ i & 1 & -i & -1\\ -1 & i & 1 & -i\\ -i & -1 & i & 1\end{pmatrix} & e^{i2\tilde{\theta}}\begin{pmatrix}1 & i & -1 & -i\\ i & -1 & -i & 1\\ -1 & -i & 1 & i\\ -i & 1 & i & -1\end{pmatrix}\\
\begin{pmatrix}-i & -1 & i & 1\\ -1 & i & 1 & -i\\ i & 1 & -i & -1\\ 1 & -i & -1 & i\end{pmatrix} & e^{i2\theta}\begin{pmatrix}-i & 1 & i & -1\\ -1 & -i & 1 & i\\ i & -1 & -i & 1\\ 1 & i & -1 & -i\end{pmatrix} & e^{i2\theta}\begin{pmatrix}-1 & i & 1 & -i\\ i & 1 & -i & -1\\ 1 & -i & -1 & i\\ -i & -1 & i & 1\end{pmatrix} & \begin{pmatrix}1 & i & -1 & -i\\ -i & 1 & i & -1\\ -1 & -i & 1 & i\\ i & -1 & -i & 1\end{pmatrix}
\end{pmatrix}$$

$$[\{Y,\,Y^{-1},\,\tilde{Y},\,\tilde{Y}^{-1}\},\{Y^{-1},\,Y,\,\tilde{Y}^{-1},\,\tilde{Y}\}]_d =$$

$$2r^2\begin{pmatrix}
0 & e^{i2\theta}\begin{pmatrix}1&0&-1&0\\0&-1&0&1\\-1&0&1&0\\0&1&0&-1\end{pmatrix} & 0 & \begin{pmatrix}i&0&-i&0\\0&-i&0&i\\-i&0&i&0\\0&i&0&-i\end{pmatrix}\\
e^{i2\tilde{\theta}}\begin{pmatrix}-1&0&1&0\\0&1&0&-1\\1&0&-1&0\\0&-1&0&1\end{pmatrix} & 0 & \begin{pmatrix}-i&0&i&0\\0&i&0&-i\\i&0&-i&0\\0&-i&0&i\end{pmatrix} & 0\\
0 & \begin{pmatrix}i&0&-i&0\\0&-i&0&i\\-i&0&i&0\\0&i&0&-i\end{pmatrix} & 0 & e^{i2\tilde{\theta}}\begin{pmatrix}1&0&-1&0\\0&-1&0&1\\-1&0&1&0\\0&1&0&-1\end{pmatrix}\\
\begin{pmatrix}-i&0&i&0\\0&i&0&-i\\i&0&-i&0\\0&-i&0&i\end{pmatrix} & 0 & e^{i2\theta}\begin{pmatrix}-1&0&1&0\\0&1&0&-1\\1&0&-1&0\\0&-1&0&1\end{pmatrix} & 0
\end{pmatrix}$$

$$S([\{Y,\,Y^{-1},\,\tilde{Y},\,\tilde{Y}^{-1}\},\{Y^{-1},\,Y,\,\tilde{Y}^{-1},\,\tilde{Y}\}]_d) = \begin{pmatrix}
0 & R_s|R_a & 0 & I_s|I_a\\
R_s|R_a & 0 & I_s|I_a & 0\\
0 & I_s|I_a & 0 & R_s|R_a\\
I_s|I_a & 0 & R_s|R_a & 0
\end{pmatrix}$$
$$= 0(8), \{R_s(4), I_s(4)\}|\{R_a(4), I_a(4)\}$$

$$\{\{Y,\,Y^{-1},\,\tilde{Y},\,\tilde{Y}^{-1}\},\{Y^{-1},\,Y,\,\tilde{Y}^{-1},\,\tilde{Y}\}\}_d =$$

$$2r^2\begin{pmatrix}
\begin{pmatrix}1&-i&-1&i\\i&1&-i&-1\\-1&i&1&-i\\-i&-1&i&1\end{pmatrix} & e^{i2\theta}\begin{pmatrix}0&i&0&-i\\i&0&-i&0\\0&-i&0&i\\-i&0&i&0\end{pmatrix} & e^{i2\theta}\begin{pmatrix}-i&-1&i&1\\1&-i&-1&i\\i&1&-i&-1\\-1&i&1&-i\end{pmatrix} & \begin{pmatrix}0&-1&0&1\\-1&0&1&0\\0&1&0&-1\\1&0&-1&0\end{pmatrix}\\
e^{i2\tilde{\theta}}\begin{pmatrix}0&i&0&-i\\i&0&-i&0\\0&-i&0&i\\-i&0&i&0\end{pmatrix} & \begin{pmatrix}1&i&-1&-i\\-i&1&i&-1\\-1&-i&1&i\\i&-1&-i&1\end{pmatrix} & \begin{pmatrix}0&-1&0&1\\-1&0&1&0\\0&1&0&-1\\1&0&-1&0\end{pmatrix} & e^{i2\tilde{\theta}}\begin{pmatrix}-i&1&i&-1\\-1&-i&1&i\\i&-1&-i&1\\1&i&-1&-i\end{pmatrix}\\
e^{i2\tilde{\theta}}\begin{pmatrix}-i&-1&i&1\\1&-i&-1&i\\i&1&-i&-1\\-1&i&1&-i\end{pmatrix} & \begin{pmatrix}0&-1&0&1\\-1&0&1&0\\0&1&0&-1\\1&0&-1&0\end{pmatrix} & \begin{pmatrix}1&-i&-1&i\\i&1&-i&-1\\-1&i&1&-i\\-i&-1&i&1\end{pmatrix} & e^{i2\tilde{\theta}}\begin{pmatrix}0&i&0&-i\\i&0&-i&0\\0&-i&0&i\\-i&0&i&0\end{pmatrix}\\
\begin{pmatrix}0&-1&0&1\\-1&0&1&0\\0&1&0&-1\\1&0&-1&0\end{pmatrix} & e^{i2\theta}\begin{pmatrix}-i&1&i&-1\\-1&-i&1&i\\i&-1&-i&1\\1&i&-1&-i\end{pmatrix} & e^{i2\theta}\begin{pmatrix}0&i&0&-i\\i&0&-i&0\\0&-i&0&i\\-i&0&i&0\end{pmatrix} & \begin{pmatrix}1&i&-1&-i\\-i&1&i&-1\\-1&-i&1&i\\i&-1&-i&1\end{pmatrix}
\end{pmatrix}$$

$$S(\{\{Y,\,Y^{-1},\,\tilde{Y},\,\tilde{Y}^{-1}\},\{Y^{-1},\,Y,\,\tilde{Y}^{-1},\,\tilde{Y}\}\}_d) = \begin{pmatrix}
R_sI_a|R_sI_s & I_s|I_s & R_aI_s|R_sI_s & R_s|R_s\\
I_s|I_s & R_sI_a|R_sI_s & R_s|R_s & R_aI_s|R_sI_s\\
R_aI_s|R_sI_s & R_s|R_s & R_sI_a|R_sI_s & I_s|I_s\\
R_s|R_s & R_aI_s|R_sI_s & I_s|I_s & R_sI_a|R_sI_s
\end{pmatrix}$$
$$= \{R_s(8), R_a(4), I_s(8), I_a(4)\}|\{R_s(12), R_a(4), I_s(16)\}$$

$$[\{Y, Y^{-1}, \tilde{Y}, \tilde{Y}^{-1}\}, \{Y^{-1}, Y, \tilde{Y}^{-1}, \tilde{Y}\}]^u =$$

$$2r^2 \begin{pmatrix} \begin{pmatrix} 0 & -i & 0 & i \\ i & 0 & -i & 0 \\ 0 & i & 0 & -i \\ -i & 0 & i & 0 \end{pmatrix} & 0 & e^{i2\theta}\begin{pmatrix} 0 & -1 & 0 & 1 \\ 1 & 0 & -1 & 0 \\ 0 & 1 & 0 & -1 \\ -1 & 0 & 1 & 0 \end{pmatrix} & 0 \\ 0 & \begin{pmatrix} 0 & i & 0 & -i \\ -i & 0 & i & 0 \\ 0 & -i & 0 & i \\ i & 0 & -i & 0 \end{pmatrix} & 0 & e^{i2\tilde{\theta}}\begin{pmatrix} 0 & 1 & 0 & -1 \\ -1 & 0 & 1 & 0 \\ 0 & -1 & 0 & 1 \\ 1 & 0 & -1 & 0 \end{pmatrix} \\ e^{i2\tilde{\theta}}\begin{pmatrix} 0 & -1 & 0 & 1 \\ 1 & 0 & -1 & 0 \\ 0 & 1 & 0 & -1 \\ -1 & 0 & 1 & 0 \end{pmatrix} & 0 & \begin{pmatrix} 0 & -i & 0 & i \\ i & 0 & -i & 0 \\ 0 & i & 0 & -i \\ -i & 0 & i & 0 \end{pmatrix} & 0 \\ 0 & e^{i2\theta}\begin{pmatrix} 0 & 1 & 0 & -1 \\ -1 & 0 & 1 & 0 \\ 0 & -1 & 0 & 1 \\ 1 & 0 & -1 & 0 \end{pmatrix} & 0 & \begin{pmatrix} 0 & i & 0 & -i \\ -i & 0 & i & 0 \\ 0 & -i & 0 & i \\ i & 0 & -i & 0 \end{pmatrix} \end{pmatrix}$$

$$S([\{Y, Y^{-1}, \tilde{Y}, \tilde{Y}^{-1}\}, \{Y^{-1}, Y, \tilde{Y}^{-1}, \tilde{Y}\}]^u) = \begin{pmatrix} I_a|I_s & 0 & R_a|R_s & 0 \\ 0 & I_a|I_s & 0 & R_a|R_s \\ R_a|R_s & 0 & I_a|I_s & 0 \\ 0 & R_a|R_s & 0 & I_a|I_s \end{pmatrix}$$
$$= 0(8), \{R_a(4), I_a(4)\}|\{R_s(4), I_s(4)\}$$

$$\{\{Y, Y^{-1}, \tilde{Y}, \tilde{Y}^{-1}\}, \{Y^{-1}, Y, \tilde{Y}^{-1}, \tilde{Y}\}\}^u =$$

$$= 2r^2 \begin{pmatrix} \begin{pmatrix} 1 & 0 & -1 & 0 \\ 0 & 1 & 0 & -1 \\ -1 & 0 & 1 & 0 \\ 0 & -1 & 0 & 1 \end{pmatrix} & e^{i2\theta}\begin{pmatrix} 1 & i & -1 & -i \\ i & -1 & -i & 1 \\ -1 & -i & 1 & i \\ -i & 1 & i & -1 \end{pmatrix} & e^{i2\theta}\begin{pmatrix} -i & 0 & i & 0 \\ 0 & -i & 0 & i \\ i & 0 & -i & 0 \\ 0 & i & 0 & -i \end{pmatrix} & \begin{pmatrix} i & -1 & -i & 1 \\ -1 & -i & 1 & i \\ -i & 1 & i & -1 \\ 1 & i & -1 & -i \end{pmatrix} \\ e^{i2\tilde{\theta}}\begin{pmatrix} -1 & i & 1 & -i \\ i & 1 & -i & -1 \\ 1 & -i & -1 & i \\ -i & -1 & i & 1 \end{pmatrix} & \begin{pmatrix} 1 & 0 & -1 & 0 \\ 0 & 1 & 0 & -1 \\ -1 & 0 & 1 & 0 \\ 0 & -1 & 0 & 1 \end{pmatrix} & \begin{pmatrix} -i & -1 & i & 1 \\ -1 & i & 1 & -i \\ i & 1 & -i & -1 \\ 1 & -i & -1 & i \end{pmatrix} & e^{i2\tilde{\theta}}\begin{pmatrix} -i & 0 & i & 0 \\ 0 & -i & 0 & i \\ i & 0 & -i & 0 \\ 0 & i & 0 & -i \end{pmatrix} \\ e^{i2\tilde{\theta}}\begin{pmatrix} -i & 0 & i & 0 \\ 0 & -i & 0 & i \\ i & 0 & -i & 0 \\ 0 & i & 0 & -i \end{pmatrix} & \begin{pmatrix} i & -1 & -i & 1 \\ -1 & -i & 1 & i \\ -i & 1 & i & -1 \\ 1 & i & -1 & -i \end{pmatrix} & \begin{pmatrix} 1 & 0 & -1 & 0 \\ 0 & 1 & 0 & -1 \\ -1 & 0 & 1 & 0 \\ 0 & -1 & 0 & 1 \end{pmatrix} & e^{i2\tilde{\theta}}\begin{pmatrix} 1 & i & -1 & -i \\ i & -1 & -i & 1 \\ -1 & -i & 1 & i \\ -i & 1 & i & -1 \end{pmatrix} \\ \begin{pmatrix} -i & -1 & i & 1 \\ -1 & i & 1 & -i \\ i & 1 & -i & -1 \\ 1 & -i & -1 & i \end{pmatrix} & e^{i2\theta}\begin{pmatrix} -i & 0 & i & 0 \\ 0 & -i & 0 & i \\ i & 0 & -i & 0 \\ 0 & i & 0 & -i \end{pmatrix} & e^{i2\theta}\begin{pmatrix} -1 & i & 1 & -i \\ i & 1 & -i & -1 \\ 1 & -i & -1 & i \\ -i & -1 & i & 1 \end{pmatrix} & \begin{pmatrix} 1 & 0 & -1 & 0 \\ 0 & 1 & 0 & -1 \\ -1 & 0 & 1 & 0 \\ 0 & -1 & 0 & 1 \end{pmatrix} \end{pmatrix}$$

$$S(\{\{Y, Y^{-1}, \tilde{Y}, \tilde{Y}^{-1}\}, \{Y^{-1}, Y, \tilde{Y}^{-1}, \tilde{Y}\}\}^u) = \begin{pmatrix} R_s|R_s & R_sI_s|R_aI_s & I_s|I_s & R_sI_s|R_sI_a \\ R_sI_s|R_aI_s & R_s|R_s & R_sI_s|R_sI_a & I_s|I_s \\ I_s|I_s & R_sI_s|R_sI_a & R_s|R_s & R_sI_s|R_aI_s \\ R_sI_s|R_sI_a & I_s|I_s & R_sI_s|R_aI_s & R_s|R_s \end{pmatrix}$$
$$= \{R_s(12), I_s(4)\}|\{R_s(6), R_a(4), I_s(8), I_a(4)\}$$

$$S(\{Y, Y^{-1}, \tilde{Y}, \tilde{Y}^{-1}\}, \{Y^{-1}, Y, \tilde{Y}^{-1}, \tilde{Y}\}) =$$

$$\begin{pmatrix} 0(8), \{R_s(4), I_s(4)\}|\{R_a(4), I_a(4)\} & 0(8), \{R_a(4), I_a(4)\}|\{R_s(4), I_s(4)\} \\ \{R_s(8), R_a(4), I_s(8), I_a(4)\}|\{R_s(12), R_a(4), I_s(16)\} & \{R_s(12), I_s(4)\}|\{R_s(6), R_a(4), I_s(8), I_a(4)\} \end{pmatrix}$$
$$= 0(16), \{R_s(24), R_a(8), I_s(16), I_a(8)\}|\{R_s(22), R_a(12), I_s(28), I_a(8)\}$$

Comparison of the results of the two groups $S(\{Y, Y^{-1}, \tilde{Y}, \tilde{Y}^{-1}\}, \{Y, Y^{-1}, \tilde{Y}, \tilde{Y}^{-1}\}) =$

$$\begin{pmatrix} 0(8), \{R_a(4), I_a(4)\} | \{R_s(4), I_s(4)\} & 0(8), \{R_s(4), I_s(4)\} | \{R_a(4), I_a(4)\} \\ \{R_s(12), R_a(2), I_s(2)\} | \{R_s(10), R_a(4), I_s(6), I_a(4)\} & \{R_s(8), R_a(4), I_s(8), I_a(4)\} | \{R_s(12), I_s(12)\} \end{pmatrix}$$

$= 0(16), \{R_s(24), R_a(10), I_s(14), I_a(8)\} | \{R_s(26), R_a(8), I_s(22), I_a(8)\}.$

When the extended part degenerates into one, the classical quantum information matrix expression pattern can be observed. However, the extended part exhibits richer combinatorial effects and various inherently rich conjugate algebraic structural features.

9. Intrinsic Time Order

Since the transformation between the four attribute regions of the complex plane must strictly satisfy the temporal order rule of synchronous action, it forms a canonical transformation sequence and related vector action operators. Let $\leftarrow$ rotate operator, vector $X = \{0, 2, 3, 1\}$, $^2\overleftarrow{X} = \{2, 3, 1, 0\}$ set, and the second vector is the initial position. Let – be the reflection operator, $\overline{X} = \{1, 3, 2, 0\}$. By applying these two operators comprehensively, eight groups of different rotation reflection states are formed.

The corresponding two sets of time feature sequences are: counter-clockwise rotation $\{1, i, -1, -i\}$, and clockwise rotation $1, -i, -1, i$. The corresponding coordinate transformation can also be distinguished according to this.

$$Y\tilde{Y} = \tilde{Y}Y = \begin{pmatrix} i & -1 & -i & 1 \\ -1 & -i & 1 & i \\ -i & 1 & i & -1 \\ 1 & i & -1 & -i \end{pmatrix}, \overline{Y}\tilde{Y} = \begin{pmatrix} 1 & i & -1 & -i \\ -i & 1 & i & -1 \\ -1 & -i & 1 & i \\ i & -1 & -i & 1 \end{pmatrix}, \tilde{Y}\overline{Y} = \begin{pmatrix} 1 & -i & -1 & i \\ i & 1 & -i & -1 \\ -1 & i & 1 & -i \\ -i & -1 & i & 1 \end{pmatrix}$$

$$\overline{Y}\overline{\tilde{Y}} = \overline{\tilde{Y}}\overline{Y} = \begin{pmatrix} -i & -1 & i & 1 \\ -1 & i & 1 & -i \\ i & 1 & -i & -1 \\ 1 & -i & -1 & i \end{pmatrix}, \tilde{Y}\overline{Y} = \overline{\tilde{Y}}Y = \begin{pmatrix} 1 & -i & -1 & i \\ i & 1 & -i & -1 \\ -1 & i & 1 & -i \\ -i & -1 & i & 1 \end{pmatrix}$$

Four sets of 4×4 product sequences $\{Y, \overline{Y}\} \times \{\tilde{Y}, \overline{\tilde{Y}}\}$

	$\tilde{Y}$	$\overline{\tilde{Y}}$
Y	$Y\tilde{Y}$	$Y\overline{\tilde{Y}}$
$\overline{Y}$	$\overline{Y}\tilde{Y}$	$\overline{Y}\overline{\tilde{Y}}$

=

	$\tilde{Y}$	$\overline{\tilde{Y}}$
Y	$Y\tilde{Y} = \begin{pmatrix} i & -1 & -i & 1 \\ -1 & -i & 1 & i \\ -i & 1 & i & -1 \\ 1 & i & -1 & -i \end{pmatrix}$	$Y\overline{\tilde{Y}} = \begin{pmatrix} 1 & i & -1 & -i \\ -i & 1 & i & -1 \\ -1 & -i & 1 & i \\ i & -1 & -i & 1 \end{pmatrix}$
$\overline{Y}$	$\overline{Y}\tilde{Y} = \begin{pmatrix} 1 & i & -1 & -i \\ -i & 1 & i & -1 \\ -1 & -i & 1 & i \\ i & -1 & -i & 1 \end{pmatrix}$	$\overline{Y}\overline{\tilde{Y}} = \begin{pmatrix} -i & -1 & i & 1 \\ -1 & i & 1 & -i \\ i & 1 & -i & -1 \\ 1 & -i & -1 & i \end{pmatrix}$

The main and auxiliary diagonals are 1, and four sets of 4×4 product sequences $\{\overline{Y}, Y\} \times \{\tilde{Y}, \overline{\tilde{Y}}\}$

	$\tilde{Y}$	$\overline{\tilde{Y}}$
$\overline{Y}$	$\overline{Y}\tilde{Y}$	$\overline{Y}\overline{\tilde{Y}}$
Y	$Y\tilde{Y}$	$Y\overline{\tilde{Y}}$

=

	$\tilde{Y}$	$\overline{\tilde{Y}}$
$\overline{Y}$	$\overline{Y}\tilde{Y}=\begin{pmatrix} 1 & i & -1 & -i \\ -i & 1 & i & -1 \\ -1 & -i & 1 & i \\ i & -1 & -i & 1 \end{pmatrix}$	$\overline{Y}\overline{\tilde{Y}}=\begin{pmatrix} -i & -1 & i & 1 \\ -1 & i & 1 & -i \\ i & 1 & -i & -1 \\ 1 & -i & -1 & i \end{pmatrix}$
Y	$Y\tilde{Y}=\begin{pmatrix} i & -1 & -i & 1 \\ -1 & -i & 1 & i \\ -i & 1 & i & -1 \\ 1 & i & -1 & -i \end{pmatrix}$	$Y\overline{\tilde{Y}}=\begin{pmatrix} 1 & i & -1 & -i \\ -i & 1 & i & -1 \\ -1 & -i & 1 & i \\ i & -1 & -i & 1 \end{pmatrix}$

Four sets of 4×4 product sequences $\{\tilde{Y}, \overline{\tilde{Y}} \times \{Y, \overline{Y}\}\}$

	Y	$\overline{Y}$
$\tilde{Y}$	$\tilde{Y}Y$	$\tilde{Y}\overline{Y}$
$\overline{\tilde{Y}}$	$\overline{\tilde{Y}}Y$	$\overline{\tilde{Y}}\overline{Y}$

=

	Y	$\overline{Y}$
$\tilde{Y}$	$\tilde{Y}Y=\begin{pmatrix} i & -1 & -i & 1 \\ -1 & -i & 1 & i \\ -i & 1 & i & -1 \\ 1 & i & -1 & -i \end{pmatrix}$	$\tilde{Y}\overline{Y}=\begin{pmatrix} 1 & -i & -1 & i \\ i & 1 & -i & -1 \\ -1 & i & 1 & -i \\ -i & -1 & i & 1 \end{pmatrix}$
$\overline{\tilde{Y}}$	$\overline{\tilde{Y}}Y=\begin{pmatrix} 1 & -i & -1 & i \\ i & 1 & -i & -1 \\ -1 & i & 1 & -i \\ -i & -1 & i & 1 \end{pmatrix}$	$\overline{\tilde{Y}}\overline{Y}=\begin{pmatrix} -i & -1 & i & 1 \\ -1 & i & 1 & -i \\ i & 1 & -i & -1 \\ 1 & -i & -1 & i \end{pmatrix}$

The main and auxiliary diagonals are 1, and four sets of 4×4 product sequences $\{\overline{\tilde{Y}}, \tilde{Y}\} \times \{Y, \overline{Y}\}$

	Y	$\overline{Y}$
$\overline{\tilde{Y}}$	$\overline{\tilde{Y}}Y$	$\overline{\tilde{Y}}\overline{Y}$
$\tilde{Y}$	$\tilde{Y}Y$	$\tilde{Y}\overline{Y}$

=

	Y	$\overline{Y}$
$\overline{\tilde{Y}}$	$\overline{\tilde{Y}}Y=\begin{pmatrix} 1 & -i & -1 & i \\ i & 1 & -i & -1 \\ -1 & i & 1 & -i \\ -i & -1 & i & 1 \end{pmatrix}$	$\overline{\tilde{Y}}\overline{Y}=\begin{pmatrix} -i & -1 & i & 1 \\ -1 & i & 1 & -i \\ i & 1 & -i & -1 \\ 1 & -i & -1 & i \end{pmatrix}$
$\tilde{Y}$	$\tilde{Y}Y=\begin{pmatrix} i & -1 & -i & 1 \\ -1 & -i & 1 & i \\ -i & 1 & i & -1 \\ 1 & i & -1 & -i \end{pmatrix}$	$\tilde{Y}\overline{Y}=\begin{pmatrix} 1 & -i & -1 & i \\ i & 1 & -i & -1 \\ -1 & i & 1 & -i \\ -i & -1 & i & 1 \end{pmatrix}$

After the rotation operator is applied, with the parameters changing from $\{0, 2, 3, 1\}$, the four sets of eigenvalues of the main diagonal/subdiagonal are converted to $\{1, i, -1, -i\}, \{1, -i, -1, i\}$ and show two circular projection effects.

$$
{}^{0}\overleftarrow{\overline{Y}}\,\tilde{Y}=\begin{pmatrix} i & -1 & -i & 1 \\ -1 & -i & 1 & i \\ -i & 1 & i & -1 \\ 1 & i & -1 & -i \end{pmatrix}, \quad {}^{2}\overleftarrow{\overline{Y}}\,\tilde{Y}=\begin{pmatrix} -1 & -i & 1 & i \\ -i & 1 & i & -1 \\ 1 & i & -1 & -i \\ i & -1 & -i & 1 \end{pmatrix}
$$

$$
{}^{3}\overleftarrow{\overline{Y}}\,\tilde{Y}=\begin{pmatrix} -i & 1 & i & -1 \\ 1 & i & -1 & -i \\ i & -1 & -i & 1 \\ -1 & -i & 1 & i \end{pmatrix}, \quad {}^{1}\overleftarrow{\overline{Y}}\,\tilde{Y}=\begin{pmatrix} 1 & i & -1 & -i \\ i & -1 & -i & 1 \\ -1 & -i & 1 & i \\ -i & 1 & i & -1 \end{pmatrix}
$$

$$
\overline{Y}^{0}\overleftarrow{\tilde{Y}} = \begin{pmatrix} 1 & i & -1 & -i \\ -i & 1 & i & -1 \\ -1 & -i & 1 & i \\ i & -1 & -i & 1 \end{pmatrix}, \overline{Y}^{2}\overleftarrow{\tilde{Y}} = \begin{pmatrix} i & -1 & -i & 1 \\ 1 & i & -1 & -i \\ -i & 1 & i & -1 \\ -1 & -i & 1 & i \end{pmatrix}
$$

$$
\overline{Y}^{3}\overleftarrow{\tilde{Y}} = \begin{pmatrix} -1 & -i & 1 & i \\ i & -1 & -i & 1 \\ 1 & i & -1 & -i \\ -i & 1 & i & -1 \end{pmatrix}, \overline{Y}^{1}\overleftarrow{\tilde{Y}} = \begin{pmatrix} -i & 1 & i & -1 \\ -1 & -i & 1 & i \\ i & -1 & -i & 1 \\ 1 & i & -1 & -i \end{pmatrix}
$$

$$
Y^{0}\overleftarrow{\tilde{Y}} = \begin{pmatrix} i & -1 & -i & 1 \\ -1 & -i & 1 & i \\ -i & 1 & i & -1 \\ 1 & i & -1 & -i \end{pmatrix}, Y^{2}\overleftarrow{\tilde{Y}} = \begin{pmatrix} -1 & -i & 1 & i \\ -i & 1 & i & -1 \\ 1 & i & -1 & -i \\ i & -1 & -i & 1 \end{pmatrix},
$$

$$
Y^{3}\overleftarrow{\tilde{Y}} = \begin{pmatrix} -i & 1 & i & -1 \\ 1 & i & -1 & -i \\ i & -1 & -i & 1 \\ -1 & -i & 1 & i \end{pmatrix}, Y^{1}\overleftarrow{\tilde{Y}} = \begin{pmatrix} 1 & i & -1 & -i \\ i & -1 & -i & 1 \\ -1 & -i & 1 & i \\ -i & 1 & i & -1 \end{pmatrix}
$$

Theorem 22 (Time Order Restriction of Operators). *The* 4×4 *feature matrix that satisfies the time order in the conjugate algebra representation represents the overall invariant structure, and there are thirty-two groups of* $\{1, i, -1, -i\} \times 4$ *feature matrix representations, involving sixteen groups of different configurations, that can represent one-third of the possible substitutions.*

Proof. In the permutation $\{0, 2, 3, 1\}, \{2, 3, 1, 0\}$ sequence, due to symmetry, the simplified eight groups of matrices contain four groups of distinguishable matrix representations. The main eigenvalue selection range of the matrix is $\{1, i, -1, -i\}$, which contains a total of sixteen groups of distinguishable matrices. $\{0, 1, 2, 3\}$ Arbitrary permutation of elements has twenty-four possibilities, which can represent $8/24 = 1/3$ of the complete set of permutations.

References

[1] E. M. Stein and R. Shakarchi. *Complex Analysis*, Princeton University Press, 2003.

[2] T. Needham. *Visual Complex Analysis*, Oxford University Press, 1997.

[3] L. K. Hua. *Collection of L. K. Hua - Multiple Complex Variables, Vol. II*, Science Press, 2013 (in Chinese).

[4] R. B. Ash and C. A. Doléans-Dade. *Probability and Measure Theory*, Elsevier, 2000.

[5] R. Durrett. *Probability: Theory and Examples*, 3rd edition, Brppks/Cole, 2005.

[6] Y. T. Zhang and K. T. Fang. *Introduction to Multiple Statistical Analysis*, Science Press, 1982 (in Chinese).

[7] J. Gribbin. *In Search of Schrodinger's Cat*, Black Swan, 1984.

[8] S. W. Hawking and L. Mlodinow. *The Grand Design*, Bantam Books International, 2010.

[9] V. I. Arnold. *Methematical Methods of Classical Mechanics*, Springer-Verlag, 1978.

[10] D. Y. Wu. *Theoretical Physics, Vol. I, Classical Dynamics*, Science Press, 2010.

[11] W. R. Hamilton. On a general method in dynamics. Philosophical Transactions of the Royal Society Part II for 1834, 247–308, 1834; Part I for 1835, 95–144, 1835.

[12] W. R. Hamilton. Second essay on a general method in dynamics. Philosophical Transactions of the Royal Society Part I for 1835, 95–144, 1835.

[13] R. Penrose. *The Road to Reality*, Vintage Books, 2005.

[14] H. Rechenbarch. *Philosophic Foundations of Quantum Mechanics*, University of California Press, 1945.
[15] S. C. Lee. *Modern Switching Theory and Digital Design*, Prentice-Hall Inc., 1978.
[16] D. E. Knuth. *The Art of Computer Programming, Vol. 4, Introduction to Combinatorial Algorithms and Boolean Functions*, Addison-Wesley, 2008.
[17] M. A. Nielsen and I. L. Chuang. *Quantum Computing and Quantum Information*, Cambridge University Press and Tsinghua University Press, 2015.
[18] C. E. Shannon. A mathematical theory of communication. *Bell Syst. Tech. J.* 27, 1948, 379–423, 623–656.
[19] G.'t Hooft. *The Cellular Automaton Interpretation of Quantum Mechanics*, Springer, 2016.
[20] S. L. Altmann. *Icons and Symmetries*, Clarendon Press, 1992.
[21] G. Birkhoff. *Lattice Theory*, 3rd edition, The American Mathematical Society, 1967.
[22] L. D. Broglie. *The Current Interpretation of Wave Mechanics*, Elsevier Publishing Company, 1964.
[23] M. G. Calkin. *Lagrangian and Hamilton Mechanics*, World Scientific, 1996.
[24] K. W. Gruenberg and A. J. Weir. *Linear Geometry*, Springer-Verlag, 1967.
[25] W. R. Hamilton. On a general method in dynamics. *Philos. Trans. R. Soc.* 1834, 347–308; in *The Mathematical Papers of Sir William Rowan Hamilton, Vol. 2, Dynamics*, Cambridge University Press, 1940, 103–316.
[26] W. Heisenberg. *The Physical Principles of the Quantum Theory*, Dover Publications, 1930.
[27] M. Jammer. *The Philosophy of Quantum Mechanics, the Interpretations of Quantum Mechanics in Historical Perspective*, John Wiley & Sons, 1974.
[28] D. Kleppner and R. Jackiw. One hundred years of quantum physics. *Science* 289, 2000, 893–898.
[29] G. Ludwig. *Wave Mechanics*, Pergamon Press, 1968.
[30] J. von Neumann. *John von Neumann Collected Works, Vol. 3, Rings of Operators*, Pergamon Press, 1961.
[31] J. von Neumann. *John von Neumann Collected Works, Vol. 4, Continuous Geometry and Other Topics*, Pergamon Press, 1962.
[32] J. Rosen. *Symmetry in Science, An Introduction to the General Theory*, Springer-Verlag, 1995.
[33] E. Schrödinger. *Four Lectures on Wave Mechanics*, Blackie & Son Limited, 1928.
[34] A. Stern. *Matrix Logic*, North-Holland, 1988.
[35] M. Tegmark and J. A. Wheeler. 100 years of quantum mysteries. *Sci. Am.* 284, 2001, 54–61.
[36] D. M. Topping. *Lectures on Von Neumann Algebras*, Van Nostrand Reinhold Company, 1971.
[37] B. L. Van der Waerden. *Group Theory and Quantum Mechanics*, Springer-Verlag, 1974.
[38] C. von Westenholz. *Differential Forms in Mathematical Physics*, North-Holland, 1978.
[39] H. Weyl. *The Theory of Groups and Quantum Mechanics*, Dover Publications Inc., 1930.
[40] S. Wolfram. *Cellular Automata and Complexity, Collected Papers*, Addison Wesley Publishing Company, 1994.
[41] S. Wolfram. *A New Kind of Science*, Wolfram Media, 2002.
[42] A. Ilachinski. *Cellular Automata - A Discrete Universe*, World Scientific, 2001.
[43] M. Kline. *Mathematical Thought from Ancient to Modern Times*, Oxford University Press, 1972.
[44] S. S. Chern. Yang-Mills equations and Yang-Baxter equations. *Chin. J. Phys.* 30(7), 1992, 949–952.
[45] Z. J. Zheng and A. J. Maeder. The conjugate classification of the kernel form of the hexagonal grid. In *Modern Geometric Computing for Visualization*, edited by T. L. Kunii and Y. Shinagawa, Springer-Verlag, 1992, 73–89.
[46] Z. J. Zheng and A. J. Maeder. The elementary equation of the conjugate transformation for hexagonal grid. In *Modoling in Computer Graphics*, edited by B. Falcidieno and T. L. Kunii, Springer-Verlag, 1993, 21–42.
[47] Z. J. Zheng and A. J. Maeder. The reversible phonomena and the conjugate classification, proceedings of 1993 Australian cognitive science conference, 1993, 43–45.
[48] Z. J. Zheng. *Conjugate Transformation of Regular Plane Lattices for Binary Images*, PhD Thesis, Monash University, 1994.
[49] Z. J. Zheng. *Variant construction from Theoretical Foundation to Applications*, Springer Nature, 2019.
[50] 郑智捷, 变值体系理论及其应用 第一卷 理论基础及其应用,科学出版社 2021. Jeffrey Z. J. Zheng. *Vairant Construction Theory and Applications, Vol. 1, Theoretical Foundation and Applications*, Science Press, 2021.

Chapter 3

Conjugate Complex Measurements of Clock Reference Signals

Conjugate Quantum Numbers of Narrow and General Projections on Complex Number Plane

Jeffrey Zheng[1]*, Qi Wang[2], Yujin Lu[2], and Youping Li[3]

[1]Conjugate System Research, Melbourne, Australia

[2]Yunnan University, Kunming, China

[3]Chinese Academy of Engineering Physics, Beijing, China

Abstract

From the studies on the time basis of modern physics, information entropy and measurement play an important role in complex dynamic systems to explore the microscopic and macroscopic effects of the universe and the evolution and variation of time. From the view of conjugate analysis, this chapter uses the clock reference signal to select the $2(m+1)$ bits for any N-bit modulo adder to determine the $2^{2(m+1)}$ states. After the full expansion of the conjugate function space on $2^{2^{m+1}+2}$, through the conjugate orthogonal reduction of the dimensional optimization, the number of distinguishable conjugate functions in the feature space is $2^{2(m+1)}$. Two projections are composed of narrow $2^{2(m+1)}$ and general $4(m+1)^2$. The created sequence has the minimum length and satisfies the optimization characteristics, such as translation invariance and maximum entropy. The typical characteristic of the narrow projection is a uniform distribution, while the typical characteristic probability and statistical distribution of the general projection is a binomial multiplication distribution. The relevant conjugate structure serves as an ideal measure of the distribution of random sequences in classical/quantum cryptography.

Keywords: Time basis, information entropy, satellite clock, adder, measurement window, conjugate partition, exhaustive distribution, binomial distribution, conjugate invariance

---

*Corresponding Author's Email: conjugatelogic@yahoo.com.

In: Conjugate Construction of Quantum Optics

Editors: Jeffrey Zheng

ISBN: 979-8-88697-651-9

Without knowledge Skill cannot be focused Without skill
Strength cannot be brought to bear Without strength
Knowledge may not be applied
– Alexander the Great

Great mathematicians, such as Archimedes, Newton, and Gauss
Always united theory and applications in equal measure
– Felix Klein

Present is saturated with past and pregnant with future
– Gottfried Leibniz

1. Aim of Research

In the exploration of satellite communication of synchronous short messages, one basic question of spacetime research needs to be considered: "Using time-based position and navigation measurements to properly handle the global time synchronous collection, how can time structure guarantee the uniqueness of the measurement under the four-dimensional spacetime coordinate representation?"

The follow-up part of this chapter shows how to use the two-dimensional conjugate measurement on the statistical distributions, from the macro-, meso-, and micro-levels, to establish a series of related research results on the time measuring foundation.

1.1 Global Positioning System

The global positioning system (GPS), also known as the "ball system," is a satellite navigation system on a medium-distance circular orbit that uses navigation satellites for timing and ranging in combination with satellites and communications.

The United States began to develop GPS in the 1970s. It took more than twenty years of research and development, and it was fully completed in 1994. It provides a new generation of satellite navigation and positioning systems to implement all-round three-dimensional navigation and positioning by sea, land, and air.

Several global satellite positioning systems have been used around the world, namely, the American GPS system, the European Union's "Galileo" system, the Russian GLONASS system, and the Chinese "BeiDou" system.

1.2 Time Measurement History

From the historical development of time measurement [1], early humans extracted models of time measurement from long-term observations of the regular periodic motions of the sun, moon, and stars.

For thirty thousand years ago, the records of the existence and development of human beings from the moon phase have indicated the practice of human time measurement.

Early sundials provided basic measuring equipment for measuring the activities of the sun and the moon and provided important support for the establishment of the solar calendar.

In the archaeological evidence of ancient relics, it can be seen that the ancestors of China used oil lamps as early as the 4th century BC to mark the elapsed time from the amount of oil consumed.

On the time measurement, a stick of incense [2] is a commonly used timekeeping unit in ancient China, approximately half an hour. Among the folks, the time period on a stick of incense is widely known. It is common from officials to civilians to use burning incense and candles to calibrate the timing through sticks in competitions and executions. In the 6th century AD, ancient Chinese used candles marked with scales to measure time.

King Alfred of England invented the candle clock, a well-known story in the 10th century AD. Ancient Chinese astronomers developed water clocks as a time-measuring device in the 2nd to 13th century AD. Sand clocks, developed in the 14th century, provided new time-measuring devices. Large-scale observatories built by astronomers from various countries accurately measure the winter solstice, summer solstice, spring equinox, autumn equinox, and other special conditions. To replace the water clock, from the mid-14th century, large mechanical clocks driven by gravity began to appear on the city's high towers. More accurate mechanical clocks appeared in the 17th century, with an error of ten seconds per day.

In the literature [7], the historical development of advanced timing, positioning, and the historical process of precise time measurement are discussed. Clock technology has changed from mechanical clocks $10^{-3}/s$ to quartz clocks $10^{-5}/s$, cesium atomic clocks $10^{-10}/s$, ultrahigh-precision polyatomic clocks array $10^{-12}/s$, and so on, and the accuracy of the satellite mounted clock is much better than $10^{-9}/s$ on nanoseconds.

Although the top clock technology is still far from the natural physical limit—Planck time $10^{-44}/s$, it is accurate enough to lead the world based on a satellite timing accuracy of $10^{-10}/s$. Internet applications, based on the local clock $10^{-6}/s$ technology of quartz clocks, provide basic support for system certification services for various social services to be confirmed in time and place around the world.

1.3 Comprehensive Application—Future Cyberspace Security

From the view of advanced applications, the series of applications are closely related to standard time measurement. Examples include standard timing clock reference sequence, new crownvirus mutation, infection, spread, disaster warning, extreme weather, secure mail, files, data blocks, network information flow, audio, video, remote sensing telemetry, lidar, holographic imaging, classical cryptography, quantum cryptography sequence, secure satellite communication, voice, temperature, ECG, EEG, EMG, vibration, air pressure, fingerprint, iris, face, complex network features, graphics, images, medical biological information, complex dynamic systems, eigenvectors, eigenvalues, and other application scenarios.

1.4 Correlation between Information Entropy and Frontier Time Research

Time is a kind of physical-based meta-parameter. In the universe, the states and fine changes of all kinds of objects from the microscopic to the macroscopic are closely related to the time process. Exploring the basis of time, Rovelli and other researchers were studied on time (past, present, future, macroscopic, microscopic), and a series of related topics such as time entropy and volume entropy [17]–[23].

1.4.1 Irreversibility from Presentism to Minkowski Space

Research from Presentism to Minkowski Space shows that four-dimensional pseudo-Riemannian manifolds are represented as local approximations of physical spacetime. The four-dimensional spacetime itself is only a series of representations of the animation forms of multiple local objects.

However, the irreversibility exhibited by the motion of the universe itself is not an assertion of the microscopic state of the universe. It is an assertion for determining a coarse-grained state: the entropy defined by a set of macroscopically observable sets, with low values in the past. The accompanying question is: What mechanism can determine the observable state of a system on a macroscopic scale?

1.4.2 Trace, Memory, and Agency

The conditions required to study the physical properties of an event trace or memory. Traces of past events are unique because these conditions are mostly met in our universe.

Satisfying (1) system separation, (2) temperature difference, and (3) long thermodynamic time is a sufficient condition for the formation of macroscopic traces.

The thermodynamic conditions of memory lead to an extremely large amount of information stored in memory as a function of the relevant thermodynamic parameters. Such mechanisms convert low entropy into usable information.

Memory contains information. This information must come from somewhere. The only possible source is an increase in entropy, since an increase in entropy corresponds to a loss of information. This mechanism converts the initial low entropy (free energy) into usable information.

The mechanism of the physical agency is explored from three aspects: (1) under what conditions a physical system is an agent, (2) why the agent has time directionality, and (3) the choice of information sources generated in an action.

Proxy and memory structural similarities were observed. In modeling an agent and tracing its events to antisymmetry, it is thermodynamically irreversible to identify the source of information generated by the agent in an entropy-increasing effect. Agents are a class of physical mechanisms capable of converting low entropy into information.

There are possible lineages of measurements to differentiate between different agents: external agents, internal agents, random agents, independent agents, and paranormal agents.

1.4.3 Statistical Discrete Geometry and Discrete Geometric Entropy

Discrete geometry entropy, as the geometric entropy accompanying the surface of a given region, plays an important role in the structure of natural spacetime. Under suitable conditions, such volumetric measurements associated with spatial regions can be calculated by the von Neumann entropy formula.

Statistical discrete geometry applies statistical mechanics to discrete geometry. Averaging and coarse-graining methods are utilized to satisfy two classes of philosophical assumptions: the concepts of atomicity and context independence.

The theory of constraints is to determine the region. In discrete geometry, such constraints make the infinite-dimensional degrees of freedom of the original theory into finite degrees of freedom, and the infinite-dimension is regarded as the limit structure.

The interaction of the two partitions satisfies the theoretical coupling intrinsic curve. It is convenient to discuss the thermodynamic limit to explore the thermodynamic properties of general relativity through the corresponding equation of state.

1.4.4 Statistical Mechanics of Life

In the statistical mechanics of life description, a microstate belongs to a low entropy macrostate maintained in a metastable state. The entropy value does not increase to its maximum value. However, after the channel is opened, the microscopic state can be promoted to high entropy, which satisfies the second law of thermodynamics. Such structures, called temporal long-range associations, have retained information over long periods of time in favor of entropy changes.

1.5 Brief Time History

For the origin and development of time and thermodynamic entropy, Ben-Naim described the beginning of time [5], to use a series of probability distribution examples to judge Rovelli, Hawking, Penrose, and so on, in related works on time evolution, time paradox, entropy description, and core difficulties to show that there is no inherent relationship between entropy and the second law of thermodynamics and time.

However, it is generally believed that there is an arrow of time. The entropy of the ancient world was small, to be related to the order and disorder of the system, and so on, and all lack substantial evidence related to the ontology of time. The relevant discussion is unfounded by the occurrence of the system structure and the measurement of the probability distribution.

Ben-Naim believed that the entropy of a thermodynamic system is related to Shannon's measure of information SMI for the system in equilibrium. Since this type of measurement is inherently related to probability and statistical distribution, in terms of entropy measurements and characteristics in a thermodynamic system, entropy is a deterministic probability of the distribution function. For any measurement under given constraints, it is extreme for the maximum and minimum values to be independent of time ontology properties.

1.6 Conjugate Measurement

For the view of phylogeny and time series measurements, this chapter does not directly address features such as thermodynamic time arrows and general evolution. The research focuses on generating models and measuring methods for time series, exploring the probability distribution by sampling sequences associated with time series, and quantifying laws based on SMI serialized information entropy measurement and merged states.

Based on a small number of initial meta-parameter conditions, the vector logic system is comprehensively used to form multiple phase spaces on variables, states, functions, and its complex representative series, and the conjugate measurement space is formed through the structure of conjugate transformation.

The complete processing process of the conjugate time series analysis framework includes multiple levels from multivariate logical variables and vector features to the multivariate probability of measurement statistics, composite entropy, and structured representation. A broad range of topics covers from basic logic to the underlying support environment for typical measurement applications.

Based on conjugate measurements, this chapter uses standard time sequences as signal sources to conduct structured analysis. From the view of state exhaustion and optimization of a probabilistic system for a standard clock signal, the analysis and projection strategy based on the conjugate principle forms a series of conjugate complex functions of measurement and analysis on the complex number plane for constructing a new type of support platform.

In the following sections, explorations are carried out on related topics, possible extended applications, and other urgently needed frontier research areas.

2. Classical Decomposition

For the convenience of comparison, the typical processing of classical decomposition is briefly summarized.

[Input Vector] → [Logical State Measurement Space] → [Narrow Projection Measurement Space] → [General Projection Visual Space]

The quantities involved in each part are:

[m Linear Vector Input] → [2^m Power Exponent Classical Logic State Space] → [2^m Power Exponent Narrow Projection Measure Space] → [$(m+1)$ Linear General Projection Measure Space]

There is a state space of 2m by exhaustively enumerating all distinguishable states to describe the classical decomposition process for the input m element vector. After converting each distinguishable state into a measure, it is natural to form the same scale narrow projection of the 2^m measure space. From the perspective of measure optimization, measures with the same value can be combined, and the resulting degenerate projection space is $(m+1)$, which constitutes a general projection measure space.

$$m \text{ Vector} \rightarrow \boxed{\text{State Exhaustion}} \rightarrow 2^m \text{ (Logical State Space)}$$

$$\rightarrow \boxed{\text{Deterministic Measure}} \rightarrow 2^m \begin{pmatrix} \text{Narrow Projection} \\ \text{Measure Space} \end{pmatrix}$$

$\rightarrow$ [Degenerate Dimension Reduction] $\rightarrow$ $m+1$ (General Measure Space Projected)

Lemma 1 (Decomposition of Classic Measurement Process). *The classical decomposition process involves four core parts, namely, the linear input variable m, the classical logic state space 2^m, the narrow projection measure space 2^m, and the general projection measure space represented by the binomial $m+1$.*

3. Data Measurement Flow

For the conjugate measurement transformation, starting from the built-in time pulse signal, a meta-data sequence is formed through a long-period accumulator conversion, sampling to form a meta-sequence, and then through exhaustive traversal to form the overall processing flow of the two variable value graphs in Fig. 1 shown in (a), which contains three core processing modules: data sequence flow, sampling sequence, and observing window.

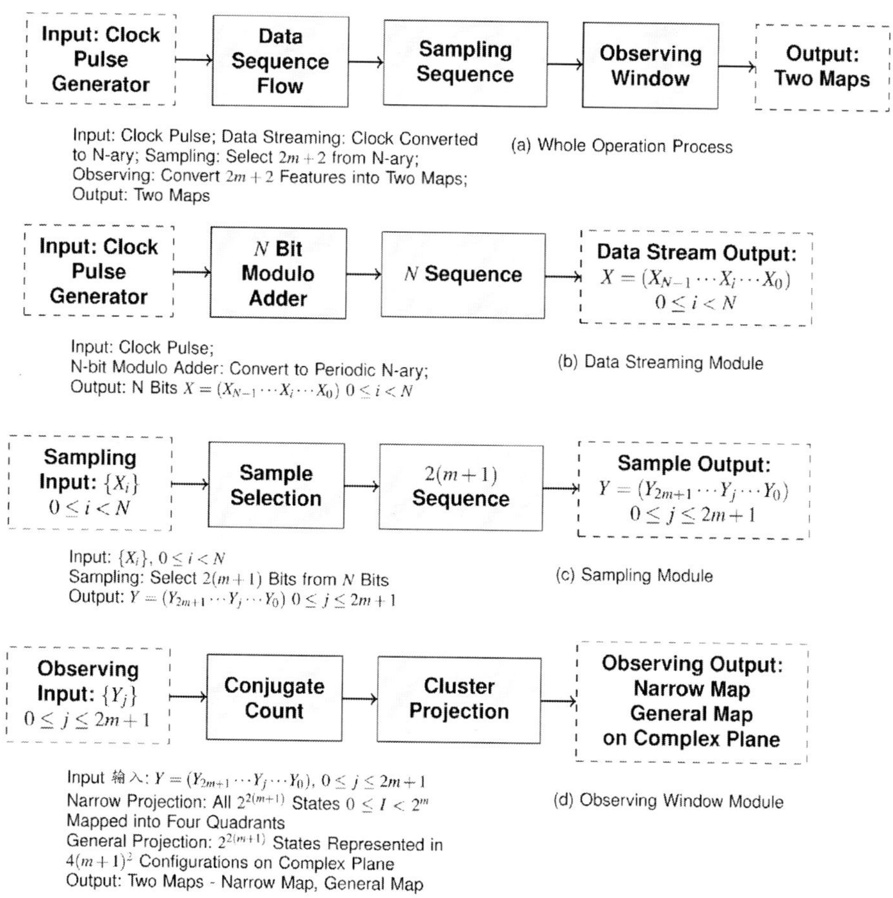

Figure 1. Clock sequence conjugate visual generation system: (a) whole operational flow; (b) data flow module; (c) sampling module; (d) observing window module.

Clock Pulse Signal	Specified Time Pulse Information Source
Data Stream	Periodically Collected Data Sequence
Sampling	Selecting Relevant Sub-Sequences
Observing Window	Selected Sequences for Specific Visual Modes
Conjugate Maps	Two Maps for Narrow and General Projections

4. Three Core Modules

There are three core processing modules: data stream, sampling, and observing window. They are shown in Figs. 1 (b)–(d), respectively.

4.1 Data Stream Module

In Fig. 1 (b), the data stream module contains two submodules: {N Bit Modulo Adder, N Sequence}. Obtaining ultrahigh-precision clock pulses from the input clock pulse sequence generator, the N bit modulo adder converts the clock pulse sequence into 2^N signal sequences whose states are periodic changes. The signal sequence is calculated by the N bit vector pattern, forming the output sequence.

4.2 Sampling Module

In Fig. 1 (c), the sampling module contains two submodules: {Sample Selection, $2(m+1)$ Sequence}. For the input sequence of N, select the adjacent $2(m+1)$ sequence as the output to form a suitable $2(m+1)$ vector.

4.3 Observing Window Module

In Fig. 1 (d), the observing window module contains two submodules: {Conjugate Count, Cluster Projection}. For a suitable $2^{2(m+1)}$ sampling sequence, the exhaustive computation submodule converts each representation into a conjugate integer value in the range of $I \in [0, 2^{2(m+1)} - 1]$ interval to form a 2D histogram distribution.

$$I = \sum_{k=0}^{2m+1} Y_k \times 2^k,\ Y_k \in \{0, 1\}, 0 \le k \le 2m+1, 0 \le I < 2^{2(m+1)} \tag{1}$$

For the clustering projection submodule, the number of 1s is selected in the statistical sampling vector for clustering, to form a histogram distribution in the range of $[0, 2^{2(m+1)} - 1]$.

5. Conjugate Measurement

Using the selected discriminant bits to form two sets of conjugate state clusters through the observing window module, a pairwise representation of the results is generated for the built-in clock sequence conjugate map.

The typical processing process is divided into the following:

Input Vector: $2m$ → Conjugate Logic State Space : 2^{2m} → Conjugate Logic Function Space : $2^{2^{m+1}}$

→ Narrow Conjugate Measure Space : 2^{2m} → General Conjugate Measure Space on Complex Plane : $(m+1)^2$

Processing the 2^{2m} exponential states collected from the selected $2m$ bit 0–1 vector, the given positions of $\{0, 1\}$ are collected in a set of 2^m state features respectively to form an orthogonal function represented by the conjugate logic function space of $2^{2^{2m}}$ on the basic logic state space. Refer to the Conjugate Transform Structures section of [40] and [42, Ch. 4] for details on related conjugate orthogonal transforms.

Two groups of conjugation means that the total number of a state space is $2^m \times 2^m = 2^{2m}$, and the total number of a conjugate function space is $2^{2^m} \times 2^{2^m} = 2^{2^{m+1}}$. Under conjugate transformation, conjugate state eigenvectors have orthogonal properties to represent all functions containing more than two similar states to be trivial representations.

From the initial $2^{2^{m+1}}$, under the orthogonal condition of conjugate 2^m, the independent states and actions themselves remain unchanged. After reducing operations, the dimensionality of the decreased function space is $2^m \times 2^m = 2^{2m}$ to form a narrow projection measure function space as the group of nontrivial orthogonal measures.

Using the degenerate mode of clustering the same measure, the total number of 2^{2m} nontrivial measures as the exponential mode is represented, which can be further reduced to a square term growth mode to be arranged in a square matrix of $(m+1)^2$ terms. This provides the general projection of the conjugate measure space on the complex number plane.

Conjugate decomposition and dimensionality reduction process schematic:

$2m$ Linear Vector Input → 2^{2m} Power Exponent Conjugate Logic State Space → $2^{2^{m+1}}$ Superpower Exponents Total Conjugate Logic Function Space Orthogonal Transform Dimensionality Reduction

→ 2^{2m} Power Exponent Narrow Measure Space Degenerate Reduction → $(m+1)^2$ Square Function General Measure Space on Complex Quadrant

$2m$ Vector → State Decomposition → 2^{2m} (Basic Logic State Space)

→ Conjugate Decomposition → $\{2^m\} : \{2^m\}$ (Conjugate Logic State Space)

$$\rightarrow \boxed{\begin{array}{c}\text{Complete}\\ \text{Function}\end{array}} \rightarrow 2^{2^{m+1}} \begin{pmatrix}\text{Full Conjugate Logic}\\ \text{Function Space}\end{pmatrix}$$

$$\rightarrow \boxed{\begin{array}{c}\text{Orthogonal}\\ \text{Reduction}\end{array}} \rightarrow 2^{2m}\ (\text{Narrow Measure Space})$$

$$\rightarrow \boxed{\begin{array}{c}\text{Degenerate}\\ \text{Reduction}\end{array}} \rightarrow (m+1)^2 \begin{pmatrix}\text{General Measure Space}\\ \text{on Complex Plane}\end{pmatrix}$$

Corollary 1 (Conjugate Measurement Process Decomposition). *The conjugate decomposition process involves five core parts, namely, the linear input variables* $2m$, *the conjugate logic state space* 2^{2m}, *the full conjugate logic function space* $2^{2^{m+1}}$, *the narrow projection measure space* 2^{2m}, *and the general projection measure space* $(m+1)^2$ *expressed by the dual-binomial.*

Proof. There is a natural procedure for a single block through multiple levels of hierarchy with a clear correspondence.

In the conjugate structure, each structure contains four independent spaces. For any state variable x, $\{\sim, \prime\}$, two groups of operators, conjugate and complementary, there are four distinguishable configurations to represent $\{x, \tilde{x}, x', \tilde{x}'\}$.

$$2(m+1)\text{Base Vector}$$

$$\rightarrow \boxed{\text{State Decomposition}} \rightarrow 2^{2(m+1)}\ (\text{Basic Logic State Space})$$

$$\rightarrow \boxed{\text{Conjugate Decomposition}} \rightarrow \{2^{m+1}\} : \{2^{m+1}\}\ (\text{Conjugate Logic State Space})$$

$$\rightarrow \boxed{\text{Complete Function}} \rightarrow 2^{2^{m+1}+2}\ (\text{Conjugate Logic Function Space})$$

$$\rightarrow \boxed{\text{Orthogonal Reduction}} \rightarrow 2^{2(m+1)}\ (\text{Narrow Measure Space})$$

$$\rightarrow \boxed{\text{Degenerate Reduction}} \rightarrow 4(m+1)^2\ \begin{array}{c}(\text{General Measure Space}\\ \text{on Complex Plane})\end{array}$$

Corollary 2 (Conjugate Complementary Measurement Process Decomposition). *The conjugate + complementary decomposition process involves five core parts, which are the linear input variable* $2(m+1)$, *the conjugate logic state space* $2^{2(m+1)}$, *the full conjugate logic function space* $2^{2^{m+1}+2}$, *the narrow projection measure space* $2^{2(m+1)}$, *and the general projection measure space* $4(m+1)^2$ *represented by the dual-binomial.*

Proof. There is a natural procedure to expand four blocks into four regions with a clear measurement.

6. Conjugate State Distribution Features

Comparing the basic and extended representation, the number of input variables is increased from $2m$ to $2(m+1)$, and the narrow and general projections on space output scales are increased by a factor of 4. The resulting structures are described in subsequent sections.

6.1 *Basic Distribution*

For any state $x, \tilde{x}, x', \tilde{x}'$, the corresponding number of connections is given by $s, \tilde{s}, s', \tilde{s}'$. The reference position is k; $i, k \in [0, m]$, and the transformation value is determined by the following formula.

$$s(x) = \sum_{\forall i \neq k} (x_i == 1)\&(x_k == 1) \equiv s \tag{2}$$

$$s(x') = -s(x) \equiv s' \tag{3}$$

$$s(\tilde{x}) = \sum_{\forall i \neq k} (x_i == 0)\&(x_k == 0) \equiv \tilde{s} \tag{4}$$

$$s(\tilde{x}') = -s(\tilde{x}) \equiv \tilde{s}' \tag{5}$$

Single state $\{-1, 1\} \in \{s, s'\}, \{-i, i\} \in \{\tilde{s}, \tilde{s}'\}$, i imaginary number $i = \sqrt{-1}$.

$(s, s' : \tilde{s}, \tilde{s}')$	s'	s
$\tilde{s}$	$(s', \tilde{s})$	$(s, \tilde{s})$
$\tilde{s}'$	$(s', \tilde{s}')$	$(s, \tilde{s}')$

6.2 *Compound Distribution*

From the viewpoint of conjugate representation, the logic function space needs to enumerate all possible state combinations. The total number of positions involved is the superpower exponent $4 \times 2^{2^m} \times 2^{2^m} = 2^{2^{m+1}+2}$. Each block contains $2^{2^{m+1}}$ configurations.

Conjugate Logic Function Space	$[2^{2^m} - 1]'$	...	$0'$	0	...	$[2^{2^m} - 1]$
$\tilde{0}$	$([2^{2^m} - 1]', \tilde{0})$	...	$(0', \tilde{0})$	$(0, \tilde{0})$	...	$([2^{2^m} - 1], \tilde{0})$
...	...		...	...		...
$[\widetilde{2^{2^m} - 1}]$	$([2^{2^m} - 1]', [\widetilde{2^{2^m} - 1}])$	...	$(0', [\widetilde{2^{2^m} - 1}])$	$(0, [\widetilde{2^{2^m} - 1}])$	...	$([2^{2^m} - 1], [\widetilde{2^{2^m} - 1}])$
$[\widetilde{2^{2^m} - 1}]'$	$([2^{2^m} - 1]', [\widetilde{2^{2^m} - 1}]')$	...	$(0', [\widetilde{2^{2^m} - 1}]')$	$(0, [\widetilde{2^{2^m} - 1}]')$	...	$([2^{2^m} - 1], [\widetilde{2^{2^m} - 1}]')$
...	...		...	...		...
$\tilde{0}'$	$([2^{2^m} - 1]', \tilde{0}')$	...	$(0', \tilde{0}')$	$(0, \tilde{0}')$	...	$([2^{2^m} - 1], \tilde{0}')$

Under the orthogonal transformation on the conjugate logic vector, all functions with more than two similar state combinations are transformed into trivial representations in the measurement projection, and all trivial representations can be removed. From the initial superpower exponential expression, each conjugate logic function block contains a total number of $2^{2^{m+1}}$ configurations, and a large dimensionality reduction forms a nontrivial measure configuration of the power exponent 2^{2m} group to be a narrow projection of the conjugate measurement space.

From the view of representation, the narrow projection exhausts all possible combinations of single-state sequence pairs. The total number of positions involved is the power exponent $4 \times 2^{2m} = 2^{2(m+1)}$. Each block contains 2^{2m} configurations.

Narrow Projection	$[2^m-1]'$	...	$0'$	0	...	$[2^m-1]$
$\tilde{0}$	$([2^m-1]',\tilde{0})$	...	$(0',\tilde{0})$	$(0,\tilde{0})$	...	$([2^m-1],\tilde{0})$
...	...		...	...		...
$\widetilde{[2^m-1]}$	$([2^m-1]',\widetilde{[2^m-1]})$	...	$(0',\widetilde{[2^m-1]})$	$(0,\widetilde{[2^m-1]})$	...	$([2^m-1],\widetilde{[2^m-1]})$
$\widetilde{[2^m-1]}'$	$([2^m-1]',\widetilde{[2^m-1]}')$	...	$(0',\widetilde{[2^m-1]}')$	$(0,\widetilde{[2^m-1]}')$	...	$([2^m-1],\widetilde{[2^m-1]}')$
...	...		...	...		...
$\tilde{0}'$	$([2^m-1]',\tilde{0}')$	...	$(0',\tilde{0}')$	$(0,\tilde{0}')$	...	$([2^m-1],\tilde{0}')$

Theorem 1 (Conjugate Numbers in One Block of Narrow Distribution). *In the conjugate decomposition process, one block of the narrow distribution contains 2^{2m} pairs of complex numbers of a $[0,2^m)^2$ region.*

Proof. Checking the narrow projection table of the upright block, there are 1–1 corresponding links for all 2^{2m} entries.

Theorem 2 (Conjugate Numbers in Four Block of Narrow Distribution). *In the conjugate decomposition process, four blocks of narrow distribution contain $2^{2(m+1)}$ pairs of complex numbers of four $[0,2^m)^2$ regions.*

Proof. Checking the narrow projection table of four blocks, there are 1–1 corresponding links for all $2^{2(m+1)}$ entries.

Corollary 3 (Number of Elements on Narrow Distribution). *In a complete narrow distribution, each entry contains one element without any degeneration.*

Proof. There are 1-1 corresponding links for all entries of 2^{2m} (one block) or $2^{2(m+1)}$ (four blocks).

In the degenerated item mode, the total number is $2^{2(m+1)}$ as the power exponential configuration. The dimensionality is greatly reduced through clustering and merging, from the power exponential growth mode to the general projection of the measurement space on $4\times(m+1)^2$ configurations.

General Projection $(s,s':\tilde{s},\tilde{s}')$	$-m$	...	-0	0	...	m
$\tilde{m}$	$(-m,\tilde{m})$	...	$(-0,\tilde{m})$	$(0,\tilde{m})$	...	$(m,\tilde{m})$
...	...		...	...		...
$\tilde{0}$	$(-m,\tilde{0})$	...	$(-0,\tilde{0})$	$(0,\tilde{0})$	...	$(m,\tilde{0})$
$-\tilde{0}$	$(-m,-\tilde{0})$	...	$(-0,-\tilde{0})$	$(0,-\tilde{0})$	...	$(m,-\tilde{m})$
...	...		...	...		...
$-\tilde{m}$	$(-m,-\tilde{m})$	...	$(-0,-\tilde{m})$	$(0,-\tilde{m})$	...	$(m,-\tilde{m})$

For row i, column j configuration, the number of collected states is $\binom{m}{i}\times\binom{m}{j}$.

Number of Degenerate States $(s, \tilde{s})$	$\binom{m}{0}$ ... $\binom{m}{i}$ ... $\binom{m}{m}$
$\binom{m}{0}$	$\binom{m}{0}\binom{m}{0}$ ... $\binom{m}{i}\binom{m}{0}$ ... $\binom{m}{m}\binom{m}{0}$
...	... ...
$\binom{m}{j}$	$\binom{m}{0}\binom{m}{j}$ ... $\binom{m}{i}\binom{m}{j}$ ... $\binom{m}{m}\binom{m}{j}$
...	... ...
$\binom{m}{m}$	$\binom{m}{0}\binom{m}{m}$ ... $\binom{m}{i}\binom{m}{m}$ ... $\binom{m}{m}\binom{m}{m}$

Theorem 3 (Conjugate Numbers in One Block of General Distribution). *One block of the general distribution contains $(m+1)^2$ pairs of complex numbers as complex basis positions in a partial complex plane of a $[0, m]^2$ region.*

Proof. Checking the general projection table of the upright block, there are 1–1 corresponding links for all $(m+1)^2$ positions. Only the first quadrant is involved.

Theorem 4 (Conjugate Numbers in Four Block of General Distribution). *Four blocks of general distribution contain $(m+1)^2$ pairs of complex numbers as complex basis positions in a complete complex plane of a $[-m, m]^2$ region.*

Proof. Checking the general projection table of four blocks, there are 1–1 corresponding links for all $4 \times (m+1)^2$ positions.

Corollary 4 (Number of Elements on General Distribution). *In a complete general distribution, each i,j position is linked to a coefficient of $\binom{m}{i}\binom{m}{j}$ degenerate states, and the total numbers of degenerate states are 2^{2m} (one block) and $2^{2(m+1)}$ (four blocks).*

Proof. In the cases of one block and four blocks, there are 2^{2m} and $2^{2(m+1)}$ total numbers.

$$\sum_{i=0}^{m}\sum_{j=0}^{m}\binom{m}{i}\binom{m}{j} = \left[\sum_{i=0}^{m}\binom{m}{i}\right]\left[\sum_{j=0}^{m}\binom{m}{j}\right]$$

$$= 2^m \times 2^m$$

$$= 2^{2m} : \text{For one block}$$

$$4\sum_{i=0}^{m}\sum_{j=0}^{m}\binom{m}{i}\binom{m}{j} = 4 \times 2^{2m}$$

$$= 2^2 \times 2^{2m} = 2^{2+2m}$$

$$= 2^{2(m+1)} : \text{For four blocks.}$$

7. Cases $m = \{0, 1, 2, 3\}$

7.1 Case $m = 0$

x	s	$\tilde{s}$	s'	$\tilde{s}'$
0		1		-1
1	1		-1	

Pair States	$1'$	1
$\tilde{0}$	$(1', \tilde{0})$	$(1, \tilde{0})$
$\tilde{0}'$	$(1', \tilde{0}')$	$(1, \tilde{0}')$

→

Pair Values	$1'$	1
$\tilde{0}$	$(-1, 1)$	$(1, 1)$
$\tilde{0}'$	$(-1, -1)$	$(1, -1)$

7.2 *Case* $m = 1$

$s^0 =$

$\downarrow$x n	s	$\tilde{s}$	s'	$\tilde{s}'$
00 0		1		-1
01 1		0		-0
10 2	0		-0	
11 3	1		-1	

; $s^1 =$

x$\downarrow$ n	s	$\tilde{s}$	s'	$\tilde{s}'$
00 0		1		-1
01 1	0		-0	
10 2		0		-0
11 3	1		-1	

; $s^0 + s^1 =$

$s^0 + s^1$	s	$\tilde{s}$	s'	$\tilde{s}'$
00 0		2		-2
01 1	0	0	-0	-0
10 2	0	0	-0	-0
11 3	2		-2	

Pair States	$1'$	$0'$	0	1
$\tilde{0}$	$(1', \tilde{0})$	$(0', \tilde{0})$	$(0, \tilde{0})$	$(1, \tilde{0})$
$\tilde{1}$	$(1', \tilde{1})$	$(0', \tilde{1})$	$(0, \tilde{1})$	$(1, \tilde{1})$
$\tilde{1}'$	$(1', \tilde{1}')$	$(0', \tilde{1}')$	$(0, \tilde{1}')$	$(1, \tilde{1}')$
$\tilde{0}'$	$(1', \tilde{0}')$	$(0', \tilde{0}')$	$(0, \tilde{0}')$	$(1, \tilde{0}')$

$\rightarrow$

Pair Values	$1'$	$0'$	0	1
$\tilde{0}$	$(-1, 1)$	$(-0, 1)$	$(0, 1)$	$(1, 1)$
$\tilde{1}$	$(-1, 0)$	$(-0, 0)$	$(0, 0)$	$(1, 0)$
$\tilde{1}'$	$(-1, -0)$	$(-0, -0)$	$(0, -0)$	$(1, -0)$
$\tilde{0}'$	$(-1, -1)$	$(-0, -1)$	$(0, -1)$	$(1, -1)$

$\rightarrow$

Merged Values	$1'$	$\{0', 0\}$	1
$\tilde{0}$	$(-1, 1)$	$2 \times (0, 1)$	$(1, 1)$
$\{\tilde{1}, \tilde{1}'\}$	$2 \times (-1, 0)$	$4 \times (0, 0)$	$2 \times (1, 0)$
$\tilde{0}'$	$(-1, -1)$	$2 \times (0, -1)$	$(1, -1)$

$\rightarrow$

Collected Number	$1'$	$\{0', 0\}$	1
$\tilde{0}$	1	2	1
$\{\tilde{1}, \tilde{1}'\}$	2	4	2
$\tilde{0}'$	1	2	1

7.3 *Case* $m = 2$

Pair Values $(s, \tilde{s})$	00	01	10	11
$\widetilde{00}$	$(0, 2)$	$(1, 2)$	$(1, 2)$	$(2, 2)$
$\widetilde{01}$	$(0, 1)$	$(1, 1)$	$(1, 1)$	$(2, 1)$
$\widetilde{10}$	$(0, 1)$	$(1, 1)$	$(1, 1)$	$(2, 1)$
$\widetilde{11}$	$(0, 0)$	$(1, 0)$	$(1, 0)$	$(2, 0)$

$\rightarrow$

Merged Values	00	$\{01, 10\}$	11
$\widetilde{00}$	$(0, 2)$	$2 \times (1, 2)$	$(2, 2)$
$\{\widetilde{01}, \widetilde{10}\}$	$2 \times (0, 1)$	$4 \times (1, 1)$	$2 \times (2, 1)$
$\widetilde{11}$	$(0, 0)$	$2 \times (1, 0)$	$(2, 0)$

$\rightarrow$

Collected Number	00	$\{01, 10\}$	11
$\widetilde{00}$	1	2	1
$\{\widetilde{01}, \widetilde{10}\}$	2	4	2
$\widetilde{11}$	1	2	1

$\rightarrow$

Collected Values	$11'$	$\{01', 10'\}$	$\{00', 00\}$	$\{01, 10\}$	11
$\widetilde{00}$	$(-2, 2)$	$2 \times (-1, 2)$	$2 \times (0, 2)$	$2 \times (1, 2)$	$(2, 2)$
$\{\widetilde{01}, \widetilde{10}\}$	$2 \times (-2, 1)$	$4 \times (-1, 1)$	$4 \times (0, 1)$	$4 \times (1, 1)$	$2 \times (2, 1)$
$\{\widetilde{11}, \widetilde{11}'\}$	$2 \times (-2, 0)$	$4 \times (-1, 0)$	$4 \times (0, 0)$	$4 \times (1, 0)$	$2 \times (2, 0)$
$\{\widetilde{01}', \widetilde{10}'\}$	$2 \times (-2, -1)$	$4 \times (-1, -1)$	$4 \times (0, -1)$	$4 \times (1, -1)$	$2 \times (2, -1)$
$\{\widetilde{00}'\}$	$(-2, -2)$	$2 \times (-1, -2)$	$2 \times (0, -2)$	$2 \times (1, -2)$	$(2, -2)$

Merged Coefficient Matrix $(5 \times 5) =$

1	2	2	2	1
2	4	4	4	2
2	4	4	4	2
2	4	4	4	2
1	2	2	2	1

7.4 *Case m* = 3

$(s, \tilde{s})$ Pair Values	000	001	010	100	011	101	110	111
$\widetilde{000}$	$(0,3)$	$(1,3)$	$(1,3)$	$(1,3)$	$(2,3)$	$(2,3)$	$(2,3)$	$(3,3)$
$\widetilde{001}$	$(0,2)$	$(1,2)$	$(1,2)$	$(1,2)$	$(2,2)$	$(2,2)$	$(2,2)$	$(3,2)$
$\widetilde{010}$	$(0,2)$	$(1,2)$	$(1,2)$	$(1,2)$	$(2,2)$	$(2,2)$	$(2,2)$	$(3,2)$
$\widetilde{100}$	$(0,2)$	$(1,2)$	$(1,2)$	$(1,2)$	$(2,2)$	$(2,2)$	$(2,2)$	$(3,2)$
$\widetilde{011}$	$(0,1)$	$(1,1)$	$(1,1)$	$(1,1)$	$(2,1)$	$(2,1)$	$(2,1)$	$(3,1)$
$\widetilde{101}$	$(0,1)$	$(1,1)$	$(1,1)$	$(1,1)$	$(2,1)$	$(2,1)$	$(2,1)$	$(3,1)$
$\widetilde{110}$	$(0,1)$	$(1,1)$	$(1,1)$	$(1,1)$	$(2,1)$	$(2,1)$	$(2,1)$	$(3,1)$
$\widetilde{111}$	$(0,0)$	$(1,0)$	$(1,0)$	$(1,0)$	$(2,0)$	$(2,0)$	$(2,0)$	$(3,0)$

$\rightarrow$

Merged Values	000	{001, 010, 100}	{011, 110, 101}	111
$\widetilde{000}$	$(0,3)$	$3 \times (1,3)$	$3 \times (2,3)$	$(3,3)$
$\{\widetilde{001}, \widetilde{010}, \widetilde{100}\}$	$3 \times (0,2)$	$9 \times (1,2)$	$9 \times (2,2)$	$3 \times (3,2)$
$\{\widetilde{011}, \widetilde{110}, \widetilde{101}\}$	$3 \times (0,1)$	$9 \times (1,1)$	$9 \times (2,1)$	$3 \times (3,1)$
$\widetilde{111}$	$(0,0)$	$3 \times (1,0)$	$3 \times (2,0)$	$(2,0)$

Merged Coefficient Matrix $(7 \times 7) =$

1	3	3	2	3	3	1
3	9	9	6	9	9	3
3	9	9	6	9	9	3
2	6	6	4	6	6	2
3	9	9	6	9	9	3
3	9	9	6	9	9	3
1	3	3	2	3	3	1

7.5 *Cases m* = {4, 5, 6, 7, 8}

$m = 4$ Merged Coefficient Matrix $(10 \times 10) =$

1	4	6	4	1	1	4	6	4	1
4	16	24	16	4	4	16	24	16	4
6	24	36	24	6	6	24	36	24	6
4	16	24	16	4	4	16	24	16	4
1	4	6	4	1	1	4	6	4	1
1	4	6	4	1	1	4	6	4	1
4	16	24	16	4	4	16	24	16	4
6	24	36	24	6	6	24	36	24	6
4	16	24	16	4	4	16	24	16	4
1	4	6	4	1	1	4	6	4	1

$m = 5$ Merged Coefficient Matrix (12×12) =

1	5	10	10	5	1	1	5	10	10	5	1
5	25	50	50	25	5	5	25	50	50	25	5
10	50	100	100	50	10	10	50	100	100	50	10
10	50	100	100	50	10	10	50	100	100	50	10
5	25	50	50	25	5	5	25	50	50	25	5
1	5	10	10	5	1	1	5	10	10	5	1
1	5	10	10	5	1	1	5	10	10	5	1
5	25	50	50	25	5	5	25	50	50	25	5
10	50	100	100	50	10	10	50	100	100	50	10
10	50	100	100	50	10	10	50	100	100	50	10
5	25	50	50	25	5	5	25	50	50	25	5
1	5	10	10	5	1	1	5	10	10	5	1

$m = 6$ Merged Coefficient Matrix (14×14) =

1	6	15	20	15	6	1	1	6	15	20	15	6	1
6	36	90	120	90	36	6	6	36	90	120	90	36	6
15	90	225	300	225	90	15	15	90	225	300	225	90	15
20	120	300	400	300	120	20	20	120	300	400	300	120	20
15	90	225	300	225	90	15	15	90	225	300	225	90	15
6	36	90	120	90	36	6	6	36	90	120	90	36	6
1	6	15	20	15	6	1	1	6	15	20	15	6	1
1	6	15	20	15	6	1	1	6	15	20	15	6	1
6	36	90	120	90	36	6	6	36	90	120	90	36	6
15	90	225	300	225	90	15	15	90	225	300	225	90	15
20	120	300	400	300	120	20	20	120	300	400	300	120	20
15	90	225	300	225	90	15	15	90	225	300	225	90	15
6	36	90	120	90	36	6	6	36	90	120	90	36	6
1	6	15	20	15	6	1	1	6	15	20	15	6	1

$m = 7$ Merged Coefficient Matrix (16×16) =

1	7	21	35	35	21	7	1	1	7	21	35	35	21	7	1
7	49	147	245	245	147	49	7	7	49	147	245	245	147	49	7
21	147	441	735	735	441	147	21	21	147	441	735	735	441	147	21
35	245	735	1225	1225	737	245	35	35	245	735	1225	1225	737	245	35
35	245	735	1225	1225	737	245	35	35	245	735	1225	1225	737	245	35
21	147	441	735	735	441	147	21	21	147	441	735	735	441	147	21
7	49	147	245	245	147	49	7	7	49	147	245	245	147	49	7
1	7	21	35	35	21	7	1	1	7	21	35	35	21	7	1
1	7	21	35	35	21	7	1	1	7	21	35	35	21	7	1
7	49	147	245	245	147	49	7	7	49	147	245	245	147	49	7
21	147	441	735	735	441	147	21	21	147	441	735	735	441	147	21
35	245	735	1225	1225	737	245	35	35	245	735	1225	1225	737	245	35
35	245	735	1225	1225	737	245	35	35	245	735	1225	1225	737	245	35
21	147	441	735	735	441	147	21	21	147	441	735	735	441	147	21
7	49	147	245	245	147	49	7	7	49	147	245	245	147	49	7
1	7	21	35	35	21	7	1	1	7	21	35	35	21	7	1

$m = 8$ Merged Coefficient Matrix (18×18) =

1	8	28	56	70	56	28	8	1	1	8	28	56	70	56	28	8	1
8	64	224	448	560	448	224	64	8	8	64	224	448	560	448	224	64	8
28	224	784	1568	1960	1568	784	224	28	28	224	784	1568	1960	1568	784	224	28
56	448	1568	3136	3920	3136	1568	448	56	56	448	1568	3136	3920	3136	1568	448	56
70	560	1960	3920	4900	3920	1960	560	70	70	560	1960	3920	4900	3920	1960	560	70
56	448	1568	3136	3920	3136	1568	448	56	56	448	1568	3136	3920	3136	1568	448	56
28	224	784	1568	1960	1568	784	224	28	28	224	784	1568	1960	1568	784	224	28
8	64	224	448	560	448	224	64	8	8	64	224	448	560	448	224	64	8
1	8	28	56	70	56	28	8	1	1	8	28	56	70	56	28	8	1
1	8	28	56	70	56	28	8	1	1	8	28	56	70	56	28	8	1
8	64	224	448	560	448	224	64	8	8	64	224	448	560	448	224	64	8
28	224	784	1568	1960	1568	784	224	28	28	224	784	1568	1960	1568	784	224	28
56	448	1568	3136	3920	3136	1568	448	56	56	448	1568	3136	3920	3136	1568	448	56
70	560	1960	3920	4900	3920	1960	560	70	70	560	1960	3920	4900	3920	1960	560	70
56	448	1568	3136	3920	3136	1568	448	56	56	448	1568	3136	3920	3136	1568	448	56
28	224	784	1568	1960	1568	784	224	28	28	224	784	1568	1960	1568	784	224	28
8	64	224	448	560	448	224	64	8	8	64	224	448	560	448	224	64	8
1	8	28	56	70	56	28	8	1	1	8	28	56	70	56	28	8	1

8. Narrow and General Projections

8.1 Associated Distributions

Examining the narrow and general distribution maps, two sets of distinguishable correlated states of representation vectors are distributed in the main diagonal and subdiagonal regions of the matrix, respectively, $\{(I,\tilde{I}),(I',\tilde{I}')\}, 0 \leq I < 2^m$ and $\{(I',\tilde{I}),(I,\tilde{I}')\}$, $0 \leq I < 2^m$.

8.2 Narrow Distribution

Two sets of vector lengths are 2^{m+1}. The narrow distributions are denoted as $S_E^a(m)$ and $A_E^a(m)$, respectively.

$$S_E^a(m) = [(0,\tilde{0}),\cdots,(I,\tilde{I}),\cdots,(2^m-1,\widetilde{2^m-1});(0',\tilde{0}'),\cdots,(I',\tilde{I}'),\cdots,([2^m-1]',\widetilde{[2^m-1]}')]$$

$$A_E^a(m) = [(0',\tilde{0}),\cdots,(I',\tilde{I}),\cdots,([2^m-1]',\widetilde{[2^m-1]});(0,\tilde{0}'),\cdots,(I,\tilde{I}'),\cdots,(2^m-1,\widetilde{[2^m-1]}')]$$

$$|S_E^a(m)| = 2 \times \sum_{j=0}^{2^m-1} 1 = 2^{m+1} \tag{6}$$

$$|A_E^a(m)| = 2 \times \sum_{j=0}^{2^m-1} 1 = 2^{m+1} \tag{7}$$

In the narrow distribution, each distinguishable state appears only once, the corresponding correlated state appears only once in the system, and there are no maximum and minimum distributions.

$$|S^{a}_{Ex}(m)| = 2 \tag{8}$$

$$|A^{a}_{Ex}(m)| = 2. \tag{9}$$

8.3 General Distribution

Two vector lengths are $2 \times (m+1)$. The general distributions are denoted as $S^{b}_{E}(m)$ and $A^{b}_{E}(m)$, respectively.

$$S^{b}_{E}(m) = [(\mathbf{0}, \tilde{\mathbf{0}}), \cdots, (\mathbf{I}, \tilde{\mathbf{I}}), \cdots, (\mathbf{m}, \tilde{\mathbf{m}}); (\mathbf{0}', \tilde{\mathbf{0}}'), \cdots, (\mathbf{I}', \tilde{\mathbf{I}}'), \cdots, (\mathbf{m}', \tilde{\mathbf{m}}')]$$

$$A^{b}_{E}(m) = [(\mathbf{0}', \tilde{\mathbf{0}}), \cdots, (\mathbf{I}', \tilde{\mathbf{I}}), \cdots, (\mathbf{m}', \tilde{\mathbf{m}}); (\mathbf{0}, \tilde{\mathbf{0}}'), \cdots, (\mathbf{I}, \tilde{\mathbf{I}}'), \cdots, (\mathbf{m}, \tilde{\mathbf{m}}')]$$

$$|S^{b}_{E}(m)| = 2 \times \sum_{j=0}^{m} \binom{m}{j}^{2} \tag{10}$$

$$|A^{b}_{E}(m)| = 2 \times \sum_{j=0}^{m} \binom{m}{j}^{2} \tag{11}$$

For correlated states, when $m = 0\ (mod\ 2)$, the position of $m/2$ provides a single maximum value; when $m = 1\ (mod\ 2)$, the $(m-1)/2$ or $(m+1)/2$ position provides pair maximum values.

$$S^{b}_{Ex}(m) = [(\mathbf{m}/\mathbf{2}, \widetilde{[\mathbf{m}/\mathbf{2}]}); ([\mathbf{m}/\mathbf{2}]', \widetilde{[\mathbf{m}/\mathbf{2}]}')], m = 0\ (mod\ 2)$$

$$A^{b}_{Ex}(m) = [(\mathbf{m}/\mathbf{2}, \widetilde{[\mathbf{m}/\mathbf{2}]}')); ([\mathbf{m}/\mathbf{2}]', \widetilde{[\mathbf{m}/\mathbf{2}]})], m = 0\ (mod\ 2)$$

$$S^{b}_{Ex}(m) = [([(\mathbf{m}-\mathbf{1})/\mathbf{2}], \widetilde{[(\mathbf{m}-\mathbf{1})/\mathbf{2}]}))|([(\mathbf{m}+\mathbf{1})/\mathbf{2}], \widetilde{[(\mathbf{m}+\mathbf{1})/\mathbf{2}]}));$$
$$([(\mathbf{m}-\mathbf{1})/\mathbf{2}]', \widetilde{[(\mathbf{m}-\mathbf{1})/\mathbf{2}]}')|([(\mathbf{m}+\mathbf{1})/\mathbf{2}]', \widetilde{[(\mathbf{m}+\mathbf{1})/\mathbf{2}]}')], m = 1\ (mod\ 2)$$

$$A^{b}_{Ex}(m) = [([(\mathbf{m}-\mathbf{1})/\mathbf{2}], \widetilde{[(\mathbf{m}-\mathbf{1})/\mathbf{2}]}'))|([(\mathbf{m}+\mathbf{1})/\mathbf{2}], \widetilde{[(\mathbf{m}+\mathbf{1})/\mathbf{2}]}'));$$
$$([(\mathbf{m}-\mathbf{1})/\mathbf{2}]', \widetilde{[(\mathbf{m}-\mathbf{1})/\mathbf{2}]})|([(\mathbf{m}+\mathbf{1})/\mathbf{2}]', \widetilde{[(\mathbf{m}+\mathbf{1})/\mathbf{2}]})], m = 1\ (mod\ 2)$$

$$|S^{b}_{Ex}(m)| = 2 \times \binom{m}{m/2}^{2}, m = 0\ (mod\ 2) \tag{12}$$

$$|A^{b}_{Ex}(m)| = 2 \times \binom{m}{m/2}^{2}, m = 0\ (mod\ 2) \tag{13}$$

$$|S^{b}_{Ex}(m)| = 2 \times \binom{m}{(m-1)/2}^{2}, m = 1\ (mod\ 2) \tag{14}$$

$$|A^{b}_{Ex}(m)| = 2 \times \binom{m}{(m-1)/2}^{2}, m = 1\ (mod\ 2) \tag{15}$$

In the associated representation for all diagonal and antidiagonal positions, the probability distribution is

m	2^{m+1}	$2^{2(m+1)}$	$S_E^a(m)$	$A_E^a(m)$	$S_E^b(m)$	$A_E^b(m)$	$\frac{S_E^a(m)+A_E^a(m)}{2^{2(m+1)}}$	$\frac{S_E^b(m)+A_E^b(m)}{2^{2(m+1)}}$
0	2	4	2	2	2	2	1	1
1	4	16	4	4	4	4	0.5	0.5
2	8	64	8	8	12	12	0.25	0.375
3	16	256	16	16	40	40	0.125	0.3125
4	32	1024	32	32	140	140	0.0625	0.2734375
5	64	4096	64	64	504	504	0.03125	0.24609375
6	128	16384	128	128	1848	1848	0.015625	0.2255859375
7	256	65536	256	256	6864	6864	0.0078125	0.20947265625
8	512	262144	512	512	25740	25740	0.00390625	0.196380615234375

The maximum probability distribution for each part is

m	2^{m+1}	$2^{2(m+1)}$	$S_{Ex}^a(m)$	$A_{Ex}^a(m)$	$S_{Ex}^b(m)$	$A_{Ex}^b(m)$	$\frac{S_{Ex}^a(m)+A_{Ex}^a(m)}{2^{2(m+1)}}$	$\frac{S_{Ex}^b(m)+A_{Ex}^b(m)}{2^{2(m+1)}}$
0	2	4	2	2	2	2	1	1
1	4	16	2	2	2	2	0.25	0.25
2	8	64	2	2	8	8	0.0625	0.25
3	16	256	2	2	18	18	0.015625	0.140625
4	32	1024	2	2	72	72	0.00390625	0.140625
5	64	4096	2	2	200	200	0.0009765625	0.09765625
6	128	16384	2	2	800	800	0.000244140625	0.09765625
7	256	65536	2	2	2450	2450	0.00006103515625	0.07476806640625
8	512	262144	2	2	9800	9800	0.0000152587890625	0.07476806640625

9. Entropy Function of Narrow and General Projection

Let E_a and E_b be the narrow and general information entropy functions, respectively.

9.1 Classical Representation

From the perspective of fundamental analysis, it is necessary to briefly summarize the classical pattern. For any m element vector, 2^m states and 2^m measures, and $m+1$ clustering, let $p_I = \frac{1}{2^m}, 0 \leq I < 2^m$, the narrow information entropy of the system can be described.

Theorem 5 (Narrow Information Entropy Function). *For a system of m variables, 2^m states and 2^m measures, the narrow information entropy function satisfies*

$$E_a(m) = m \tag{16}$$

Proof.

$$
\begin{aligned}
E_a(m) &= -\sum_{I=0}^{2^m-1} \frac{1}{2^m} \times \log_2 \frac{1}{2^m} \\
&= -\log_2 \frac{1}{2^m} \\
&= \log_2 2^m \\
&= m
\end{aligned}
$$

Let $p_i = \binom{m}{i}/2^m, 0 \leq i \leq m$; then, the general information entropy of the system can be described.

Theorem 6 (General Information Entropy Function). *For a system of m variables, 2^m states, and 2^m measures, the general information entropy function satisfies*

$$E_b(m) \leq \log_2(m+1) \tag{17}$$

Proof.

$$
\begin{aligned}
E_b(m) &= -\sum_{i=0}^{m} \frac{\binom{m}{i}}{2^m} \times \log_2 \frac{\binom{m}{i}}{2^m} \\
&\leq -\sum_{i=0}^{m} \frac{2^m}{(m+1) \times 2^m} \times \log_2 \frac{2^m}{(m+1) \times 2^m} \\
&= -\sum_{i=0}^{m} \frac{1}{m+1} \times \log_2 \frac{1}{m+1} \\
&= \log_2(m+1)
\end{aligned}
$$

9.2 Conjugate Representation

9.2.1 Narrow Projection

From the sampling viewpoint, when the selected input state reaches 2^{m+1}, the two sets of conjugate states associated with the selected k will appear once, and $(s, \tilde{s})$ will appear on the horizontal and vertical orientations of the region. All conjugate state projections of this region can be formed by combining horizontal and vertical positions. The full number of narrow projection sequence pairs is $2^m \times 2^m = 2^{2m}$. The number of pairs in the $(s, s' : \tilde{s}, \tilde{s}')$ area is $4 \times 2^m \times 2^m = 2^{2(m+1)}$.

Theorem 7 (Single Block of Narrow Information Entropy Function). *For $m+1$ variables, 2^{2m} measures and $(s, \tilde{s})$ conjugate function, the narrow system entropy function satisfies*

$$E_a((s, \tilde{s})(m)) = 2m \tag{18}$$

Proof.

$$E_a((s, \tilde{s})(m)) = \log_2(2^{2m}) = 2m$$

Corollary 5 (Four-Block Narrow Information Entropy Function). *For $m+1$ variables, $2^{2(m+1)}$ measures, $(s, s' : \tilde{s}, \tilde{s}')$ extended conjugate function system, the narrow information entropy function satisfies*

$$E_a((s, s' : \tilde{s}, \tilde{s}')(m)) = 2(m+1) \tag{19}$$

Proof.

$$E_a((s, s' : \tilde{s}, \tilde{s}')(m)) = \log_2(2^{2(m+1)}) = 2(m+1)$$

9.2.2 General Projection

It is essential to calculate the 2^{2m} configurations of the single block. The probability of the i-th row j-th column is $p(i,j) = \frac{\binom{m}{i}\binom{m}{j}}{2^{2m}}$, and the entropy function of the system satisfies the following theorem.

Theorem 8 (Single Block General Entropy Function). *For $m+1$ variables, 2^{2m} measures on the $(s, \tilde{s})$ conjugate function, and the general entropy function satisfies*

$$E_b((s, \tilde{s})(m)) \leq 2\log_2(m+1) \tag{20}$$

Proof.

$$\begin{aligned}
E_b((s, \tilde{s})(m)) &= -\sum_{i=0}^{m}\sum_{j=0}^{m} \frac{\binom{m}{i}\binom{m}{j}}{2^{2m}} \times \log_2 \frac{\binom{m}{i}\binom{m}{j}}{2^{2m}} \\
&= -\sum_{i=0}^{m}\sum_{j=0}^{m} \frac{\binom{m}{i}\binom{m}{j}}{2^m \times 2^m} \times \log_2 \frac{\binom{m}{i}\binom{m}{j}}{2^m \times 2^m} \\
&= -\sum_{i=0}^{m}\sum_{j=0}^{m} \frac{\binom{m}{i}\binom{m}{j}}{2^m \times 2^m} \times \left[\log_2 \frac{\binom{m}{i}}{2^m} + \log_2 \frac{\binom{m}{j}}{2^m}\right] \\
&= -\sum_{i=0}^{m}\sum_{j=0}^{m} \frac{\binom{m}{i}\binom{m}{j}}{2^m \times 2^m} \times \log_2 \frac{\binom{m}{i}}{2^m} - \sum_{i=0}^{m}\sum_{j=0}^{m} \frac{\binom{m}{i}\binom{m}{j}}{2^m \times 2^m} \times \log_2 \frac{\binom{m}{j}}{2^m} \\
&= -2\sum_{i=0}^{m} \frac{\binom{m}{i}}{2^m} \times \log_2 \frac{\binom{m}{i}}{2^m} \\
&= 2E_b(m) \\
&\leq 2\log_2(m+1)
\end{aligned}$$

9.2.3 Any Two Blocks

When any two blocks need to be processed, $p(i,j) = \binom{m}{i}\binom{m}{j}/2^{2m+1}$, the system entropy function satisfies following relationship.

Corollary 6 (Two-Block General Entropy Function). *For the $m+1$ variables, choose any two 2^{2m} measures $(s, s' : \tilde{s}, \tilde{s}')$ conjugate function, and the general entropy function satisfies*

$$E_b((s, \tilde{s})(m)) \leq 1 + 2\log_2(m+1) \tag{21}$$

Proof.

$$\begin{aligned}
E_b((s, s' : \tilde{s}, \tilde{s}')(m)) &= -2\sum_{i=0}^{m}\sum_{j=0}^{m} \frac{\binom{m}{i}\binom{m}{j}}{2^{2m+1}} \times \log_2 \frac{\binom{m}{i}\binom{m}{j}}{2^{2m+1}} \\
&= -\sum_{i=0}^{m}\sum_{j=0}^{m} \frac{\binom{m}{i}\binom{m}{j}}{2^{2m}} \times \log_2 \frac{\binom{m}{i}\binom{m}{j}}{2^{2m+1}} \\
&= -\sum_{i=0}^{m}\sum_{j=0}^{m} \frac{\binom{m}{i}\binom{m}{j}}{2^{2m}} \times \left[\log_2 \frac{\binom{m}{i}\binom{m}{j}}{2^{2m}} - 1\right] \\
&= 1 - 2\sum_{i=0}^{m} \frac{\binom{m}{i}}{2^{m}} \times \log_2 \frac{\binom{m}{i}}{2^{m}} \\
&= 1 + 2E_b(m) \\
&\leq 1 + 2\log_2(m+1)
\end{aligned}$$

9.2.4 Full Four Blocks

When all four blocks need to be processed, $p(i,j) = \binom{m}{i}\binom{m}{j}/2^{2(m+1)}$, the system entropy function satisfies following relationship.

Corollary 7 (Four-Block General Entropy Function). *For the $m+1$ variables, select the four-block 2^{2m} regions $(s, s' : \tilde{s}, \tilde{s}')$ conjugate function, and the general entropy function satisfies*

$$E_b((s, \tilde{s})(m)) \leq 2 + 2\log_2(m+1) \tag{22}$$

Proof.

$$\begin{aligned}
E_b((s, s' : \tilde{s}, \tilde{s}')(m)) &= -4\sum_{i=0}^{m}\sum_{j=0}^{m} \frac{\binom{m}{i}\binom{m}{j}}{2^{2(m+1)}} \times \log_2 \frac{\binom{m}{i}\binom{m}{j}}{2^{2(m+1)}} \\
&= -\sum_{i=0}^{m}\sum_{j=0}^{m} \frac{\binom{m}{i}\binom{m}{j}}{2^{2m}} \times \log_2 \frac{\binom{m}{i}\binom{m}{j}}{2^{2(m+1)}} \\
&= -\sum_{i=0}^{m}\sum_{j=0}^{m} \frac{\binom{m}{i}\binom{m}{j}}{2^{2m}} \times \left[\log_2 \frac{\binom{m}{i}\binom{m}{j}}{2^{2m}} - 2\right] \\
&= 2 - 2\sum_{i=0}^{m} \frac{\binom{m}{i}}{2^{m}} \times \log_2 \frac{\binom{m}{i}}{2^{m}} \\
&= 2[1 + E_b(m)] \\
&\leq 2[1 + \log_2(m+1)].
\end{aligned}$$

9.3 *Merged Entropy Values*

9.3.1 Two Blocks with 0 Points Merged

When $(s, \tilde{s})$ and $(s', \tilde{s}')$ or $(s', \tilde{s})$ and $(s, \tilde{s}')$ are selected. If the two blocks are merged, then the corresponding two pairs of 0 points $\{0, \tilde{0}\}$ and $\{-0, -\tilde{0}\}$ or $\{-0, \tilde{0}\}$ and $\{0, -\tilde{0}\}$ are involved in merging. For the two points before the operation, one point occupies two values after merging, and the other point is empty. The probability of a single point is $p(0, \tilde{0}) = p(-0, -\tilde{0}) = 1/2^{2m+1}$ $p'(0, \tilde{0}) = 2/2^{2m+1}, p(-0, -\tilde{0}) = 0$.

Theorem 9 (Difference between Two 0-Point Combined General Entropy Functions). *When $(s, \tilde{s})$ and $(s', \tilde{s}')$ or $(s', \tilde{s})$ and $(s, \tilde{s}')$ are selected, if the two 0 points of the two blocks are merged, then the difference of the general entropy function satisfies*

$$E(m,m) - E'(m,m) = \frac{1}{2^{2m}} \tag{23}$$

Proof.

$$\begin{aligned} E(m,m) - E'(m,m) &= -\frac{2}{2^{2m+1}} \log_2 \frac{1}{2^{2m+1}} + \frac{2}{2^{2m+1}} \log_2 \frac{2}{2^{2m+1}} \\ &= \frac{2}{2^{2m+1}} \log_2(2) \\ &= \frac{2}{2^{2m+1}} \\ &= \frac{1}{2^{2m}} > 0 \end{aligned}$$

9.3.2 Two Blocks of 0 Lines Merged

The points of the two blocks are involved, $p(i) = \binom{m}{i}/2^{2m+1}, p'(i) = 2\binom{m}{i}/2^{2m+1}, 0 \leq i \leq m$.

Theorem 10 (Difference between Two 0-Lines Combined General Entropy Functions). *When $(s, \tilde{s})$ and $(s', \tilde{s})$ or $(s', \tilde{s})$ and $(s', \tilde{s}')$ or $(s, \tilde{s}')$ and $(s', \tilde{s}')$ or $(s, \tilde{s})$ and $(s, \tilde{s}')$ are selected, if two blocks are combined, then the difference of the general entropy function satisfies*

$$E(m,m) - E'(m,m) = \frac{1}{2^m} \tag{24}$$

Proof.

$$\begin{aligned} E(m,m) - E'(m,m) &= -2\sum_{i=0}^{m} \frac{\binom{m}{i}}{2^{2m+1}} \log_2 \frac{\binom{m}{i}}{2^{2m+1}} + \frac{2\binom{m}{i}}{2^{2m+1}} \log_2 \frac{2\binom{m}{i}}{2^{2m+1}} \\ &= \frac{2}{2^{2m+1}} \sum_{i=0}^{m} \binom{m}{i} \left[\log_2 \frac{2\binom{m}{i}}{2^{2m+1}} - \log_2 \frac{\binom{m}{i}}{2^{2m+1}} \right] \\ &= \frac{2}{2^{2m+1}} \sum_{i=0}^{m} \binom{m}{i} \\ &= \frac{1}{2^m} > 0 \end{aligned}$$

9.3.3 Four Blocks of 0 Lines Merged

Four blocks are $p(i) = \frac{\binom{m}{i}}{2^{2m+2}}, 0 \le i \le m; p'(j) = \frac{2\binom{m}{j}}{2^{2m+2}}, 0 \le j < m, p'(m) = \frac{4}{2^{2m+2}}$

The points of the four blocks are involved, $p(i) = \frac{\binom{m}{i}}{2^{2m+2}}, 0 \le i \le m; p'(j) = \frac{2\binom{m}{j}}{2^{2m+2}}, 0 \le j < m, p'(m) = \frac{4}{2^{2m+2}}$.

Corollary 8 (Four-Block 0-Lines Combined General Entropy Function). *When* $\{(s,\tilde{s}), (s',\tilde{s}), (s,\tilde{s}'), (s',\tilde{s}')\}$ *are selected, if the four blocks are combined, then the difference of the general entropy function satisfies*

$$E(m,m) - E'(m,m) > \frac{1}{2^{m-1}} \tag{25}$$

Proof.

$$\begin{aligned}
E(m,m) - E'(m,m) &= -8\sum_{i=0}^{m} \frac{\binom{m}{i}}{2^{2m+2}} \log_2 \frac{\binom{m}{i}}{2^{2m+2}} + 4\frac{1}{2^{2m+2}} log_2 \frac{1}{2^{2m+2}} \\
&\quad +4\sum_{i=0}^{m-1} \frac{2\binom{m}{i}}{2^{2m+2}} \log_2 \frac{2\binom{m}{i}}{2^{2m+2}} - \frac{4}{2^{2m+2}} \log_2 \frac{4}{2^{2m+2}} \\
&= 8\sum_{i=0}^{m-1} \frac{\binom{m}{i}}{2^{2m+2}} \left[\log_2 \frac{2\binom{m}{i}}{2^{2m+2}} - \log_2 \frac{\binom{m}{i}}{2^{2m+2}}\right] \\
&\quad - \frac{4}{2^{2m+2}} \left[log_2 \frac{1}{2^{2m+2}} + \log_2 \frac{4}{2^{2m+2}}\right] \\
&= 8\sum_{i=0}^{m-1} \frac{\binom{m}{i}}{2^{2m+2}} + \frac{4}{2^{2m+2}} [2m+2+2m] \\
&= \frac{8(2^m-1)}{2^{2m+2}} + \frac{4(4m+2)}{2^{2m+2}} \\
&= \frac{8(2^m+2m)}{2^{2m+2}} \\
&> \frac{1}{2^{m-1}} > 0
\end{aligned}$$

The increment is approximately twice that of the two-block merged, and from the perspective of the variation in the entropy function, the magnitude is small.

10. Feature Analysis

Comparing **Lemma 1** and **Corollary 1**, there is a clear difference between the classical and conjugated measurement procedures.

10.1 Classic Representation Analysis

In the four steps of the classical analysis, it is worth noting that the narrow projection measurement space is equal to the scale of the logical state space, and the general projection measurement space is equivalent to the symmetric function cluster. The number of functions involved is equivalent to the general projected measurement space. From the logic function representation, m variables, and 2^m states, the 2^{2^m} functions are established.

Due to the superexponential growth property of the function space, the analysis steps associated with information entropy are usually associated with symmetric functions to simplify the analysis complexity.

10.1.1 Classic Sequence Narrow Information Entropy

From the sequence exhaustion, the 2^m states associated with the analysis can be obtained from m positions in the standard clock signal as a total of 2^m count values $I,\ 0 \leq I < 2^m$. When each distinguishable state occurs once, such a sequence can form all logical state measurement spaces, guaranteeing each position with an element and being the total number of narrow projection measurement spaces on 2^m.

From the information entropy level, the uniform distribution structure provides the narrow projection measure information entropy $E_a(m) = m$. Since the information entropy reaches the maximum value in this type of system, from the time basis of physical meaning, it is the longest coordinate of the basic position of linear measurements. A special distribution is formed in this linear region to support arbitrary agent working modes. When the sequence of total length $m \times 2^m$ forms a loop and the shift operator moves the $t \times m, 0 \leq t < 2^m$ position, the distribution and entropy value in the phase space will not change.

Such shift-invariance effects exhibit the energy-conserving properties of the system from the physical measurements that are invariant with time.

10.1.2 Classic Sequence General Information Entropy

For the general projection clustering mode, the clustering mode characterized by a binomial distribution makes the information entropy satisfy $E_b(m) \leq \log_2(m + 1)$. It is not difficult to verify that when the information entropy of this type of clustering is $m = \{0, 1\}$, the equation holds.

For the whole structure analysis, the important difference between general projection and narrow projection is that the basis of general projection is the intrinsic measure with the ordered properties of maxima and minima. However, this kind of measurement feature on ordered clusters cannot be represented in the narrow projection of the distribution. From the order and disorder measurement, the value of the system information entropy decreases from the initial m to $\log_2(m + 1)$, and the accompanying changes are the comprehensive effects of nonequilibrium statistical clustering and bifurcation.

10.1.3 Micro and Macro State Representation

If the corresponding states in the same cluster change interactively, the local cyclic change pattern within the cluster will not cause the entire probability distribution to be changed, or it will not change the information entropy. Compared with the classical time-information entropy characteristics, it is possible to determine this type of change pattern as a microstate mode.

However, the state movement between different clusters inevitably leads to a change in the probability distribution. Such patterns of change can be identified as patterns of macrostates.

10.2 Conjugate Representation

In the conjugate measurement, the total number of the five main modules involved is not equal. Using the conjugate partition condition, the state space between the conjugate logic state space and the input variable $2(m+1)$ maintains the same number of states $2^{2(m+1)}$ as in the classical model, but through two sets of conjugate clustering, each forms an independent combined cluster to form a superpower exponential function space $2^{2^{2m}}$, and the whole conjugate logistic function space has a huge number of superpower exponentials $2^{2^{2m+1}+2}$. Unlike classical logic functions, which can maintain the large scale of nontrivial superpower exponents, the conjugate system reduces the total number of distinguishable vectors from the superpower exponentials $2^{2^{2m+1}+2}$ to the exponentials $2^{2(m+1)}$ range. Through the symmetric function clustering mode, the corresponding number of clusters is decreased as the product of binomial coefficients, and the number of distinguishable clusters after merging is $4(m+1)^2$.

10.2.1 Conjugate Sequence Narrow Information Entropy

For the distinguishable function space, the narrow projection function space is on $2^{2(m+1)}$, far exceeding the scale of the classical narrow projection measure space on 2^m.

From the levels of information entropy, it represents an $(s, \tilde{s})$ type data sequence, and the smallest mode forms a 2^{2m} uniform distribution to ensure narrow projection information entropy $E_a(m) = 2m$.

This information entropy reaches the maximum value of the system, and in terms of the physical meaning of time representation, it is the maximum value coordinate of the square measurements on base positions. Arbitrary agent working modes can be supported to form special distributions in the area.

When the sequence of total length $2m \times 2^{2m}$ forms a loop, and the shift operator moves $t \times 2m$, $0 \leq t < 2^{2m}$, the distribution in the phase space and the entropy value will not be changed.

If it is necessary to form a $(s, s' : \tilde{s}, \tilde{s}')$ type data sequence, the smallest mode is to form a $2^{2(m+1)}$ uniform distribution to guarantee that the narrow projection measures information entropy $E_a(m+1) = 2(m+1)$. The information entropy reaches the maximum value of the system, and in terms of the physical meaning of time representation, it is the maximum value coordinate of the square measurements on the base positions. Arbitrary agent working modes are supported to form special distributions in the area. When the total length is $2(m+1) \times 2^{2(m+1)}$, the sequence forms a loop, and the shift operator moves the $t \times 2(m+1)$, $0 \leq t < 2^{2(m+1)}$ position; neither the distribution nor the entropy value in the phase space will change. Satisfying this type of data sequence, four interrelated conjugate projection regions can be generated on the complex number plane. The required four distinguishable conjugate and complementary combination blocks are distinguished using the specified state bits to fully satisfy the functional representation to support all potential measure coordinate positions.

For the physical measurement invariant with time, the conjugate shift invariant effect also exhibits the energy conservation property of the system.

10.2.2 Conjugate Sequence General Information Entropy

For the general projection of the clustering measurement mode, the clustering mode characterized by pairwise product binomial distribution makes the information entropy satisfy $E_b(m) \leq 2\log_2(m+1)$. It is not difficult to verify that when the information entropy of this type of clustering is $m = \{0, 1\}$, the equation holds.

The complete four-block information entropy satisfies $E_b(m) \leq 2 + 2\log_2(m+1)$.

For the structure analysis, the important difference between general projection and narrow projection is that the basis of general projection is the intrinsic measure to have the ordered properties of maxima and minima. However, this kind of measure feature on ordered clustering cannot be represented in the narrow projection distribution. For the ordered and disordered measurements, the value of information entropy decreases from the initial $2m$ to $2\log_2(m+1)$, and the changes are nonequilibrium statistical conjugate clustering and bifurcation synthesis effects.

10.2.3 Conjugate Micro and Macro Representation

In the conjugate system, if the related states in the same cluster change interactively, the local cyclic change pattern within the cluster will not influence the probability distribution, nor will it change the information entropy value. Compared with the classical time-information entropy features, it is possible to determine this type of change pattern as a conjugate microstate pattern. From the perspective of intrinsic morphology, four feature representations can be distinguished.

However, state movement between different clusters will cause the probability distribution to change. Such change patterns can be identified as conjugate macrostate patterns, and from the perspective of external morphological representation, four characteristics can be distinguished.

10.2.4 Various Observations

From the two modes of classical and conjugate measurements, the biggest difference is that the number of distinguishable functions in the logic function space of the classical m element is 2^{2^m}. For more variables, it is difficult to classify and exhaustively.

However, the total number of distinguishable functions in conjugate logic is 2^{2m}. Considering the expansion of conjugate combinations, the space of conjugate logic functions contains huge volumes and may significantly reduce the dimension of trivial functions.

What observable measurable effects correspond to such reduction effects? In natural physical space, it is worth exploring in depth.

10.3 Narrow and General Conjugate Measures of Ideal Sampling Sequences

The distribution is determined by the value of I, and the narrow-sense conjugate measurement of the sampling sequence occupies the position of 2^{2m} for a 2D uniform distribution. Taking the number of included 0/1 as the measure, the cluster corresponding to the general conjugate measurement of the sampling sequence occupies $(m + 1)^2$ positions as the binomial product distribution; for the complete sampling sequence, the entire area can be occupied without leaving a blank. From the perspective of data, the sampling sequence has a very small length, and from the perspective of clustering, the clustering distribution of the sampling sequence forms a greater coverage.

Narrow Measurement: Each distinguishable state corresponds to a unique location
General Measurement: Each distinguishable cluster corresponds to a unique location
Parallel Shift Features Invariant: Energy conservation
Maximum Clustering: Maximum information entropy.

10.4 Random Sequence of Equal Length

The value of each I determines the distribution, and the narrow conjugate measurement may occupy the position of 2^{2m} for a $2D$ nonuniform distribution. The same measurement sequence contains $\{0, 1\}$ number of measures, and the general conjugate measurements of the sampling sequence may occupy $(m+1)^2$ positions to form a discrete distribution. When a multivariate distribution is represented, the occupancy of the entire distribution area may be left blank. Along with the characteristics of the random sequence, the corresponding data of the sampling sequence may contain repeated passages. For the clustering and sampling sequences, specific cluster distributions of coverage features are formed. For the sequence of the time and space translations, a random sequence has the following characteristics:

Parallel Shift Characteristic Variation—Energy is not conserved
Far Away from Equilibrium Clustering—Information entropy decreases.

10.5 Classical Symmetric Distribution

For any random sequence using the extended symmetry, the given $m \times M$ length $\{0, 1\}$ sequence can be expanded two times to form a reflection-symmetric 1D distribution. The expanding mode is described by the following algorithm.

Reflection Symmetric Distribution Algorithm
Input: $m \times M$ long $\{0, 1\}$ sequence.
Output: $\boxed{\{-x\}}\boxed{\{x\}}$.
Process: For each m long segment, it can be computed as the x measure; then, the related measure expands to $-x$. Do the same actions be applied to the M segment data in the input sequence, and two sets of sequence pairs are projected to relevant regions as a 1D histogram.

Under the action of the algorithm, it can ensure that when the measure x is nonzero, both positions are associated with the same measure. For the data measurement, the random sequence can confirm the feature distribution to retain the 1D reflection symmetry under the two-sequence expansion.

10.6 Conjugate Symmetric Distribution

For any random sequence using the extended symmetry, the given $2 \times (m+1) \times M$ length 0-1 sequence can be extended by eight times the measure to form a conjugate characteristic symmetrical distribution. The expanding mode is described by the following algorithm.

Conjugate Measure Symmetric Distribution Algorithm
Input: $2 \times (m+1) \times M$ long $\{0, 1\}$ sequence.

Output: $\begin{array}{|c|c|}\hline \{(-x,y),(-y,x)\} & \{(x,y),(y,x)\} \\ \hline \{(-x,-y),(-y,-x)\} & \{(x,-y),(y,-x)\} \\ \hline\end{array}$.

Processing: For each $2\times(m+1)$ segment, calculate the (x, y) measure, and then expand to $(x,y) \rightarrow \{(x,y),(-x,y),(x,-y),(-x,-y)\} \rightarrow \{(y,x),(-y,x),(y,-x),(-y,-x)\}$. The same processing is performed on the M segment data in the input sequence, and the eight groups of sequence pairs are used to project to related regions to form a histogram.

Under the action of the algorithm, it can be ensured that when the measure (a, b) is nonzero, then all eight positions are associated with the same measure. For the data measurement, the random sequence can confirm the feature distribution to retain the symmetry under the eightfold sequence expansion. This kind of algorithm is equivalent to using multiple reflections and other processing mechanisms to obtain the full invariance of the 2D four-area distribution on the system measurement.

10.7 Distribution Features

To identify the essential difference between ideal and random distributions, classical and conjugate probability distribution characteristics are compared in Table 1.

Table 1. Comparison of conjugate probability distribution between ideal sequence and random sequence.

Classical Signal $\{N, m\}$	**Narrow 1D Distribution** 2^m	**General 1D Cluster Map** $(m+1)$
Ideal Sequence Clock Signal	1D Uniform Distribution 2^m 2^m Equal Probability	Binomial Distribution $\{\binom{m}{i}\}, 0 \le i, \le m$
Random Sequence Classical/Quantum Signal Double Expansion	Nonuniform Distribution 1D Unequal Probability Reflection Symmetry	Sub-Poisson~Poisson Map 1D Probability 1D Reflection Symmetry Map
Conjugate Signal $\{N, 2(m+1)\}$	**Narrow 2D Map** $2^{2(m+1)}$	**General 2D Polygon Class** $4(m+1)^2$
Ideal Sequence Standard Clock Signal	4-Block 2D Uniform $2^m \times 2^m$ $2^{2(m+1)}$ Equal Probability	Four Region Binomial Products $\{\binom{m}{i}\binom{m}{j}\}, 0 \le i,j \le m$
Random Sequence Classical/Quantum Signal Eightfold Extension	Nonuniform Distribution 2D Unequal Probability Conjugate Symmetry	Sub-Poisson~Poisson 2D Probability 2D Quadruple Conjugate Symmetry

11. Ideal Sequence and Conjugate Distribution Comparison

11.1 Composite Distribution Example

For a more complex example, it is extremely complicated to express numerically, and Fig. 2 illustrates a typical $m = 50$ distribution.

11.2 Superposition Effects of Selected Features

More complex interactions are shown in Figs. 3(a)–(h). Three types of results are included: (1) Separate projections (a), (2) Separate projections (b), and (3) Enlarged overlay modes (c)–(h).

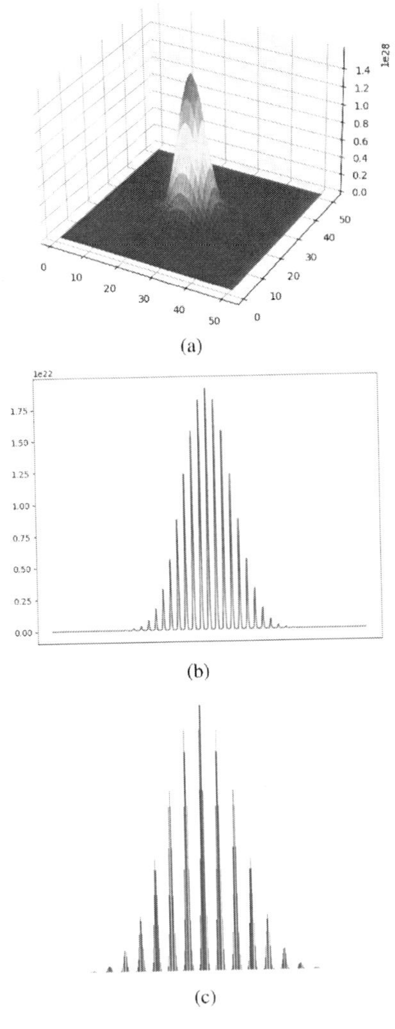

Figure 2. $m = 50$ 3D distribution map of single general projection: (a) 3D surface; (b) 1D wave distribution on envelope projection; (c) 1D wave distribution on histogram projection.

Interactive interference effects can be observed from various distributions. The four maps {(c),(d),(g),(h)} show obvious fluctuation patterns, and there are two sets of distributions {(e),(f)} showing classic superposition characteristics. These illustrations of the interactive properties are associated with quantum superposition activities.

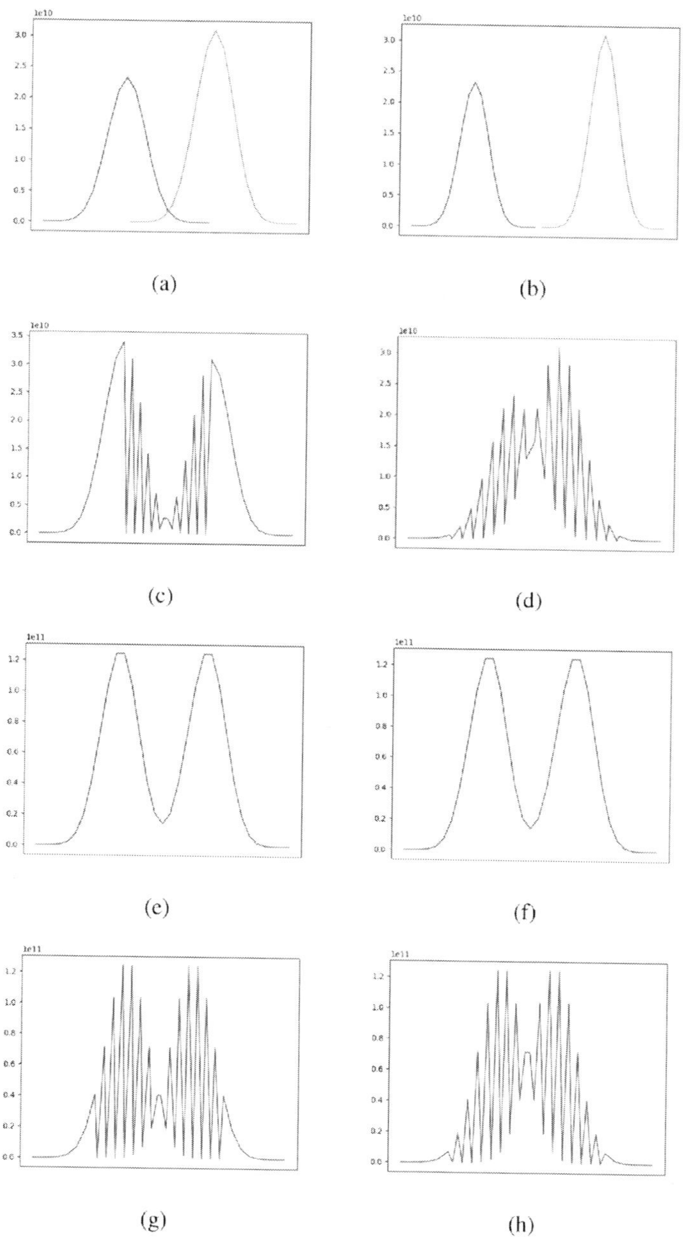

Figure 3. Two Sets of 1D distribution maps on various interaction lengths: (a) $m = 20$ twice histogram; (b) $m = 20$ twice separation histogram; (c) $m = 20$ double superposition distribution; (d) $m = 20$ five times superposition distribution; (e) $m = 21$ double histogram distribution; (f) $m = 21$ double separation histogram distribution; (g) $m = 21$ triple superposition distribution; (h) $m = 21$ quadruple superposition distribution.

12. Comparison between Ideal and Random Sequences

12.1 Typical Distribution

The narrow distribution corresponds to each distinguishable state only once, and it is easy to imagine a horizontal straight line for any parameter.

The general projection has two modes: all distinguishable combinatorial states appear once, and the distribution is determined by a composite binomial product distribution.

However, if only part of the states is present, the resulting distribution is a random distribution, which is difficult to describe by the binomial formula, and the distribution can be formed by piecewise statistical analysis.

Pairwise conjugate selection determines a location first. For example, from a complex number (a, b), the selected quadruple is $\{(a, b), (-a, b), (a, -b), (-a, -b)\}$ as a Klein group. Then the conjugate quadruple is $\{(b, a), (b, -a), (-b, a), (-b, -a)\}$ as a conjugate Klein group. When $a \neq b$, there are eight columns observed in the distribution.

12.1.1 Formula Distribution Maps

The results are calculated using the compound binomial formula shown in Fig. 4.

(a):

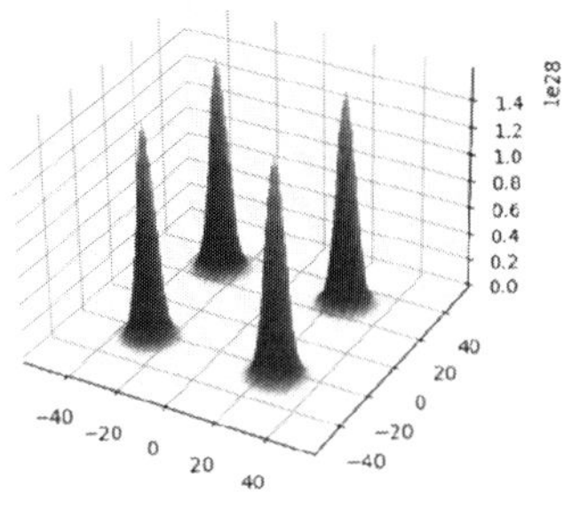

(b):

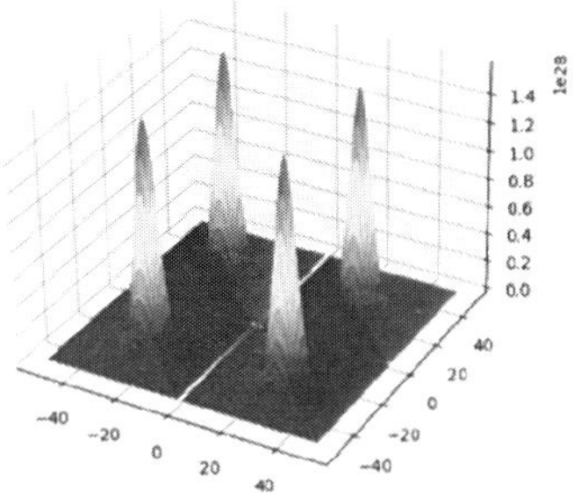

(c):

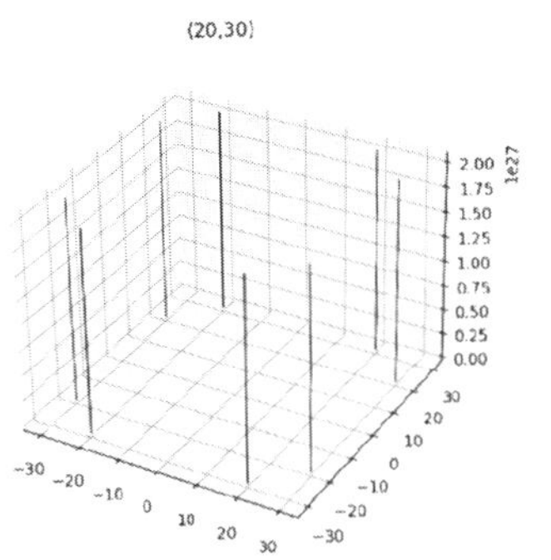

(d):

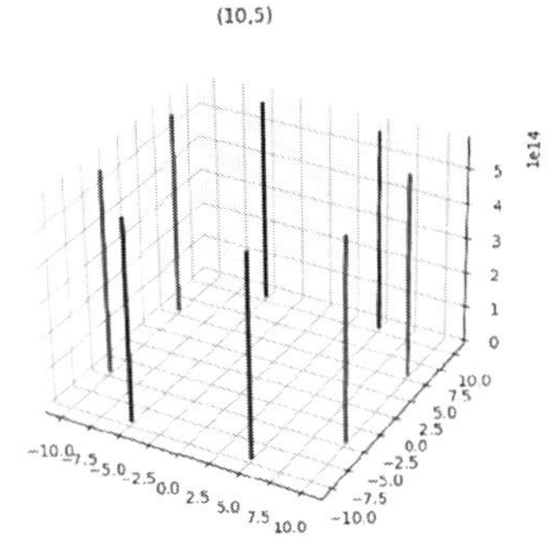

Figure 4. 3D Distributions of general projection: (a) $m = 50$ histogram: (b) $m = 50$ thermal distribution; (c) select conjugate feature (20.30); (d) select conjugate feature (10.5).

12.1.2 Random Data Sequence Distribution Maps

The results using random sequences are shown in Fig. 5.

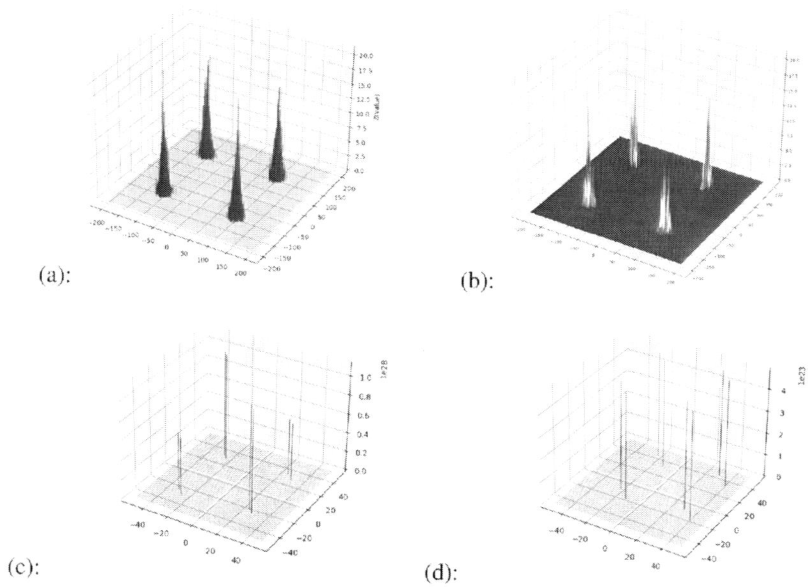

Figure 5. 3D random data sequence distributions of general projection: (a) $m = 200$ histogram; (b) $m = 200$ thermal distribution; (c) select conjugate features (51.52); (d) select conjugate features (15.25).

12.1.3 Random Data Series Symmetrical Distribution Maps

The results of extended corrections using random sequences are shown in Fig. 6.

After correction, the random distribution can be observed for the feature position with the value, showing the symmetric property. The random distribution characteristics exhibit specific distributions.

12.2 Fiber Bundle Distribution

Referring to the Yang-Mills equation, it is interesting to show the simulation maps.

12.3 Pairwise Distribution

Three sets of sequences are selected with lengths of $N \in \{10^4, 10^5, 10^6\}$, segment on $m = 100$ for statistics, and project to the first and third quadrants, so the distributions are illustrated in Figs. 8–9.

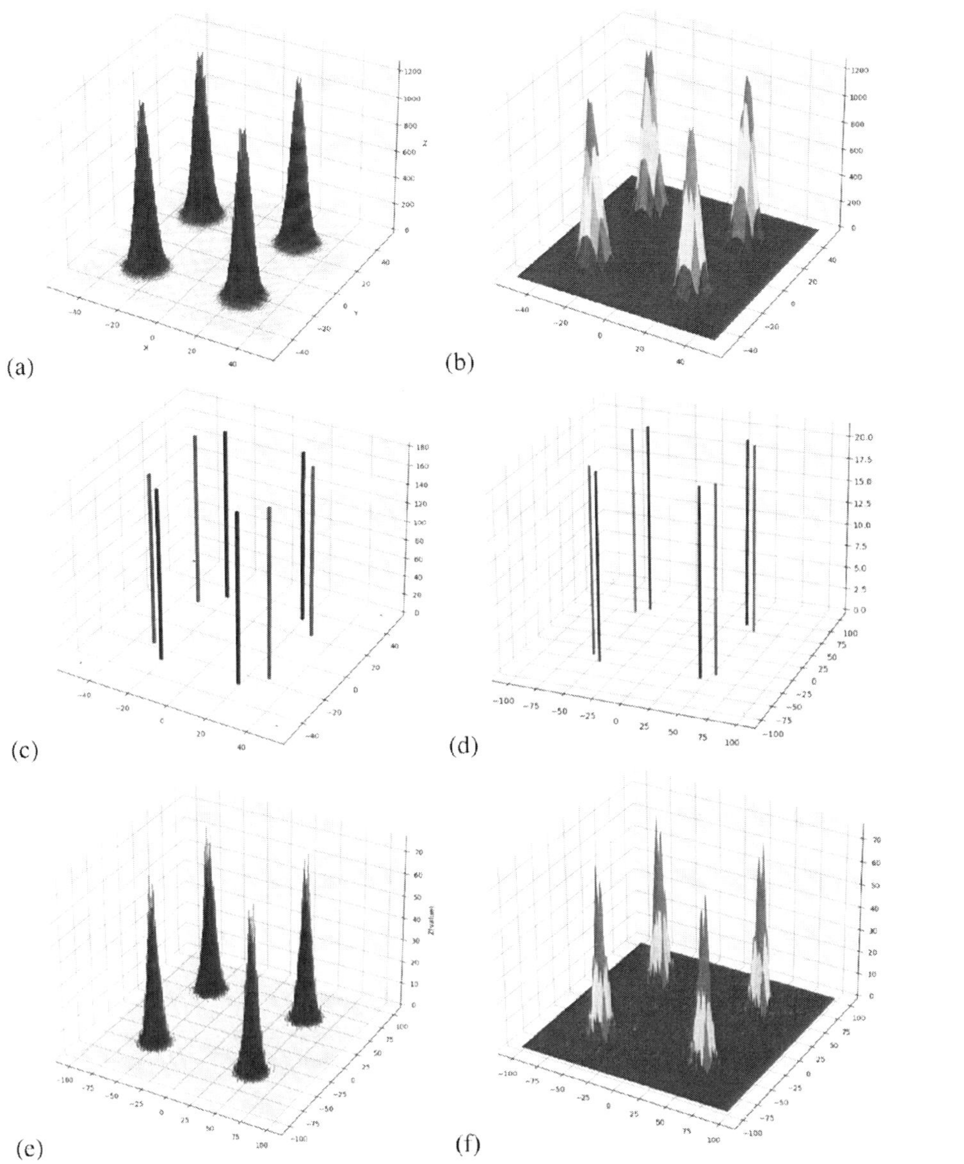

Figure 6. 3D random data sequence on corrected symmetry distributions of general projection: (a) $m = 40$ histogram; (b) $m = 40$ thermal distribution; (c) select features $m = 40, (20, 30)$; (d) select features $m = 100, (50, 45)$ (e) $m = 100$ histogram; (f) $m = 100$ thermal distribution.

Two distributions are included in map Figure 8(b). Along with the complex plane, the base description forms a conjugate pair from $(x, y) \rightarrow (y, x)$, and the distributions have reflection symmetry characteristics. Increasing the total sequence length in Figs. 9(a)–(b), the distributions in Fig. (b) are smoother than those in Fig. (a) in detail. A closer look reveals the reflection symmetry effect. For the pairwise distribution, how to accurately establish the correspondence between maps Figure 8(a),(b) and maps Figures 8 and 9 can be described using the conjugate transformation mechanism on the fiber bundle.

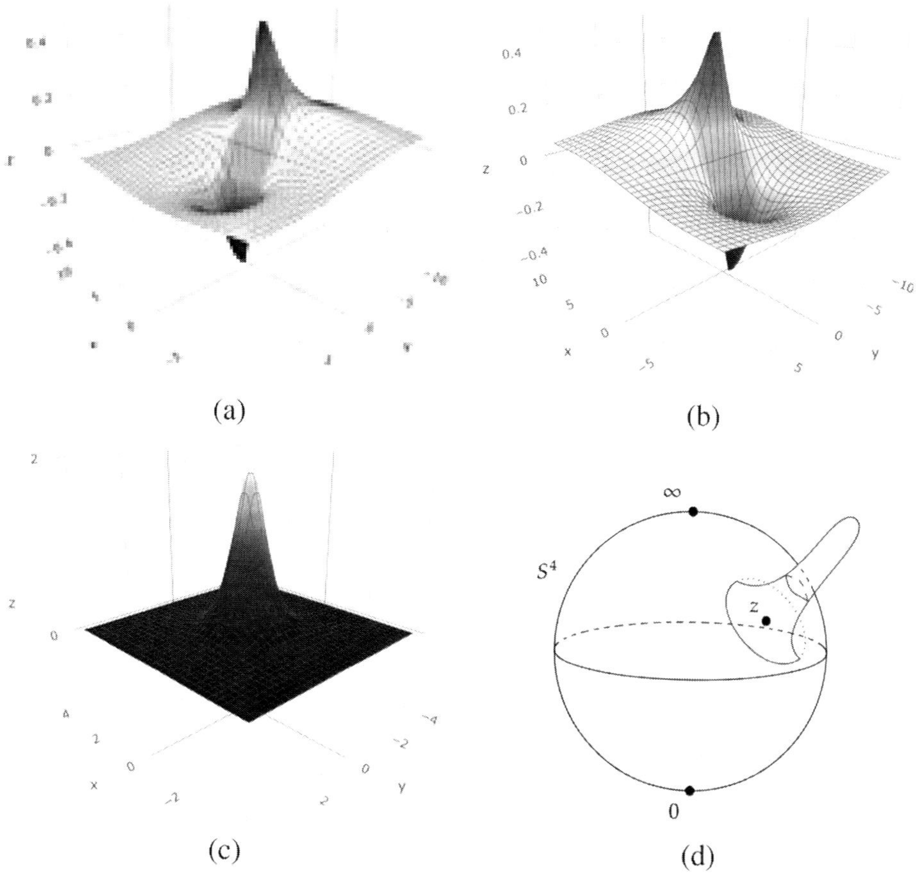

Figure 7. Typical BSPT of yang-mills instanton action distributions: (a) $dx^1 \otimes \sigma_3$; (b) $dx^2 \otimes \sigma_3$; (c) $g = 2, \rho = 1, z = 0$; (d) $z \in S^4 \subset R^4$. [37, 38].

13. Conclusion

Under conjugate coding, the standard clock counting sequence forms an ideal structure with a 2D probability distribution. Through the narrow and general measurement of the conjugate transformation, the standard time signal exhibits more precise analysis characteristics than classical time series analysis and can be observed to identify a series of ideal traversal features: minimum sequence length, maximum coverage area, complete distribution, energy conservation, maximum entropy, and so on.

For other random sequences or quantum random sequences, the statistical projection of the conjugate measurement is a 2D subset of the ideal distribution. Under combinatorial explosion, the feature distribution is constrained in the complete distribution.

Using the 2D conjugate function representation, it can be expressed quantitatively. The region where quantum interactions occur in entangled states can be used to estimate the proportion of entangled states in the whole sequence.

For random sequences, the location-specific measurements of the selected projections under the correction operation have a well-balanced distribution.

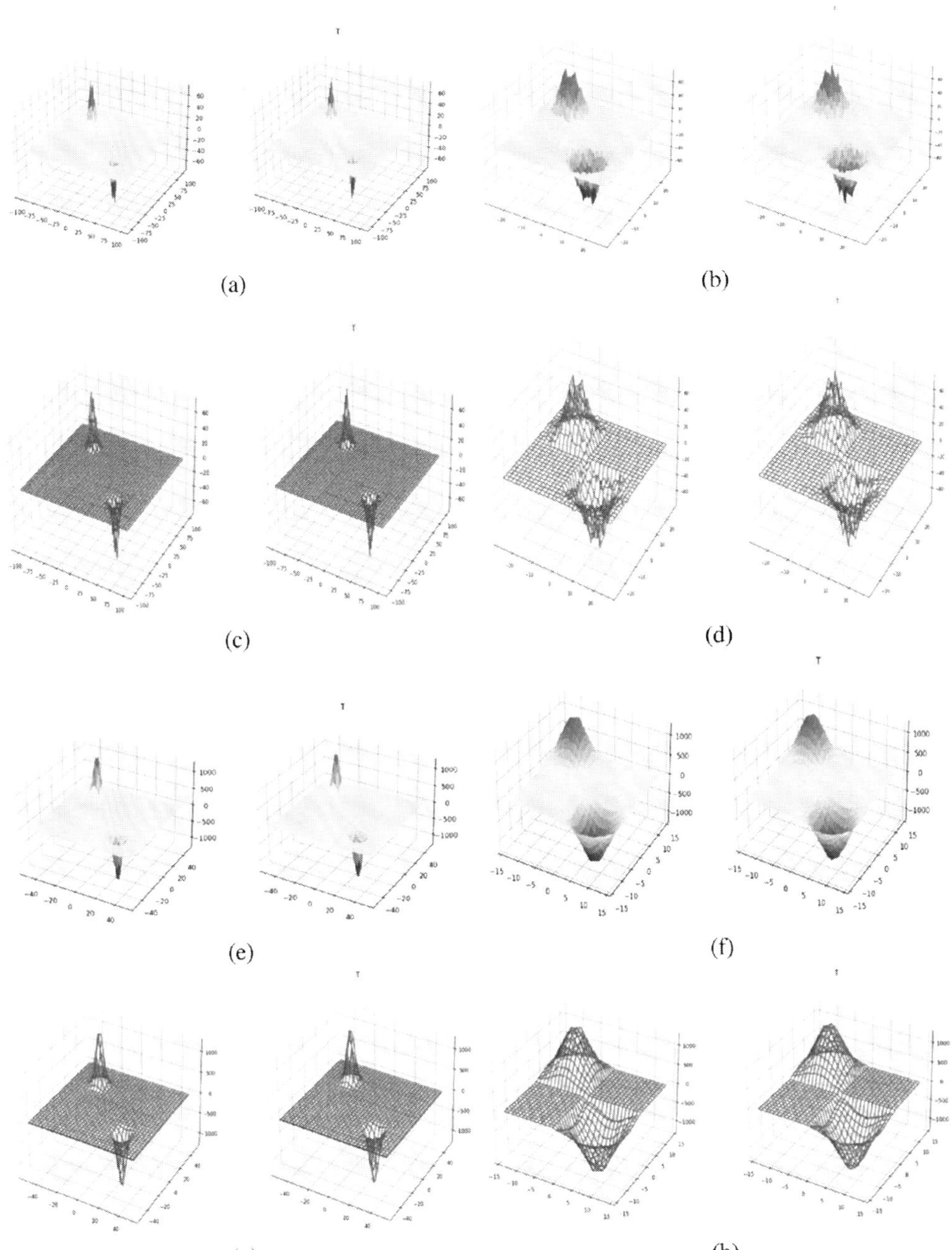

Figure 8. Conjugate distributions $N = 10^4, m = 100$: (a) pairwise distribution; (b) partial enlargement; (c) whole map; (d) enlarged map $N = 10^5, m = 100$; (e) pairwise distribution; (f) local zoom; (g) whole map; (h) zoomed map.

Using an ideal sequence to provide an optimized 2D conjugate measurement space, the self-consistency of spacetime is a dimensionality reduction representation of quantum entanglement that is established in the variation of a clock-accurate sampling sequence, and a rigorous demonstration will be provided in Chapter 6 of the book.

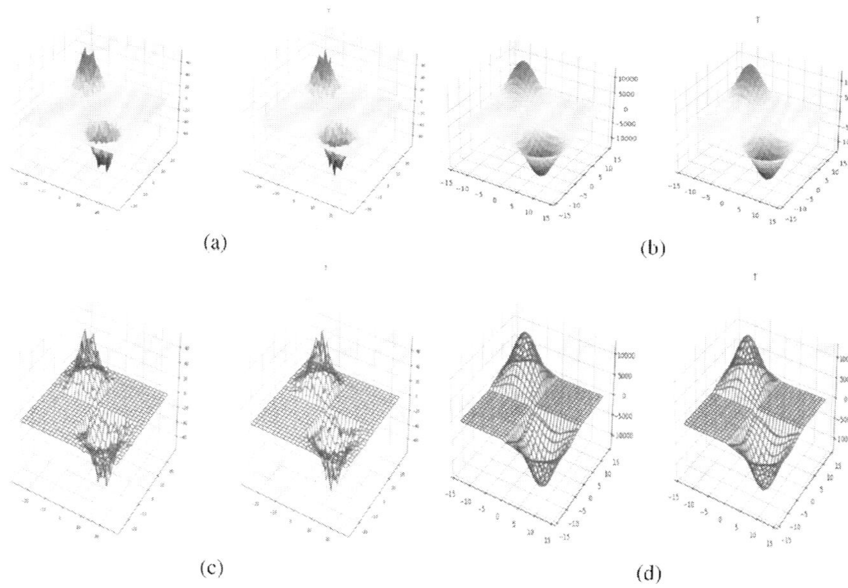

Figure 9. Conjugate Action Distributions $m = 100$: (a) $N = 10^4$ Partial Enlargement; (b) $N = 10^6$ Partial Enlargement; (c) $N = 10^4$ Enlarged Map; (d) $N = 10^6$ Enlarged Map

References

[1] Leo Rogers. *A Brief History of Time Measurement*. http://nrich.maths.org/6070.
[2] 一炷香 http://baike.baidu.com/item/一炷香/18892185.
[3] 艾小文, 数论问题 - 指数和的均值分析与研究,电子工业出版社 2021。
[4] D. Z. Albert. *Time and Chance*, Harvard University Press, 2000.
[5] 阿里耶 本-纳伊姆著;李永学译,时间的起点,北京联合出版社 2021 A. Ben-Naim. *The Briefest History of Time: The History of Histories of Time and the Miscondtrued Association between Entropy and Time*, World Scientific Publishing Co. Pte. Ltd., 2021.
[6] 吉姆 巴戈特 著;徐彬 于秀秀译, 量子通史 - 量子物理史上的40个重大时刻,中信出版社 2021。J. Baggott. *The Quantum Story in 40 Moments*, Oxford University Press, 2011.
[7] 艾伦 柏狄克 著; 李英松译,时间的质量,贵州人民出版社 2020。A. Burdick. *Why Time Flies: A Mostly Scientific Investigation*, Dterling Lord Literistic, 2017.
[8] P. C. W. Davis. *The Physics of Time Assymetry*, University of California Press, 1974.
[9] A. Eddington. *The Nature of Physical World*, CambridgeUniversity Press, 1928.
[10] P. H. Frampton. *Did Time Begin? Will Time End?* World Scientific Singapore, 2010.
[11] B. Green. *The Fabric of the Cosmos: Space, Time and the Texture of Reality*, Alfred A. Knopf, 2004.
[12] S. Hawking. *A Brief History of Time: From the Big Bang to Black Holes*, Bantam, 1988.
[13] 史蒂芬 霍金, 埃利斯著 王文浩译,李泳校, 时空的大尺度结构,湖南科学技术出版社 2021。S. Hawking and G. Ellis. *The Large Scale Structure of Space-Time*, Cambridge University Press, 1973.
[14] M. C. Mackey. *Time'sArrow: The Origins of Thermodynamic Behavior*, Dover, 1992.
[15] 瓦妮莎 奥格尔著; 郭科,章柳怡译, 时间的全球史, 浙江大学出版社 2021 *The Global Transformation of Time: 1970–1950*, Harvard University Press, 2015.
[16] *Roger Penrose. The Emperor's Mind: Concerning Computers, Minds and the Law of Physics*, Penguin, 1989.
[17] Carlo Rovelli. Neither presentism nor eternalism, 2020, Arxiv 1910.02474.
[18] Carlo Rovelli. Where was past low-entropy? 2019, Arxiv 1812.03578.
[19] Carlo Rovelli. Memory and entropy, 2020, Arxiv 2003.06687.
[20] Carlo Rovelli. Agency in physics, 2020, Arxiv 2007.05200.

[21] Valerio Astuti, Marios Christodoulou and Carlo Rovelli. Volume entropy, 2016, Arxiv 1603.01561.

[22] S. Ariwahjoedi, V. Astuti, J. S. Kosasih, C. Rovelli and F. P. Zen. Statistical discrete geometry, 2016, Arxiv 1607.08629.

[23] Kate Jeffery, Robert Pollack and Carlo Rovelli. On the statistical mechanics of life: Schrödinger revisited, 2019, Arxiv 1908.08374.

[24] R. M. Sainsbury. *Paradoxes*, Cambridge University Press, 2009.

[25] C. Seife. *Decoding the Universe: How the Science of Information is Explaining Everything in the Cosmos, from our Brains to Black Holes*, Penguin, 2006.

[26] T. Siegfried. *The Bit and the Pendulum: From Quantum Computing to M Theory - The New Physics of Information*, John Wiley & Sons, 2000.

[27] L. Smolin. *Time Reborn: From the Crisis in Physics to he Future of the Universe*, Penguin, 2004.

[28] G. Smoot and K. Davidson. *Wrinkles in Time: Witness to the Birth of the Universe*, Harper Perennial, 1993.

[29] G. 't Hooft and S. Vandoren. *Time in Powers of Ten: Natural Phenomena and Their Timescales*, World Scientific Singapore, 2014.

[30] S. Weinberg. *The First Three Minutes: A Mordern View of the Origin of the Universe*, Basic, 1977.

[31] H. Weyl. *Space Time Matter* (Translated by H. L. Brose), Dover, 1950.

[32] M. M. Woolfson. *Time and Age: Time Machines, Relativity and Fossils*, Imperial College Press, 2015.

[33] J. H. van Lint and R. M. Wilson. *A Course in Combinatorics*, 2nd edition, Cambridge University Press, 2001.

[34] L. X. Wang. *An Elementary Treatise on Combinations*, Harbin Institute of Technology Press, 2012 (in Chinese).

[35] L. Z. Xu, M. S. Jiang and Z. Q. Zhu. *Combinatorial Mathematics of Computation*, Shanghai Science & Technology Press, 1983 (in Chinese).

[36] 菲利普 津巴多, 约翰 博伊德著;张迪衡译,中信出版社 2018oP. Zimbardo and J. Boyd. *The Paradox: The New Psychology of Time That Will Change Your Life*, The Marsh Agency Ltd, 2008.

[37] Yang-Mills equations. http://en.wikipedia.org/wiki/Yang-Mills_equations.

[38] BSPT instanton. http://en.wikipedia.org/wiki/BSPT_instanton.

[39] Z. J. Zheng and A. Maeder. The conjugate classification of the kernel form of the hexagonal grid. In *Modern Geometric Computing for Visualization*, edited by T. L. Kunii and Y. Shinagawa, Springer-Verlag, 1992, 73–89.

[40] Z. J. Zheng. *Conjugate Transformation of Regular Plan Lattices for Binary Images*, PhD Thesis, Monash University, 1994.

[41] Jeffrey Z. J. Zheng. *Variant Construction from Theoretical Foundation to Applications*, Springer Nature, 2019.

[42] 郑智捷,变值体系理论及其应用 第一卷 理论基础及其应用,科学出版社 2021. Jeffrey Z. J. Zheng. *Vairant Construction Theory and Applications, Vol. 1, Theoretical Foundation and Applications*, Science Press, 2021.

Chapter 4

Combinatorial Lattice Clusters of Conjugate Complex Vector Functions

Quantitative Distributions and Refined Measurements of Strong Interactions on Complex Planes

Jeffrey Zheng[1]*, Yangyang Tao[2], and Wanyi Zhou[2]
[1]Conjugate System Research, Melbourne, Australia
[2]Yunnan University, Kunming, China

Abstract

In the research of cutting-edge quantum information, quantum entangled states appear in pairs in a variety of fundamental theories of quantum information and quantum interference and other cutting-edge physical experimental scenarios appear in incredible patterns. The state space of complex dynamic systems is a frontier research topic that is worth exploring. In this chapter, the model and method of the hierarchical structural analysis of clock standard signals and conjugate transforms are used and the conjugate function feature space 2^{2m} is used, and its two projection parts of the overall conjugate measurement, narrow 2^{2m} and generalized $(m+1)^2$, are the basic measurement space. According to the symmetric and antisymmetric characteristics of the qualified representation of the conjugate Klein group, the conjugate measurement space is divided into correlated and uncorrelated state representation clusters, and then the probability of each cluster appearing in the conjugate measurement space is calculated to determine the conjugate. Quantitative probability-statistical distribution properties provide the qualified Klein coherent representation. From the perspective of structure measurement, the qualified distribution structure of the conjugated Klein group is used as a reference measurement standard for the fine identification of quantum coherent states.

Keywords: Time basis, conjugate partition, exhaustive distribution, binomial product distribution, conjugate invariance, quantum coherent state, narrow measure, general measure, conjugate klein group, combinatorial lattice, measurement space

---

*Corresponding Author's Email: conjugatelogic@yahoo.com.

In: Conjugate Construction of Quantum Optics
Editors: Jeffrey Zheng
ISBN: 979-8-88697-651-9

Theory attracts practice as magnet attracts iron.
– Carl Friedrich Gauss

Mathematics in general is fundamentally
science of self-evident things.
– Felix Klein

Look deep into nature and then
you will understand everything better.
– Albert Einstein

1. Conjugate Representation and Coherent Measurement

In the classic literature, Weyl's research on space-time matter [1] establishes the foundation for the framework of modern mathematical physics. The modern theory of multivariate complex functions is closely related to the series of studies by Hua Luogeng (Loo-Keng Hua) [2] and Chen Xingshen (Shiing-Shen Chern) [3]. Researches of cutting-edge quantum physics and related systems can be referred in holistic mathematical physics [4]–[14], special number theory research and combinatorial exploration [15]–[18], and lattice theory and their advanced applications [19]–[24].

In Chapter 3, the overall distribution patterns, such as conjugate complementary measurement space, narrow projection conjugate configuration, and general projection, are described. In Chapter 5, the frontier quantum theory system and its corresponding various representations of the quantum entangled state and their typical complex geometric invariants satisfying the characteristics of global transformation are reviewed. Related series of materials refer to [25]–[28].

This section summarizes the above to facilitate the understanding of definitions and descriptions for conjugate group qualification.

1.1 Pair of Complex Function and Wave Functions

In the subsequent description, following Hamilton's parenthesis notation, complex numbers are uniformly represented as binary parentheses: $a + ib = (a, b)$, a simplified representation of typical paired wave functions.

$$\Psi = \frac{1}{\sqrt{2}}(\psi_L + i\psi_R) = \frac{1}{\sqrt{2}}(\psi_L, \psi_R)$$

$$\psi_0 = \frac{1}{\sqrt{2}}(\psi_{1L} + i\psi_{1R}) = \frac{1}{\sqrt{2}}(\psi_{1L}, \psi_{1R})$$

$$\psi_\pi = \frac{1}{\sqrt{2}}(\psi_{1L} - i\psi_{1R}) = \frac{1}{\sqrt{2}}(\psi_{1L}, -\psi_{1R})$$

From the perspective of fully describing the conjugate state complex variable function, it is necessary to distinguish four kinds of complex primitive bracket representations $\{(a, b), (a, -b), (-a, b), (-a, -b)\}$, and their conjugate states represent $\{(b, a), (-b, a), (b, -a), (-b, -a)\}$. Each group can be represented by a combination of selected and unselected modes to form an output comprehensive complex function, which can be combined to form $2^4 \times 2^4 = 256$ expressions.

1.2 Conjugate Measurement Process

Using the selected discriminant bits to form two sets of conjugate state clusters through the observation window module, a pairwise representation of the system results is generated for the built-in clock sequence conjugate graph.

The typical processing process is divided into the following:

Input Vector $: m+1$ → Conjugate Logic State Space $: 2^{m+1}$

→ Conjugate Logic Function Space $: 2^{2^{m+1}}$

→ Narrow Conjugate Measure Space $: 2^{2m}$

→ General Conjugate Measure Space $: (m+1)^2$

Process the 2^{m+1} power exponent states collected from the selected $m+1$ bit 0–1 vector, distinguish the given positions 0 and 1, and collect a set of 2^m state features to form a positive conjugate logic function space represented by the intersection function. The hyper-power exponential conjugate logic function space formed from the basic logic state space is $2^{2^{m+1}}$. For details of the related conjugate classification and orthogonal transformation, please refer to [25], [26], Measurement System [27], and Conjugate Transformation Structure [28, chapter 4].

The two groups of conjugates represent the total number of power exponential state spaces formed by $2^m + 2^m = 2^{m+1}$, which can be converted into a superpower exponential conjugate function by arbitrary selection to represent the space of $2^{2^m} \times 2^{2^m} = 2^{2^{m+1}}$. Associated with conjugate orthogonal properties of conjugate state eigenvectors under conjugate transformation, all functions containing more than two similar states are trivial representations. From the initial $2^{2^{m+1}}$, under the orthogonal condition of conjugate logic, 2^m independent states and their own actions remain unchanged, and the system satisfies the nontrivial conjugate logic function representation, from a superpower exponential logical function to a power exponential measure function, and the dimensionality reduction projection is $2^m \times 2^m = 2^{2m}$ group of nontrivial orthogonal measures, forming a narrow projection measure function space. Then, using the degenerate mode of clustering the same measure, the total number of 2^{2m} nontrivial measures is represented as a power exponential mode, which is reduced to a square term growth mode, which is arranged on a square matrix and contains $(m+1)^2$ terms, which constitute the generalized projection conjugate measure space.

Conjugate decomposition and dimensionality reduction process schematic:

$m+1$ Linear Vector Input → 2^{m+1} Exponent Conjugate Logic State Space → $2^{2^{m+1}}$ Hyper Exponent Conjugate Logic Function Space Conjugate Orthogonal Transform Removed Trivial Reduction → 2^{2m} Exponent Narrow Conjugate Measure Space Degenerate Reduction → $(m+1)^2$ Squared General Conjugate Measure Space

For any state variable x, let $\{\sim, \prime\}$ be conjugate and complementary operators; there are four differentiable extended configuration representations $\{x, \tilde{x}, x', \tilde{x}'\}$.

$2(m+1)$ {Meta Variable Space} → State Decomposition → $2^{2(m+1)}$ (Basic Logical State Space)

→ Conjugate Decomposition → $\{2^{(m+1)}\} : \{2^{(m+1)}\}$ (Conjugate Logic State Space)

→ Complete Function → $2^{2^{m+1}+2}$ (Full Conjugate Logic Function Space)

→ Orthogonal Reduction → $2^{2(m+1)}$ (Narrow Measure Function Space)

→ Degenerate Reduction Conjugate, Complement → $4 \times (m+1)^2$ (General Measure Function Space).

Lemma 1 (Conjugate Complementary Measurement Process). *The conjugate + complementary decomposition process involves five core parts, which are the linear input variable* $2(m+1)$, *the conjugate logic state space* $2^{2(m+1)}$, *the full conjugate logic function space* $2^{2^{m+1}+2}$, *the extended narrow projection measure space* $2^{2(m+1)}$, *and its extended generalized projection measure space* $4(m+1)^2$.

For any state $x, \tilde{x}, x', \tilde{x}'$, the corresponding number of connections is given by $s, \tilde{s}, s', \tilde{s}'$. The reference position is k, $j, k \in [0, m]$, and the corresponding measure transformation value is determined by the following formula:

$$s(x) = \sum_{\forall j \neq k} (x_j == 1) \& (x_k == 1) \equiv s \tag{1}$$

$$s(x') = -s(x) \equiv s' \tag{2}$$

$$s(\tilde{x}) = \sum_{\forall j \neq k} (x_j == 0) \& (x_k == 0) \equiv \tilde{s} \tag{3}$$

$$s(\tilde{x}') = -s(\tilde{x}) \equiv \tilde{s}' \tag{4}$$

Among them, the monomorphic $\{-1, 1\} \in \{s, s'\}, \{-i, i\} \in \{\tilde{s}, \tilde{s}'\}, i$ is an imaginary number.

$(s, s' : \tilde{s}, \tilde{s}')$	s'	s
$\tilde{s}$	$(s', \tilde{s})$	$(s, \tilde{s})$
$\tilde{s}'$	$(s', \tilde{s}')$	$(s, \tilde{s}')$

1.3 Conjugate State Distribution

From the viewpoint of conjugate logistic function representation, the logistic function space needs to enumerate all possible state combinations. The total number of positions involved is the superpower exponent $4 \times 2^{2^m} \times 2^{2^m} = 2^{2^{m+1}+2}$. Each block contains a $2^{2^{m+1}}$ configuration.

Conjugate Logic Function Space	$[2^{2^m}-1]'$	...	$0'$	0	...	$[2^{2^m}-1]$
$\tilde{0}$	$([2^{2^m}-1]', \tilde{0})$	...	$(0', \tilde{0})$	$(0, \tilde{0})$	...	$([2^{2^m}-1], \tilde{0})$
...	...		...	...		...
$\widetilde{[2^{2^m}-1]}$	$([2^{2^m}-1]', \widetilde{[2^{2^m}-1]})$	...	$(0', \widetilde{[2^{2^m}-1]})$	$(0, \widetilde{[2^{2^m}-1]})$	...	$([2^{2^m}-1], \widetilde{[2^{2^m}-1]})$
$\widetilde{[2^{2^m}-1]}'$	$([2^{2^m}-1]', \widetilde{[2^{2^m}-1]}')$	...	$(0', \widetilde{[2^{2^m}-1]}')$	$(0, \widetilde{[2^{2^m}-1]}')$	...	$([2^{2^m}-1], \widetilde{[2^{2^m}-1]}')$
...	...		...	...		...
$\tilde{0}'$	$([2^{2^m}-1]', \tilde{0}')$	...	$(0', \tilde{0}')$	$(0, \tilde{0}')$	...	$([2^{2^m}-1], \tilde{0}')$

Under the action of the peculiar orthogonal transformation of the eigenfunction of the conjugate logic vector, all functions with more than two similar state combinations are transformed into trivial representations under the condition of measure projection, and all trivial representations are removed. From the initial superpower exponent, each conjugate logic function block contains a $2^{2^{m+1}}$ configuration, and a large dimensionality reduction forms a nontrivial measure configuration of the power exponent 2^{2m} group, which constitutes a narrow projection in conjugate measures of a representative space.

From the point of view of representation, the narrow projection exhausts all possible combinations of single-state sequence pairs. The total number of positions involved is the power exponent $4 \times 2^{2m} = 2^{2(m+1)}$. Each block contains 2^{2m} configuration.

Narrow Projection Configuration	$[2^m-1]'$	...	$0'$	0	...	$[2^m-1]$
$\tilde{0}$	$([2^m-1]', \tilde{0})$	...	$(0', \tilde{0})$	$(0, \tilde{0})$	...	$([2^m-1], \tilde{0})$
...	...		...	...		...
$\widetilde{[2^m-1]}$	$([2^m-1]', \widetilde{[2^m-1]})$	...	$(0', \widetilde{[2^m-1]})$	$(0, \widetilde{[2^m-1]})$	...	$([2^m-1], \widetilde{[2^m-1]})$
$\widetilde{[2^m-1]}'$	$([2^m-1]', \widetilde{[2^m-1]}')$	...	$(0', \widetilde{[2^m-1]}')$	$(0, \widetilde{[2^m-1]}')$	...	$([2^m-1], \widetilde{[2^m-1]}')$
...	...		...	...		...
$\tilde{0}'$	$([2^m-1]', \tilde{0}')$	...	$(0', \tilde{0}')$	$(0, \tilde{0}')$	...	$([2^m-1], \tilde{0}')$

In the degenerate similar item mode, the total number is the power exponent $2^{2(m+1)}$ configuration, and the dimensionality is greatly reduced through clustering and merging, from the power exponential function growth mode to the generalized projection total of the square function growth Yoke measure space. The extended mode includes the $4 \times (m+1)^2$ configuration.

General Projection $(s, s' : \tilde{s}, \tilde{s}')$	$-m$	...	-0	0	...	m
$\tilde{m}$	$(-m, \tilde{m})$	...	$(-0, \tilde{m})$	$(0, \tilde{m})$	...	$(m, \tilde{m})$
...	...		...	...		...
$\tilde{0}$	$(-m, \tilde{0})$	...	$(-0, \tilde{0})$	$(0, \tilde{0}$	...	$(m, \tilde{0})$
$-\tilde{0}$	$(-m, -\tilde{0})$	...	$(-0, -\tilde{0})$	$(0, -\tilde{0})$	...	$(m, -\tilde{m})$
...	...		...	...		...
$-\tilde{m}$	$(-m, -\tilde{m})$	...	$(-0, -\tilde{m})$	$(0, -\tilde{m})$	...	$(m, -\tilde{m})$

For row i, column j configuration, the number of states collected is $\binom{m}{i} \times \binom{m}{j}$

Number of Degenerate States $(s, \tilde{s})$	$\binom{m}{0}$	...	$\binom{m}{i}$	...	$\binom{m}{m}$
$\binom{m}{0}$	$\binom{m}{0}\binom{m}{0}$	...	$\binom{m}{i}\binom{m}{0}$	...	$\binom{m}{m}\binom{m}{0}$
...	...		...		...
$\binom{m}{j}$	$\binom{m}{0}\binom{m}{j}$	...	$\binom{m}{i}\binom{m}{j}$	...	$\binom{m}{m}\binom{m}{j}$
...	...		...		...
$\binom{m}{m}$	$\binom{m}{0}\binom{m}{m}$	...	$\binom{m}{i}\binom{m}{m}$	...	$\binom{m}{m}\binom{m}{m}$

1.4 Conjugate Basis Representation

For single-term conjugate representation, the conjugation operator $\tilde{}$ converts binary vectors to conjugate numbers, which are then combined into complex parentheses. The complementary operator $'$ works with the conjugate operator $\tilde{}$ to represent the negative value of the corresponding number.

$$\mathbf{I}' = -\mathbf{I} \tag{5}$$

$$\tilde{\mathbf{I}}' = -\tilde{\mathbf{I}} \tag{6}$$

$$\mathbf{I} = \sum_{j=0}^{m} b_j \times 2^j, b_j \in \{0, 1\}, 0 \leq \mathbf{I} < 2^m \tag{7}$$

The smallest quaternion is represented as a 2×2 matrix

$$|I\rangle \equiv \begin{pmatrix} 1 & 0 \\ 0 & 0 \end{pmatrix} \equiv \mathbf{I} \tag{8}$$

$$|I'\rangle \equiv \mathbf{I}' \equiv \begin{pmatrix} 0 & 0 \\ 0 & 1 \end{pmatrix} \tag{9}$$

$$|\tilde{I}\rangle \equiv \begin{pmatrix} 0 & 1 \\ 0 & 0 \end{pmatrix} \equiv \tilde{\mathbf{I}} \tag{10}$$

$$|\tilde{I}'\rangle \equiv \tilde{\mathbf{I}}' \equiv \begin{pmatrix} 0 & 0 \\ 1 & 0 \end{pmatrix} \tag{11}$$

$$\{|I\rangle, |I'\rangle|, \tilde{I}\rangle, |\tilde{I}'\rangle\} \rightarrow \{(\mathbf{I}, \tilde{\mathbf{I}}), (\mathbf{I}, \tilde{\mathbf{I}}'), (\mathbf{I}', \tilde{\mathbf{I}}), (\mathbf{I}', \tilde{\mathbf{I}}')\} \tag{12}$$

$$\sim: \{(\mathbf{I}, \tilde{\mathbf{I}}), (\mathbf{I}, \tilde{\mathbf{I}}'), (\mathbf{I}', \tilde{\mathbf{I}}), (\mathbf{I}', \tilde{\mathbf{I}}')\} \rightarrow \{(\mathbf{I}, \tilde{\mathbf{I}})^{\sim}, (\mathbf{I}, \tilde{\mathbf{I}}')^{\sim}, (\mathbf{I}', \tilde{\mathbf{I}})^{\sim}, (\mathbf{I}', \tilde{\mathbf{I}}')^{\sim}\} \tag{13}$$

Correspondence using measures: $\{(\mathbf{I}, \tilde{\mathbf{I}}) = (a, b), (\mathbf{I}, \tilde{\mathbf{I}}') = (a, -b), (\mathbf{I}', \tilde{\mathbf{I}}) = (-a, b), (\mathbf{I}', \tilde{\mathbf{I}}') = (-a, -b)\}$. Conjugate measure correspondence: $\{(\mathbf{I}, \tilde{\mathbf{I}})^{\sim} = (b, a), (\mathbf{I}, \tilde{\mathbf{I}}')^{\sim} = (-b, a), (\mathbf{I}', \tilde{\mathbf{I}})^{\sim} = (b, -a), (\mathbf{I}', \tilde{\mathbf{I}}')^{\sim} = (-b, -a)\}$.

No.	(a, b)	$(a, -b)$	$(-a, b)$	$(-a, -b)$	$\sum =$
0000					$(0, 0)$
0001				$(-a, -b)$	$(-a, -b)$
0010			$(-a, b)$		$(-a, b)$
0100		$(a, -b)$			$(a, -b)$
1000	(a, b)				(a, b)
0011			$(-a, b)$	$(-a, -b)$	$(-2a, 0)$
0101		$(a, -b)$		$(-a, -b)$	$(0, -2b)$
0110		$(a, -b)$	$(-a, b)$		$(0, 0)$
1001	(a, b)			$(-a, -b)$	$(0, 0)$
1010	(a, b)		$(-a, b)$		$(0, 2b)$
1100	(a, b)	$(a, -b)$			$(2a, 0)$
0111		$(a, -b)$	$(-a, b)$	$(-a, -b)$	$(-a, -b)$
1011	(a, b)		$(-a, b)$	$(-a, -b)$	$(-a, b)$
1101	(a, b)	$(a, -b)$		$(-a, -b)$	$(a, -b)$
1110	(a, b)	$(a, -b)$	$(-a, b)$		(a, b)
1111	(a, b)	$(a, -b)$	$(-a, b)$	$(-a, -b)$	$(0, 0)$

Differentiable complex numbers are represented as nine groups:
$\{(0, 0), (-a, -b), (-a, b), (a, -b), (a, b), (-2a, 0), (0, -2b), (0, 2b), (2a, 0)\}$

Conjugate state: $\{(b, a), (-b, a), (b, -a), (-b, -a)\}$

No.	(b, a)	$(-b, a)$	$(b, -a)$	$(-b, -a)$	$\sum =$
$\tilde{0}$: 0000					$(0, 0)$
$\tilde{1}$: 0001				$(-b, -a)$	$(-b, -a)$
$\tilde{2}$: 0010			$(b, -a)$		$(b, -a)$
$\tilde{4}$: 0100		$(-b, a)$			$(-b, a)$
$\tilde{8}$: 1000	(b, a)				(b, a)
$\tilde{3}$: 0011			$(b, -a)$	$(-b, -a)$	$(0, -2a)$
$\tilde{5}$: 0101		$(-b, a)$		$(-b, -a)$	$(-2b, 0)$
$\tilde{6}$: 0110		$(-b, a)$	$(b, -a)$		$(0, 0)$
$\tilde{9}$: 1001	(b, a)			$(-b, -a)$	$(0, 0)$
$\widetilde{10}$: 1010	(b, a)		$(b, -a)$		$(2b, 0)$
$\widetilde{12}$: 1100	(b, a)	$(-b, a)$			$(0, 2a)$
$\tilde{7}$: 0111		$(-b, a)$	$(b, -a)$	$(-b, -a)$	$(-b, -a)$
$\widetilde{11}$: 1011	(b, a)		$(b, -a)$	$(-b, -a)$	$(b, -a)$
$\widetilde{13}$: 1101	(b, a)	$(-b, a)$		$(-b, -a)$	$(-b, a)$
$\widetilde{14}$: 1110	(b, a)	$(-b, a)$	$(b, -a)$		(b, a)
$\widetilde{15}$: 1111	(b, a)	$(-b, a)$	$(b, -a)$	$(-b, -a)$	$(0, 0)$

There are also nine groups of complex conjugate representations that can be distinguished:
$\{(0,0),(-b,-a),(b,-a),(-b,a),(b,a),(0,-2a),(-2b,0),(2b,0),(0,2a)\}$.

By comprehensively using the conjugate and complementary operators, combined with any number, the following quaternary complex number representation relationship can be formed:

$(,)$	$\mathbf{I}'$	$\mathbf{I}$
$\tilde{\mathbf{I}}$	$(\mathbf{I}',\tilde{\mathbf{I}})$	$(\mathbf{I},\tilde{\mathbf{I}})$
$\tilde{\mathbf{I}}'$	$(\mathbf{I}',\tilde{\mathbf{I}}')$	$(\mathbf{I},\tilde{\mathbf{I}}')$

$=$

$(,)$	$-\mathbf{I}$	$\mathbf{I}$
$\tilde{\mathbf{I}}$	$(-\mathbf{I},\tilde{\mathbf{I}})$	$(\mathbf{I},\tilde{\mathbf{I}})$
$-\tilde{\mathbf{I}}$	$(-\mathbf{I},-\tilde{\mathbf{I}})$	$(\mathbf{I},-\tilde{\mathbf{I}})$

; $\sim:$

$(,)$	$\mathbf{I}'$	$\mathbf{I}$
$\tilde{\mathbf{I}}$	$(\mathbf{I}',\tilde{\mathbf{I}})$	$(\mathbf{I},\tilde{\mathbf{I}})$
$\tilde{\mathbf{I}}'$	$(\mathbf{I}',\tilde{\mathbf{I}}')$	$(\mathbf{I},\tilde{\mathbf{I}}')$

$\rightarrow$

$(,)^{\sim}$	$\mathbf{I}'$	$\mathbf{I}$
$\tilde{\mathbf{I}}$	$(\tilde{\mathbf{I}},\mathbf{I}')$	$(\tilde{\mathbf{I}},\mathbf{I})$
$\tilde{\mathbf{I}}'$	$(\tilde{\mathbf{I}}',\mathbf{I}')$	$(\tilde{\mathbf{I}}',\mathbf{I})$

For the eight-element primitive of the same number, the product can be further expanded:

$(,)$	$\mathbf{I}'$	$\mathbf{I}$
$\tilde{\mathbf{I}}$	$(\mathbf{I}',\tilde{\mathbf{I}})$	$(\mathbf{I},\tilde{\mathbf{I}})$
$\tilde{\mathbf{I}}'$	$(\mathbf{I}',\tilde{\mathbf{I}}')$	$(\mathbf{I},\tilde{\mathbf{I}}')$

$\otimes$

$(,)^{\sim}$	$\mathbf{I}'$	$\mathbf{I}$
$\tilde{\mathbf{I}}$	$(\tilde{\mathbf{I}},\mathbf{I}')$	$(\tilde{\mathbf{I}},\mathbf{I})$
$\tilde{\mathbf{I}}'$	$(\tilde{\mathbf{I}}',\mathbf{I}')$	$(\tilde{\mathbf{I}}',\mathbf{I})$

1.5 Conjugate Algebraic Representation

Set $\{\mathbf{0},\mathbf{0}',\tilde{\mathbf{0}},\tilde{\mathbf{0}}'\} = \{|0\rangle,|0'\rangle,|\tilde{0}\rangle,|\tilde{0}'\rangle\}$ contains a generative structure whose four elements are zero-element conjugate vector spaces and generates four sets of complex representations.

Set $\{\mathbf{0},\mathbf{0}',\tilde{\mathbf{0}},\tilde{\mathbf{0}}'\} = \{|0\rangle,|0'\rangle,|\tilde{0}\rangle,|\tilde{0}'\rangle\}$ containing eight elements is a generative structure of a one-element conjugate vector space, generating sixteen sets of complex representations.

$(,)$	$\mathbf{0}'$	$\mathbf{0}$
$\tilde{\mathbf{0}}$	$(\mathbf{0}',\tilde{\mathbf{0}})$	$(\mathbf{0},\tilde{\mathbf{0}})$
$\tilde{\mathbf{0}}'$	$(\mathbf{0}',\tilde{\mathbf{0}}')$	$(\mathbf{0},\tilde{\mathbf{0}}')$

Set $\{\mathbf{0},\mathbf{1},\mathbf{0}',\mathbf{1}',\tilde{\mathbf{0}},\tilde{\mathbf{1}},\tilde{\mathbf{0}}',\tilde{\mathbf{1}}'\} = \{|0\rangle,|0'\rangle,|1\rangle,|1'\rangle,|\tilde{0}\rangle,|\tilde{0}'\rangle,|\tilde{1}\rangle,|\tilde{1}'\rangle\}$ containing eight elements is a generative structure of a one-element conjugate vector space, generating sixteen sets of complex representations.

$(,)$	$\mathbf{1}'$	$\mathbf{0}'$	$\mathbf{0}$	$\mathbf{1}$
$\tilde{\mathbf{1}}$	$(\mathbf{1}',\tilde{\mathbf{1}})$	$(\mathbf{0}',\tilde{\mathbf{1}})$	$(\mathbf{0},\tilde{\mathbf{1}})$	$(\mathbf{1},\tilde{\mathbf{1}})$
$\tilde{\mathbf{0}}$	$(\mathbf{1}',\tilde{\mathbf{0}})$	$(\mathbf{0}',\tilde{\mathbf{0}})$	$(\mathbf{0},\tilde{\mathbf{0}})$	$(\mathbf{1},\tilde{\mathbf{0}})$
$\tilde{\mathbf{0}}'$	$(\mathbf{1}',\tilde{\mathbf{0}}')$	$(\mathbf{0}',\tilde{\mathbf{0}}')$	$(\mathbf{0},\tilde{\mathbf{0}}')$	$(\mathbf{1},\tilde{\mathbf{0}}')$
$\tilde{\mathbf{1}}'$	$(\mathbf{1}',\tilde{\mathbf{1}}')$	$(\mathbf{0}',\tilde{\mathbf{1}}')$	$(\mathbf{0},\tilde{\mathbf{1}}')$	$(\mathbf{1},\tilde{\mathbf{1}}')$

Set $\{\mathbf{0},\mathbf{1},\mathbf{2},\mathbf{3},\mathbf{0}',\mathbf{1}',\mathbf{2}',\mathbf{3}',\tilde{\mathbf{0}},\tilde{\mathbf{1}},\tilde{\mathbf{2}},\tilde{\mathbf{3}},\tilde{\mathbf{0}}',\tilde{\mathbf{1}}',\tilde{\mathbf{2}}',\tilde{\mathbf{3}}'\} = \{|00\rangle,|01\rangle,|10\rangle,|11\rangle,|00'\rangle,|01'\rangle,|10'\rangle,|11'\rangle,|\tilde{00}\rangle,|\tilde{01}\rangle,|\tilde{10}\rangle,|\tilde{11}\rangle|\tilde{00}'\rangle,|\tilde{01}'\rangle,|\tilde{10}'\rangle,|\tilde{11}'\rangle\}$ containing sixteen elements is a generative structure of a two-element conjugate vector space, generating sixty-four sets of complex representations.

Using the conjugation relationship of these numbers, the following sixteen basic combinations are formed:

(,)	**0**	**1**	**2**	**3**
$\tilde{0}$	$(0,\tilde{0})$	$(1,\tilde{0})$	$(2,\tilde{0})$	$(3,\tilde{0})$
$\tilde{1}$	$(0,\tilde{1})$	$(1,\tilde{1})$	$(2,\tilde{1})$	$(3,\tilde{1})$
$\tilde{2}$	$(0,\tilde{2})$	$(1,\tilde{2})$	$(2,\tilde{2})$	$(3,\tilde{2})$
$\tilde{3}$	$(0,\tilde{3})$	$(1,\tilde{3})$	$(2,\tilde{3})$	$(3,\tilde{3})$

Using the complement operator, the sixty-four groups of complex function relations are expressed as

(,)	$3'$	$2'$	$1'$	$0'$	**0**	**1**	**2**	**3**
$\tilde{0}$	$(3',\tilde{0})$	$(2',\tilde{0})$	$(1',\tilde{0})$	$(0',\tilde{0})$	$(0,\tilde{0})$	$(1,\tilde{0})$	$(2,\tilde{0})$	$(3,\tilde{0})$
$\tilde{1}$	$(3',\tilde{1})$	$(2',\tilde{1})$	$(1',\tilde{1})$	$(0',\tilde{1})$	$(0,\tilde{1})$	$(1,\tilde{1})$	$(2,\tilde{1})$	$(3,\tilde{1})$
$\tilde{2}$	$(3',\tilde{2})$	$(2',\tilde{2})$	$(1',\tilde{2})$	$(0',\tilde{2})$	$(0,\tilde{2})$	$(1,\tilde{2})$	$(2,\tilde{2})$	$(3,\tilde{2})$
$\tilde{3}$	$(3',\tilde{3})$	$(2',\tilde{3})$	$(1',\tilde{3})$	$(0',\tilde{3})$	$(0,\tilde{3})$	$(1,\tilde{3})$	$(2,\tilde{3})$	$(3,\tilde{3})$
$\tilde{3}'$	$(3',\tilde{3}')$	$(2',\tilde{3}')$	$(1',\tilde{3}')$	$(0',\tilde{3}')$	$(0,\tilde{3}')$	$(1,\tilde{3}')$	$(2,\tilde{3}')$	$(3,\tilde{3}')$
$\tilde{2}'$	$(3',\tilde{2}')$	$(2',\tilde{2}')$	$(1',\tilde{2}')$	$(0',\tilde{2}')$	$(0,\tilde{2}')$	$(1,\tilde{2}')$	$(2,\tilde{2}')$	$(3,\tilde{2}')$
$\tilde{1}'$	$(3',\tilde{1}')$	$(2',\tilde{1}')$	$(1',\tilde{1}')$	$(0',\tilde{1}')$	$(0,\tilde{1}')$	$(1,\tilde{1}')$	$(2,\tilde{1}')$	$(3,\tilde{1}')$
$\tilde{0}'$	$(3',\tilde{0}')$	$(2',\tilde{0}')$	$(1',\tilde{0}')$	$(0',\tilde{0}')$	$(0,\tilde{0}')$	$(1,\tilde{0}')$	$(2,\tilde{0}')$	$(3,\tilde{0}')$

Reverting the complement operator to the negative sign, we can obtain the following representation:

(,)	**−3**	**−2**	**−1**	**−0**	**0**	**1**	**2**	**3**
$\tilde{0}$	$(-3,\tilde{0})$	$(-2,\tilde{0})$	$(-1,\tilde{0})$	$(-0,\tilde{0})$	$(0,\tilde{0})$	$(1,\tilde{0})$	$(2,\tilde{0})$	$(3,\tilde{0})$
$\tilde{1}$	$(-3,\tilde{1})$	$(-2,\tilde{1})$	$(-1,\tilde{1})$	$(-0,\tilde{1})$	$(0,\tilde{1})$	$(1,\tilde{1})$	$(2,\tilde{1})$	$(3,\tilde{1})$
$\tilde{2}$	$(-3,\tilde{2})$	$(-2,\tilde{2})$	$(-1,\tilde{2})$	$(-0,\tilde{2})$	$(0,\tilde{2})$	$(1,\tilde{2})$	$(2,\tilde{2})$	$(3,\tilde{2})$
$\tilde{3}$	$(-3,\tilde{3})$	$(-2,\tilde{3})$	$(-1,\tilde{3})$	$(-0,\tilde{3})$	$(0,\tilde{3})$	$(1,\tilde{3})$	$(2,\tilde{3})$	$(3,\tilde{3})$
$-\tilde{3}$	$(-3,-\tilde{3})$	$(-2,-\tilde{3})$	$(-1,-\tilde{3})$	$(-0,-\tilde{3})$	$(0,-\tilde{3})$	$(1,-\tilde{3})$	$(2,-\tilde{3})$	$(3,-\tilde{3})$
$-\tilde{2}$	$(-3,-\tilde{2})$	$(-2,-\tilde{2})$	$(-1,-\tilde{2})$	$(-0,-\tilde{2})$	$(0,-\tilde{2})$	$(1,-\tilde{2})$	$(2,-\tilde{2})$	$(3,-\tilde{2})$
$-\tilde{1}$	$(-3,-\tilde{1})$	$(-2,-\tilde{1})$	$(-1,-\tilde{1})$	$(-0,-\tilde{1})$	$(0,-\tilde{1})$	$(1,-\tilde{1})$	$(2,-\tilde{1})$	$(3,-\tilde{1})$
$-\tilde{0}$	$(-3,-\tilde{0})$	$(-2,-\tilde{0})$	$(-1,-\tilde{0})$	$(-0,-\tilde{0})$	$(0,-\tilde{0})$	$(1,-\tilde{0})$	$(2,-\tilde{0})$	$(3,-\tilde{0})$

1.6 Typical Case m = 3

$(s,\tilde{s})$ Conjugate Value	000	001	010	100	011	101	110	111
$\widetilde{000}$	(0, 3)	(1, 3)	(1, 3)	(1, 3)	(2, 3)	(2, 3)	(2, 3)	(3, 3)
$\widetilde{001}$	(0, 2)	(1, 2)	(1, 2)	(1, 2)	(2, 2)	(2, 2)	(2, 2)	(3, 2)
$\widetilde{010}$	(0, 2)	(1, 2)	(1, 2)	(1, 2)	(2, 2)	(2, 2)	(2, 2)	(3, 2)
$\widetilde{100}$	(0, 2)	(1, 2)	(1, 2)	(1, 2)	(2, 2)	(2, 2)	(2, 2)	(3, 2)
$\widetilde{011}$	(0, 1)	(1, 1)	(1, 1)	(1, 1)	(2, 1)	(2, 1)	(2, 1)	(3, 1)
$\widetilde{101}$	(0, 1)	(1, 1)	(1, 1)	(1, 1)	(2, 1)	(2, 1)	(2, 1)	(3, 1)
$\widetilde{110}$	(0, 1)	(1, 1)	(1, 1)	(1, 1)	(2, 1)	(2, 1)	(2, 1)	(3, 1)
$\widetilde{111}$	(0, 0)	(1, 0)	(1, 0)	(1, 0)	(2, 0)	(2, 0)	(2, 0)	(3, 0)

→

Degenerate Value	000	{001, 010, 100}	{011, 110, 101}	111
$\widetilde{000}$	$(0,3)$	$3\times(1,3)$	$3\times(2,3)$	$(3,3)$
$\{\widetilde{001},\widetilde{010},\widetilde{100}\}$	$3\times(0,2)$	$9\times(1,2)$	$9\times(2,2)$	$3\times(3,2)$
$\{\widetilde{011},\widetilde{110},\widetilde{101}\}$	$3\times(0,1)$	$9\times(1,1)$	$9\times(2,1)$	$3\times(3,1)$
$\widetilde{111}$	$(0,0)$	$3\times(1,0)$	$3\times(2,0)$	$(3,0)$

$$m = 3 \text{ Degenerate Coefficient Matrix } (8\times 8) = \begin{array}{|cccc|cccc|}\hline 1 & 3 & 3 & 1 & 1 & 3 & 3 & 1 \\ 3 & 9 & 9 & 3 & 3 & 9 & 9 & 3 \\ 3 & 9 & 9 & 3 & 3 & 9 & 9 & 3 \\ 1 & 3 & 3 & 1 & 1 & 3 & 3 & 1 \\ \hline 1 & 3 & 3 & 1 & 1 & 3 & 3 & 1 \\ 3 & 9 & 9 & 3 & 3 & 9 & 9 & 3 \\ 3 & 9 & 9 & 3 & 3 & 9 & 9 & 3 \\ 1 & 3 & 3 & 1 & 1 & 3 & 3 & 1 \\ \hline \end{array}$$

2. Conjugate Combinatorial Lattice

Existing conjugated structures represent the results, and very complex combinatorial features can be seen. From the conjugate characteristics, the complementary operational structures are different from typical algebraic representations such as groups, rings, and fields. The aim of this chapter explores deeply group-qualified structures derived from vector logic to properly describe the interaction of quantum coherent states.

2.1 Conjugate Combinatorial Lattice Generation Process

To facilitate the description of the subsequent complex interaction distributions, the constructions were adapted using the models and methods already documented in "Fundamentals of Representation of Entangled States in Classical Quantum Measurements and Global Geometric Transformations" further information referred Chapters 5 & 6 of the book.

Let the structure in question be called "**Conjugate Klein Group Combinatorial Lattice**" and abbreviated as "**Conjugate Combinatorial Lattice**" or "**Combinatorial Lattice**".

From selecting the ground state to the conjugated combination lattice, the generation process of the lattice structure is illustrated by the {input, expansion, refinement, generation, output} five-part process:

[Input: Choose Ground State B] → [Extension: Klein Group $K = \{e, u, v, w\}$, Conjugate Klein Group $\tilde{K} = \{\tilde{e}, \tilde{u}, \tilde{v}, \tilde{w}\}$]

→ [Extract: Conjugate Combination Primitives: $K^1(B), \tilde{K}^1(B)$] → [Generate: Combinatorial Clusters: $K^1(B) \bigoplus \tilde{K}^1(B)$]

→ [Output: Conjugate Combinatorial Lattice: 9×9]

2.2 Conjugate Combinatorial Lattice Step by Step

2.2.1 Input Ground State

Choose any complex order pair $(a, b) \in C$ as the ground state $B = \{(a, b)\}$.

2.2.2 Extended Klein Group and Conjugate Structure

Let the Klein four-group matrix be expressed as $K = \{e, u, v, w\}$, and let the conjugate state be

$$K = \left\{ e = \begin{pmatrix} 1 & 0 \\ 0 & 1 \end{pmatrix}, u = \begin{pmatrix} -1 & 0 \\ 0 & 1 \end{pmatrix}, v = \begin{pmatrix} 1 & 0 \\ 0 & -1 \end{pmatrix}, w = \begin{pmatrix} -1 & 0 \\ 0 & -1 \end{pmatrix} \right\} \tag{14}$$

$$\tilde{K} = \left\{ \tilde{e} = \begin{pmatrix} 0 & 1 \\ 1 & 0 \end{pmatrix}, \tilde{u} = \begin{pmatrix} 0 & 1 \\ -1 & 0 \end{pmatrix}, \tilde{v} = \begin{pmatrix} 0 & -1 \\ 1 & 0 \end{pmatrix}, \tilde{w} = \begin{pmatrix} 0 & -1 \\ -1 & 0 \end{pmatrix} \right\} \tag{15}$$

$K, \tilde{K}$ act on the ground state B to form two sets of feature clusters.

$$K(B) = \{(a, b), (-a, b), (a, -b), (-a, -b)\} \tag{16}$$

$$\tilde{K}(B) = \{(b, a), (b, -a), (-b, a), (-b, -a)\}. \tag{17}$$

2.2.3 Refinement Combinatorial Generator

In the two sets of matrices, each matrix is a nonzero representation, and the four matrices are a set of generating base representations. The corresponding quaternary binary number is

$$K = \{K_{0001} = e, K_{0010} = u, K_{0100} = v, K_{1000} = w\} \tag{18}$$

$$\tilde{K} = \{\tilde{K}_{0001} = \tilde{e}, \tilde{K}_{0010} = \tilde{u}, K_{0100} = \tilde{v}, \tilde{K}_{1000} = \tilde{w}\} \tag{19}$$

$K^1, \tilde{K}^1$ represent a combination of four primitives, each containing sixteen cases:

$$K^1 = \{K_{0000}, \ldots, K_I, \ldots, K_{1111}\}, 0 \le I < 16 \tag{20}$$

$$\tilde{K}^1 = \{\tilde{K}_{0000}, \ldots, \tilde{K}_J, \ldots, \tilde{K}_{1111}\}, 0 \le J < 16 \tag{21}$$

i.e.

$$K_{0000} = 0 \cdot w + 0 \cdot v + 0 \cdot u + 0 \cdot e = \begin{pmatrix} 0 & 0 \\ 0 & 0 \end{pmatrix}$$

$$K_{0101} = 0 \cdot w + 1 \cdot v + 0 \cdot u + 1 \cdot e = \begin{pmatrix} 1 & 0 \\ 0 & -1 \end{pmatrix} + \begin{pmatrix} 1 & 0 \\ 0 & 1 \end{pmatrix} = \begin{pmatrix} 2 & 0 \\ 0 & 0 \end{pmatrix}$$

$$\begin{aligned} K_{1111} &= 1 \cdot w + 1 \cdot v + 1 \cdot u + 1 \cdot e \\ &= K_{0110} = 0 \cdot w + 1 \cdot v + 1 \cdot u + 0 \cdot e \\ &= K_{1001} = 1 \cdot w + 0 \cdot v + 0 \cdot u + 1 \cdot e \\ &= \begin{pmatrix} 0 & 0 \\ 0 & 0 \end{pmatrix} \end{aligned}$$

Other numbers can be calculated with the corresponding add operator, and it can be observed that different numbers generate the same matrix, accompanied by degeneration. Such results indicate that the selected base element clusters have intrinsic correlation characteristics.

The resulting feature clusters can be extended by a combinatorial lattice of sum-accumulator operators. The first combined expansion is denoted as $K^1, \tilde{K}^1$, and different distinguishable representations are combined to form an extended basis vector, each with nine elements.

$$\begin{aligned}
K^1(B) = \{&K_0^1(B) = K_{0000}(B) = (0,0), K_1^1(B) = K_{0001}(B) = (a,b),\\
&K_2^1(B) = K_{0010}(B) = (-a,b), K_3^1(B) = K_{0100}(B) = (a,-b),\\
&K_4^1(B) = K_{1000}(B) = (-a,-b), K_5^1(B) = K_{0011}(B) = (0,2b),\\
&K_6^1(B) = K_{0101}(B) = (2a,0), K_0^1(B) = K_{1010}(B) = (0,0),\\
&K_0^1(B) = K_{0110}(B) = (0,0), K_7^1(B) = K_{1010}(B) = (-2a,0),\\
&K_8^1(B) = K_{1100}(B) = (0,-2b), K_1^1(B) = K_{0111}(B) = (a,b),\\
&K_2^1(B) = K_{1011}(B) = (-a,b), K_3^1(B) = K_{1101}(B) = (a,-b),\\
&K_4^1(B) = K_{1110}(B) = (-a,-b), K_0^1(B) = K_{1111}(B) = (0,0)\}\\
= \{&(0,0),(a,b),(-a,b),(a,-b),(-a,-b),\\
&(0,2b),(2a,0),(-2a,0),(0,-2b)\}\\
= \{&K_j^1\}_{j=0}^{8}
\end{aligned}$$

$$\begin{aligned}
\tilde{K}^1(B) = \{&\tilde{K}_0^1(B) = \tilde{K}_{0000}(B) = (0,0), \tilde{K}_1^1(B) = K_{0001}(B) = (b,a),\\
&\tilde{K}_2^1(B) = \tilde{K}_{0010}(B) = (b,-a), \tilde{K}_3^1(B) = \tilde{K}_{0100}(B) = (-b,a),\\
&\tilde{K}_4^1(B) = \tilde{K}_{1000}(B) = (-b,-a), \tilde{K}_5^1(B) = \tilde{K}_{0011}(B) = (2b,0),\\
&\tilde{K}_6^1(B) = \tilde{K}_{0101}(B) = (0,2a), \tilde{K}_0^1(B) = \tilde{K}_{1010}(B) = (0,0),\\
&\tilde{K}_0^1(B) = \tilde{K}_{0110}(B) = (0,0), \tilde{K}_7^1(B) = \tilde{K}_{1010}(B) = (0,-2a),\\
&\tilde{K}_8^1(B) = \tilde{K}_{1100}(B) = (-2b,0), \tilde{K}_1^1(B) = \tilde{K}_{0111}(B) = (b,a),\\
&\tilde{K}_2^1(B) = \tilde{K}_{1011}(B) = (b,-a), \tilde{K}_3^1(B) = \tilde{K}_{1101}(B) = (-b,a),\\
&\tilde{K}_4^1(B) = \tilde{K}_{1110}(B) = (-b,-a), \tilde{K}_0^1(B) = \tilde{K}_{1111}(B) = (0,0)\}\\
= \{&(0,0),(b,a),(b,-a),(-b,a),(-b,-a),\\
&(2b,0),(0,2a),(0,-2a),(-2b,0)\}\\
= \{&\tilde{K}_j^1\}_{j=0}^{8}.
\end{aligned}$$

2.2.4 Generate Combinatorial Lattice Clusters

If two sets of eigenvectors are used to perform the addition operation between different elements, the generated result is equivalent to selecting all primitives of $\{K^1(B), \tilde{K}^1(B)\}$ and combining the conjugated lattices of states. The element at each (i,j) position satisfies the formula.

$$K^1(B) \oplus \tilde{K}^1(B) = \left\{\forall i,j | [K^1(B) \oplus \tilde{K}^1(B)]_{i,j} = K_I^1(B) + \tilde{K}_j^1(B), 0 \leq i,j \leq 8\right\} \tag{22}$$

For the corresponding 9×9 matrix, the conjugate state combination lattice can be expressed as

$K^1(B) \bigoplus \tilde{K}^1(B)$	$K_0^1(B)$	$\cdots$	$K_I^1(B)$	$\cdots$	$K_8^1(B)$
$\tilde{K}_0^1(B)$	$K_0^1(B) + \tilde{K}_0^1(B)$	$\cdots$	$K_I^1(B) + \tilde{K}_0^1(B)$	$\cdots$	$K_8^1(B) + \tilde{K}_0^1(B)$
$\vdots$	$\vdots$		$\vdots$		$\vdots$
$\tilde{K}_J^1(B)$	$K_0^1(B) + \tilde{K}_J^1(B)$	$\cdots$	$K_I^1(B) + \tilde{K}_J^1(B)$	$\cdots$	$K_8^1(B) + \tilde{K}_J^1(B)$
$\vdots$	$\vdots$		$\vdots$		$\vdots$
$\tilde{K}_8^1(B)$	$K_0^1(B) + \tilde{K}_8^1(B)$	$\cdots$	$K_I^1(B) + \tilde{K}_8^1(B)$	$\cdots$	$K_8^1(B) + \tilde{K}_8^1(B)$

2.2.5 Output Conjugate Combinatorial Lattice

Once the $K^1(B) \oplus \tilde{K}^1(B)$ operation is completed, the entire processing flow ends and the corresponding output result is obtained.

3. Properties of Conjugate Lattice Quantitative Distributions

3.1 Basic Features

For the constructed conjugated lattices, through a series of basic theoretical characteristic analyses, the unique distinguishable structural distribution characteristics of conjugated lattices are shown.

Lemma 2 (Conjugate Feature Group + Multiplicative Identity Element). *K4 group* $K = \{e, u, v, w\}$ *contains four elements;* $e = \begin{pmatrix} 1 & 0 \\ 0 & 1 \end{pmatrix}$ *is the multiplication unit; adding the extended conjugate structure* $\tilde{K} = \{\tilde{e}, \tilde{u}, \tilde{v}, \tilde{w}\}$, $\{K, \tilde{K}\}$ *form* D_4 *group including eight elements* $\{K, \tilde{K}\}$.

Corollary 1 (Conjugate Global Invariance). *For the* D_4 *group, the eight-element cluster under the action of the conjugate operator,* $D_4 = \tilde{D}_4$, *as a whole remains unchanged.*

Proof.

$$\begin{aligned} D_4 &= \{e, u, v, w, \tilde{e}, \tilde{u}, \tilde{v}, \tilde{w}\} \\ &= \{\tilde{e}, \tilde{u}, \tilde{v}, \tilde{w}, e, u, v, w\} \\ &= \tilde{D}_4 \end{aligned}$$

Theorem 1 (Eigenvector Length Constraint). *For arbitrary states* $B = (a, b)$, $K(B)$, *and* $\tilde{K}(B)$ *The distinguishable eigenvector length of* ≤ 4.

Proof. When $a > 0, b = 0$, the eigenvector length is 2. When $a, b > 0$, the eigenvector length is 4.

Theorem 2 (Additive Group + Inverse/Zero Combination). *For arbitrary states* $B = (a, b)$, $K(B)$, $\tilde{K}(B)$, *and the combined state of the components* $\{K, \tilde{K}\}$, *each contains three sets of* *zero elements of the zero matrix corresponding to the additive group, and the four elements are two pairs of eigenvalues, which provide inverse features of paired values. Reverse paired eigenvalues ensure that after the eigenvalue clusters are accumulated as a whole, the sum is expressed as a zero element* $= (0,0)$.

Proof.

$$(a,b)\ :\ e+w=u+v=e+u+v+w=\begin{pmatrix}0\,0\\0\,0\end{pmatrix}\ K \text{ Three Zero Matrices}$$
$$=\tilde{e}+\tilde{w}=\tilde{u}+\tilde{v}=\tilde{e}+\tilde{u}+\tilde{v}+\tilde{w}=\begin{pmatrix}0\,0\\0\,0\end{pmatrix}\ \tilde{K} \text{ Three Zero Matrices}$$
$$\rightarrow (0,0) \text{ Zero.}$$

Corollary 2 (Complementary Digestion Properties). *Choose any combination of three primitives from the four nonzero primitives of* $K(B)$ *or* $\tilde{K}(B)$ *addition, equivalent to selecting a single primitive.*

Proof. According to **Theorem 2**, if three primitives are arbitrarily selected, a pair of features must be cancelled after the accumulation operation, leaving only one primitive.

3.2 Combination Properties

Theorem 3 (Combination Eigenvalues). *According to* **Theorem 2** *and* **Corollary 2**, *for any state* $B=(a,b)$, $K^1(B)$ *or* $\tilde{K}^1(B)$, $2^4=16$ *combined eigenvalues of, respectively, including four zeros and four pairs of identical features value, the length of distinguishable eigenvectors after merging is* ≤ 9.

Proof. When $a>0, b=0$, there are more identical parts in the combined state, the length of the distinguishable feature vector is less than 9, and it is equal to 9 in general.

Corollary 3 (Partial Sum Zero). *Accumulate from nine combined primitives of* $K^1(B)$ *or* $\tilde{K}^1(B)$. *The sum is zero.*

$$\sum_{0\leq i\leq 8}\left[K^1(B)\right]_i=\sum_{i=0}^{8}K_i^1(B)=(0,0) \tag{23}$$

$$\sum_{0\leq j\leq 8}\left[\tilde{K}^1(B)\right]_j=\sum_{j=0}^{8}\tilde{K}_j^1(B)=(0,0) \tag{24}$$

Proof. Except for one intrinsic zero element, eight combined primitives form four pairs of mutually opposite eigenvalues, and the sum is a zero element after accumulating and summing up.

Theorem 4 (Combined Lattice Eigenvalues Sum Zero). *All* $K^1(B)\bigoplus\tilde{K}^1(B)$ *elements sum to zero after accumulation.*

$$\sum_{0\leq i,j\leq 8}\left[K^1(B)\oplus\tilde{K}^1(B)\right]_{i,j}=\sum_{i=0}^{8}\sum_{j=0}^{8}\left[K_i^1(B)+\tilde{K}_j^1(B)\right]=(0,0) \tag{25}$$

Proof. Use **Corollary 3** to combine the row and column summation parts, and the sum is zero after the cumulative sum.

Theorem 5 (Combination Lattice Equivalence). *Using $\{K(B), \tilde{K}(B)\}$ eigenvectors constructed $K^1(B) \bigoplus \tilde{K}^1(B)$, which is equivalent to the combined lattice structure formed by the eight element feature clustering contained in $D_4(B)$.*

$$K^1(B) \oplus \tilde{K}^1(B) = D_4^1(B) \tag{26}$$

Proof.

$$\begin{aligned}
D_4^1(B)_I &= \sum_{j=0}^{7} I_j \times D_4(B)_j, 0 \le I < 256 \\
&= I_0 \times e + I_1 \times u + I_2 \times v + I_3 \times w + I_4 \times \tilde{e} + I_5 \times \tilde{u} + I_6 \times \tilde{v} + I_7 \times \tilde{w} \\
&= [K^1(B) \oplus \tilde{K}^1(B)]_{J,K} \\
&\quad J = \sum_{j=0}^{3} I_j \times 2^j, K = \sum_{k=4}^{7} I_j \times 2^{k-4}, 0 \le J, K \le 16 \\
&= [K^1(B) \oplus \tilde{K}^1(B)]_{J',K'} \\
&\quad 0 \le J', K' \le 8 \text{ satisfies } \textbf{Theorem 4} \text{ condition, renumbering}
\end{aligned}$$

Corollary 4 (Conjugates satisfy combinatorial conditions). *For any choice state B, conjugates $K^1(B) \bigoplus \tilde{K}^1(B)$ satisfy all the basic conditions defined by abstract lattice theory.*

Proof. According to **Lemma 1, Theorem 2**, the conjugate lattice has the identity element of the multiplicative group and the negative and zero elements of the additive group. The additive group involved itself is an Abelian group, and it is not difficult to verify that the conjugate lattice also satisfies the associative law of addition, the commutative law of addition, the associative law of multiplication, and the distributive law of multiplication.

Corollary 5 (Maximum Number of Distinguishable Conjugated Combinatorial Lattice Points). *For any choice state B, conjugated combinatorial lattice $K^1(B) \bigoplus \tilde{K}^1(B)$. The feature distribution contains at most eighty-one distinguishable elements in the 9×9 table.*

Corollary 6 (Number of Nonzero Conjugate Combinatorial Lattice Points). *Under general conditions, the conjugated combinatorial lattice $K^1(B) \bigoplus \tilde{K}^1(B)$ provides eighty nonzero eigenvalues to distinguish lattice distributions.*

Theorem 6 (Combinatorial Eigenvalue Structure Extension). *For any state $B = (a, b)$, if $T(B) = \{u, v, \ldots, x\}$ is a cluster containing m distinguishable eigenvalues, $T'(B) = \{u, v, \ldots, x, y\}$ is an extension of $T(B)$ containing $m + 1$ distinguishable eigenvalues, then the combined cell $T^1(B)$ contains at most 2^m combined eigenvalues, and $T'^1(B)$ contains at most 2^{m+1} combined eigenvalues, all lattice points contained in $T^1(B)$ are completely contained in $T'^1(B)$. Additional features do not reduce the existing grid distribution.*

Proof. In $T'^1(B)$, select the new eigenvalue y projection coefficient to be 0, and the combined lattice point of $T^1(B)$ can be generated. Then, the projection coefficient of the new eigenvalue y is selected to be 1, and $|T^1(B)|$ distinguishable points are newly generated, which may coincide with the old grid points or generate new grid points. Thus, $|T^1(B)| \leq |T'^1(B)|$ holds.

Corollary 7 (Conjugate Combination Lattice Condition). *For any state $B = (a, b)$, if $T(B) = \{u, v, \cdots, x\}$ is a cluster containing m distinguishable eigenvalues, $\exists l \in Z, 0 \leq l \leq m$ can divide the combined lattice into two sublattices, including $\{l, m-l\}$generators corresponding to $\{2^l, 2^{m-l}\}$ lattice points.*

Proof.

$$
\begin{aligned}
T^1(B)_I &= \sum_{j=0}^{m} I_j \times T_j(B), 0 \leq I < 2^m \\
&= I_0 \times T_0(B) + I_1 \times T_1(B) + \cdots + I_j \times T_j(B) + \cdots + I_m \times T_m(B) \\
&= [T^1(B) \bigoplus T^1(B)]_{J,K} \\
& \quad J = \sum_{j=0}^{l} I_j \times 2^j, K = \sum_{k=l}^{m} I_k \times 2^{k-l}, 0 \leq J < 2^l, 0 \leq K < 2^{m-l} \\
&= [T^1(B) \bigoplus T^1(B)]_{J',K'} \\
& \quad 0 \leq J' < 2^l, 0 \leq K < 2^{m-l} \text{ Relabeling after merged}
\end{aligned}
$$

3.3 Optimized Features

Theorem 7 (Combinatorial Lattice Partition Optimization). *For any state $B = (a, b)$, if $T(B) = \{u, v, \ldots, x\}$ is a cluster containing m distinguishable eigenvalues, $\exists l \in Z, 0 \leq l \leq m$ divides the combined lattice into two sublattices, including $\{l, m-l\}$ generator correspondences $\{2^l, 2^{m-l}\}$ lattice points. When $m = 0(mod\ 2), l = \frac{m}{2}$ or $m = 1(mod\ 2), l = \{\frac{m+1}{2}, \frac{m-1}{2}\}$, the combinatorial lattice has overall optimization characteristics.*

Proof. In the differentiated substructure, the relevant computational complexity is $l \times 2^l$ and $(m-l) \times 2^{m-l}$, respectively. The complexity parameter is $F(l) = l \times 2^l + (m-l) \times 2^{m-l}$, and the extreme value condition is determined by differential $\frac{dF(l)}{dl} = 0$.

$$
\begin{aligned}
F(l) &= l \times 2^l + (m-l) \times 2^{m-l} \\
\frac{dF(l)}{dl} &= dl \times 2^l + l \times \frac{d2^l}{dl} + \frac{d(m-l)}{dl} \times 2^{m-l} + (m-l) \times \frac{d2^{m-l}}{dl} \\
&= 2^l + l \times 2^l \times ln\ 2 - 2^{m-l} - (m-l) \times 2^{m-l} \times ln\ 2 \\
&= (2^l - 2^{m-l}) + ln\ 2 \times (l \times 2^l - (m-l) \times 2^{m-l}) = 0 \\
&\Rightarrow \{2^l = 2^{m-l},\ l \times 2^l = (m-l) \times 2^{m-l}\} \\
&\Rightarrow l = m - l \\
&\Rightarrow m = 2l \text{ Minimum Value}
\end{aligned}
$$

Theorem 8 (Conjugate Lattice Global Optimization). *For any state $B = (a, b)$, $T(B) = \{u, v, \ldots, x\}$ is a cluster containing m distinguishable eigenvalues, and the number of distinguishable eigenvalues of the conjugate combination lattice $\|[T^1(B) \oplus \tilde{T}^1(B)]\| \leq 2^{2m}$ balances the $\{T^1(B), \tilde{T}^1(B)\}$ feature clusters to meet the overall computational optimization characteristics.*

Proof. According to **Theorem 7**, both clusters contain the same number of eigenvalues $|T^1(B)| = |\tilde{T}^1(B)| \leq 2^m$, and the overall number of eigenvalues of the conjugate combination lattice is the product of the two parts $\leq 2^{2m}$. Based on the balance of the two clusters, the calculation amount of the lattice generation of the conjugate group reaches the overall minimum value.

3.4 Symmetrical Distribution

Theorem 9 (Conjugate Symmetric Permutation). *If the feature numbers for any m terms are $\{P(1), P(2), \ldots, P(i), \ldots, P(m)\}, 1 \leq i \leq m$, the corresponding m term conjugate number is $\{\tilde{P}(1), \tilde{P}(2), \ldots, \tilde{P}(j), \ldots, \tilde{P}(m)\}, 1 \leq j \leq m$. Choose any permutation P and its conjugate permutation $\tilde{P}$ is the horizontal and vertical alignment coordinates, because $P(i,j) = P(i) + \tilde{P}(j)$; then the formed $m \times m$ indicates that the matrix has conjugate symmetry.*

$$P(i,j) = (x, y) \& \tilde{P}(i,j) = P(j, i) = (y, x)$$

Proof. Let $P(i) = (a, b)$, $\tilde{P}(j) = (c, d)$, $P(i,j) = P(i) + \tilde{P}(j) = (a, b) + (c, d) = (x, y)$, $\tilde{P}(i,j) = \tilde{P}(i) + P(j) = (b.a) + (d, c) = (y, x) \rightarrow x = a + c, y = b + d$

Theorem 10 (Antisymmetric Permutation). *If the feature numbers for any m terms are $\{P(1), P(2), \ldots, P(i), \ldots, P(m)\}, 1 \leq i \leq m$, the corresponding m item inverse item number is $\{P^{-1}(1), P^{-1}(2), \ldots, P^{-1}(j), \ldots, P^{-1}(m)\}, 1 \leq j \leq m$, choose any permutation P for the number and its inverse transformation P^{-1} are horizontal and vertical alignment coordinates $P(i,j) = P(i) + P^{-1}(j)$, then the formed m timesm indicates that the matrix is antisymmetric.*

$$P(i,j) = (x, y) \rightarrow P^{-1}(i,j) = P(j, i) = -(x, y)$$

Proof. Let $P(i) = (a, b)$, $P^{-1}(j) = -(c, d)$, $P(i,j) = P(i) + P^{-1}(j) = (a, b) - (c, d)$, $P^{-1}(i,j) = P^{-1}(i) + P(j) = -(a, b) + (c, d) = -P(i,j) \rightarrow x = a - c, y = b - d$

Corollary 8 (Conjugate Symmetry and Antisymmetric Association Conditions). *If $a = -b, (b, a) = \tilde{P}(i) = P^{-1}(i) = -(a, b)$ conjugate symmetric and antisymmetric matrices have the same value, then $P(i, i) = (0, 0)$.*

Proof. $P(i, i) = P(i) + \tilde{P}(i) = (a, b) + (b, a) = (a, -a) + (-a, a) = (0, 0)$; $P(i, i) = P(i) + P^{-1}(i) = (a, b) + [-(a, b)] = (a, b) + (-a, -b) = (0, 0)$.

Two theorems describe that the related feature clusters can be expressed as two kinds of symmetric feature representations, corresponding to the conjugate symmetry and antisymmetric cases. The corollary establishes under what conditions the conjugation and inverse operators have the same value. All diagonal elements in an antisymmetric matrix are $(0, 0)$.

3.5 Four Groups of Typical Cluster Topology Categories

Looking at the conjugate lattice representation, the extended structure associated with each eigenvalue is accompanied by a difference in value. However, under different parameter values, the specific grid distributions are different. However, from the perspective of topological classification, clusters with the same number can be classified into one category. For a given value, the following inferences hold.

Theorem 11 (Four Groups of Conjugate Combination Lattice Points). *For any order pair (a, b), the accompanying conjugate combination lattice has four distinguishable numbers:*

$$\{(0,0), \{(1,0),(1,1)\}, \{(2,1),(3,1)\},\ Other\}$$

Value	$(0,0)$	$\{(1,0),(1,1)\}$	$\{(2,1),(3,1)\}$	*Other*
Cluster No.	1	25	73	81
Value condition	$(0,0)$	$\{(1,0),(1,1)\}$	$\{(2,1),(3,1)\}$	*other*
Number of clusters	1	25	73	81

Proof. Here are two sets of $\{(0,0),(a,b)\}$ proofs, and the other two sets of conditions refer to the following examples. If $(a,b) = (0,0)$, no other expansion operators will form a new state. The state does not change after expansion. For any $a \neq b \neq 0$, the structure formed under the condition of quaternary conjugate combination lattice expansion is a 9×9 conjugate matrix; usually, each position has a unique representation and distinguishable point. The number of group states is equal to 81.

Corollary 9 (Number of Distinguishable Conjugate Combination Lattices). *Corresponding to the value of any order pair (a, b), there are six distinguishable state distributions.*

Proof. In $\{(0,0), \{(1,0),(1,1)\}, \{(2,1),(3,1)\}, \text{other}\}$, although the number of distinguishable grid points is the same, the specific distribution pattern can be distinguished. $\{(1,0),(1,1)\}, \{(2,1),(3,1)\}$ can be further distinguished into four clusters.

3.6 Local Expansion Quantization Feature

The local features of the conjugate group lattice are associated with the corresponding value positions of the complex plane. From the level of overall description, six typical local category features under the condition of small parameters are observed. $(x, y), x, y \in \{0, \ldots, 12\}$ parametric conditions for local expansion. $\odot : (0,0)$; $\circ : (0,1)$; $\bullet : (1,1)$; $\times : (1,2)$; $\varDelta : (1,3)$ condition.

The corresponding number of lattice points is

Lattice Type	$(0,0)$	$\{(0,1),(1,0)\}$	$(1,1)$	$\{(1,2),(2,1)\}$	$\{(1,3),(3,1)\}$	Other
Notation	$\odot$	$\circ$	$\bullet$	$\times$	$\varDelta$	
Clustered Lattice	1	25	25	73	73	81

The first quadrant small area $n \in \{0, \ldots, 12\}$ is distributed as

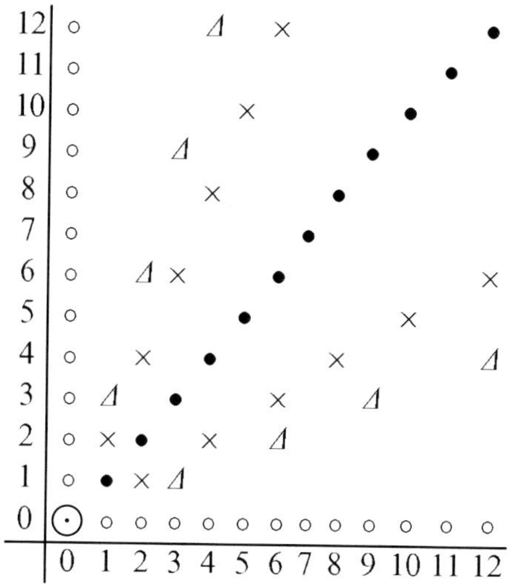

The number of corresponding quantization nodes is

n	(0, 0)	(0, 1)	(1, 1)	(1, 2)	(1, 3)	Other	$(n+1)^2$
0	1						1
1	1	2	1				4
2	1	4	2	2			9
3	1	6	3	2	2	2	16
4	1	8	4	4	2	6	25
5	1	10	5	4	2	14	36
6	1	12	6	6	4	20	49
7	1	14	7	6	4	32	64
8	1	16	8	8	4	44	81
n	1	$2 \times n$	n	$2 \times \lfloor \frac{n}{2} \rfloor$	$2 \times \lfloor \frac{n}{3} \rfloor$	$O(n^2 - \frac{8}{3}n)$	$(n+1)^2$

Combining the four quadrants, the corresponding number of quantization nodes is

n	(0, 0)	(0, 1)	(1, 1)	(1, 2)	(1, 3)	Other	$4 \times (n+1)^2$
0	4						4
1	4	8	4				16
2	4	16	8	8			36
3	4	24	12	8	8	8	64
4	4	32	16	16	8	24	100
5	4	40	20	16	8	56	144
6	4	48	24	24	16	80	196
7	4	56	28	24	16	128	256
8	4	64	32	32	16	176	324
n	4	$8 \times n$	$4 \times n$	$8 \times \lfloor \frac{n}{2} \rfloor$	$8 \times \lfloor \frac{n}{3} \rfloor$	$O(4 \times n^2 - \frac{32}{3} \times n)$	$4 \times (n+1)^2$

As seen from the table, when $n = \{0, 1, 2\}$, there are corresponding marked points in the area, starting from $n = 3$, with other clusters appearing, to $n = 7$. The number of marked points is equal to that of others. Beyond this range, the sum of the number of other points exceeds the sum of the number of marked point groups.

Corollary 10 (Regional Convergence Condition of Label Lattice). *For a given n and n > 7, the sum of the three states can be distinguished by the conjugate combination lattice topology and the number of lattice points in the given region. The ratio is a descending sequence.*

Proof. The ratio of the sum of the three states to the number of points in the area is

$$\lim_{n\to\infty} \frac{\frac{14}{3} \times n + 1}{(n+1)^2} \to 0$$

For example, for $n = 12, 57/169 = 0.337 = 33.7\%; n = 99, 155/10^4 = 0.0155 = 1.55\%$, it can be observed that after the area increases, the number of other states increases with the effect. The marked part is only a small percentage.

Theorem 12 (Conjugate Combination Lattice Mean Limit). *Choose a larger n, and the number of conjugate combination lattices at any position on the complex plane is 81.*

Proof. According to **Corollary 10**, the three marked parts decay rapidly with the growth of the spatial scale of the complex plane solution.

$$\lim_{n\to\infty} \frac{81 \times [(n+1)^2 - (\frac{14}{3} \times n + 1)]}{(n+1)^2} = 81$$

Corollary 11 (Labeled Area of Conjugate Combination Grid). *Select the area where the number of conjugated combination grid nodes is not equal to 81, gathered near the* $(0,0)$ *point, and its level* $(1,0)$*/vertical* $(0,1)$*, two diagonals* $(1,1)$*, paired bisections* $(1,2),(2,1)$*, paired three-point line* $(1,3),(3,1)$*.*

Theorem 13 (Local Entropy Measure). *For any order pair* (a,b)*, the accompanying conjugate lattice has four distinguishable numbers:*
$\{(0,0), \{(1,0),(1,1)\}, \{(2,1),(3,1)\}, other\}$*, corresponding to four kinds of local entropy values, respectively.*

Value condition	$(0,0)$	$\{(1,0),(1,1)\}$	$\{(2,1),(3,1)\}$	*other*
Local entropy	$= 0$	$\le \log_2(25)$	$\le \log_2(73)$	$\le \log_2(81)$

Proof. Since the number of four local lattice points is the maximum number of corresponding lattice points,

$$-\sum_{i=1}^{m} p_i \log_2(p_i) \le -\sum_{i=1}^{m} \frac{1}{m} \log_2(\frac{1}{m}) = \log_2(m)$$

From a probabilistic measurement point of view, a given entropy value provides the largest partition upper bound value, so the inequality holds.

Corollary 12 (Maximum Entropy of Conjugate Lattices). *On the complex plane, the maximum entropy of conjugated lattices at any position is measured as* $\log_2(81)$*.*

4. Conjugate Case—Typical Example

For the convenience of describing the structure of this additive group, a typical conjugate transform solution is constructed step by step for the case of unary, 2-, and 4-variable variables. The conjugation operator $\sim$ flips the real and imaginary measures of complex pairs.

4.1 Non-Zero Sum

$\sim: \{(a,b)\} \rightarrow \{(b,a)\}$

	0 $(0,0)$	1 (a,b)
$\tilde{0}$ $(0,0)$	$(0,0)$	(a,b)
$\tilde{1}$ (b,a)	(b,a)	$(a+b,a+b)$

$\rightarrow$

Complex Function Solutions	Numbers
$(0,0)$	1
(a,b)	1
(b,a)	1
$(a+b,a+b)$	1
4 Groups, Total	4

In the related illustration, $\bigodot$ marks the position of the $(0,0)$ point, the $\odot$ grid point indicates the position occupied by other generators, and the $\bullet$ grid point indicates the position occupied by the newly generated feature.

In general, there are four grid points distributed in the first quadrant.

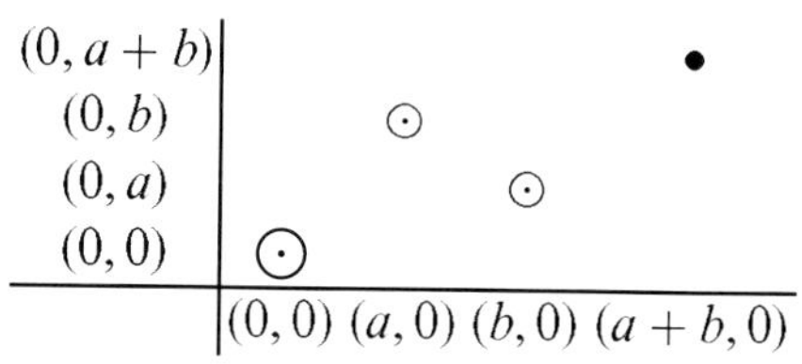

$a=b$ shrinks to antidiagonal three lattice points.

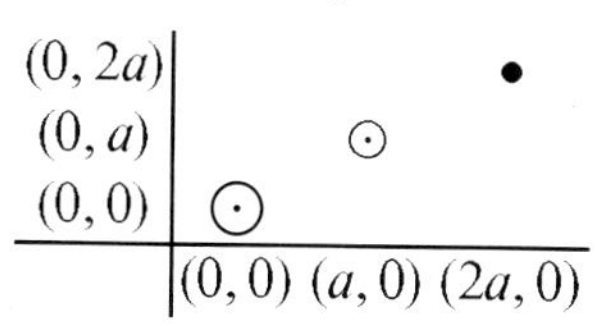

4.2 Zero Sum

$\sim: \{(a,b),(-a,-b)\} \rightarrow \{(b,a),(-b,-a)\}$

	00 $(0,0)$	01 (a,b)	10 $(-a,-b)$	11 $(0,0)$
$\widetilde{00}$ $(0,0)$	$(0,0)$	(a,b)	$(-a,-b)$	$(0,0)$
$\widetilde{01}$ (b,a)	(b,a)	$(a+b,a+b)$	$(b-a,a-b)$	(b,a)
$\widetilde{01}$ $(-b,-a)$	$(-b,-a)$	$(a-b,b-a)$	$(-a-b,-a-b)$	$(-b,-a)$
$\widetilde{11}$ $(0,0)$	$(0,0)$	(a,b)	$(-a,-b)$	$(0,0)$

$\rightarrow$

Complex Function Solutions	Numbers
$(0,0)$	4
(a,b)	2
(b,a)	2
$(-a,-b)$	2
$(-b,-a)$	2
Middle	4×1
9 Groups, Total	16

$a \neq b$, nine distinguishable grid points distributed diagonally in the grid.

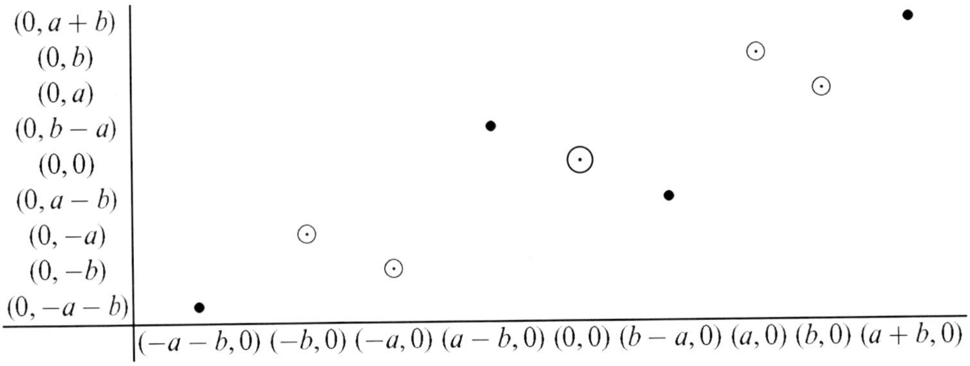

$a = b$, five distinguishable grid points, distributed diagonally in the grid.

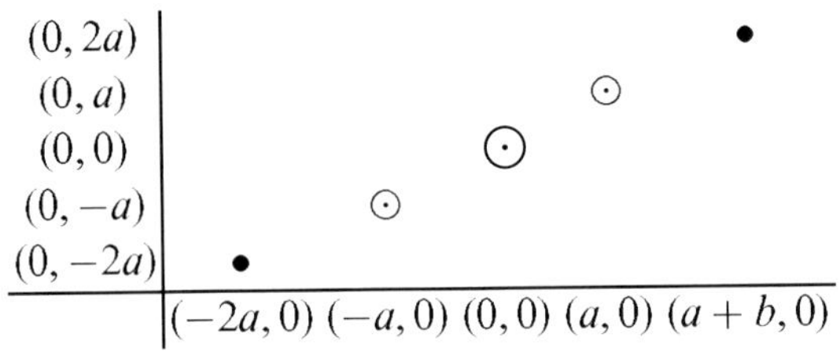

$\sim: \{(a,b),(-a,b)\} \rightarrow \{(b,a),(b,-a)\}$

	00 $(0,0)$	01 (a,b)	10 $(a,-b)$	11 $(2a,0)$
$\widetilde{00}$ $(0,0)$	$(0,0)$	(a,b)	$(a,-b)$	$(2a,0)$
$\widetilde{01}$ (b,a)	(b,a)	$(a+b,a+b)$	$(b+a,a-b)$	$(2a+b,a)$
$\widetilde{01}$ $(-b,a)$	$(-b,a)$	$(a-b,a+b)$	$(a-b,a-b)$	$(2a-b,-a)$
$\widetilde{11}$ $(0,2a)$	$(0,2a)$	$(a,2a+b)$	$(a,2a-b)$	$(2a,2a)$

$\rightarrow$

Complex Function Solutions	Numbers
$(0,0)$	1
(a,b)	1
(b,a)	1
$(a,-b)$	1
$(-b,a)$	1
Other	11×1
16 groups, total	16

$\sim: \{(a,b),(-a,b),(a,-b)\} \rightarrow \{(b,a),(b,-a),(-b,a)\}$

	000 $(0,0)$	001 (a,b)	010 $(a,-b)$	011 $(2a,0)$	100 $(-a,b)$	101 $(0,2b)$	110 $(0,0)$	111 (a,b)
$\widetilde{000}$ $(0,0)$	$(0,0)$	(a,b)	$(a,-b)$	$(2a,0)$	$(-a,b)$	$(0,2b)$	$(0,0)$	(a,b)
$\widetilde{001}$ (b,a)	(b,a)	$(a+b,a+b)$	$(a+b,a-b)$	$(2a+b,a)$	$(b-a,a+b)$	$(b,a+2b)$	(b,a)	$(a+b,a+b)$
$\widetilde{010}$ $(b,-a)$	$(b,-a)$	$(a+b,b-a)$	$(a+b,-a-b)$	$(2a+b,-a)$	$(-a+b,b-a)$	$(b,2b-a)$	$(b,-a)$	$(a+b,b-a)$
$\widetilde{011}$ $(2b,0)$	$(2b,0)$	$(a+2b,b)$	$(a+2b,-b)$	$(2(a+b),0)$	$(2b-a,b)$	$(2b,2b)$	$(2b,0)$	$(a+2b,b)$
$\widetilde{100}$ $(-b,a)$	$(-b,a)$	$(a-b,a+b)$	$(a-b,a-b)$	$(2a-b,a)$	$(-a-b,a+b)$	$(-b,a+2b)$	$(-b,a)$	$(a-b,a+b)$
$\widetilde{101}$ $(0,2a)$	$(0,2a)$	$(a,2a+b)$	$(-a,2a+b)$	$(0,2a)$	$(a,2a-b)$	$(2a,2b)$	$(0,2(a+b))$	$(a,2a+b)$
$\widetilde{110}$ $(0,0)$	$(0,0)$	(a,b)	$(a,-b)$	$(2a,0)$	$(-a,b)$	$(0,2b)$	$(0,0)$	(a,b)
$\widetilde{111}$ (b,a)	(b,a)	$(a+b,a+b)$	$(a+b,a-b)$	$(2a+b,a)$	$(b-a,a+b)$	$(b,a+2b)$	(b,a)	$(a+b,a+b)$

	Complex Function Solutions	Numbers
	$\{(0,0),(a,b),(b,a),(a+b,a+b)\}$	$4\times 4=16$
	$\{(-a,b),(-b,a),(a,-b),(b,-a)\}$	$4\times 2=8$
	$\{(2a,0),(0,2b),(2b,0),(0,2a)\}$	$4\times 2=8$
	$\{(a+b,b-a),(a-b,a+b),(a+b,a-b),(b-a,a+b)\}$	$4\times 2=8$
$\rightarrow$	$\{(a+2b,b),(a,2a+b),(2a+b,a),(b,a+2b)\}$	$4\times 2=8$
	$\{(b-a,b-a),(-a-b,a+b),(a+b,-a-b),(a-b,a-b)\}$	$4\times 1=4$
	$\{(2a,2a),(2(a+b),0),(0,2(a+b)),(2b,2b)\}$	$4\times 1=4$
	$\{(2b-a,b),(-b,2a+b),(a+2b,-b),(b,2b-a)\}$	$4\times 1=4$
	$\{(-b,2b-a),(-a,2a+b),(2a+b,-a),(2b-a,-b)\}$	$4\times 1=4$
	36 groups, total	64

4.3 Conjugate Combinatorial Lattice—Complete Representation

No.	(a,b)	$(a,-b)$	$(-a,b)$	$(-a,-b)$	$\sum =$
$0:0000$					$(0,0)$
$1:0001$				$(-a,-b)$	$(-a,-b)$
$2:0010$			$(-a,b)$		$(-a,b)$
$4:0100$		$(a,-b)$			$(a,-b)$
$8:1000$	(a,b)				(a,b)
$3:0011$			$(-a,b)$	$(-a,-b)$	$(-2a,0)$
$5:0101$		$(a,-b)$		$(-a,-b)$	$(0,-2b)$
$6:0110$		$(a,-b)$	$(-a,b)$		$(0,0)$
$9:1001$	(a,b)			$(-a,-b)$	$(0,0)$
$10:1010$	(a,b)		$(-a,b)$		$(0,2b)$
$12:1100$	(a,b)	$(a,-b)$			$(2a,0)$
$7:0111$		$(a,-b)$	$(-a,b)$	$(-a,-b)$	$(-a,-b)$
$11:1011$	(a,b)		$(-a,b)$	$(-a,-b)$	$(-a,b)$
$13:1101$	(a,b)	$(a,-b)$		$(-a,-b)$	$(a,-b)$
$14:1110$	(a,b)	$(a,-b)$	$(-a,b)$		(a,b)
$15:1111$	(a,b)	$(a,-b)$	$(-a,b)$	$(-a,-b)$	$(0,0)$

	Complex Function Solutions	Numbers
	$\{(0,0)\}$	$1\times 4=4$
$\rightarrow$	$\{(a,b),(-a,b),(a,-b),(-a,-b)\}$	$4\times 2=8$
	$\{(-2a,0),(0,-2b),(0,2b),(2a,0)\}$	$4\times 1=4$
	9 Groups, Total	16

Nine distinguishable grid points are symmetrically distributed in the grid.

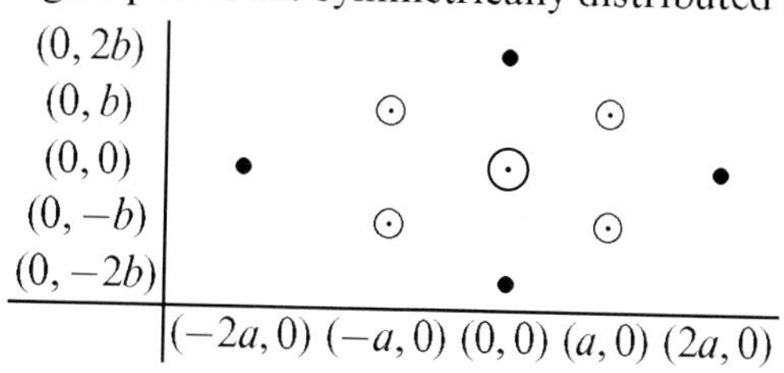

Conjugate state: $\{(b,a),(-b,a),(b,-a),(-b,-a)\}$.

No.	(b,a)	$(-b,a)$	$(b,-a)$	$(-b,-a)$	$\sum =$
$\tilde{0}:0000$					$(0,0)$
$\tilde{1}:0001$				$(-b,-a)$	$(-b,-a)$
$\tilde{2}:0010$			$(b,-a)$		$(b,-a)$
$\tilde{4}:0100$		$(-b,a)$			$(-b,a)$
$\tilde{8}:1000$	(b,a)				(b,a)
$\tilde{3}:0011$			$(b,-a)$	$(-b,-a)$	$(0,-2a)$
$\tilde{5}:0101$		$(-b,a)$		$(-b,-a)$	$(-2b,0)$
$\tilde{6}:0110$		$(-b,a)$	$(b,-a)$		$(0,0)$
$\tilde{9}:1001$	(b,a)			$(-b,-a)$	$(0,0)$
$\widetilde{10}:1010$	(b,a)		$(b,-a)$		$(2b,0)$
$\widetilde{12}:1100$	(b,a)	$(-b,a)$			$(0,2a)$
$\tilde{7}:0111$		$(-b,a)$	$(b,-a)$	$(-b,-a)$	$(-b,-a)$
$\widetilde{11}:1011$	(b,a)		$(b,-a)$	$(-b,-a)$	$(b,-a)$
$\widetilde{13}:1101$	(b,a)	$(-b,a)$		$(-b,-a)$	$(-b,a)$
$\widetilde{14}:1110$	(b,a)	$(-b,a)$	$(b,-a)$		(b,a)
$\widetilde{15}:1111$	(b,a)	$(-b,a)$	$(b,-a)$	$(-b,-a)$	$(0,0)$

	Complex Function Solution	Number
	$\{(0,0)\}$	$1\times 4=4$
$\rightarrow$	$\{(b,a),(-b,a),(b,-a),(-b,-a)\}$	$4\times 2=8$
	$\{(0,-2a),(-2b,0),(2b,0),(0,2a)\}$	$4\times 1=4$
	9 Groups, Total	16

There are nine lattice points on a slightly symmetric distribution.

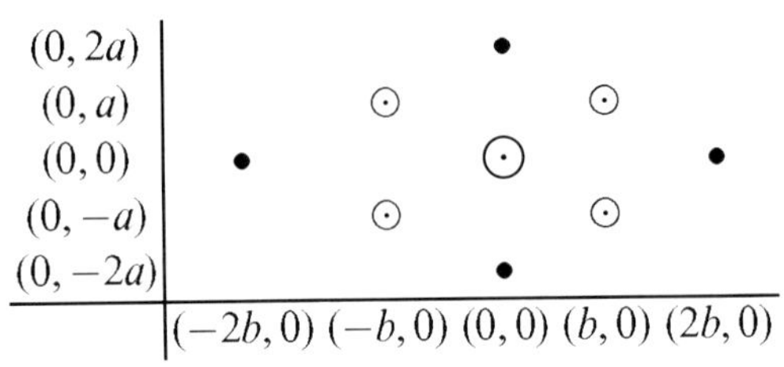

There are seventeen lattice points, the center $\odot$ coincides, and the remaining sixteen points are

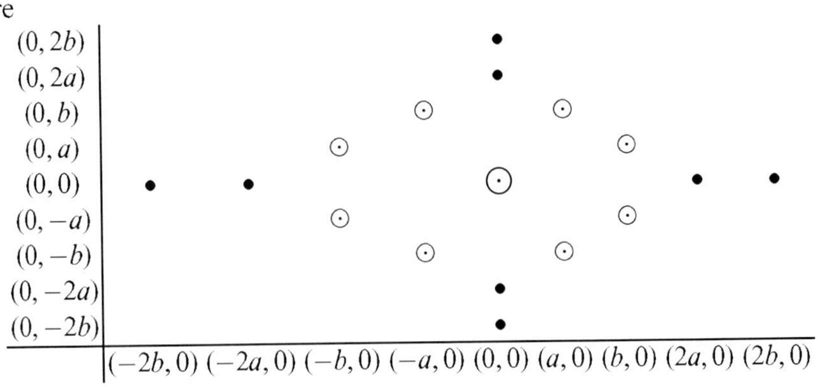

$\sum =$	0 $(0,0)$	1 $(-a,-b)$	2 $(-a,b)$	3 $(-2a,0)$	4 $(a,-b)$	5 $(0,-2b)$	6 $(0,0)$	7 $(-a,-b)$	8 (a,b)	9 $(0,0)$	10 $(0,2b)$	11 $(-a,b)$	12 $(2a,0)$	13 $(a,-b)$	14 (a,b)	15 $(0,0)$
$\tilde{0}(0,0)$	$(0,0)$	$(-a,-b)$	$(-a,b)$	$(-2a,0)$	$(a,-b)$	$(0,-2b)$	$(0,0)$	$(-a,-b)$	(a,b)	$(0,0)$	$(0,2b)$	$(-a,b)$	$(2a,0)$	$(a,-b)$	(a,b)	$(0,0)$
$\tilde{1}(-b,-a)$	$(-b,-a)$	$(-a-b,-a-b)$	$(-a-b,b-a)$	$(-2a-b,-a)$	$(a-b,-a-b)$	$(-b,-2b-a)$	$(-b,-a)$	$(-a-b,-a-b)$	$(a-b,b-a)$	$(-b,-a)$	$(-b,2b-a)$	$(-a-b,b-a)$	$(2a-b,-a)$	$(a-b,-a-b)$	$(a-b,b-a)$	$(-b,-a)$
$\tilde{2}(b,-a)$	$(b,-a)$	$(b-a,-a-b)$	$(b-a,b-a)$	$(b-2a,-a)$	$(a+b,-a-b)$	$(b,-2b-a)$	$(b,-a)$	$(b-a,-a-b)$	$(a+b,b-a)$	$(b,-a)$	$(b,2b-a)$	$(b-a,b-a)$	$(2a+b,-a)$	$(a+b,-a-b)$	$(a+b,b-a)$	$(b,-a)$
$\tilde{3}(0,-2a)$	$(0,-2a)$	$(-a,-2a-b)$	$(-a,-2a+b)$	$(-2a,-2a)$	$(a,-2a-b)$	$(0,-2(a+b))$	$(0,-2a)$	$(-a,-2a-b)$	$(-a,-2a+b)$	$(0,-2a)$	$(0,2(b-a))$	$(-a,-2a+b)$	$(2a,-2a)$	$(a,-b-2a)$	$(a,b-2a)$	$(0,-2a)$
$\tilde{4}(-b,a)$	$(-b,a)$	$(-a-b,a-b)$	$(-a-b,a+b)$	$(-2a-b,a)$	$(a-b,a-b)$	$(-b,a-2b)$	$(-b,a)$	$(-a-b,a-b)$	$(-a-b,a+b)$	$(-b,a)$	$(-b,a+2b)$	$(-a-b,a+b)$	$(2a-b,a)$	$(a-b,a-b)$	$(a-b,a+b)$	$(-b,a)$
$\tilde{5}(-2b,0)$	$(-2b,0)$	$(-2b-a,-b)$	$(-2b-a,b)$	$(-2(a+b),0)$	$(-2b+a,-b)$	$(-2b,-2b)$	$(-2b,0)$	$(-a-2b,-b)$	$(a-2b,b)$	$(-2b,0)$	$(-2b,2b)$	$(-2b-a,b)$	$(2(a-b),0)$	$(-2b+a,-b)$	$(a-2b,b)$	$(-2b,0)$
$\tilde{6}(0,0)$	$(0,0)$	$(-a,-b)$	$(-a,b)$	$(-2a,0)$	$(a,-b)$	$(0,-2b)$	$(0,0)$	$(-a,-b)$	(a,b)	$(0,0)$	$(0,2b)$	$(-a,b)$	$(2a,0)$	$(a,-b)$	(a,b)	$(0,0)$
$\tilde{7}(-b,-a)$	$(-b,-a)$	$(-a-b,-a-b)$	$(-a-b,b-a)$	$(-2a-b,-a)$	$(a-b,-a-b)$	$(-b,-2b-a)$	$(-b,-a)$	$(-a-b,-a-b)$	$(a-b,b-a)$	$(-b,-a)$	$(-b,2b-a)$	$(-a-b,b-a)$	$(2a-b,-a)$	$(a-b,-a-b)$	$(a-b,b-a)$	$(-b,-a)$
$\tilde{8}(b,a)$	(b,a)	$(b-a,a-b)$	$(b-a,a+b)$	$(b-2a,a)$	$(a+b,a-b)$	$(b,a-2b)$	(b,a)	$(b-a,a-b)$	$(b+a,a+b)$	(b,a)	$(b,a+2b)$	$(b-a,a+b)$	$(2a+b,a)$	$(a+b,a-b)$	$(a+b,a+b)$	(b,a)
$\tilde{9}(0,0)$	$(0,0)$	$(-a,-b)$	$(-a,b)$	$(-2a,0)$	$(a,-b)$	$(0,-2b)$	$(0,0)$	$(-a,-b)$	(a,b)	$(0,0)$	$(0,2b)$	$(-a,b)$	$(2a,0)$	$(a,-b)$	(a,b)	$(0,0)$
$\tilde{10}(2b,0)$	$(2b,0)$	$(2b-a,-b)$	$(2b-a,b)$	$(2(b-a),0)$	$(2b+a,-b)$	$(2b,-2b)$	$(2b,0)$	$(2b-a,-b)$	$(a+2b,b)$	$(2b,0)$	$(2b,2b)$	$(2b-a,b)$	$(2(a+b),0)$	$(2b+a,-b)$	$(a+2b,b)$	$(2b,0)$
$\tilde{11}(b,-a)$	$(b,-a)$	$(b-a,-a-b)$	$(b-a,b-a)$	$(b-2a,-a)$	$(a+b,-a-b)$	$(b,-2b-a)$	$(b,-a)$	$(b-a,-a-b)$	$(a+b,b-a)$	$(b,-a)$	$(b,2b-a)$	$(b-a,b-a)$	$(2a+b,-a)$	$(a+b,-a-b)$	$(a+b,b-a)$	$(b,-a)$
$\tilde{12}(0,2a)$	$(0,2a)$	$(-a,2a-b)$	$(-a,2a+b)$	$(-2a,2a)$	$(a,2a-b)$	$(0,2(a-b))$	$(0,2a)$	$(-a,2a-b)$	$(-a,2a+b)$	$(0,2a)$	$(0,2(a+b))$	$(-a,2a+b)$	$(2a,2a)$	$(a,2a-b)$	$(a,2a+b)$	$(0,2a)$
$\tilde{13}(-b,a)$	$(-b,a)$	$(-a-b,a-b)$	$(-a-b,a+b)$	$(-2a-b,a)$	$(a-b,a-b)$	$(-b,a-2b)$	$(-b,a)$	$(-a-b,a-b)$	$(-a-b,a+b)$	$(-b,a)$	$(-b,a+2b)$	$(-a-b,a+b)$	$(2a-b,a)$	$(a-b,a-b)$	$(a-b,a+b)$	$(-b,a)$
$\tilde{14}(b,a)$	(b,a)	$(b-a,a-b)$	$(b-a,a+b)$	$(b-2a,a)$	$(a+b,a-b)$	$(b,a-2b)$	(b,a)	$(b-a,a-b)$	$(b+a,a+b)$	(b,a)	$(b,a+2b)$	$(b-a,a+b)$	$(2a+b,a)$	$(a+b,a-b)$	$(a+b,a+b)$	(b,a)
$\tilde{15}(0,0)$	$(0,0)$	$(-a,-b)$	$(-a,b)$	$(-2a,0)$	$(a,-b)$	$(0,-2b)$	$(0,0)$	$(-a,-b)$	(a,b)	$(0,0)$	$(0,2b)$	$(-a,b)$	$(2a,0)$	$(a,-b)$	(a,b)	$(0,0)$

$\sum =$	0 $(0,0)$	1 $(-a,-b)$	2 $(-a,b)$	3 $(-2a,0)$	4 $(a,-b)$	5 $(0,-2b)$	6 $(0,0)$	7 $(-a,-b)$	8 (a,b)	9 $(0,0)$	10 $(0,2b)$	11 $(-a,b)$	12 $(2a,0)$	13 $(a,-b)$	14 (a,b)	15 $(0,0)$
$\tilde{0}(0,0)$	$(0,0)$	$(-a,-b)$	$(-a,b)$	$(-2a,0)$	$(a,-b)$	$(0,-2b)$	$(0,0)$	$(-a,-b)$	(a,b)	$(0,0)$	$(0,2b)$	$(-a,b)$	$(2a,0)$	$(a,-b)$	(a,b)	$(0,0)$
$\tilde{1}(-b,-a)$	$(-b,-a)$	$(-a-b,-a-b)$	$(-a-b,b-a)$	$(-2a-b,-a)$	$(a-b,-a-b)$	$(-b,-2b-a)$	$(-b,-a)$	$(-a-b,-a-b)$	$(a-b,b-a)$	$(-b,-a)$	$(-b,2b-a)$	$(-a-b,b-a)$	$(2a-b,-a)$	$(a-b,-a-b)$	$(a-b,b-a)$	$(-b,-a)$
$\tilde{2}(b,-a)$	$(b,-a)$	$(b-a,-a-b)$	$(b-a,b-a)$	$(b-2a,-a)$	$(a+b,-a-b)$	$(b,-2b-a)$	$(b,-a)$	$(b-a,-a-b)$	$(a+b,b-a)$	$(b,-a)$	$(b,2b-a)$	$(b-a,b-a)$	$(2a+b,-a)$	$(a+b,-a-b)$	$(a+b,b-a)$	$(b,-a)$
$\tilde{4}(-b,a)$	$(-b,a)$	$(-a-b,a-b)$	$(-a-b,a+b)$	$(-2a-b,a)$	$(a-b,a-b)$	$(-b,a-2b)$	$(-b,a)$	$(-a-b,a-b)$	$(-a-b,a+b)$	$(-b,a)$	$(-b,a+2b)$	$(-a-b,a+b)$	$(2a-b,a)$	$(a-b,a-b)$	$(a-b,a+b)$	$(-b,a)$
$\tilde{8}(b,a)$	(b,a)	$(b-a,a-b)$	$(b-a,a+b)$	$(b-2a,a)$	$(a+b,a-b)$	$(b,a-2b)$	(b,a)	$(b-a,a-b)$	$(b+a,a+b)$	(b,a)	$(b,a+2b)$	$(b-a,a+b)$	$(2a+b,a)$	$(a+b,a-b)$	$(a+b,a+b)$	(b,a)
$\tilde{3}(0,-2a)$	$(0,-2a)$	$(-a,-2a-b)$	$(-a,-2a+b)$	$(-2a,-2a)$	$(a,-2a-b)$	$(0,-2(a+b))$	$(0,-2a)$	$(-a,-2a-b)$	$(-a,-2a+b)$	$(0,-2a)$	$(0,2(b-a))$	$(-a,-2a+b)$	$(2a,-2a)$	$(a,-b-2a)$	$(a,b-2a)$	$(0,-2a)$
$\tilde{5}(-2b,0)$	$(-2b,0)$	$(-2b-a,-b)$	$(-2b-a,b)$	$(-2(a+b),0)$	$(-2b+a,-b)$	$(-2b,-2b)$	$(-2b,0)$	$(-a-2b,-b)$	$(a-2b,b)$	$(-2b,0)$	$(-2b,2b)$	$(-2b-a,b)$	$(2(a-b),0)$	$(-2b+a,-b)$	$(a-2b,b)$	$(-2b,0)$
$\tilde{6}(0,0)$	$(0,0)$	$(-a,-b)$	$(-a,b)$	$(-2a,0)$	$(a,-b)$	$(0,-2b)$	$(0,0)$	$(-a,-b)$	(a,b)	$(0,0)$	$(0,2b)$	$(-a,b)$	$(2a,0)$	$(a,-b)$	(a,b)	$(0,0)$
$\tilde{9}(0,0)$	$(0,0)$	$(-a,-b)$	$(-a,b)$	$(-2a,0)$	$(a,-b)$	$(0,-2b)$	$(0,0)$	$(-a,-b)$	(a,b)	$(0,0)$	$(0,2b)$	$(-a,b)$	$(2a,0)$	$(a,-b)$	(a,b)	$(0,0)$
$\tilde{10}(2b,0)$	$(2b,0)$	$(2b-a,-b)$	$(2b-a,b)$	$(2(b-a),0)$	$(2b+a,-b)$	$(2b,-2b)$	$(2b,0)$	$(2b-a,-b)$	$(a+2b,b)$	$(2b,0)$	$(2b,2b)$	$(2b-a,b)$	$(2(a+b),0)$	$(2b+a,-b)$	$(a+2b,b)$	$(2b,0)$
$\tilde{12}(0,2a)$	$(0,2a)$	$(-a,2a-b)$	$(-a,2a+b)$	$(-2a,2a)$	$(a,2a-b)$	$(0,2(a-b))$	$(0,2a)$	$(-a,2a-b)$	$(-a,2a+b)$	$(0,2a)$	$(0,2(a+b))$	$(-a,2a+b)$	$(2a,2a)$	$(a,2a-b)$	$(a,2a+b)$	$(0,2a)$
$\tilde{7}(-b,-a)$	$(-b,-a)$	$(-a-b,-a-b)$	$(-a-b,b-a)$	$(-2a-b,-a)$	$(a-b,-a-b)$	$(-b,-2b-a)$	$(-b,-a)$	$(-a-b,-a-b)$	$(a-b,b-a)$	$(-b,-a)$	$(-b,2b-a)$	$(-a-b,b-a)$	$(2a-b,-a)$	$(a-b,-a-b)$	$(a-b,b-a)$	$(-b,-a)$
$\tilde{11}(b,-a)$	$(b,-a)$	$(b-a,-a-b)$	$(b-a,b-a)$	$(b-2a,-a)$	$(a+b,-a-b)$	$(b,-2b-a)$	$(b,-a)$	$(b-a,-a-b)$	$(a+b,b-a)$	$(b,-a)$	$(b,2b-a)$	$(b-a,b-a)$	$(2a+b,-a)$	$(a+b,-a-b)$	$(a+b,b-a)$	$(b,-a)$
$\tilde{13}(-b,a)$	$(-b,a)$	$(-a-b,a-b)$	$(-a-b,a+b)$	$(-2a-b,a)$	$(a-b,a-b)$	$(-b,a-2b)$	$(-b,a)$	$(-a-b,a-b)$	$(-a-b,a+b)$	$(-b,a)$	$(-b,a+2b)$	$(-a-b,a+b)$	$(2a-b,a)$	$(a-b,a-b)$	$(a-b,a+b)$	$(-b,a)$
$\tilde{14}(b,a)$	(b,a)	$(b-a,a-b)$	$(b-a,a+b)$	$(b-2a,a)$	$(a+b,a-b)$	$(b,a-2b)$	(b,a)	$(b-a,a-b)$	$(b+a,a+b)$	(b,a)	$(b,a+2b)$	$(b-a,a+b)$	$(2a+b,a)$	$(a+b,a-b)$	$(a+b,a+b)$	(b,a)
$\tilde{15}(0,0)$	$(0,0)$	$(-a,-b)$	$(-a,b)$	$(-2a,0)$	$(a,-b)$	$(0,-2b)$	$(0,0)$	$(-a,-b)$	(a,b)	$(0,0)$	$(0,2b)$	$(-a,b)$	$(2a,0)$	$(a,-b)$	(a,b)	$(0,0)$

16 × 16 Action matrix

$\sum =$	0 $(0,0)$	1 $(-a,-b)$	2 $(-a,b)$	4 $(a,-b)$	8 (a,b)	12 $(2a,0)$	5 $(0,-2b)$	6 $(0,0)$
$\tilde{0}(0,0)$	$(0,0)$	$(-a,-b)$	$(-a,b)$	$(a,-b)$	(a,b)	$(2a,0)$	$(0,-2b)$	$(0,0)$
$\tilde{1}(-b,-a)$	$(-b,-a)$	$(-a-b,-a-b)$	$(-a-b,b-a)$	$(a-b,-b-a)$	$(a-b,b-a)$	$(2a-b,-a)$	$(-b,-2b-a)$	$(-b,-a)$
$\tilde{2}(b,-a)$	$(b,-a)$	$(b-a,-a-b)$	$(b-a,b-a)$	$(a+b,-a-b)$	$(a+b,b-a)$	$(b+2a,-a)$	$(b,-2b-a)$	$(b,-a)$
$\tilde{4}(-b,a)$	$(-b,a)$	$(-a-b,a-b)$	$(-a-b,a+b)$	$(a-b,a-b)$	$(-a-b,a+b)$	$(2a-b,a)$	$(-b,a-2b)$	$(-b,a)$
$\tilde{8}(b,a)$	(b,a)	$(b-a,a-b)$	$(b-a,a+b)$	$(a+b,a-b)$	$(a+b,a+b)$	$(b+2a,a)$	$(b,a-2b)$	(b,a)
$\tilde{12}(0,2a)$	$(0,2a)$	$(-a,2a-b)$	$(-a,2a+b)$	$(a,2a-b)$	$(-a,2a+b)$	$(2a,2a)$	$(0,2(a-b))$	$(0,2a)$
$\tilde{5}(-2b,0)$	$(-2b,0)$	$(-2b-a,-b)$	$(-2b-a,b)$	$(-2b+a,-b)$	$(a-2b,b)$	$(2(a-b),0)$	$(-2b,-2b)$	$(-2b,0)$
$\tilde{6}(0,0)$	$(0,0)$	$(-a,-b)$	$(-a,b)$	$(a,-b)$	(a,b)	$(2a,0)$	$(0,-2b)$	$(0,0)$
$\tilde{9}(0,0)$	$(0,0)$	$(-a,-b)$	$(-a,b)$	$(a,-b)$	(a,b)	$(2a,0)$	$(0,-2b)$	$(0,0)$
$\tilde{10}(2b,0)$	$(2b,0)$	$(2b-a,-b)$	$(2b-a,b)$	$(2b+a,-b)$	$(a+2b,b)$	$(2(b+a),0)$	$(2b,-2b)$	$(2b,0)$
$\tilde{3}(0,-2a)$	$(0,-2a)$	$(-a,-2a-b)$	$(-a,-2a+b)$	$(a,-2a-b)$	$(-a,b-2a)$	$(2a,-2a)$	$(0,-2(a+b))$	$(0,-2a)$
$\tilde{7}(-b,-a)$	$(-b,-a)$	$(-a-b,-a-b)$	$(-a-b,b-a)$	$(a-b,-a-b)$	$(a-b,b-a)$	$(2a-b,-a)$	$(-b,-2b-a)$	$(-b,-a)$
$\tilde{11}(b,-a)$	$(b,-a)$	$(b-a,-a-b)$	$(b-a,b-a)$	$(a+b,-a-b)$	$(a+b,b-a)$	$(b+2a,-a)$	$(b,-2b-a)$	$(b,-a)$
$\tilde{13}(-b,a)$	$(-b,a)$	$(-a-b,a-b)$	$(-a-b,a+b)$	$(a-b,a-b)$	$(-a-b,a+b)$	$(2a-b,a)$	$(-b,a-2b)$	$(-b,a)$
$\tilde{14}(b,a)$	(b,a)	$(b-a,a-b)$	$(b-a,a+b)$	$(a+b,a-b)$	$(a+b,a+b)$	$(b+2a,a)$	$(b,a-2b)$	(b,a)
$\tilde{15}(0,0)$	$(0,0)$	$(-a,-b)$	$(-a,b)$	$(a,-b)$	(a,b)	$(2a,0)$	$(0,-2b)$	$(0,0)$

$\sum =$	9 $(0,0)$	10 $(0,2b)$	3 $(-2a,0)$	7 $(-a,-b)$	11 $(-a,b)$	13 $(a,-b)$	14 (a,b)	15 $(0,0)$
$\tilde{0}(0,0)$	$(0,0)$	$(0,2b)$	$(-2a,0)$	$(-a,-b)$	$(-a,b)$	$(a,-b)$	(a,b)	$(0,0)$
$\tilde{1}(-b,-a)$	$(-b,-a)$	$(-b,2b-a)$	$(-2a-b,-a)$	$(-a-b,-b-a)$	$(-a-b,b-a)$	$(a-b,-a-b)$	$(a-b,b-a)$	$(-b,-a)$
$\tilde{2}(b,-a)$	$(b,-a)$	$(b,2b-a)$	$(b-2a,-a)$	$(b-a,-a-b)$	$(b-a,b-a)$	$(a+b,-a-b)$	$(a+b,b-a)$	$(b,-a)$
$\tilde{4}(-b,a)$	$(-b,a)$	$(-b,a+2b)$	$(-2a-b,a)$	$(-a-b,a-b)$	$(-a-b,a+b)$	$(a-b,a-b)$	$(a-b,a+b)$	$(-b,a)$
$\tilde{8}(b,a)$	(b,a)	$(b,a+2b)$	$(b-2a,a)$	$(b-a,a-b)$	$(b-a,a+b)$	$(a+b,a-b)$	$(a+b,a+b)$	(b,a)
$\tilde{12}(0,2a)$	$(0,2a)$	$(0,2(a+b))$	$(-2a,2a)$	$(-a,2a-b)$	$(-a,2a+b)$	$(a,2a-b)$	$(a,2a+b)$	$(0,2a)$
$\tilde{5}(-2b,0)$	$(-2b,0)$	$(-2b,2b)$	$(-2(a+b),0)$	$(-a-2b,-b)$	$(-2b-a,b)$	$(-2b+a,-b)$	$(a-2b,b)$	$(-2b,0)$
$\tilde{6}(0,0)$	$(0,0)$	$(0,2b)$	$(-2a,0)$	$(-a,-b)$	$(-a,b)$	$(a,-b)$	(a,b)	$(0,0)$
$\tilde{9}(0,0)$	$(0,0)$	$(0,2b)$	$(-2a,0)$	$(-a,-b)$	$(-a,b)$	$(a,-b)$	(a,b)	$(0,0)$
$\tilde{10}(2b,0)$	$(2b,0)$	$(2b,2b)$	$(2(b-a),0)$	$(2b-a,-b)$	$(2b-a,b)$	$(2b+a,-b)$	$(a+2b,b)$	$(2b,0)$
$\tilde{3}(0,-2a)$	$(0,-2a)$	$(0,2(b-a))$	$(-2a,-2a)$	$(-a,-2a-b)$	$(-a,-2a+b)$	$(a,-b-2a)$	$(a,b-2a)$	$(0,-2a)$
$\tilde{7}(-b,-a)$	$(-b,-a)$	$(-b,2b-a)$	$(-2a-b,-a)$	$(-a-b,-a-b)$	$(-a-b,b-a)$	$(a-b,-a-b)$	$(a-b,b-a)$	$(-b,-a)$
$\tilde{11}(b,-a)$	$(b,-a)$	$(b,2b-a)$	$(b-2a,-a)$	$(b-a,-a-b)$	$(b-a,b-a)$	$(a+b,-a-b)$	$(a+b,b-a)$	$(b,-a)$
$\tilde{13}(-b,a)$	$(-b,a)$	$(-b,a+2b)$	$(-2a-b,a)$	$(-a-b,a-b)$	$(-a-b,a+b)$	$(a-b,a-b)$	$(a-b,a+b)$	$(-b,a)$
$\tilde{14}(b,a)$	(b,a)	$(b,a+2b)$	$(b-2a,a)$	$(b-a,a-b)$	$(b-a,a+b)$	$(a+b,a-b)$	$(a+b,a+b)$	(b,a)
$\tilde{15}(0,0)$	$(0,0)$	$(0,2b)$	$(-2a,0)$	$(-a,-b)$	$(-a,b)$	$(a,-b)$	(a,b)	$(0,0)$

Complex Function Solutions	Numbers
$\{(0,0)\}$	$1 \times 16 = 16$
$\{(a,b),(-a,b),(a,-b),(-a,-b)\}$	$4 \times 8 = 32$
$\{(b,a),(-b,a),(b,-a),(-b,-a)\}$	$4 \times 8 = 32$
$\{(0,-2a),(-2b,0),(2b,0),(0,2a)\}$	$4 \times 4 = 16$
$\{(-2a,0),(0,-2b),(0,2b),(2a,0)\}$	$4 \times 4 = 16$
$\{(-a-b,-a-b),(b-a,-a-b),(-a-b,a-b),(b-a,a-b)\}$	$4 \times 4 = 16$
$\rightarrow$ $\{(-a-b,b-a),(b-a,b-a),(-a-b,a+b),(b-a,a+b)\}$	$4 \times 4 = 16$
$\{(a-b,-a-b),(b+a,-a-b),(a-b,a-b),(b+a,a-b)\}$	$4 \times 4 = 16$
$\{(a-b,b-a),(b+a,b-a),(a-b,a+b),(b+a,a+b)\}$	$4 \times 4 = 16$
$\{(-a,2a-b),(-2b-a,-b),(2b-a,-b),(-a,-2a-b)\}$	$4 \times 2 = 8$
$\{(-a,2a+b),(-2b-a,b),(2b-a,b),(-a,-2a+b)\}$	$4 \times 2 = 8$
$\{(a,2a-b),(a-2b,-b),(a+2b,-b),(a,-2a-b)\}$	$4 \times 2 = 8$
$\{(a,2a+b),(a-2b,b),(a+2b,b),(a,-2a+b)\}$	$4 \times 2 = 8$
$\{(-b,-a+2b),(b,-a+2b),(-b,a+2b),(b,a+2b)\}$	$4 \times 2 = 8$

$\{(-b,-a-2b),(b,-a-2b),(-b,a-2b),(b,a-2b)\}$	$4\times 2=8$
$\{(-2a-b,-a),(-2a+b,-a),(-2a-b,a),(-2a+b,a)\}$	$4\times 2=8$
$\{(2a-b,-a),(2a+b,-a),(2a-b,a),(2a+b,a)\}$	$4\times 2=8$
$\{(2a,2a),(2(a-b),0),(2(a+b),0),(2a,-2a)\}$	$4\times 1=4$
$\{(-2a,2a),(-2(a+b),0),(2(b-a),0),(-2a,-2a)\}$	$4\times 1=4$
$\{(-2b,-2b),(0,2(a-b)),(0,-2(a+b)),(2b,-2b)\}$	$4\times 1=4$
$\{(-2b,2b),(0,2(a+b)),(0,2(b-a)),(2b,2b)\}$	$4\times 1=4$
81 Groups, Total	256

4.4 Conjugate Combinatorial Lattice—Optimized Representation

Reduced redundant parts to be an active matrix 9×9

$\sum =$	0 $(0,0)$	1 $(-a,-b)$	2 $(-a,b)$	4 $(a,-b)$	8 (a,b)	12 $(2a,0)$	5 $(0,-2b)$	10 $(0,2b)$	3 $(-2a,0)$
$\tilde{0}(0,0)$	$(0,0)$	$(-a,-b)$	$(-a,b)$	$(a,-b)$	(a,b)	$(2a,0)$	$(0,-2b)$	$(0,2b)$	$(-2a,0)$
$\tilde{1}(-b,-a)$	$(-b,-a)$	$(-a-b,-a-b)$	$(-a-b,b-a)$	$(a-b,-b-a)$	$(a-b,b-a)$	$(-2a-b,-a)$	$(-b,-2b-a)$	$(-b,2b-a)$	$(-2a-b,-a)$
$\tilde{2}(b,-a)$	$(b,-a)$	$(b-a,-a-b)$	$(b-a,b-a)$	$(a+b,-a-b)$	$(a+b,b-a)$	$(b+2a,-a)$	$(b,-2b-a)$	$(b,2b-a)$	$(b-2a,-a)$
$\tilde{4}(-b,a)$	$(-b,a)$	$(-a-b,a-b)$	$(-a-b,a+b)$	$(a-b,a-b)$	$(-a-b,a+b)$	$(2a-b,a)$	$(-b,a-2b)$	$(-b,a+2b)$	$(-2a-b,a)$
$\tilde{8}(b,a)$	(b,a)	$(b-a,a-b)$	$(b-a,a+b)$	$(a+b,a-b)$	$(a+b,a+b)$	$(b+2a,a)$	$(b,a-2b)$	$(b,a+2b)$	$(b-2a,a)$
$\tilde{12}(0,2a)$	$(0,2a)$	$(-a,2a-b)$	$(-a,2a+b)$	$(a,2a-b)$	$(-a,2a+b)$	$(2a,2a)$	$(0,2(a-b))$	$(0,2(a+b))$	$(-2a,2a))$
$\tilde{5}(-2b,0)$	$(-2b,0)$	$(-2b-a,-b)$	$(-2b-a,b)$	$(-2b+a,-b)$	$(a-2b,b)$	$(2(a-b),0)$	$(-2b,-2b)$	$(-2b,2b)$	$(-2(a+b),0)$
$\tilde{10}(2b,0)$	$(2b,0)$	$(2b-a,-b)$	$(2b-a,b)$	$(2b+a,-b)$	$(a+2b,b)$	$(2(b+a),0)$	$(2b,-2b)$	$(2b,2b)$	$(2(b-a),0)$
$\tilde{3}(0,-2a)$	$(0,-2a)$	$(-a,-2a-b)$	$(-a,-2a+b)$	$(a,-2a-b)$	$(-a,b-2a)$	$(2a,-2a)$	$(0,-2(a+b))$	$(0,2(b-a))$	$(-2a,-2a)$

	Complex Function Solutions	Numbers
	$\{(0,0)\}$	$1\times 1=1$
	$\{(a,b),(-a,b),(a,-b),(-a,-b)\}$	$4\times 1=4$
	$\{(b,a),(-b,a),(b,-a),(-b,-a)\}$	$4\times 1=4$
	$\{(0,-2a),(-2b,0),(2b,0),(0,2a)\}$	$4\times 1=4$
	$\{(-2a,0),(0,-2b),(0,2b),(2a,0)\}$	$4\times 1=4$
	$\{(-a-b,-a-b),(b-a,-a-b),(-a-b,a-b),(b-a,a-b)\}$	$4\times 1=4$
	$\{(-a-b,b-a),(b-a,b-a),(-a-b,a+b),(b-a,a+b)\}$	$4\times 1=4$
	$\{(a-b,-a-b),(b+a,-a-b),(a-b,a-b),(b+a,a-b)\}$	$4\times 1=4$
	$\{(a-b,b-a),(b+a,b-a),(a-b,a+b),(b+a,a+b)\}$	$4\times 1=4$
	$\{(-a,2a-b),(-2b-a,-b),(2b-a,-b),(-a,-2a-b)\}$	$4\times 1=4$
$\rightarrow$	$\{(-a,2a+b),(-2b-a,b),(2b-a,b),(-a,-2a+b)\}$	$4\times 1=4$
	$\{(a,2a-b),(a-2b,-b),(a+2b,-b),(a,-2a-b)\}$	$4\times 1=4$
	$\{(a,2a+b),(a-2b,b),(a+2b,b),(a,-2a+b)\}$	$4\times 1=4$
	$\{(-b,-a+2b),(b,-a+2b),(-b,a+2b),(b,a+2b)\}$	$4\times 1=4$
	$\{(-b,-a-2b),(b,-a-2b),(-b,a-2b),(b,a-2b)\}$	$4\times 1=4$
	$\{(-2a-b,-a),(-2a+b,-a),(-2a-b,a),(-2a+b,a)\}$	$4\times 1=4$
	$\{(2a-b,-a),(2a+b,-a),(2a-b,a),(2a+b,a)\}$	$4\times 1=4$
	$\{(2a,2a),(2(a-b),0),(2(a+b),0),(2a,-2a)\}$	$4\times 1=4$
	$\{(-2a,2a),(-2(a+b),0),(2(b-a),0),(-2a,-2a)\}$	$4\times 1=4$
	$\{(-2b,-2b),(0,2(a-b)),(0,-2(a+b)),(2b,-2b)\}$	$4\times 1=4$
	$\{(-2b,2b),(0,2(a+b)),(0,2(b-a)),(2b,2b)\}$	$4\times 1=4$
	81 Groups, Total	81

Optimization $OP : (0, 1, 2, 4, 8, 12, 5, 10, 3) \to (0, 1, 2, 4, 8, 12, 5, 10, 3) \times (0, 8, 2, 4, 1, 5, 12, 3, 10)$

$\sum =$	0 $(0,0)$	1 $(-a,-b)$	2 $(-a,b)$	4 $(a,-b)$	8 (a,b)	12 $(2a,0)$	5 $(0,-2b)$	10 $(0,2b)$	3 $(-2a,0)$
$\tilde{0}(0,0)$	$(0,0)$	$(-a,-b)$	$(-a,b)$	$(a,-b)$	(a,b)	$(2a,0)$	$(0,-2b)$	$(0,2b)$	$(-2a,0)$
$\tilde{8}(b,a)$	(b,a)	$(b-a,a-b)$	$(b-a,a+b)$	$(a+b,a-b)$	$(a+b,a+b)$	$(b+2a,a)$	$(b,a-2b)$	$(b,a+2b)$	$(b-2a,a)$
$\tilde{2}(b,-a)$	$(b,-a)$	$(b-a,-a-b)$	$(b-a,b-a)$	$(a+b,-a-b)$	$(a+b,b-a)$	$(b+2a,-a)$	$(b,-2b-a)$	$(b,2b-a)$	$(b-2a,-a)$
$\tilde{4}(-b,a)$	$(-b,a)$	$(-a-b,a-b)$	$(-a-b,a+b)$	$(a-b,a-b)$	$(-a-b,a+b)$	$(2a-b,a)$	$(-b,a-2b)$	$(-b,a+2b)$	$(-2a-b,a)$
$\tilde{1}(-b,-a)$	$(-b,-a)$	$(-a-b,-a-b)$	$(-a-b,b-a)$	$(a-b,-b-a)$	$(a-b,b-a)$	$(2a-b,-a)$	$(-b,-2b-a)$	$(-b,2b-a)$	$(-2a-b,-a)$
$\tilde{5}(-2b,0)$	$(-2b,0)$	$(-2b-a,-b)$	$(-2b-a,b)$	$(-2b+a,-b)$	$(a-2b,b)$	$(2(a-b),0)$	$(-2b,-2b)$	$(-2b,2b)$	$(-2(a+b),0)$
$\tilde{12}(0,2a)$	$(0,2a)$	$(-a,2a-b)$	$(-a,2a+b)$	$(a,2a-b)$	$(-a,2a+b)$	$(2a,2a)$	$(0,2(a-b))$	$(0,2(a+b))$	$(-2a,2a))$
$\tilde{3}(0,-2a)$	$(0,-2a)$	$(-a,-2a-b)$	$(-a,-2a+b)$	$(a,-2a-b)$	$(-a,b-2a)$	$(2a,-2a)$	$(0,-2(a+b))$	$(0,2(b-a))$	$(-2a,-2a)$
$\tilde{10}(2b,0)$	$(2b,0)$	$(2b-a,-b)$	$(2b-a,b)$	$(2b+a,-b)$	$(a+2b,b)$	$(2(b+a),0)$	$(2b,-2b)$	$(2b,2b)$	$(2(b-a),0)$

4.5 *Conjugate Combination Lattice—Antisymmetric and Conjugate Symmetric Matrix Representation*

When $a = b$, the action matrix of 9×9 is

$\sum =$	0 $(0,0)$	1 $(-a,-a)$	2 $(-a,a)$	4 $(a,-a)$	8 (a,a)	12 $(2a,0)$	5 $(0,-2a)$	10 $(0,2a)$	3 $(-2a,0)$
$\tilde{0}(0,0)$	$(0,0)$	$(-a,-a)$	$(-a,a)$	$(a,-a)$	(a,a)	$(2a,0)$	$(0,-2a)$	$(0,2a)$	$(-2a,0)$
$\tilde{1}(-a,-a)$	$(-a,-a)$	$(-a-a,-a-a)$	$(-a-a,a-a)$	$(a-a,-a-a)$	$(a-a,a-a)$	$(2a-a,-a)$	$(-a,-2a-a)$	$(-a,2a-a)$	$(-2a-a,-a)$
$\tilde{2}(a,-a)$	$(a,-a)$	$(a-a,-a-a)$	$(a-a,a-a)$	$(a+a,-a-a)$	$(a+a,a-a)$	$(a+2a,-a)$	$(a,-2a-a)$	$(a,2a-a)$	$(a-2a,-a)$
$\tilde{4}(-a,a)$	$(-a,a)$	$(-a-a,a-a)$	$(-a-a,a+a)$	$(a-a,a-a)$	$(-a-a,a+a)$	$(2a-a,a)$	$(-a,a-2a)$	$(-a,a+2a)$	$(-2a-a,a)$
$\tilde{8}(a,a)$	(a,a)	$(a-a,a-a)$	$(a-a,a+a)$	$(a+a,a-a)$	$(a+a,a+a)$	$(a+2a,a)$	$(a,a-2a)$	$(a,a+2a)$	$(a-2a,a)$
$\tilde{12}(0,2a)$	$(0,2a)$	$(-a,2a-a)$	$(-a,2a+a)$	$(a,2a-a)$	$(-a,2a+a)$	$(2a,2a)$	$(0,2(a-a))$	$(0,2(a+a)$	$(-2a,2a)$
$\tilde{5}(-2a,0)$	$(-2a,0)$	$(-2a-a,-a)$	$(-2a-a,a)$	$(-2a+a,-a)$	$(a-2a,a)$	$(2(a-a),0)$	$(-2a,-2a)$	$(-2a,2a)$	$(-2(a+a),0)$
$\tilde{10}(2a,0)$	$(2a,0)$	$(2a-a,-a)$	$(2a-a,a)$	$(2a+a,-a)$	$(a+2a,a)$	$(2(a+a),0)$	$(2a,-2a)$	$(2a,2a)$	$(2(a-a),0)$
$\tilde{3}(0,-2a)$	$(0,-2a)$	$(-a,-2a-a)$	$(-a,-2a+a)$	$(a,-2a-a)$	$(-a,a-2a)$	$(2a,-2a)$	$(0,-2(a+a))$	$(0,2(a-a))$	$(-2a,-2a)$

Complex Function Solutions	Numbers
$\{(0,0),(0,0),(0,0),(0,0),(0,0),(0,0),(0,0),(0,0),(0,0)\}$	$1 \times 9 = 9$
$\{(a,a),(a,a),(a,a),(a,a),(a,a),(a,a)\}$	$1 \times 6 = 6$
$\{(-a,a),(-a,a),(-a,a),(-a,a),(-a,a),(-a,a)\}$	$1 \times 6 = 6$
$\{(a,-a),(a,-a),(a,-a),(a,-a),(a,-a),(a,-a)\}$	$1 \times 6 = 6$
$\{(-a,-a),(-a,-a),(-a,-a),(-a,-a),(-a,-a),(-a,-a)\}$	$1 \times 6 = 6$
$\{(0,-2a),(0,-2a),(0,-2a),(0,-2a)\}$	$1 \times 4 = 4$
$\{(-2a,0),(-2a,0),(-2a,0),(-2a,0)\}$	$1 \times 4 = 4$
$\{(2a,0),(2a,0),(2a,0),(2a,0)\}$	$1 \times 4 = 4$
$\to$ $\{(0,2a),(0,2a),(0,2a),(0,2a)\}$	$1 \times 4 = 4$
$\{(-2a,-2a),(-2a,-2a),(-2a,-2a)\}$	$1 \times 3 = 3$
$\{(-2a,2a),(-2a,2a),(-2a,2a)\}$	$1 \times 3 = 3$
$\{(2a,-2a),(2a,-2a),(2a,-2a)\}$	$1 \times 3 = 3$
$\{(2a,2a),(2a,2a),(2a,2a)\}$	$1 \times 3 = 3$
$\{(-3a,-a),(-3a,-a),(-a,3a),(-a,3a),(-a,-3a),(-a,-3a)\}$	$3 \times 2 = 6$
$\{(-3a,a),(-3a,a),(3a,-a),(3a,-a),(a,-3a),(a,-3a)\}$	$3 \times 2 = 6$
$\{(a,3a),(a,3a),(3a,a),(3a,a)\}$	$2 \times 2 = 4$
$\{(4a,0),(-4a,0),(0,-4a),(0,4a)\}$	$4 \times 1 = 4$
25 Groups, Total	81

In optimization, select the same label horizontally and vertically as $OP : (0,1,2,4,8,12,5,10,3) \to (0,1,2,4,8,12,5,10,3) \times (0,1,2,4,8,12,5,10,3)$, (Conjugate Symmetric Matrix $(a,b) \to (b,a)$)

$\sum =$	0 $(0,0)$	1 $(-a,-a)$	2 $(-a,a)$	4 $(a,-a)$	8 (a,a)	12 $(2a,0)$	5 $(0,-2a)$	10 $(0,2a)$	3 $(-2a,0)$
$\tilde{0}(0,0)$	$(0,0)$	$(-a,-a)$	$(-a,a)$	$(a,-a)$	(a,a)	$(2a,0)$	$(0,-2a)$	$(0,2a)$	$(-2a,0)$
$\tilde{1}(-a,-a)$	$(-a,-a)$	$(-2a,-2a)$	$(-2a,0)$	$(0,-2a)$	$(0,0)$	$(a,-a)$	$(-a,-3a)$	$(-a,a)$	$(-3a,-a)$
$\tilde{2}(a,-a)$	$(a,-a)$	$(0,-2a)$	$(0,0)$	$(2a,-2a)$	$(2a,0)$	$(3a,-a)$	$(a,-3a)$	(a,a)	$(-a,-a)$
$\tilde{4}(-a,a)$	$(-a,a)$	$(-2a,0)$	$(-2a,2a)$	$(0,0)$	$(-2a,2a)$	(a,a)	$(-a,-a)$	$(-a,3a)$	$(-3a,a)$
$\tilde{8}(a,a)$	(a,a)	$(0,0)$	$(0,2a)$	$(2a,0)$	$(2a,2a)$	$(3a,a)$	$(a,-a)$	$(a,3a)$	$(-a,a)$
$\tilde{12}(0,2a)$	$(0,2a)$	$(-a,a)$	$(-a,3a)$	(a,a)	$(-a,3a)$	$(2a,2a)$	$(0,0)$	$(0,4a)$	$(-2a,2a))$
$\tilde{5}(-2a,0)$	$(-2a,0)$	$(-3a,-a)$	$(-3a,a)$	$(-a,-a)$	$(-a,a)$	$(0,0)$	$(-2a,-2a)$	$(-2a,2a)$	$(-4a,0)$
$\tilde{10}(2a,0)$	$(2a,0)$	$(a,-a)$	(a,a)	$(3a,-a)$	$(3a,a)$	$(4a,0)$	$(2a,-2a)$	$(2a,2a)$	$(0,0)$
$\tilde{3}(0,-2a)$	$(0,-2a)$	$(-a,-3a)$	$(-a,-a)$	$(a,-3a)$	$(-a,-a)$	$(2a,-2a)$	$(0,-4a)$	$(0,0)$	$(-2a,-2a)$

Optimization $OP : (0,1,2,4,8,12,5,10,3) \to (0,1,2,4,8,12,5,10,3) \times (0,8,2,4,1,5,12,3,10)$, (Anti-symmetric Matrix $(a,b) \to -(a,b)$)

$\sum =$	0 $(0,0)$	1 $(-a,-a)$	2 $(-a,a)$	4 $(a,-a)$	8 (a,a)	12 $(2a,0)$	5 $(0,-2a)$	10 $(0,2a)$	3 $(-2a,0)$
$\tilde{0}(0,0)$	$(0,0)$	$(-a,-a)$	$(-a,a)$	$(a,-a)$	(a,a)	$(2a,0)$	$(0,-2a)$	$(0,2a)$	$(-2a,0)$
$\tilde{8}(a,a)$	(a,a)	$(0,0)$	$(0,2a)$	$(2a,0)$	$(2a,2a)$	$(3a,a)$	$(a,-a)$	$(a,3a)$	$(-a,a)$
$\tilde{2}(a,-a)$	$(a,-a)$	$(0,-2a)$	$(0,0)$	$(2a,-2a)$	$(2a,0)$	$(3a,-a)$	$(a,-3a)$	(a,a)	$(-a,-a)$
$\tilde{4}(-a,a)$	$(-a,a)$	$(-2a,0)$	$(-2a,2a)$	$(0,0)$	$(0,2a)$	(a,a)	$(-a,-a)$	$(-a,3a)$	$(-3a,a)$
$\tilde{1}(-a,-a)$	$(-a,-a)$	$(-2a,-2a)$	$(-2a,0)$	$(0,-2a)$	$(0,0)$	$(a,-a)$	$(-a,-3a)$	$(-a,a)$	$(-3a,-a)$
$\tilde{5}(-2a,0)$	$(-2a,0)$	$(-3a,-a)$	$(-3a,a)$	$(-a,-a)$	$(-a,a)$	$(0,0)$	$(-2a,-2a)$	$(-2a,2a)$	$(-4a,0)$
$\tilde{12}(0,2a)$	$(0,2a)$	$(-a,a)$	$(-a,3a)$	(a,a)	$(a,3a)$	$(2a,2a)$	$(0,0)$	$(0,4a)$	$(-2a,2a)$
$\tilde{3}(0,-2a)$	$(0,-2a)$	$(-a,-3a)$	$(-a,-a)$	$(a,-3a)$	$(a,-a)$	$(2a,-2a)$	$(0,-4a)$	$(0,0)$	$(-2a,-2a)$
$\tilde{10}(2a,0)$	$(2a,0)$	$(a,-a)$	(a,a)	$(3a,-a)$	$(3a,a)$	$(4a,0)$	$(2a,-2a)$	$(2a,2a)$	$(0,0)$

Distinguishable twenty-five grid points are symmetrically distributed in the grid.

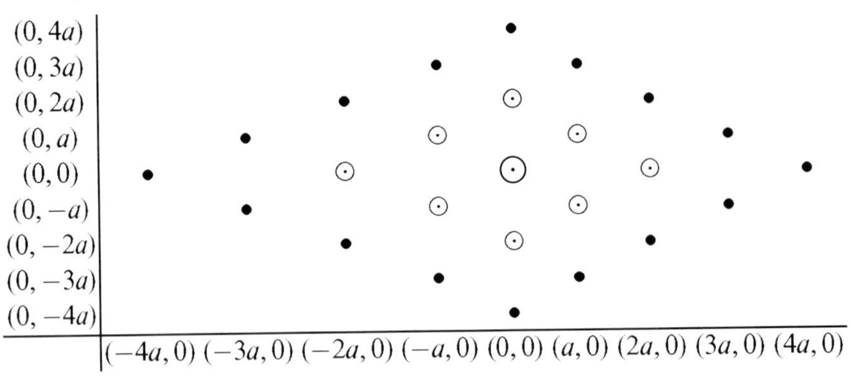

In the core part of $a = b$, the nine distinguishable lattice points of the four generator sets are symmetrically distributed in the grid.

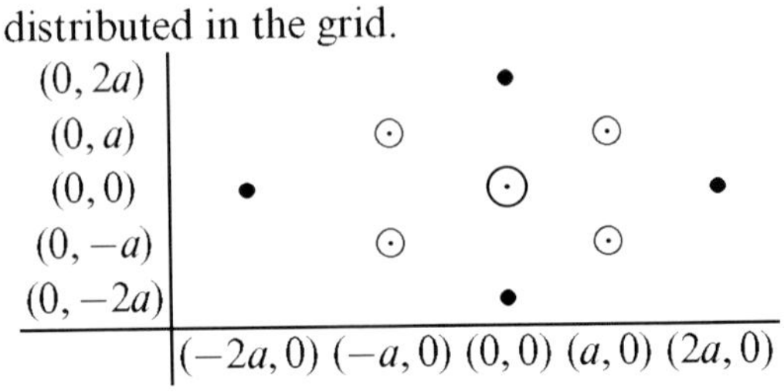

In the core part of $a \neq b$, the two combined grid points each containing the nine grid points overlap into seventeen grid points, two grid points overlap at the center $\odot$, and the remaining sixteen points are, respectively, distinguishable.

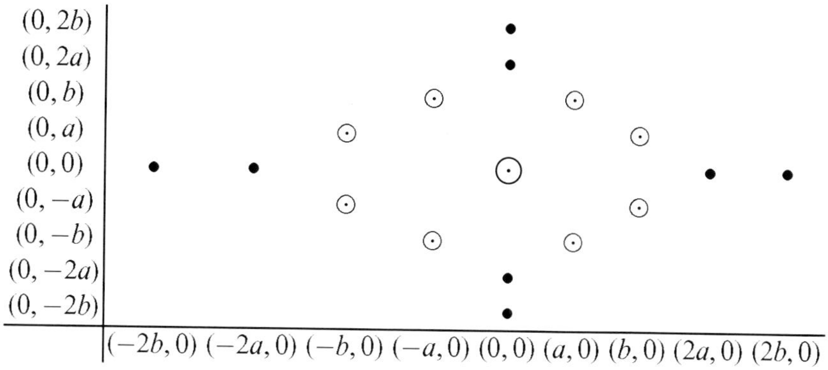

Looking at the series lattice distribution, the substructure does not have symmetry. The generator usually occupies the central area, and the newly generated points occupy the outer area. Under different conditions, the accompanying merging process is determined by the location of the lattice points. For the complete conjugated combinatorial lattice structure, the overall distribution formed after combinatorial optimization has good symmetry properties.

The five groups of distribution diagrams are given in Figure 1, and although the number of lattice points appearing in (a) and (b) are the same, their main distribution shapes are completely different. A similar situation occurs in (c) and (d) with different parameter conditions, resulting in the same number of lattice points but in completely different distributions. For the more general case of $a \neq b$, shown in (e), the distribution of different selection parameters is also completely different, but the total number of lattice points is the same, which is eighty-one.

5. Conclusion

Using **Lemma 1**, through the conjugate state representation, based on the extended $K4$ group and conjugate structure, the eight-element representation basis and its extended combined structure are formed, and the nine-element irreducible representation is optimized to be conjugated. The primitives form a lattice structure of conjugated states.

From **Lemma 2** to **Theorem 13**, a series of lemmas, theorems, and corollaries describe the algebraic and measurement properties of conjugated combinatorial lattices from different perspectives. Describe the combination theorem of additive group + inverse element/zero element, the sum of eigenvalues is zero, etc.

Through eigenvalue structure expansion and combinatorial lattice optimization segmentation, the overall optimization conditions of conjugated combinatorial lattices are confirmed. The two types of permutations formed by the transformation rules are labeled conjugate symmetric permutation and antisymmetric permutation structures (**Theorem 9, 10**).

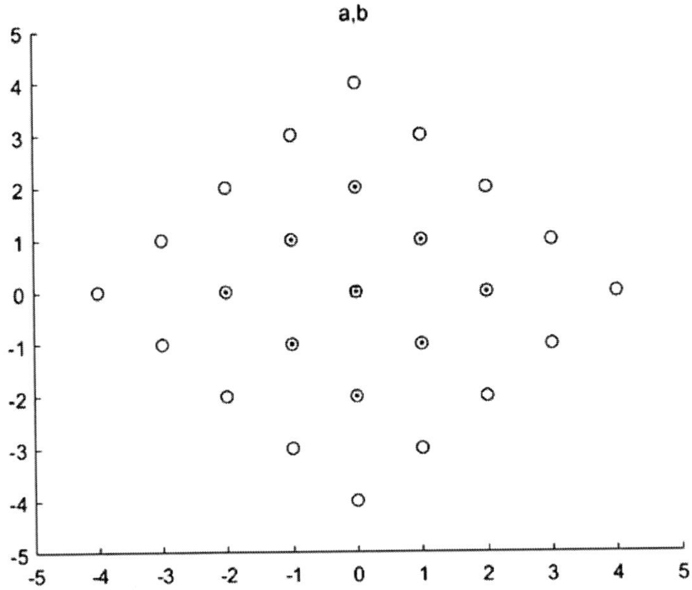

(a) 25 Lattice Points $(a, b) \in \{(1,1), (1,-1), (-1,1), (-1,-1)\}$

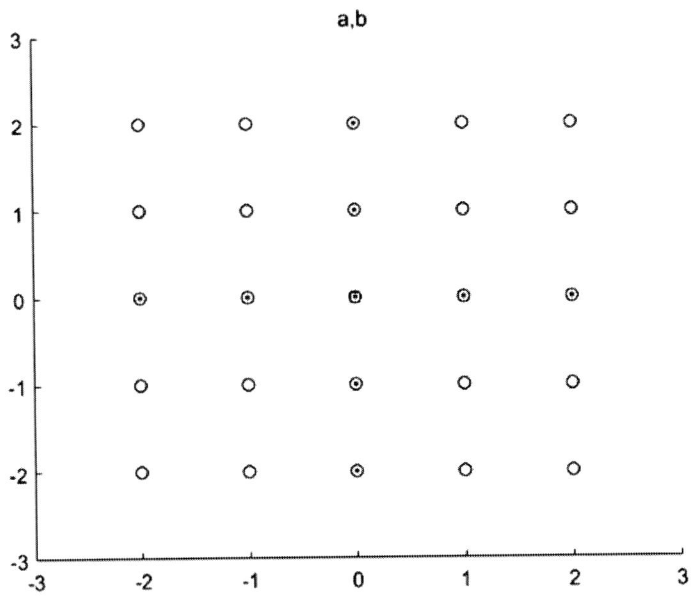

(b) 25 Lattice Points $(a, b) \in \{(1,0), (-1,0), (0,1), (0,-1)\}$

The typical cluster topological properties are expressed by four types of conjugated combination lattice points and six groups of distinguishable modes, and the distribution of samples near the zero point is observed in graphical mode using the complex plane.

Whether it is from the average measure or the information entropy measure, it can be seen that from the perspective of any measure, the structure has the characteristics of dimension and entropy value in the constraint interval of the maximum and minimum values.

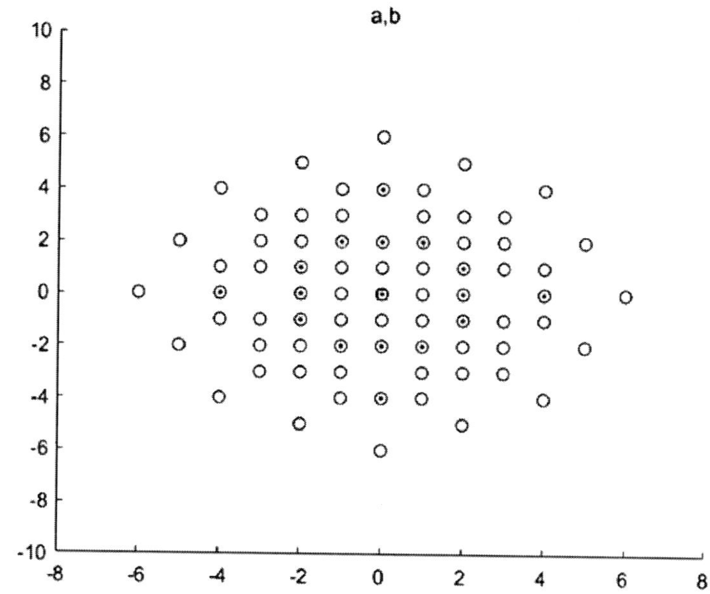

(c) 73 Lattice Points $(a,b) \in \{(1,2),(-1,2),(1,-2),(-1,-2)\}$

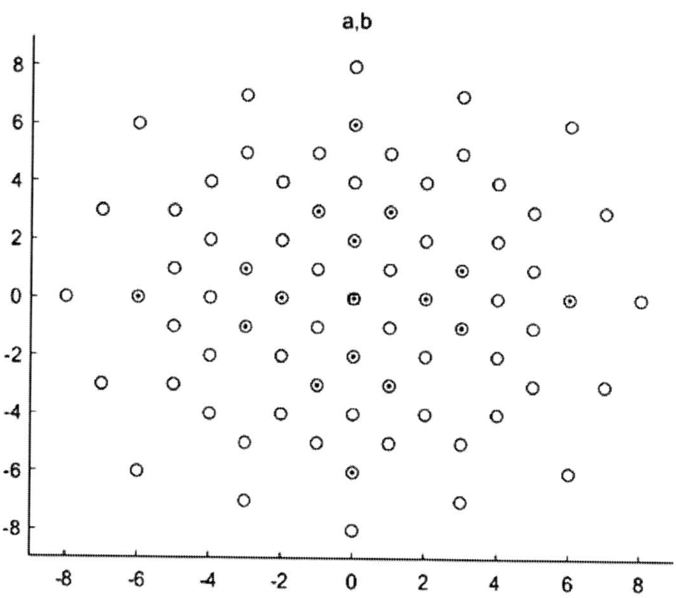

(d) 73 Lattice Points $(a,b) \in \{(1,3),(-1,3),(1,-3),(-1,-3)\}$

From a series of typical examples, the spatial distribution of different combinatorial lattice structures is shown. Selecting a specific region, the corresponding structure has nonzero sum representation features. However, for a series of complete representation examples, it is not difficult to observe the symmetric properties of the corresponding group lattice distribution.

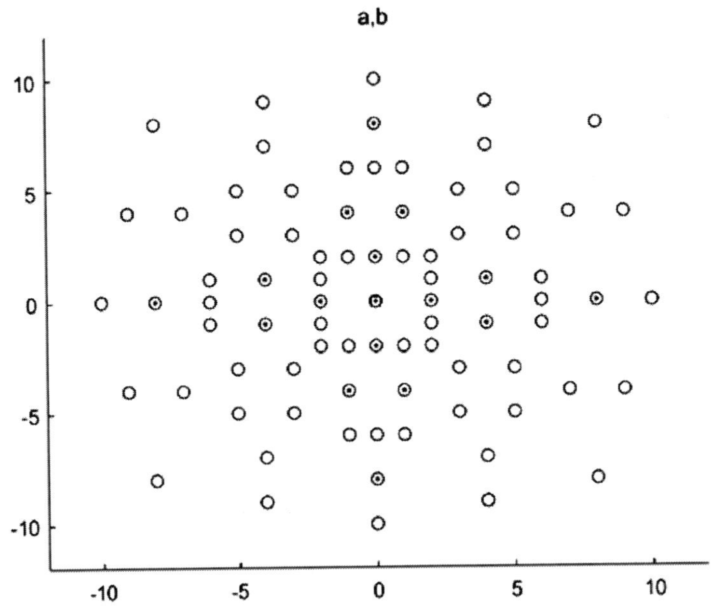

(e) 81 Lattice Points $(a,b) \in \{(1,4),(-1,4),(1,-4),(-1,-4)\}$

Figure 1. Conjugate lattice distributions with different conditions: (a) twenty-five lattice points $(a,b) = (1,1)$; (b) twenty-five lattice points $(a,b) = (0,1)$; (c) seventy-three lattice points $(a,b) = (1,2)$; (d) seventy-three lattice points $(a,b) = (1,3)$; (e) eighty-one lattice points $a \neq b, i.e.(a,b) = (1,4)$.

The 16×16 conjugated combinatorial representation matrix, with several identical parts, can be simplified to the 9×9 optimized representation. A distinction can be observed between two symmetric representation matrices, the accompanying conjugate symmetry, and the antisymmetry.

A more detailed description can be observed in the subsequent distribution diagrams of the lattice point clustering of a series of conjugate combinations under different conditions of (a,b).

Related canonical quantum coherent-entangled state applications will be explored in subsequent chapters.

References

[1] H. Weyl. *Space Time Matter* (Translated by H. L. Brose), Dover, 1950.

[2] 华罗庚, 华罗庚文集 - 多复变函数论卷 *II*, 科学出版社 2013.

[3] 张奠宙 王善平, 陈省身文集,华东师范大学出版社 2002.

[4] R. S. Varga. *Matrix Iterative Analysis*, Prentice-Hall Inc., 1962.

[5] V. I. Arnold. *Mathematical Methods of Classical Mechanics*, Springer-Verlag, 1978.

[6] R. Penrose. *The Emperor's Mind: Concerning Computers, Minds and the Law of Physics*, Penguin, 1989.

[7] T. Siegfried. *The Bit and the Pendulum: From Quantum Computing to M Theory - The New Physics of Information*, John Wiley & Sons, 2000.

[8] R. Penrose. *The Road to Reality - A Complete Guide to the Laws of the Universe*, Vintage Book, 2005.

[9] C. Seife. *Decoding the Universe: How the Science of Information is Explaining Everything in the Cosmos, from our Brains to Black Holes*, Penguin, 2006.
[10] 吉姆 巴戈特 著;徐彬 于秀秀译, 量子通史 - 量子物理史上的40个重大时刻,中信出版社 2021。J. Baggott. *The Quantum Story in 40 Moments*, Oxford University Press, 2011.
[11] 许天周 李炳照, 线性正则变换及其应用,科学出版社 2013.
[12] 马天, 从数学观点看物理世界 - 基本粒子与统一场理论, 科学出版社 2015.
[13] G.'t Hooft. *The Cellular Automaton Interpretation of Quantum Mechanics*, Springer, 2016.
[14] 杨振宁,葛墨林,何杨辉编, 拓扑与物理,江西科学技术出版社 2021.
[15] 艾小文, 数论问题 - 指数和的均值分析与研究,电子工业出版社 2021。
[16] L. Z. Xu, M. S. Jiang and Z. Q. Zhu. *Combinatorial Mathematics of Computation*, Shanghai Science & Technology Press, 1983 (in Chinese).
[17] J. H. van Lint and R. M. Wilson. *A Course in Combinatorics*, 2nd edition, Cambridge University Press, 2001.
[18] L. X. Wang. *An Elementary Treatise on Combinations*, Harbin Institute of Technology Press, 2012 (in Chinese).
[19] G. Birkhoof. *Lattice Theory*, The American Mathematical Society, 1948.
[20] G. Birkhoof. Von Neumann and lattice theory, momories of the American mathematical society, no. 10, 1955.
[21] R. J. Simpson. Convex lattice polygons on mimimum area. *Bull. Austral. Math. Soc.* 42, 1990, 353–367.
[22] A. Day. Doubling constructions in lattice theory. *Can. J. Math.* 44(2), 1992, 252–269.
[23] M. Stern. *Semimodular Lattices Theory and Applications*, Cambridge University Press, 1999.
[24] 许光午,王小云, 格的计算和密码学应用, 中国科学:数学 2020年 第50卷第10期 1417–1436。
[25] Z. J. Zheng and A. Maeder. The conjugate classification of the kernel form of the hexagonal grid. In *Modern Geometric Computing for Visualization*, edited by T. L. Kunii and Y. Shinagawa, Springer-Verlag, 1992, 73–89.
[26] Z. J. Zheng. *Conjugate Transformation of Regular Plane Lattices for Binary Images*, PhD Thesis, Monash University, 1994.
[27] Jeffrey Z. J. Zheng. *Variant Construction from Theoretical Foundation to Applications*, Springer Nature, 2019.
[28] 郑智捷, 变值体系理论及其应用 第一卷 理论基础及其应用,科学出版社 2021. Jeffrey Z. J. Zheng. *Vairant Construction Theory and Applications, Vol. 1, Theoretical Foundation and Applications*, Science Press, 2021.

Part II

Applications on Entangled Measurements of Quantum Optics

More you know
more you know you do not know.
– Aristotle

From the special to the general and it is extremely peculiar that
such a procedure has led to so few of the so-called paradoxes.
– Niels Henrik Abel

I consider that I understand an equation
when I can predict the properties of its solutions
without actually solving it.
– Paul Dirac

In Part II, three chapters are arranged as Chapters 5–7.

Chapter 5. Modern Progress in Entangled Measurements of Quantum Optics The historic development of complex conjugate systems, complex geometry, multiple variable complex functions, and entangled measurements of quantum optics are reviewed. The description of the whole structure is composed of twenty-two lemmas and twenty corollaries.

Wootters have listed various measurement schemes to indicate entanglement vectors without orthogonal properties in quantitative measuring conditions.

The proving procedure of the **Hardy Theorem** is a perfect case to illustrate a series of complex operators applied as the basic reasons in the derivation.

Chapter 6. Entanglement Foundation of Quantum Optics on Conjugate Phase Space The Entanglement Foundation of Quantum Optics is composed of three parts. The first part discusses the core matrix structure of quantum interaction. The second part summarizes the most important results of quantum entanglements during 2002–2022. The third part describes the necessary and sufficient conditions of quantum entanglement to establish a self-consistent description of space and time in the quanta optical measurement system, under a formal description of the measurement being paradox-free. The whole structure is composed of five lemmas, six assumptions, six predictions, sixteen theorems, and sixteen corollaries.

From Lie algebra to conjugate algebra, the classical quantum formula simplifies the eight-variable vectors into two vectors on a pair of complex conjugate representations. The complete quantum solution needs to be equipped with principal and auxiliary double diagonal complex eigenvalues, and the n qubit phase space representation expands as the n cobit (cobit : complete or conjugate qubit) phase space representation from 2^n to 2^{2n}.

Chapter 7. Conjugate Construction in Future Explorations Future explorations explore a series of problems related to conjugate vector algebra operations from several aspects, including classical paradox, modern logic, topological geometry, visual paradox, classical Tai Chi, resolving Mobius paradox, two-body entanglement and time-reversal symmetry, symmetry + supersymmetry criteria, and variant constructions + knowledge model.

The 0-1 vector logic leads to the multivariable complex function system. Future challenges are predicted for further exploration. The description of the whole structure is composed of ten lemmas, twelve theorems, eleven corollaries, and three predictions.

Chapter 5

Modern Progress in Entangled Measurements of Quantum Optics

Entanglement States of Quantum Measurements on Geometric Transformations

Jeffrey Zheng* and Jianzhong Liu
Conjugate System Research, Melbourne, Australia

Abstract

For a proper understanding of the complex functions and entanglement of quantum optics, it is necessary to review the key milestones in the historical and modern developments of the scientific world. This chapter is composed of three parts: (1) historical development of complex conjugate schemes from geometric viewpoints in Sections 2–6; (2) modern representations of complex functions in Sections 7–9; and (3) basic representation of quantum states, quantum paradoxes, detection of quantum entangled states, formal measurement, quantum state measurement and reflection representation, and Hardy's theorem in Sections 10–12. Section 11 discusses measurement representations of quantum paradoxes, including measurement and reflection representations, EPR paradox, Bell representations, inequalities, single-quantum entangled states, double-quantum entangled states, noninequalities Bell theorem, multiphoton entangled states, **Hardy Theorem**, cluster state representation, and classical models and methods of quantum information. Tables 1–3 summarize the relevant important conclusions. In Table 1, the correspondence between some symmetries and conserved quantities is listed. The relationship between transformation group fields and disciplines is given in Table 2. Typical representations of quantum states and quantum entangled states are given in Table 3. Important results gathered from the documentation are summarized by **Lemmas 1–21 and Corollaries 1–22**.

Keywords: Multivariate complex functions, geometric classification of complex numbers, fundamentals of quantum states, quantum measurement paradox

*Corresponding Author's Email: conjugatelogic@yahoo.com

In: Conjugate Construction of Quantum Optics
Editors: Jeffrey Zheng
ISBN: 979-8-88697-651-9

There have been only three epoch-making mathematicians
Archimedes Newton Eisenstein.
– Carl Friedrich Gauss

Knowledge is realizing that the street is one way
Wisdom is looking in both directions anyway
– Albert Einstein.

I see it but I do not believe it that which is provable
Ought not to be believed in science without proof.
– Richard Dedekind

1. Aim of Studies

For a proper understanding of the complex functions and entanglement of quantum optics, it is necessary to review the key milestones in the historical and modern developments of the scientific world. Most materials are collected on mathematics [1, 2, 3], quantum information [4], multiple websites at https://enwikipedia.org/wiki/, and other interesting materials cited.

This chapter is composed of three parts: 1) historical development of complex conjugate schemes from geometric viewpoints; 2) modern representations of complex functions; and 3) basic representation of quantum states, quantum paradoxes, detection of quantum entangled states, formal measurement, quantum state measurement and reflection representation, and Hardy theorem.

In Sections 2–6, the first part traces the early development of complex functions from the 16th-century conjugate description, complex geometry, fundamental theorem of algebra, exponential representation, modern representation, theory of complex functions, Riemann surface, elliptic functions, complex equations on mathematical physics, Green functions, automorphic functions, differential equations, variations, Hamilton–Jacobi formula, and the 20th-century Hilbert spaces.

In Sections 7–9, the second part describes the main development from the 20th-century Hermite matrix, Pauli/Dirac matrices, Jones vectors and matrix, gauge invariance, lattices, complex geometry classification, topological gravity, non-Euclidean geometry, symmetry principle, unitary group, Langlands program, dark matter, and past/future quantum explanations.

In Sections 10–12, the third part collects from the 1930s on quantum entanglement, representation of quantum states, detection of quantum entangled states, quantum state complex functions, two body entanglement, time-reversal symmetry, symmetry+supersymmetry, vector representation of quantum states, Hadamard matrix, complex matrix, pan-complex variable functions, formal measurement, quantum paradoxes, reflection representation, EPR paradox, Bell states, CHSH–Bell inequality, single qubit entanglement, Bell's theorem of noninequalities, multiphoton entangled states, Hardy theorem, and cluster states.

Typical representations of quantum states/quantum entanglements are listed in Table 3.

2. Brief Reviews on Geometric Representation of Complex Functions

In the classic literature, Weyl established the foundation for the modern mathematical physics framework for the study of space-time matter [27]. The modern theory of multivariate complex functions is closely related to the series of studies by Hua Luogeng (Loo-Keng Hua) [10] and Chen Xingshen (Shiing-Shen Chern) [11]. Researches of cutting-edge quantum physics and related systems can be referred in holistic mathematical physics [12]–[21], special number theory research and combinatorial exploration [23]–[30], and lattice theory and their advanced applications [34]–[39].

2.1 *A Few Anecdotes about Complex Functions*

A few short stories about the origin of complex functions are given below:

(A) In 1799, Gauss proved the Fundamental Theorem of Algebraic Equations, playing a central role as the main mover of functions of complex variables. Based on basic research in the early years, Gauss repeatedly proved that any algebraic equation of order n can be determined to be one of the n groups of complex number pairs as its exact solution. After the solution is determined, the dimensionality is reduced successively, and another $n-1$ group of solutions is solved until all n groups of complex solutions are completed.

(B) In 1830, Hamilton expressed complex numbers in the form of binary pairs, namely, $(a, b) = a + ib$, and systematically explored the algebraic properties under the conditions of addition, subtraction, multiplication, and division of quaternary arithmetic operations. It lays the foundation for the all-around application of modern complex functions.

(C) In 1850, Riemann, as a student of Gauss, developed the theory of multivariable complex functions, and a few years later, he established Riemann geometry. Half a century later, Riemannian geometry became famous all over the world. It is fortunate to be the foundational support core of the space-time primitives of Einstein's general theory of relativity.

2.2 *Conjugate Orthogonal Features*

The study of the conjugation of the nature of space and time can be traced back to the period when Einstein published his special theory of relativity.

In 1905, Minkowski saw the research paper published by Einstein. As Einstein's mathematics teacher, he knew that he lacked advanced mathematics skills and suggested using the four-dimensional space-time complex variable function structure, indicating that the special theory of relativity involves a unified description of time and space.

Hilbert is rumored to have said, "Every kid on the street in Gottingen knows more about four-dimensional geometry than Einstein. However, even so, it was Einstein who built the system of relativity, not mathematicians."

In 1915, Einstein successfully established the general theory of relativity using the Riemann geometric complex function representation framework. In the basic theoretical system of quantum mechanics, position and momentum are the core conjugates. von Neumann suggested that the inner product space of a function of multiple complex variables is a Hilbert space. The inner product invariant associated with quantum mechanics is a quantitative feature of the conjugate property of a quantum system.

For the multivariate vector description system, multiple sets of eigenvalues and eigenvector clusters exhibit orthogonality, which are the basic measurement characteristics of complex dynamic systems. From any two sets of complex function eigenvalues, the corresponding eigenvectors are orthogonal. Orthogonal conditions provide basic support for the description of different eigenvalues and eigenvector primitives.

2.3 Invariant of Multivariate Complex Function

From the invention of calculus, the germs of differential geometry were born [1]. Euler, Clairaut, and Monge made differential geometry an independent discipline. Euler has obtained important results in geodesy work, and the calculation of normal curvature obtains Euler's formula. Clairaut studied the curvature and torsion of curves, while Monge expressed the important properties of curves and surfaces in differential equations, leading to a peak in the development of classical differential geometry.

In the study of geodesy in 1827, Gauss found through complicated calculations that the product of the two principal curvatures of a surface has nothing to do with its shape in Euclidean space and created intrinsic geometry. Stripped from the peripheral space, the curved surface itself is studied as a space.

In 1854, Riemann popularized Gauss's two-dimensional surface intrinsic geometry with "Assumptions on Geometric Foundations" and developed n-dimensional Riemann geometry with the development of multiple complex functions. The research object of differential geometry is extended to complex manifolds and the theory of complex analytic spaces, including singularities. Differential geometry interacts with the theory of functions of multiple complex variables, Lie group theory, algebraic geometry, and PDE, which promote the continuous differentiation and integration of modern mathematics.

Multicomplex variable function theory is combined with differential geometry. From the Poincare metric defined on the unit circle and the upper half-plane (both of which can establish conformal mapping), the connection between the single-complex variable function theory and differential geometry is clearly visible. The Poincare metric is a conformal invariant. After the introduction of the Poincare metric, Schwarz's theorem is interpreted as follows: the Poincare metric does not increase for analytical mappings on the unit circle if and only if the Poincare metric does not change in the piecewise linear transformation of the mapping.

In the theory of functions of multiple complex variables, after the metric is defined in the complex affine space region, the curvature calculation of real differential geometry and other series of calculations are carried out. For a single complex variable function, all singularities satisfy the discrete distribution. In the case of multiple complex variables, all isolated singularities are engulfed due to the Hartogs exploiting phenomenon, the continuous region formed by singularities is also engulfed, and only the actual codimension is one. The manifold can be avoided. In 1922, Bergman introduced the Bergman kernel function, the Bergman metric on the multicomplex function field, and the one-dimensional case is the Poincare metric on the unit circle and the Poincare upper half-plane.

The research object of algebraic geometry is the properties of common zeros (algebraic varieties) of algebraic equations (defining equations) in any-dimensional affine space or projective space. The domain where the correlation coefficient and algebraic cluster points are located is the base domain. The irreducible algebraic variety is a finite extension field of the base field. Linear space is an extension field with the number field as the base field, and the dimension of the linear space is the number of extensions. Algebraic geometry studies finite extended fields, and the relationship between algebraic varieties and base fields is extremely close.

Based on algebraic varieties of complex affine spaces or complex projective spaces over fields, many concepts overlap with differential geometry and the theory of functions of multiple complex variables using similar tools. The progress of complex manifolds and their complex analytic spaces has influenced the research of these disciplines.

In 1944, Lookeng Hua, a master of modern mathematics, proposed matrix theory for the study of multicomplex variable functions [10]. The motion groups of the four typical field theories are rectangular matrix geometry, symmetric matrix geometry, obliquely symmetric matrix geometry, and complex sphere geometry, combined with the unit circle transformation system to lay the foundation for the complex conjugate matrix and its symplectic geometric representation. Based on the Riemann curvature, the curvature judgment conditions of hyperbolic space and ellipse space are established, the complete classification of symplectic group hypercircles is given, and the positive and negative characteristics of the label of spherical geometry are used to distinguish definite positive (ellipse), reduced rank (parabolic), and one positive and one negative (hyperbolic). For the Lorentz group study $(\det P)^2 = 1$, the determinant of P is the extended group structure of ± 1.

In 1944, Chern (Shiing-shen Chern) studied [11]'s contribution to the Gauss–Bonnet intrinsic proof in his early years. Using the Lie algebra of the special orthogonal group $SO(2n)$, that is, the space formed by the antisymmetric real $2n$ order matrix, from the recursive expression

$$d\Pi = \frac{1}{2^{2n}\pi^n n!}\psi_{n-1} = \Theta \tag{1}$$

shows a high-dimensional complex manifold. The global invariant feature is intrinsically associated with the 2^{2n} order $2n$-element combination group and the $n!$ order n-element permutation group. Indicative classes of subsequent Hermite spaces have being involved algebraic geometry, differential geometry, and the theory of functions of multiple complex variables. Hironaka studied algebraic geometry singularity resolution, working from complex manifolds to complex analytic spaces to form the theory of complex analytic spaces. Yau (Shing-Tung Yau) proved the Calabi conjecture using algebraic and differential geometry in classical general relativity and the standard model of elementary particles.

3. Historical Development of Pairwise Complex Numbers

From the perspective of solving equations, the strict representation of pairwise conjugate solutions of complex functions is explored. The process of solving quadratic equations can be traced back to the early history of mathematics. To meet the challenge of solving cubic equations, Cardano, in his 1545 book *The Important Art* Ars Magna, extended the solution of missing cubic equations to all cubic equations.

3.1 Complex Representation

In Algebra, published in 1572, Bombelli identified that the solution given by Cardan's formula is a pair of conjugate complex numbers expressed in the form $\{a + ib, a - ib\}$. Vieta gives all the roots of an irreducible cubic equation using the cosine and arc cosine functions. His solution to trigonometric functions was published in 1615.

In Descartes' "La Geometrie" published in 1637, $i = \sqrt{-1}$ was named an imaginary number, and an imaginary number was associated with the impossibility of geometric drawing. Before it was officially named, the square root of a negative number was called esoteric sophisticated or subtle.

In a letter to Huygens in 1673, Leibniz boasted about the results obtained, that he was the first person to reduce irrational radicals in the form of false numbers to real numbers. However, Bombelli was the first person to be worthy of his name. The same result was obtained a century ago.

Wallis formally presented his analysis of imaginary numbers in Treatise on Algebra in 1685. However, his work is remembered today only by historians as a pioneer in representing plurals as points on a plane.

Corollary 1 (Imaginary Representation). *Paired complex numbers have been around since the 16th century. The imaginary number is officially taken seriously and has gone through a long-term historical evolution process.*

3.2 Complex Geometry

Using geometry to explore complex numbers, the chapter "On the Analytical Representation of Direction" was submitted by Wessel in 1797 to the Royal Danish Academy of Sciences, published in Danish in the journal Memories in 1799, due to publication that are rarely read outside of Denmark. The paper was not rediscovered until 1895 and is considered a pioneer of complex geometry.

Wessel's interpretation of complex numbers is straightforward. A complex number can be represented as a point $a + ib$ in the complex plane or as a directed radial vector from the origin to this point. If $\theta = \arctan\left(\frac{b}{a}\right)$, then

$$a + ib = \sqrt{a^2 + b^2}(\cos\theta + i\sin\theta) \tag{2}$$

The length of the $\sqrt{a^2 + b^2}$ radial vector is called the complex modulus of $a + ib$, and the value of the polar angle $\arctan\left(\frac{b}{a}\right)$ is called the argument of $a+ib$, denoted as $\arg(a+ib)$

$$(a + ib) = \sqrt{a^2 + b^2} \angle \arctan\left(\frac{b}{a}\right) \tag{3}$$

Using the Wessel representation, the imaginary number is

$$\boxed{i = \sqrt{-1} = 1\angle\tfrac{\pi}{2}(\text{radians}) = 1\angle 90^\circ\ (\text{angle})}$$

Although Gauss implied in his work that he had obtained the same result in 1796, it was not published at that time. In Cauchy's 1848 book, there is a scribble description of a humble scholar Truel, who has used a vertical axis to represent imaginary values.

$$1\angle\theta = \cos\theta + i\sin\theta$$
$$1\angle\beta = \cos\beta + i\sin\beta$$
$$1\angle(\theta+\beta) = \cos(\theta+\beta) + i\sin(\theta+\beta)$$

Because $(1\angle\theta)(1\angle\beta) = 1\angle(\theta+\beta)$, consider $(\cos\theta + i\sin\theta)(\cos\beta + i\sin\beta) = \cos(\theta+\beta) + i\sin(\theta+\beta)$. Expand left section,

$$(\cos\theta + i\sin\theta)(\cos\beta + i\sin\beta) = (\cos\theta\cos\beta - \sin\theta\sin\beta) + i(\sin\theta\cos\beta + \cos\theta\sin\beta)$$
$$\cos(\theta+\beta) = \cos\theta\cos\beta - \sin\theta\sin\beta$$
$$\sin(\theta+\beta) = \sin\theta\cos\beta + \cos\theta\sin\beta$$

For a more concise example, observe

$$\left(1\angle\frac{\theta}{m}\right)^m = 1\angle\theta = \cos\theta + i\sin\theta$$
$$(1\angle\theta)^{\frac{1}{m}} = 1\angle\frac{\theta}{m} = \cos\frac{\theta}{m} + i\sin\frac{\theta}{m}$$

Lemma 1 (Complex Geometry). *A complex number can be represented as a point $a+ib$ in the complex plane or as a directed radial vector from the origin to this point.*

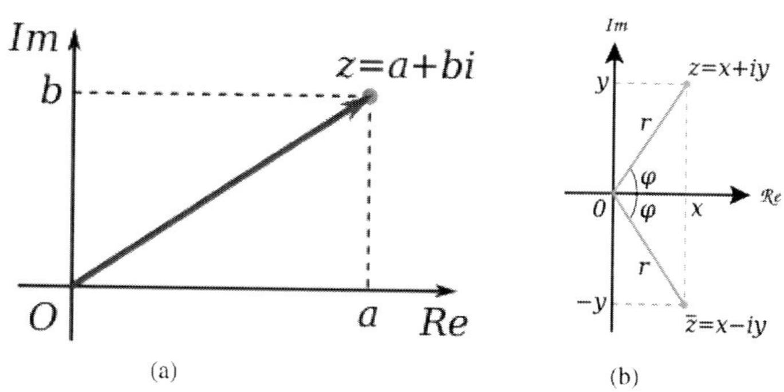

Figure 1. Complex vector representation: (a) single complex number; (b) complex conjugate pair representation.

3.3 Fundamental Theorem of Algebra

From the perspective of rigorously proving the fundamental theorem of algebra, it shows the painstaking exploration of world-class mathematicians for the basic structure of complex functions.

Fundamental Theorem of Algebra: Every nonzero, univariate polynomial of degree n with complex coefficients has n complex roots.

The first incomplete proof was provided by d'Alembert in 1746. Based on the assumption of the existence of such roots, other explorers include Euler (1749), de Foncenex (1759), Lagrange (1772), and Laplace (1795). At the end of the 18th century, two new proofs were published that did not assume root existence, but neither was complete. James Wood explored it from an algebraic point of view in 1798. His argument was completely proven by distance, with an algebraic gap, and the result was completely ignored by later generations. Another famous proof was made by Gauss in 1799 using the geometrical mode of argument, the first substantive proof of the Fundamental Theorem. His method is not to calculate a root but to prove the existence of such roots. Gauss' approach to the fundamental theorems of algebra opens up new ways of studying existential problems in mathematics. By modern standards, his proof is still lax, with a topological gap from a complete proof. Gauss provided two more proofs in 1816 and another modified version of the 1799 proof in 1849. The remaining topological gap was completely eliminated by Alexander Ostrowski in 1920.

In terms of computability, the first rigorous proof of the theorem was made by Argand in 1806 in his revised 1813 book. Since then, the Fundamental Theorem of Algebra explicitly asserts that polynomials of degree n have n complex coefficients, not just real coefficients.

Lemma 2 (Fundamental Theorem of Algebra). *Every nonzero, univariate polynomial of degree n with complex coefficients has n complex roots.*

3.4 Complex Exponential Representation

These formulas are often referred to as de Moivre's theorems, and he published a paper in Philosophical Transactions of the Royal Society in 1698, mentioning that in 1676, Newton knew the equivalent representation of de Moivre's theorem. De Moivre wrote *The Doctrine of Chances* in 1718, describing the normal distribution, which is often named after Gauss.

From the point of view of function transformation,

$$f(\theta)f(\beta) = f(\theta + \beta) \tag{4}$$

$$f^n(\theta) = f(n\theta) \tag{5}$$

$f(\theta) = e^{K\theta}, f(\beta) = e^{K\beta}$,

$$e^{K\theta}e^{K\beta} = e^{K(\theta+\beta)} \tag{6}$$

$$(e^{K\theta})^n = e^{nK\theta} \tag{7}$$

Let $f(\theta) = \cos\theta + i\sin\theta = e^{K\theta}$, for $\theta = 0, 1 = e^0$, but for the general case of θ K cannot be any value. Using differential form checking,

$$\frac{df(\theta)}{d\theta} = -\sin\theta + i\cos\theta = Ke^{K\theta} = i(\cos\theta + i\sin\theta) = ie^{K\theta} \rightarrow K = i$$

$$f(\theta) = \cos\theta + i\sin\theta = 1\angle\theta = e^{i\theta}$$

This representation is often called Euler's formula. Derived from the formula published by Euler in his book *Introduction in Analysis Infinitorum* in 1748,

$$e^{\pm ix} = \cos x \pm i \sin x \tag{8}$$

Lemma 3 (Complex Exponential Representation). *Elegant and concise complex function representation, opening up the complex magic door for complex exponential function operations.*

3.5 Multiplayer Exploration

Despite Wessel's outstanding contributions to the geometry of imaginary numbers, the results were not discovered until a century later. During this time, these quintessential ideas were rediscovered. In 1806, two treatises appeared, relating the complex plane to the imaginary numerical coordinates with the vertical axis.

Argand published "Essay on the Geometrical Interpretation of Imaginary Quantities" on the Geometrical Interpretation of Imaginary Quantities, which introduced the concept of modulus for the first time. Legender, a prominent mathematician at the time, received the work and wrote to a friend Francois Francais briefly describing the relevant important results. The letter was found when Francois Francais's brother Jacques Francais was sorting through his belongings after his death. In 1813, Jacques published an article in the *Annales de Mathematiques* giving the basic principles of complex geometry, asking the unsigned authors of the results to appear. Fortunately, Argand was informed of the request. The responding article is published in the next issue of the journal. In a brief note by Jacques accompanying the article, it was announced that Argand was the first to study the geometry of imaginary numbers. Argand happily embraces this creative priority. In 1876, the mathematician Houel reprinted Argand's work, and in the preface to the reprint, he described Argand as the true founder of the theory of plane complex numbers. None of them knew about Wessel's work until 20 years later, when Wessel's earlier creative work was discovered.

Another paper published in 1806 was that of French priest Buee, a 65-page article published in French in the Royal Society's *Philosophical Transactions*. Unlike the finer styles of Wessel and Argand, the paper has a decidedly mysterious air.

4. Complex Modern Representation

Hamilton submitted the paper "Theory of Conjugate Functions or Algebraic Couples" to the Irish Academy of Sciences in 1835, using the algebraic structure to establish the order pair representation of complex functions. The paper defines an ordered pair of real numbers of (a, b) as an even couple, and the addition, subtraction, multiplication, and division of two even numbers are defined as:

$$(a, b) \pm (c, d) = (a \pm c, b \pm d) \tag{9}$$

$$(a, b)(c, d) = (ac - bd, bc + ad) \tag{10}$$

$$\frac{(a, b)}{(c, d)} = \left(\frac{ac + bd}{c^2 + d^2}, \frac{bc - ad}{c^2 + d^2}\right) \tag{11}$$

Although Hamilton did not focus on complex geometry, the generalization of the exploration of two-dimensional complex plane rotation vectors to three-dimensional space rotation vectors inspired Hamilton to discover quaternions, that is, hypercomplex numbers.

After more than 30 years of research precipitation, in 1831, Gauss submitted the paper "The Theory of Biquadratic Residues" to the Royal Society in Gottingen. Gauss has reached maturity in complex number theory, formally naming complex numbers with i instead of $\sqrt{-1}$. Based on complex geometry, it represents $a+ib$ as a point on the complex plane and describes complex geometry addition and multiplication. Gauss mastered these complex geometric concepts before Wessel in 1797, but he was accustomed to publishing all his research after he had perfected it. In the paper, he wrote: "The mystery of complex numbers is largely due to inappropriate terminology. If $+1, -1, \sqrt{-1}$ are not called positive, negative, and imaginary numbers, but are succinctly named straight, inverse, and lateral units, so that people do not have a dark, mysterious feeling and negative impression of these numbers."

Lemma 4 (Modern Representation of Complex Numbers). *The algebraic number-even theory of conjugate functions uses algebraic structures to establish ordinal pairs of complex variable functions.*

4.1 Theory of Complex Variable Functions

In 1752, d'Alembert essay on a New Theory of the Resistance of Fluids considered an object moving through an isotropic ideal fluid and determined that two functions p, q satisfy the differential equation (Cauchy–Riemann),

$$dq = Mdx + Ndy \tag{12}$$

$$dp = Ndx - Mdy \tag{13}$$

$$\frac{\partial p}{\partial x} = \frac{\partial q}{\partial y} \tag{14}$$

$$\frac{\partial p}{\partial y} = -\frac{\partial q}{\partial x} \tag{15}$$

Cauchy's first research paper on the theory of complex functions, "Report on the Theory of Definite Integrals," was read at the Paris Academy of Sciences in 1814 and was submitted for publication in 1825 in 1827. Explore the problem of changing the order of integration and introduce two functions $P(x, y), Q(x, y)$

$$\int_{x_0}^{X}\int_{y_0}^{Y} f(x,y)dxdy = \int_{y_0}^{Y}\int_{x_0}^{X} f(x,y)dydx \tag{16}$$

$$\frac{\partial P}{\partial y} = \frac{\partial Q}{\partial x} \tag{17}$$

$$\frac{\partial P}{\partial x} = -\frac{\partial Q}{\partial y} \tag{18}$$

Substitute $f = \frac{\partial P}{\partial y}$ to enter the left, $f = \frac{\partial Q}{\partial x}$ enter the right formula,

$$\int_{x_0}^{X}\int_{y_0}^{Y}\frac{\partial Q}{\partial x}dxdy = \int_{y_0}^{Y}\int_{x_0}^{X}\frac{\partial P}{\partial y}dydx$$

$$\int_{x_0}^{X}\int_{y_0}^{Y}\frac{\partial P}{\partial x}dxdy = -\int_{y_0}^{Y}\int_{x_0}^{X}\frac{\partial Q}{\partial y}dydx$$

In 1825, Cauchy's paper "On the Study of Definite Integrals with Integral Limits to Imaginary Numbers" brought constants and variables into $z = x + iy$ with complex values, and the integral of complex functions was the limit of the sum formula.

$$\int_{x_0+iy_0}^{X+iY} f(z)dz = \lim_{n\to\infty}\sum_{k=0}^{n-1} f(x_k + iy_k)[(x_{k+1} - x_k) + i(y_{k+1} - y_k)] \tag{19}$$

The paper was only officially published in 1974, carrying Cauchy's important contribution to the fundamental research of complex function integrals.

From 25 years after 1821, when Cauchy developed the theory of complex functions alone, until 1843, when Laurent proved that when a function is discontinuous at an isolated point, it is necessary to replace the Taylor expansion formula with the expansion of the variable to the ascending and descending powers.

$$f(z) = \sum_{n=-\infty}^{\infty} a_n(z-a)^n \tag{20}$$

Lemma 5 (Complex Number Function Theory). *Cauchy–Riemann differential equations provide local differential optimization conditions for complex functions, and Cauchy definite integral equations provide global integral optimization conditions for complex functions from the integrated optimization level.*

4.2 Riemann Surface

The core of Riemann's exploration of multivalued functions is to propose the concept of the Riemann surface. An n-valued function requires an n leaf Riemann surface. For example, the function $w^2 = z$ is multivalued. For one value of z, $w = \pm\sqrt{z}$ has two values. A plane is introduced as a leaf for each value of z.

The single-valued analytic function $f(z) = u + iv$ is analytic and continuously differentiable within a point and its neighborhood, satisfying

$$\frac{\partial u}{\partial x} = \frac{\partial v}{\partial y}, \quad \frac{\partial u}{\partial y} = -\frac{\partial v}{\partial x} \tag{21}$$

For a certain u,

$$v = \int\left(-\frac{\partial u}{\partial y}dx + \frac{\partial u}{\partial x}dy\right)$$

Furthermore, it is determined that the minimization Dirichlet integration condition is satisfied

$$\iint \left\{ \left(\frac{\partial u}{\partial x}\right)^2 + \left(\frac{\partial u}{\partial y}\right)^2 \right\} dxdy \tag{22}$$

Corollary 2 (Riemann Surface). *The Riemann face shows the highest state of the differential conjugate analytical local constraint. To satisfy the Dirichlet principle, this condition ensures that the complex variable function is a minimum value after being integrated through the whole space.*

4.3 *Theory of Elliptic Functions*

Abel and Jacobi are the founders of elliptic functions. In 1828, Abel published the paper "Research on Elliptic Functions," starting from the inverse function of elliptic integrals. Abel died young in 1829, while Jacobi lived until 1851, publishing more research results. Fundamenta Nova Theoriae Functionum Ellipticarum, published by Jacobi in 1829, is the core work of elliptic function research.

To study the binomial theorem in which both variables and exponents take complex numbers, Abel introduced

$$\sin\theta = i\operatorname{tg}\phi,\ \cos\theta = \frac{1}{\cos\phi},\ \Delta(\theta,k) = \frac{\Delta(\phi,k')}{\cos\phi} \tag{23}$$

where $\theta = \text{am } iu$

$$sn(iu,k) = i\frac{sn(u,k')}{cn(u,k')} \tag{24}$$

$$cn(iu,k) = \frac{1}{cn(u,k')} \tag{25}$$

$$dn(iu,k) = \frac{dn(u,k')}{cn(u,k')} \tag{26}$$

$$u = A(x) = \int_0^x \frac{1}{\sqrt{1-x^2}}dx \tag{27}$$

$$A(x) = arc\sin x \tag{28}$$

$$A(x_1) + A(x_2) = A(x_1y_2 + x_2y_1) \tag{29}$$

$$y_1 = \sqrt{1-x_1^2},\ y_2 = \sqrt{1-x_2^2} \tag{30}$$

Abel introduced the inversion operation into elliptic integration as a key step in exploring elliptic integration. If $z = u + iv$, it is interesting to use the addition theorem to generalize $sn(z) = sn(u+iv)$ to complex numbers, which can be expressed by the formula of u, iv, sn, cn, dn.

$$sn(iu + 2iK', k) = sn(iu, k)$$
$$cn(iu + 4iK', k) = cn(iu, k)$$

$$dn(iu + 4iK', k) = dn(iu, k)$$

These elliptic functions are biperiodic, a major finding of Abel's exploratory research.

Jacobi uses the elliptic function to establish the θ function and uses four θ functions to construct

$$\theta(z) = \sum_{n=\infty}^{\infty} e^{-n^2 t + 2niz}; \quad z, t \in C, Re(t) > 0 \tag{31}$$

Corollary 3 (Elliptic Function). *A series of studies on elliptic functions are carried out from the inverse functions of elliptic integrals and have the characteristics of a series of periodic functions.*

5. Complex Function Equation on Mathematical Physics

5.1 *Fourier Mathematical Physics Equations*

Fourier, in his award-winning paper in 1811, derived the formula from the solution of the bounded region heat equation,

$$u = \sum_{n=1}^{\infty} a_n e^{-kq_n^2 t} \cos q_n x \tag{32}$$

where q_n is determined by the boundary conditions and a_n is determined by the initial conditions.

When $a_n = Q(q_n)$, Q is a function of q

$$u = \int_0^{\infty} Q(q_n) e^{-kq_n^2 t} \cos qx \, dq$$

$$a_n = \frac{2}{\pi} \int_0^{\pi} \phi(x) \cos nx \, dx$$

$$\phi(x) \text{ initial function}, a_n \to Q, n \to q$$

$$Q = \frac{2}{\pi} \int_0^{\infty} F(x) \cos qx \, dx$$

$$u = \frac{2}{\pi} \int_0^{\infty} F(\alpha) d\alpha \int_0^{\infty} e^{-kq^2 t} \cos qx \cos q\alpha \, dq$$

$$u = \frac{1}{\pi} \int_{-\infty}^{\infty} F(\alpha) d\alpha \int_0^{\infty} e^{-kq^2 t} \cos q(x - \alpha) dq$$

Write the Fourier integral in exponential form using Euler's formula $e^{ix} = \cos x + i \sin x$

$$F(x) = \frac{1}{2\pi} \int_{-\infty}^{\infty} e^{iqx} dq \int_{-\infty}^{\infty} F(\alpha) e^{-iq\alpha} d\alpha \tag{33}$$

Cauchy's 1816 "Theory of Propagation of Waves" discusses the Fourier integral. Starting from $\frac{\partial^2 q}{\partial x^2} + \frac{\partial^2 q}{\partial^2 y} = 0$, where q is the velocity potential and x, y are space coordinates, the equation solution is expressed as

$$q = \int_0^\infty \cos\ mxe^{-ym} f(m)dm$$

$$q|_{y=0} \to F(x)$$

$$F(x) = \int_0^\infty \cos\ mx\, f(m)dm$$

$$f(m) = \frac{2}{\pi}\int_0^\infty \cos mx\ F(u)du$$

$$F(x) = \frac{2}{\pi}\int_0^\infty\int_0^\infty \cos\ mx\ \cos muF(u)dudm$$

Through the series integral transformation, Cauchy obtains the Fourier double integral representation of $F(x)$ and the Fourier transform from $f(m)$ to $F(x)$ and its inverse transformation.

Lemma 6 (Fourier Mathematical Physics Equations). *The Fourier Mathematical Physics equations strictly establish the representation of the step integral between the 2n eigenfunction and the same number of eigencoefficient values, exhibiting parsimony effects under orthogonal conditions.*

5.2 *Green Math Physics Equations*

Green studied the gravitational potential problem of variable density ellipsoids in 1833. U, V are any two continuous functions of x, y, z, and the derivative is not infinite at any point. Given V on the boundary of the object, there is a function on the object that satisfies $\Delta V = 0$. No singularity has the given boundary. Applying Dirichlet's principle, the following formula has a minimization function integral:

$$\iiint \left[\left(\frac{\partial V}{\partial x}\right)^2 + \left(\frac{\partial V}{\partial y}\right)^2 + \left(\frac{\partial V}{\partial z}\right)^2\right] dv \tag{34}$$

The most important type of partial differential equation is the wave equation, and its basic form in three-dimensional space is

$$\frac{\partial^2 u}{\partial x^2} + \frac{\partial^2 u}{\partial y^2} + \frac{\partial^2 u}{\partial z^2} = a^2 \frac{\partial^2 u}{\partial t^2} \tag{35}$$

For simple harmonic motion, let $u = w(x, y, z)e^{ikt}$, Helmholtz obtains the degenerate wave equation, $\Delta w + k^2 w = 0$ represents all harmonic, acoustic, elastic, and electromagnetic and other series of simple harmonics,

$$\Delta w + k^2 w = \frac{\partial^2 w}{\partial x^2} + \frac{\partial^2 w}{\partial y^2} + \frac{\partial^2 w}{\partial z^2} + k^2 w = 0 \tag{36}$$

Kirchhoff uses Helmholtz's solution method to solve the initial value problem of the wave equation, assuming that $\Delta w + k^2 w = 0$ comes from

$$\frac{\partial^2 u}{\partial t^2} = c^2 \Delta u,\ u = we^{i\sigma t}, k = \frac{\sigma}{c}$$

$$u(P,t) = -\frac{1}{4\pi}\iint \frac{e^{i\sigma(t-r/c)}}{r}\frac{\partial u}{\partial n}dS + \frac{1}{4\pi}\iint u\frac{\partial}{\partial n}\left(\frac{e^{i\sigma(t-r/c)}}{r}\right)dS$$

Corollary 4 (Green Mathematical Physics Equations). *Green mathematical physics equations show a typical mode of differential conjugate analytical local constraints, which partially satisfy the Dirichlet principle, ensuring that the Green function is a minimum value after spatial integration.*

5.3 *Automorphic Function*

Klein and Poincare are the founders of automorphic functions. Automorphic functions are the generalization of circular functions, hyperbolic functions, and elliptic functions and are analyzed from the perspective of transformation invariance, transformation group

$$z' = \frac{az+b}{cz+d};\ a,b,c,d \in \{\text{real, complex}\}, ad - bc = 1 \tag{37}$$

Poincare proposed the θ series based on the construction of autodeformed functions in 1882.

$$z' = \frac{a_i z + b_i}{c_i z + d_i};\ a_i d_i - b_i c_i = 1, i = 0,1,2,\ldots \tag{38}$$

Let $\{z_1, z_2 \ldots\}$ be the image of z in the group, $H(z)$ rational function, θ series.

$$\theta(z) = \sum_{i=0}^{\infty}(c_i z + d_i)^{-2m} H(z_i), m > 1 \tag{39}$$

$$\theta(z_j) = (c_j z + d_j)^{-2m}\theta(z) \tag{40}$$

Let $\theta_1(z), \theta_2(z)$ be the same m and two θ series, which are both single-valued and integral functions. Therefore, $F(z)$ is an autodefense function.

$$F(z) = \frac{\theta_1(z)}{\theta_2(z)} \tag{41}$$

To distinguish the Fuchs group from the Klein group, Poincare named the θ-Fuchs series (real coefficients) and θ-Klein series (complex coefficients), respectively.

A more general autodeserving function studied second-order linear differential equations

$$\frac{\partial^2 \eta}{\partial z^2} + p_1\frac{\partial \eta}{\partial z} + p_2\eta = 0$$

A special case of hypergeometric equations is associated with three singularities $\{0, 1, \infty\}$

$$\frac{\partial^2 \eta}{\partial z^2} + \frac{\gamma - (\alpha + \beta + 1)}{z(1 - z)}\frac{\partial \eta}{\partial z} + \frac{\alpha\beta}{z(1 - z)}\eta = 0$$

Lemma 7 (Automorphic Function). *As a generalization of many types of periodic functions, the autologous function exhibits complex transformation effects under orthogonal conditions for supporting complex periodic transformation modes.*

5.4 Partial Differential Equations

Navier 1821 obtained the fundamental equations of hydrodynamics, which are now known as Navier–Stokes equations.

$$\rho\frac{Du}{Dt} = \rho X - \frac{\partial p}{\partial x} + \frac{1}{3}\mu\frac{\partial \theta}{\partial x} + \mu\,\Delta u \tag{42}$$

$$\rho\frac{Dv}{Dt} = \rho Y - \frac{\partial p}{\partial y} + \frac{1}{3}\mu\frac{\partial \theta}{\partial yx} + \mu\,\Delta v \tag{43}$$

$$\rho\frac{Dw}{Dt} = \rho Z - \frac{\partial p}{\partial z} + \frac{1}{3}\mu\frac{\partial \theta}{\partial z} + \mu\,\Delta w \tag{44}$$

$$\theta = \frac{\partial u}{\partial x} + \frac{\partial v}{\partial y} + \frac{\partial z}{\partial z} \tag{45}$$

u, v, w displacement components, θ expansion coefficients, and λ, μ body or medium constants.

Maxwell 1864 derived the laws of electromagnetism, expressed in four equations:

$$\text{curl } H = \frac{1}{c}\frac{\partial(\epsilon E)}{\partial t} \tag{46}$$

$$\text{curl } E = -\frac{1}{c}\frac{\partial(\mu H)}{\partial t} \tag{47}$$

$$\text{div } \epsilon E = \rho \tag{48}$$

$$\text{div } \mu H = 0 \tag{49}$$

Corollary 5 (Partial Differential Equations). *The Navier–Stokes equations and Maxwell equations are both PDEs associated with typical frontier applications. Such representations of partial differential equations form a central part of modern mathematical physics applications.*

6. Variation Method

6.1 Lagrange

Euler and Lagrange established the variational method in the 18th century. The principle of least action is the basis for its construction. Poisson uses Lagrange's generalized coordinates, starting from the Lagrange equation:

$$\frac{d}{dt}\left(\frac{\partial T}{\partial \dot{q}_i}\right) - \frac{\partial T}{\partial q_i} + \frac{\partial V}{\partial q_i} = 0, i = 1, 2, \ldots, n; \quad V \text{ Potential Energy}, T \text{ Kinetic Energy}, \tag{50}$$

$$T = \frac{1}{2}\sum_{i,j=1}^{n} a_{ij}\dot{q}_i\dot{q}_j \tag{51}$$

$$L = T - V \tag{52}$$

$$\frac{\partial L}{\partial q_i} = \frac{\partial T}{\partial \dot{q}_i}; \quad V \text{ depends on } q_i, \text{ not on } \dot{q}_i \text{ launch}$$

$$\frac{d}{dt}\left(\frac{\partial L}{\partial \dot{q}_i}\right) - \frac{\partial L}{\partial q_i} = 0, i = 1, 2, \ldots, n$$

$$p_i = \frac{\partial L}{\partial \dot{q}_i} = \frac{\partial T}{\partial q_i}; \quad \text{After the introduction of the relationship}$$

$$\dot{p}_i = \frac{\partial L}{\partial q_i}, i = 1, 2, \ldots, n$$

where p_i is the momentum component and q_i is the coordinate component of the position.

Corollary 6 (Lagrange Equation). *The Lagrange equation is a system of second-order differential equations with n elements $q_i(t)$.*

6.2 Hamilton–Jacobi Formula

A series of papers published by Hamilton from 1824 to 1832 established transformation theory. Starting from the difference between kinetic energy and potential energy as a function of time,

$$S = \int_{P_1,t_1}^{P_2,t_2} (T - V)dt \tag{53}$$

$T - V$ is called the Lagrange function, $P_1 = \{q_1^1, q_2^1, \ldots, q_n^1\}$ $P_2 = \{q_1^2, q_2^2, \ldots, q_n^2\}$.

Hamilton's generalized model starts at time t_1 at position P_1 and ends at time t_2 at position P_2, while the law of conservation of energy does not need to hold.

Under the assertion of the principle of least action, $T + V$ =constant, $T - V = 2T-$constant. In generalized coordinates

$$S = \int_{t_1}^{t_2} L(q_1, \ldots, q_n, \dot{q}_1, \ldots, \dot{q}_n, t)dt \tag{54}$$

The equation becomes a second-order ordinary differential equation system under the condition that the first-order variation is zero,

$$\frac{\partial L}{\partial q_k} - \frac{d}{dt}\left(\frac{\partial L}{\partial \dot{q}_k}\right) = 0, k = 1, 2, \ldots, n \tag{55}$$

Introducing $p_i = \frac{\partial L}{\partial \dot{q}_i}$ and $\dot{p}_i = \frac{\partial L}{\partial q_i}$ provides a set of symmetric differential equations

$$p_i = \frac{\partial L}{\partial \dot{q}_i}, \dot{p}_i = \frac{\partial L}{\partial q_i}, i = 1, 2, \ldots, n$$

In 1835, Hamilton introduced a new function H, which is the sum of the total energy $2T$

$$H(p_i, q_i, t) = -L + \sum_{i=1}^{n} p_i q_i \tag{56}$$

H is a function of $\{p_i, q_i, t\}$, and $L = T - V$ is a function of $\{q_i.\dot{q}_i, t\}$. Because the function of $p_i = \frac{\partial L}{\partial \dot{q}_i}$ can be solved and brought into L after solving $\dot{q}_i$. The differential equation of motion can be shown to have

$$\dot{q}_i = \frac{\partial H}{\partial p_i}, \dot{p}_i = -\frac{\partial H}{\partial q_i}, i = 1, 2, \ldots, n \tag{57}$$

From the perspective of formal expression, Hamilton's equation is a first-order ordinary differential equation with $2n$ element variable p_i, q_i is a time t function, while the Lagrange equation is a second-order ordinary differential equation with n element variable $q_i(t)$ system of differential equations.

Jacobi called the Hamilton equation a canonical differential equation, a variational equation of the S integral (the Euler equation)

$$s = \int_{P_1,t_1}^{P_2,t_2} (T - V)dt = \int Ldt = \int \left\{ \sum_{j=1}^{n} p_j \dot{q}_j - H(p_j, q_j, t) \right\} dt \tag{58}$$

Jacobi's 1837 paper generalizing the Hamilton equation is explained as follows.
Find a transformation $Q_j = Q_j(p_j, q_j, t), P_j = P_j(p_j, q_j, t)$ such that

$$\delta \int_{t_1}^{t_2} \left(\sum_{j=1}^{n} p_j \dot{q}_j - H(p_j, q_j, t) \right) dt = 0 \tag{59}$$

Transform the equation into

$$\delta \int_{t_1}^{t_2} \left(\sum_{j=1}^{n} P_j \dot{Q}_j - K(P_j, Q_j, t) \right) dt = 0$$

Then, the Hamilton equation is transformed to be expressed as

$$\dot{Q}_j = \frac{\partial K}{\partial P_j}, \dot{P}_j = -\frac{\partial K}{\partial Q_j}, j = 1, 2, \ldots, n$$

where $K(P_j, Q_j, t)$ is the new Hamilton operator such that

$$K = H(p_j, q_j, t) + \frac{\partial \Omega}{\partial t}(Q_j, q_j, t)$$

Among them, Ω represents the new function, which is the transformed generating function. Select $K = 0$ to obtain $\dot{Q}_j = \dot{P}_j = 0$, that is, $\{Q_j, P_j\}$ is a constant.

Since $H + \frac{\partial \Omega}{\partial t} = 0$, it can be shown that $p_j = \frac{\partial \Omega}{\partial q_j}$ substituting the H variable yields

$$H\left(\frac{\partial \Omega}{\partial q_j}, q_j, t\right) + \frac{\partial \Omega}{\partial t}(Q_j, q_j, t) = 0 \tag{60}$$

Since $\dot{Q}_j = 0$, $Q_j = \alpha_j$, the solution containing n arbitrary constants in Ω is

$$\Omega(\alpha_1, \ldots, \alpha_n, q_1, \ldots, q_n, t) \tag{61}$$

Since $P_j = \frac{\partial \Omega}{\partial \alpha_j}$, and $\dot{P}_j = 0$, $P_j = \beta_j$ is constant. Solve each q_j sequentially from the equation $\frac{\partial \Omega}{\partial \alpha_j} = -\beta_j, j = 1, 2, \ldots, n$.

Therefore, these solutions $q_j = f_j(\alpha_1, \ldots, \alpha_n, \beta_1, \ldots, \beta_n, t), j = 1, 2, \ldots, n$ constitute a systematic solution to Hamilton's canonical equations.

From the perspective of cutting-edge research, the work system of Hamilton and Jacobi has promoted the in-depth development of variational methods and the systematic exploration of ordinary differential equations and partial differential equations.

Corollary 7 (Hamilton–Jacobi Equation). *The Hamilton–Jacobi equation is a system of first-order differential equations with 2n metavariables p_i, q_i being a function of time t. The variational condition is that the typical differential optimization condition satisfies the Dirichlet principle, which ensures that the minimum value of the corresponding equation after the full-space integration is accompanied by the evaluation condition that each corresponding solution satisfies 2n constant coefficients.*

6.3 Axiomatization of Hilbert Spaces

von Neumann proposed the Hilbert space and operator axiomatic method [41]. For the space L^2 formed by the square-integrable complex variable functions on any measurable set E on the complex plane, the complex sequence space is the set formed by all complex series $a_1, a_2, \ldots$, satisfying

$$\sum_{j=1}^{\infty} |a_j|^2 < \infty \tag{62}$$

Furthermore, there is a 1-1 correspondence between functions in the function space and sequences in the sequence space. Select a set of standard orthogonal function systems $\{\psi_n\}$ in the function space. If f is an element of the function space, then the expansion of f is about $\{\psi_n\}$. The Fourier coefficient is a sequence in the sequence space; in contrast, for such a sequence, there is only one function in the L^2 space that uses the sequence as a function of the Fourier coefficient of $\{\psi_n\}$.

Starting from the integral equation, two square-integrable real variables $f, g \in R$ have an inner product in the interval $[a, b]$

$$\langle f, g\rangle = \int_a^b f(x)g(x)dx \tag{63}$$

From the perspective of orthogonal spectral decomposition, K is a continuous function symmetric in x, y

$$f(x) \to \int_a^b K(x, y) f(y) dy$$

The function K is expressed in the form of a series, corresponding to the extended representation of the characteristic function, and the ϕ_n function is orthogonal, $\langle \phi_n, \phi_m \rangle = 0$, $\forall n \neq m$

$$K(x, y) = \sum_n \lambda_n \phi_n(x) \phi_n(y) \tag{64}$$

The inner product operator (f, g) in the space of complex functions is

$$(f, g) = \int_E f(z) \overline{g(z)} dz \tag{65}$$

Here, $\overline{g(z)}$ is the complex conjugate of $g(z)$; the inner product of the sequence in the sequence space is

$$(a, b) = \sum_{j=1}^{\infty} a_j \overline{b}_j \tag{66}$$

When f corresponds to a and g corresponds to b, $(f, g) = (a, b)$. In the L^2 space, the Hermite operator R is a linear operator satisfying $(Rf, g) = (f, Rg)$, and in sequence space $(Ra, b) = (a, Rb)$.

H is the Hilbert space, C is the complex number space, and the five axioms (A)–(E) proposed by von Neumann are:

(A) If $f_a, f_2 \in H; a_1, a_2 \in C$, then $a_1 f_1 + a_2 f_2 \in H$
(B) If $f, g, f_1, f_2 \in H, a \in C$, then the inner product (f, g) satisfies
(a) $(af, g) = a(f, g)$; (b) $(f_1 + f_2, g) = (f_1, g) + (f_2, g)$;
(c) $(f, g) = \overline{(g, f)}$; (d) $(f, f) \geq 0$; (e) $(f, f) = 0$ iff $f = 0$
If $(f, g) = 0$, then f and g are said to be orthogonal.
The norm of f is expressed as $\| f \|$,
$\| f \| = \sqrt{(f, f)}$, norm $\| f - g \|$ is a spatial measure.
(C) If H is separable, then there is a countable dense set in H with respect to $||f - g||$
(D) For a given positive integer n, there are n linearly independent elements in H
(E) if $\{f_n\} \in H$ is such that $\lim_{\{n,m \to \infty\}} ||f_n - f_m|| \to 0$ then exists $f \in H$ such that $\lim_{\{n \to \infty\}} \| f - f_m \| \to 0$

For the second element of the inner product, the complex inner product representation is inverse linear, called the complex conjugate linear,

$$\langle x, ay_1 + by_2 \rangle = \overline{a} \langle x, y_1 \rangle + \overline{b} \langle x, y_2 \rangle$$

The normal form $\| x \|$ is a real-valued function that satisfies $\| x \| = \sqrt{(x, x)}$. Definable distance d between two points $x, y \in H$,

$$d(x, y) = \| x - y \| = \sqrt{(x - y, x - y)}$$

The distance measure satisfies the triangle inequality,

$$d(x, z) \leq d(x, y) + d(y, z) \tag{67}$$

The paradigm also satisfies the following relation if and only if x, y are linearly independent:

$$|\langle x, y \rangle| \leq \| x \| \| y \| \tag{68}$$

There is an adjoint operator R^*, and the Hermite operator $R = R^*$ is called a self-adjoint operator. If $RR^* = R^*R$, then R is normal. If $RR^* = R^*R = I$, and I is an identity operator, R and orthogonal transformation are similarly called unitary operators. For unitary operators R, $||R(f)|| = ||f||$.

The linearly bounded Hermite operator R has two operators E_-, E_+ with properties

(a) $E_-E_- = E_-, E_+E_+ = E_+, I = E_- + E_+$
(b) E_- is commutative with E_+ and is commutative with any operator that is commutative with R
(c) RE_- and RE_+ are negative and positive, respectively, for example, $RE_+(f) \geq 0$ holds for all f
(d) For all f that make $Rf = 0$, $E_-f = 0$ and $E_+f = f$

For $U = e^{iR}$, U is the unitary operator and R is the Hermite operator.

Lemma 8 (Hilbert Space). *Hilbert space is the product space L^2 of multivariable complex variable functions, which constitutes the core part of the foundation of modern mathematical physics, and promotes mathematics with modes such as inner product, orthogonal decomposition, and normal form in advanced physics.*

7. Vector and Matrix of Complex Functions

7.1 Positive Definite and Nonnegative Vector and Matrix of Complex Functions

Complex conjugation operations in the quantum mechanical formula [2]—Hermite conjugation operations—are of central importance, denoted $\{*, \dagger\}$. For the 2×2 Hermite matrix

$$\begin{pmatrix} a & b \\ c & d \end{pmatrix}^* = \begin{pmatrix} \overline{a} & \overline{c} \\ \overline{b} & \overline{d} \end{pmatrix} \tag{69}$$

The two elements $v, w \in V$ can be expressed as $\langle v|w \rangle = v^* \cdot w = h(v, w)$.

$$\langle v|w \rangle = \overline{\langle w|v \rangle}$$

for $w = v$,

$$||v|| = \langle v|v \rangle$$

yields $*$ with respect to the norm of v.

A set of bases $(e_1, e_2, \ldots, e_n)$ of V is selected, and under the representation of this set of bases, the $h_{a'b}$ components are simple n^2 complex numbers.

$$_{a'b}= \mathbf{h}(e_a, e_b) = \langle e_a | e_b \rangle \tag{70}$$

The elements that make up the Hermite matrix, the basis $(e_1, e_2, \ldots, e_n)$ is the pseudo-orthogonal basis of $*$

$$\langle e_i | e_j \rangle = \begin{cases} \pm 1 & \text{if} \quad i = j \\ 0 & \text{if} \quad i \neq j \end{cases} \tag{71}$$

When every ± 1 is 1, the set of bases is said to be an orthonormal basis. In the pseudo-orthogonal basis representation, $h_{a'b}$ is a diagonal matrix, and the diagonal elements are only the number of 1 p and the number of -1 q, the number of the sign difference (p, q) remains unchanged under the action of the operator $*$.

If $q = 0$, then the $*$ operator is positive definite (> 0, nonnegative ≥ 0) [12]; that is, the norm of a nonzero vector is always positive.

$$v \neq 0 \rightarrow \text{ positive definite: } ||v|| > 0 \text{ or nonnegative: } ||v|| \geq 0$$

n dimension U transformation group, or $n \times n$ unitary matrix group, unitary group $U(n)$. When there is a sign difference in (p, q), record the pseudounitary group $U(p, q)$. If the transformation is accompanied by the unit determinant $\det U = 1$, it is recorded as $SU(n)$ and $SU(p, q)$.

Lemma 9 (Hermite Complex Function 2 x 2 Matrix). *In complex function matrix representation 2×2, the Hermite matrix conjugate transpose operation has core importance, and the norm is a multivariate vector converted to a scalar. The inner product operation of the pseudo-orthogonal basis is accompanied by the conjugate operator in the selected n-element basis vector representation. When the transformation group is accompanied by the unit determinant* $\det\ U = 1$, *the corresponding transformation groups are $SU(n)$ and $SU(p, q)$.*

7.2 *Real/Complex Linear Regular Transformation*

Real linear regular transformation line canonical transform, LCT [18], typical special applications are generalized integral transform, Fourier transform, fractional Fourier transform, Fresnel transform, Hilbert transform, etc.

Let $A = \begin{pmatrix} a & b \\ c & d \end{pmatrix}$ $A = \{a, b, c, d\} \in SL(2, \mathbf{R}), a, b, c, d \in \mathbf{R}, \det A = 1$ is a 2×2 matrix with a determinant of 1, and the matrix is all two-dimensional real A special linear group of spaces in which all regular transformation geometries form a noncommutative group.

The complex linear canonical transform (CLCT) and typical special applications are the Laplace transform, Bargmann transform, and Gauss–Weierstrass transform.

Let $A_C = \begin{pmatrix} a & b \\ c & d \end{pmatrix} = \begin{pmatrix} a_r + ia_c & b_r + ib_c \\ c_r + ic_c & d_r + id_c \end{pmatrix}$ $A_C = \{a, b, c, d\}, a_r, a_c, b_r, b_c, c_r, c_c, d_r, d_c \in \mathbf{R}, \det A_C = 1$ All unit complex matrices form a symplectic group $SP(2, \mathbf{C})$. The complex linear regular transformation is not a unitary operator and maps the real square-integrable function to the analytic function of the Hilbert space.

Lemma 10 (Real/Complex Linear Regular Transformation Matrix). *The real linear regular transformation LCT represents the special linear group SP(2, R) of the 2 × 2 real symmetric matrix, and the regular transformation constitutes a noncommutative group. The complex linear regular transformation CLCT forms a symplectic group SP(2, C), which maps the real square-integrable function to the analytic function of the Hilbert space.*

7.3 Klein Group

In 1884, Klein named the four-element Abel group the four-group V, which constitutes the smallest acyclic group, which can be expressed as a double-copy second-order cyclic group $Z_2 \times Z_2$

$$V = \langle u, v | u^2 = v^2 = (uv)^2 = e \rangle$$

×	e	u	v	w
e	e	u	v	w
u	u	e	w	v
v	v	w	e	u
w	w	v	u	e

Using 0, 1 element representation, the Klein four group can be represented as a 2 × 2 matrix:

$$\left\{ e = \begin{pmatrix} 1 & 0 \\ 0 & 1 \end{pmatrix}, u = \begin{pmatrix} 1 & 0 \\ 0 & -1 \end{pmatrix}, v = \begin{pmatrix} -1 & 0 \\ 0 & 1 \end{pmatrix}, w = \begin{pmatrix} -1 & 0 \\ 0 & -1 \end{pmatrix} \right\} \tag{72}$$

Lemma 11 (Klein Quaternary Matrix). *The Klein four group is the smallest acyclic group, which can be represented as a 2 × 2 matrix with four distinguishable elements.*

7.4 Pauli Matrix

In mathematical physics tools, a Pauli matrix is a set of three matrices.

$$\sigma_1 = \sigma_x = \begin{pmatrix} 0 & 1 \\ 1 & 0 \end{pmatrix}, \sigma_2 = \sigma_y = \begin{pmatrix} 0 & -i \\ i & 0 \end{pmatrix}, \sigma_3 = \sigma_z = \begin{pmatrix} 1 & 0 \\ 0 & -1 \end{pmatrix} \tag{73}$$

Through the repeated action of three basic operators, the following sixteen matrices can be formed:

$$\begin{array}{l} \text{Eight } 2 \times 2 \\ \text{Real Matrices} \end{array} : \left\{ \begin{pmatrix} 1 & 0 \\ 0 & 1 \end{pmatrix}, \sigma_z = \begin{pmatrix} 1 & 0 \\ 0 & -1 \end{pmatrix}, \begin{pmatrix} -1 & 0 \\ 0 & 1 \end{pmatrix}, \begin{pmatrix} -1 & 0 \\ 0 & -1 \end{pmatrix}, \right.$$
$$\left. \sigma_x = \begin{pmatrix} 0 & 1 \\ 1 & 0 \end{pmatrix}, \begin{pmatrix} 0 & -1 \\ 1 & 0 \end{pmatrix}, \begin{pmatrix} 0 & 1 \\ -1 & 0 \end{pmatrix}, \begin{pmatrix} 0 & -1 \\ -1 & 0 \end{pmatrix} \right\}$$

$$\begin{array}{l} \text{Eight } 2 \times 2 \\ \text{Imaginary Matrices} \end{array} : \left\{ \begin{pmatrix} i & 0 \\ 0 & i \end{pmatrix}, \begin{pmatrix} i & 0 \\ 0 & -i \end{pmatrix}, \begin{pmatrix} -i & 0 \\ 0 & i \end{pmatrix}, \begin{pmatrix} -i & 0 \\ 0 & -i \end{pmatrix}, \right.$$
$$\left. \begin{pmatrix} 0 & i \\ i & 0 \end{pmatrix}, \sigma_y = \begin{pmatrix} 0 & -i \\ i & 0 \end{pmatrix}, \begin{pmatrix} 0 & i \\ -i & 0 \end{pmatrix}, \begin{pmatrix} 0 & -i \\ -i & 0 \end{pmatrix} \right\}$$

It can be observed that the first four matrices among the eight real matrices are elements of the Klein quartet.

Lemma 12 (Pauli Matrix). *The irreducible representation of a Pauli matrix is a set of three matrices. The transformation group contains a total of sixteen 2×2 matrices. Eight are real matrices, and the other eight are imaginary matrices. The first four of the real matrices are Klein quaternions.*

7.5 Dirac Representation

Among the simplified representations of the quantum mechanical equations circulating in Dirac, the following equations deserve attention.

$$(\partial - m)\psi = 0 \tag{74}$$

The natural unit is expressed as

$$(i\partial\!\!\!/ - m)\psi = 0 \tag{75}$$

Strictly expressed as

$$(i\hbar\gamma^{\mu}\partial_{\mu} - mc)\psi = 0, \mu \in \{0, 1, 2, 3\} \tag{76}$$

The γ matrix is constructed based on the Pauli 2×2 matrix plus the identity matrix I_2, which is strictly expressed as

$$\beta = \gamma^0 = \begin{pmatrix} I_2 & 0 \\ 0 & -I_2 \end{pmatrix}, \alpha_1 = \gamma^1 = \begin{pmatrix} 0 & \sigma_x \\ -\sigma_x & 0 \end{pmatrix}$$
$$\alpha_2 = \gamma^2 = \begin{pmatrix} 0 & \sigma_y \\ -\sigma_y & 0 \end{pmatrix}, \alpha_3 = \gamma^3 = \begin{pmatrix} 0 & \sigma_z \\ -\sigma_z & 0 \end{pmatrix}$$
$$\{\gamma^{\mu}, \gamma^{\nu}\} = 2\eta^{\mu\nu} I_4$$
$$\{a, b\} = ab + ba$$

Defining $\partial\!\!\!/ \equiv \gamma^{\mu}\partial_{\mu}$ and $\hbar = c = 1$ as the natural unit, we obtain the equation (75).

The four 4×4 matrices are $\alpha_1, \alpha_2, \alpha_3, \beta$ satisfying

$$\alpha_i^2 = \beta^2 = I_4 \tag{77}$$

$$\alpha_i\alpha_j + \alpha_j\alpha_i = 0\ (i \neq j), \alpha_i\beta + \beta\alpha_i = 0 \tag{78}$$

Lemma 13 (Dirac Equation). *The Dirac equation has three forms: simplified representation, natural unit representation, and strict representation. In strict representation, the four $4 \times 4\gamma$ matrices are three 2×2 Pauli. The matrix plus an identity matrix 4×4 represents the structure.*

7.6 Jones Vector and Jones Matrix

The tools for describing polarized light and polarizers in physical optics are represented by Jones vectors and Jones matrices. The relevant corresponding representations are listed below.

Polarization	Jones vector
Along x-axis	$E = \begin{pmatrix} 1 \\ 0 \end{pmatrix}$
Along y-axis	$E = \begin{pmatrix} 1 \\ 0 \end{pmatrix}$
Along x-axis $\pm\theta$	$E = \begin{pmatrix} cos\theta \\ sin\theta \end{pmatrix}$
Left Rotate	$E = \frac{1}{\sqrt{2}} \begin{pmatrix} 1 \\ i \end{pmatrix}$
Right Rotate	$E = \frac{1}{\sqrt{2}} \begin{pmatrix} 1 \\ -i \end{pmatrix}$

Polarization	Jones Matrix
Along x-axis	$\begin{pmatrix} 1 & 0 \\ 0 & 0 \end{pmatrix}$
Along y-axis	$\begin{pmatrix} 0 & 0 \\ 0 & 1 \end{pmatrix}$
Along x-axis $\pm 45°$	$\begin{pmatrix} 1 & \pm 1 \\ \pm 1 & 1 \end{pmatrix}$
Along x-axis $\pm\theta$	$\begin{pmatrix} \cos^2\theta & \frac{1}{2}\sin 2\theta \\ \frac{1}{2}\sin 2\theta & \sin^2\theta \end{pmatrix}$
x-axis 1/4 WP	$\begin{pmatrix} 1 & 0 \\ 0 & i \end{pmatrix}$
y-axis 1/4 WP	$\begin{pmatrix} 1 & 0 \\ 0 & -i \end{pmatrix}$
x-axis 1/2 WP	$\begin{pmatrix} 1 & 0 \\ 0 & -1 \end{pmatrix}$
y-axis 1/2 WP	$\begin{pmatrix} 0 & 1 \\ 1 & 0 \end{pmatrix}$
x-axis $\pm 45°$ 1/4 WP	$\begin{pmatrix} 1 & \pm i \\ \pm i & 1 \end{pmatrix}$
x-axis WP	$\begin{pmatrix} 1 & 0 \\ 0 & \exp(i \cdot \delta) \end{pmatrix}$
y-axis WP	$\begin{pmatrix} 1 & 0 \\ 0 & \exp(-i \cdot \delta) \end{pmatrix}$
x-axis $\pm 45°$ WP	$\cos\frac{\delta}{2} \begin{pmatrix} 1 & \pm i \cdot \tan\frac{delta}{2} \\ \pm i \cdot \tan\frac{\delta}{2} & 1 \end{pmatrix}$
Left Polarization	$\frac{1}{2}\begin{pmatrix} 1 & -i \\ i & 1 \end{pmatrix}$
Right Polarization	$\frac{1}{2}\begin{pmatrix} 1 & i \\ -i & 1 \end{pmatrix}$

Corollary 8 (Jones Vector and Matrix). *In physical optics, various optical signal measurement metrics, in the complex form of the Jones vector and Jones matrix, systematically deal with complex optical system measurement metrics that can be represented by complex variable functions.*

7.7 Majorana Fermion Represents

In terms of symmetry representation, the Majorana Fermion indicates that its antiparticle is itself [5], which has the magical invariance of self-reflexivity. The Majorana Fermion operator $\{a, b\}$ can be represented by the Fermion creation and annihilation operators $\{\psi^\dagger, \psi\}$.

$$
\begin{aligned}
(\psi^\dagger)^2 = \psi^2 &= 0 \\
\psi\psi^\dagger + \psi^\dagger\psi &= 1
\end{aligned}
$$

$$\{\psi, \psi^{\dagger}\} \rightarrow \begin{cases} \psi = (a+ib)/2 \\ \psi^{\dagger} = (a-ib)/2 \end{cases}$$
$$\Downarrow$$
$$a^{\dagger} = a, \quad b^{\dagger} = b$$
$$a^2 = b^2 = 1$$
$$ab + ba = 0$$

Corollary 9 (Antiparticle Representation). *Majorana Fermion Antiparticle represents itself, with reflex invariance.*

7.8 Gauge Invariance

Chen-Ning Yang's [6] early research on gauge invariance starts from simple formulas.

$$F_{\mu\nu} = A_{\mu,\nu} - A_{\nu,\mu}$$
$$F_{\mu\nu} = B_{\mu,\nu} - B_{\nu,\mu}$$

However, in the follow-up derivation, when encountering insurmountable obstacles, the only option is to give up.

In 1954, in collaboration with Robert Mills, two additions were made to transform the formula into

$$F_{\mu\nu} = B_{\mu,\nu} - B_{\nu,\mu} + B_{\mu}B_{\nu} - B_{\nu}B_{\mu}$$

The subsequent derivation is simpler and simpler and becomes the basis of Yang–Mills theory.

$$F_{\mu\nu} = B_{\mu,\nu} - B_{\nu,\mu} + B_{\mu}B_{\nu} - B_{\nu}B_{\mu}.$$

Using curvature to express the [7] Yang–Mills equation as

$$F_A = dA + \frac{1}{2}[A, A]$$

From the perspective of geometric distribution, the BSPT instanton distribution [7, 8] corresponding to the Yang–Mills equation is shown in Fig. 2.

Corollary 10 (Yang–Mills Equation). *The Yang–Mills equation consists of two parts: the first part involves the differential operator of the primitives, and the second part, the binary nonreciprocal condition via Poisson brackets, determines the difference feature under the bivariate exchange product.*

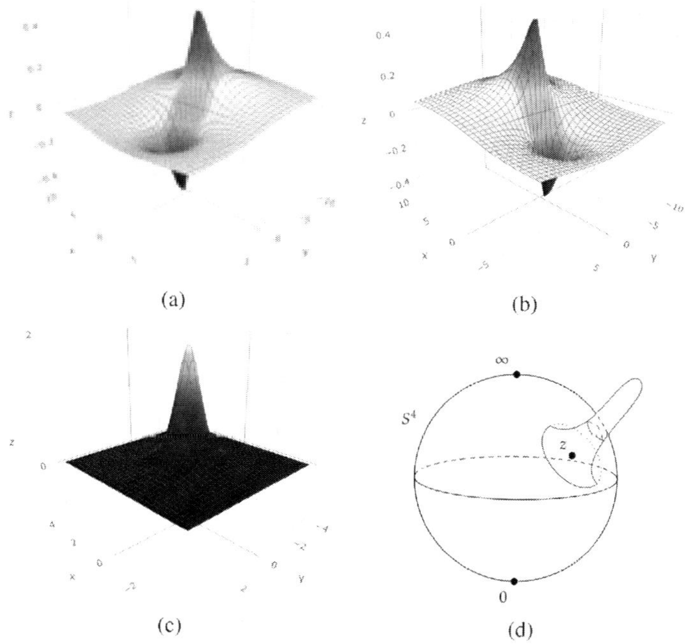

Figure 2. Yang–Mills typical BSPT instanton action distributions: (a) $dx^1 \otimes \sigma_3$; (b) $dx^2 \otimes \sigma_3$; (c) $g = 2, \rho = 1, z = 0$; (d) $z \in S^4 \subset R^4$. Reference [7, 8].

8. Lattice Theory and System Feature Discrimination

8.1 Core Part of Lattice Theory

As a basic theoretical system for the preselection of postquantum cryptographic algorithms that can effectively resist quantum cryptographic algorithm attacks, the **lattice theory** model and method have attracted international attention in the frontier research of quantum/antiquantum cryptography in recent years. For the convenience of subsequent expansion and comparison, a brief summary of the contents related to lattice theory is as follows.

From the perspective of the operator function, the lattice structure provides two sets of unit elements $\{0, 1\}$ for $\{+, \times\}$, as well as the corresponding inverse operations and inverse elements. In addition to this type of structure, a transcendental structure containing the characteristics of the group ring domain synthesis system can be constructed.

Let V be nonempty, P domain $\forall \alpha, \beta, \gamma \in V, \forall k, l \in P$

Additive Associativity: $\alpha + (\beta + \gamma) = (\alpha + \beta) + \gamma$

Commutative Law of Addition: $\alpha + \beta = \beta + \alpha$

Additive Zero: $\exists 0 \in V, \forall \alpha \in V \Rightarrow \alpha + 0 = \alpha$

Additive Negative: $\forall \alpha \in V, \exists \beta \in V \Rightarrow \alpha + \beta = 0, \beta = -\alpha$

Multiplicative Unit: $1 \in P, \alpha \in V \Rightarrow 1 \times \alpha = \alpha$

Multiplicative Associativity: $\forall k, l \in P, \alpha \in V \Rightarrow (kl)\alpha = k(l\alpha)$

Distributive Law of Multiplication: $\forall k, l \in P, \alpha \in V \Rightarrow (k + l)\alpha = k\alpha + l\alpha$

Distributive Law of Multiplication: $\forall k \in P, \alpha, \beta \in V \Rightarrow k(\alpha + \beta) = k\alpha + k\beta$

Basis Vectors: Special vectors that are **linearly independent** from each other, and the rest of the space can be linearly represented by basis vectors through scaling, addition, subtraction, and generation operations.

V is the field P vector space, and P is the real number field. For the base set $\{\alpha_1, \alpha_2, \ldots, \alpha_n\}, \exists$ is not all zero, the constant set $\{k_1, k_2, \ldots, k_n\}$, if

$$k_1\alpha_1 + k_2\alpha_2 + \cdots + k_n\alpha_n = 0 \tag{79}$$

Then, the primitive set $\{\alpha_1, \alpha_2, \ldots, \alpha_n\}$ **is linearly related**;

$$k_1 \neq 0, \alpha_1 = -\left(\frac{k_2}{k_1}\alpha_2 + \cdots + \frac{k_n}{k_1}\alpha_n\right) \tag{80}$$

Otherwise, the set is said to be **Linearly Independent**.

Let L denote a classical lattice, given a set of linearly independent vectors $(\vec{b}_1, \vec{b}_2, \ldots, \vec{b}_n) \in R^n$, for any integer $z_1, z_2, \ldots, z_n \in Z$, lattice L integer coefficient based on $\vec{b}_1, \vec{b}_2, \ldots, \vec{b}_n$ linear combination generation.

i.e., $A\vec{x} = \begin{pmatrix} 1\ 2\ 3 \\ 6\ 5\ 4 \\ 7\ 8\ 9 \end{pmatrix} \begin{pmatrix} x_1 \\ x_2 \\ x_3 \end{pmatrix} = \begin{pmatrix} 1 \\ 1 \\ 0 \end{pmatrix} = \vec{b}$

Lemma 14 (Lattice Operator Specification). *The core part of lattice theory is $\{+, \times\}$, which provides two sets of unit elements $\{0, 1\}$ and corresponding inverse operations and inverse operations elements. Operator operations satisfy the associative law of addition, the commutative law of addition, the zero element of addition, and the negative element of addition, as well as the identity element of multiplication, the associative law of multiplication, and the distribution rate of multiplication.*

8.2 *System Stability Using Eigenvalues*

For a one-dimensional system of n elements with small amplitude periodic oscillations [13] p. 104,

$$\det|B - \lambda A| = 0 \tag{81}$$

from the batch eigenvalues. From the perspective of determining the stability of the system, there are three situations:

1) $\lambda = \omega^2 > 0$, the periodic solution
2) $\lambda = 0$, neutral equilibrium solution
3) $\lambda = \omega^2 < 0$, unstable solution

Corollary 11 (Complex System Stability). *A one-dimensional system with a small amplitude periodic oscillation of the n element determines the system stability from the batch eigenvalues. There are three kinds of system solutions: periodic, neutral equilibrium, and uncertainty.*

8.3 Topological Gravity Study

Witten's frontier topological physics research [21] p. 41 describes how anomalous measurement effects appear on Riemann surfaces with g holes in 2D topological gravitational pairs.

On a Riemann surface of genus g, there are $\frac{1}{2}(2^{2g}+2^g)$ even spins and $\frac{1}{2}(2^{2g}-2^g)$ odd spin structures. The genus g partition function of **T** is

$$Z_g = \frac{1}{2}\left(\left(\frac{1}{2}(2^{2g}+2^g)\right) - \left(\frac{1}{2}(2^{2g}-2^g)\right)\right) = 2^{g-1} \tag{82}$$

It can be seen from the formula structure of the partition function that although the 2^{2g} parameter is included in the description of the parity spin itself, after the corresponding components are mutually resolved, only the 2^{g-1} component can be divided into observed components. This type of topological gravitational coupling has no effect on the measurement when only closed Riemann surface conditions are considered.

Corollary 12 (Identification of Topological Gravitational Variables). *In the research environment of topological gravitation, there is an obvious difference between the number of components of descriptive variables and observable variables, which needs to be identified in specific exploration.*

9. Complex Geometry Structure Classification

From the perspective of geometric classification and representation [1], complex geometry can be divided into Euclidean geometry and non-Euclidean geometry, while non-Euclidean geometry is divided into Roche geometry and Riemann geometry; other typical geometric representations include fractal geometry, computational geometry, algebraic geometry, differential geometry, affine geometry, and topological geometry. Three types of forms are mainly reviewed here: Euclidean geometry, Riemann geometry, and Roche geometry.

9.1 Euclidean Geometry

Euclidean geometry studies the geometric structure and metric properties (length, area, angle, etc.) of straight lines and quadratic curves (conics: ellipse, hyperbola, and parabola). The natural transition from plane geometry to solid geometry in three-dimensional space requires calculus operations for complex area and volume problems. After Descartes introduced the coordinate system, the relationship between algebra and geometry became increasingly close. From the point of view of analytic geometry, the properties of geometric figures can be reduced to the analytical and algebraic properties that describe the equations. The structure of Euclidean geometry is examined in the context of flat space structures, which are not concerned with the geometry in curved space.

9.2 Non-Euclidean Geometry

Non-Euclidean geometry is mainly divided into Riemannian geometry and Roche geometry. The fifth postulate of Euclidean geometry (parallel axiom) is: If two straight lines intersect a third straight line and the sum of the interior angles on the same side is less than two right angles, then the two straight lines must intersect this side shown in Fig. 3. Simply put, there is only one straight line that is parallel to the known straight line through a point outside the straight line, which is the basis of *n* Euclidean geometry; see Fig. 4(c).

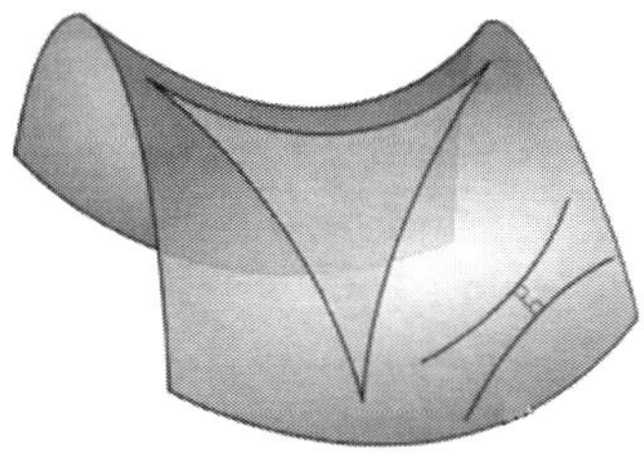

Figure 3. Non-euclid geometry maps.

9.2.1 Riemann Geometry

Mathematician Riemann founded Riemann geometry—elliptic geometry. Under the system of Riemannian geometric axioms, the existence of parallel lines is not recognized. Two straight lines in any plane must intersect. In the plane, the straight lines can be extended infinitely, but the total length is limited. The Riemann geometry can be thought of as a stretched sphere, forming a structure with positive curvature (see Fig. 4(a)).

Associated with Riemann geometry, many branches of mathematics have developed algebraic topology, partial differential equations, and multiple variable complex functions as foundations of differential geometry and general relativity.

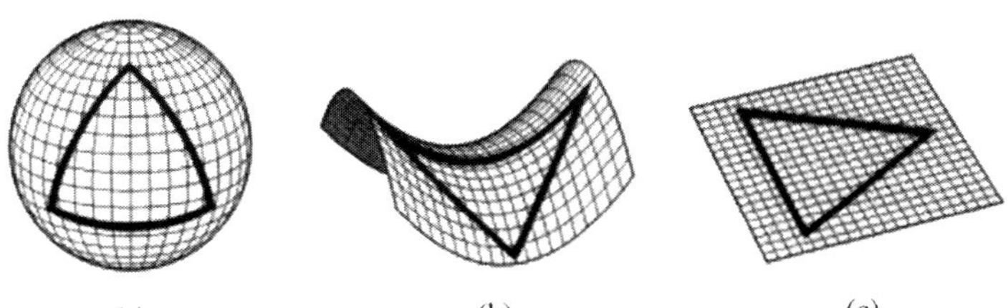

Figure 4. Three geometric illustrations: (a) positive curvature; (b) negative curvature; (c) flat curvature.

9.2.2 Lobatchevsky Geometry

Mathematician Lobatchevsky founded Lobatchevsky geometry—hyperbolic geometry, a system of axioms independent of Euclidean geometry. The fifth axiom of Euclidean geometry is replaced by the "hyperbolic parallel axiom": at least two lines passing through a point outside the line are parallel to the known lines. Basic triangles have negative curvature structures; see Fig. 4(b). In the Roche geometry axiom system, it is possible to prove completely different propositions from Euclidean geometry. For example, the sum of the interior angles of a triangle is less than 180 degrees. Any conclusions involving parallel axioms are not valid in Lobatchevsky geometry.

Table 1. Correspondence between partial symmetry and conserved quantity

Symmetry	Invariance	Conservation	Range
Space	Translation	Momentum (P)	All
Time	Translation	Energy (E)	All
Orientation	Rotation	Angular Momentum (J)	All
Particle Phase	Charge Gauge Transform	Charge (Q)	All
Baryon Phase	Baryon Gauge Transform	Baryon No. (B)	All
e, ν_e Phase	Electron Gauge	e LeptonNo. (L_e)	All
μ, ν_μ Phase	μ Subnumber Transform	μ Lepton No. (L_μ)	All
τ, ν_τ Phase	τ Subnumber Transform	τ Lepton No. (L_τ)	All
Left-Right Symmetry	Spatial Reflection	Parity (π)	Except Weak
Isospin Multiplet	Isospin Space Rotation	Isospin(I, I_3)	Strong Interaction I_3 Fits EM
π Meson Symmetry	G Conjugate Transformation	G Parity (G)	Strong interaction
Past/Future	Time Reflex	T Conservation	Weak Action Destruction
Antiparticles	Charge Conjugation	C Parity (C)	Weak Partial Destruction
CPT Symmetry	Charge Conjugation (C) Space Reflection (P) Time Reflection (T) Simultaneous Transformation	CPT Conservation	All

9.3 Symmetry Principles and Modes

Einstein's most important contributions to the fundamentals of physics [19] in two respects:

(1) physics is founded on fundamental principles and (2) the role of symmetry in it. Symmetry plays a role in three aspects:

(I) Symmetry determines the form of the Lagrange action, which is equivalent to the form of differential equations;
(II) Symmetry determines the structure of the universe;
(III) Through Neother's theorem, symmetry has a corresponding relationship with the conserved quantity of physics, and one kind of symmetry determines one kind of conserved quantity.

(I) [principle of symmetry] → [differential equation form]
(II) [principle of symmetry] → [space structure]
(III) [symmetry] → [conservation law]

Notably, the symmetries of (I) and (II) do not affect the symmetry of (III); that is, the symmetry that determines the form of the equation is not necessarily conserved, and the symmetry corresponding to the conservation does not necessarily affect the differential equation form.

Table 2. Transformation groups and domain disciplines

Principles	Groups	Symmetry	State Functions	Differential Equations	Fields of Application
Galileo	Galileo Group	Spacetime Translation Inertial System	$(x, \dot{x})$	Hamilton System	Celestial Motion
Lorentz	Lorentz Group L^4	Inertial System	Dirac spinor Scalar	Dirac equation Klein– Gordon	Quantum Mechanics
Space Rotation	$SO(3)$	Space Orientation	Spinor Wave	Spinor BEC Eq.	BE Condensation
General Relativity	Coordinate Group M^4	Non-Inertial	Tensor $\{g_{\mu\nu}\}$	Einstein Field	Gravitation
Invariance	$U(1) \times L^4$	$U(1)$	EM field$\{A_\mu\}$	Maxwell Eq.	EM Interaction
Lorentz	$SU(2) \times L^4$	$SU(2)$	Weak $\{W^\alpha_{mu}\}$	Weak Field Eq.	Weak Interaction
Invariance	$SU(3) \times L^4$	$SU(3)$	Strong $\{S^k_\mu\}$	Strong Field Eq.	Strong Interaction
Gauge Field	$U(1) \times L^2$	Gauge Invariance	$dA + \frac{1}{2}A, A$	Yang–Mills Eq.	Multiple Interactions
General Relativity Gauge Invariance Group Invariance	$M^4 \times U(1)$ $\times SU(2)$ $\times SU(3)$ $\times GCL(2)$ $\times GCL(3)$	Non-Inertial $U(1) \times SU(1)$ $\times SU(3)$ Canonical Symmetric with PRI	$\{g_{\mu\nu}$, A_μ, W^α_μ, $S^k_\mu\}$	Unified Field Eq.	Four Interactions

9.4 Unified Field Geometry for PID and PRI

A new symmetry is developed based on $SU(N)$ gauge theory [19]: representation invariance. Two fundamental physical principles are established based on the constrained variation theory and Riemann's geometric theory on $SU(N)$: the principle of interaction dynamics (PID) and the principle of representation invariance (PRI).

Each symmetry principle determines the geometry and topology of the manifold M. Unified field geometry is a geometry that uses all symmetry principles to comprehensively determine the geometry and topology of M. The four symmetry principles involved are general relativity, Lorentz invariance, gauge invariance, and representation invariance. Lagrange action satisfying four symmetries simultaneously.

$$\mathbf{L} = \mathbf{L}_{EH} + \mathbf{L}_{YM} + \mathbf{L}_{D} + \mathbf{L}_{KG} \tag{83}$$

where $\mathbf{L}_{EH}$ is the Einstein–Hilbert general relativity action, $\mathbf{L}_{YM}$ is the Yang–Mills gauge field action, $\mathbf{L}_{D}$ is the Dirac spinner, and $\mathbf{L}_{KG}$ is the Klein–Gordon field.

9.5 From General Linear Group to Unitary Group

From the perspective of matrix structure representation [47], the general linear group $GL(n, R)$ of the real number field is a real Lie group, and the dimension is n^2; the general linear group $GL(n, C)$ of the complex number field represents a complex Lie group with a dimension of n^2, and a complex Lie group matrix based on real numbers has a dimension of $2n^2$. $GL(2n, R)$ is a real Lie group with dimension $(2n)^2$. Several linear group matrices represent the dimensional correspondence as

$$GL(n, R) < GL(n, C) < GL(2n, R)$$

The corresponding real matrix dimensions are $\{n^2, 2n^2, 4n^2\}$.

As a manifold, $GL(n, R)$ is nonconnected and divided into two-leaf connected components: the corresponding matrix determinants are ± 1. The $GL(n, R)$ group is a noncompact structure, and the subgroup with compact properties is the orthogonal group $O(n)$. The largest compact subgroup of $GL^+(n, R)$ is the special orthogonal group $SO(n)$.

The Lie algebra corresponding to $GL(n, C)$ consists of all $n \times n$ complex matrices with Lie brackets as transposition operators. Unlike $GL(n, R)$, $GL(n, C)$ is connected due to the connectivity of the complex multiplicative group C^*. While the group manifold of $GL(n, C)$ is not compact, the maximal compact group is the unitary group $U(n)$.

9.6 Langlands Program

The comprehensive use of symmetry principles to develop modern theory is the Langlands program, linking number theory, algebraic geometry, and reduction group representation theory. Starting from the generalization of Artin's reciprocity law, a Galois group is given an L function for the number field of a commutative group, and it is asserted that this kind of L function is equal to some Dirichlet L function.

$$\text{Dirichlet } L \text{ function: } L(s, \chi) = \sum_{n=1}^{\infty} \frac{\chi(n)}{n^s} \tag{84}$$

$$\text{Riemann } \zeta \text{ function: } \zeta(s) = \sum_{n=1}^{\infty} \frac{1}{n^s} \tag{85}$$

$$\text{Group Conjugate Operator: } x \to gxg^{-1} \tag{86}$$

Stable conjugate classes can be decomposed into a finite number of general conjugate classes.

A schematic diagram describing the Langlands program in brief structural relationships is shown in Fig. 5.

The proposed program includes the following:

Reciprocal Conjecture The finite-dimensional representation of every Galois group is the Artin L function equal to the self-defeating cusp representation of the L function.

Functoriality Conjecture If two reduced groups allow homomorphism between their correspondences, then there should be some kind of relation compatible with the L function between the automorphic functions of the two groups.

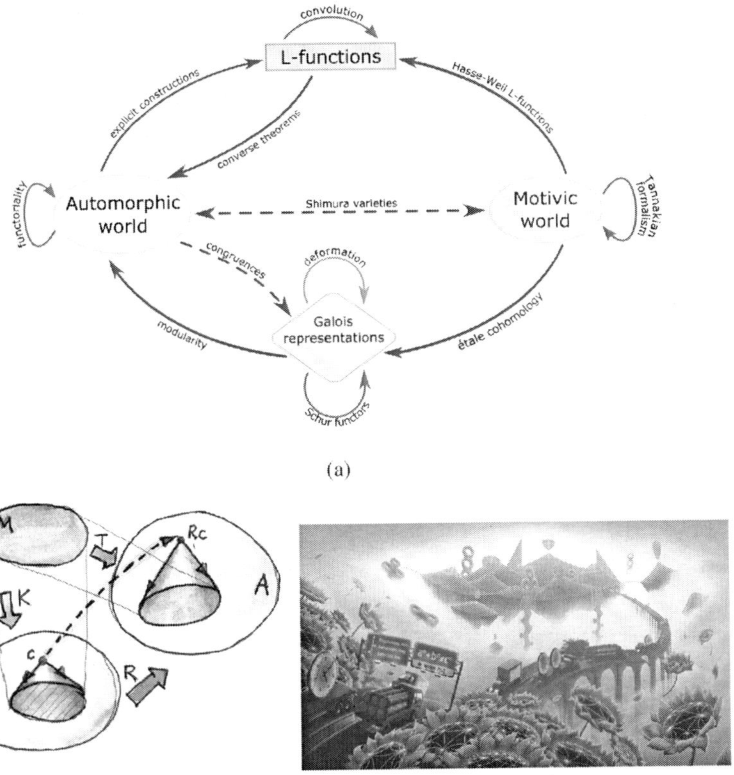

(a)

(b) (c)

Figure 5. Langlands program: (a) core module; (b) transformation mode; (c) bridge between geometry and number theory.

Functoriality Principle Properties that may hold for semisimple (reduced) Lie groups should hold for all extended cases.

For example, conjecture the general $GL(n)$ case by observing the representation of $GL(2)$ in modular form and $GL(1)$ in class domain theory.

Further promotion:

1. Replace the general linear group $GL(n)$ with the reduction group G;
2. Construct the complex Lie group L_G (dual group or L group);
3. Replace the self-defensive representation with the self-defeating L package representation, and each L package is a finite set of self-defensive representations as an equivalence class;
4. For each L self-defeating cusp representation and each finite-dimensional L_G representation, an L function is associated; representations in the same L package have the same L function and ϵ factor. Conjecture: Two functions satisfy some kind of functional equation.

The Langlands program on the number field can be transformed into a geometric framework. The steps are as follows:

1. Replace the number field with the meromorphic function field of the compact Riemann surface;

2. Replace the Galois group with the fundamental group;
3. To replace the Galois representation with a local system;
4. Replace $GL(n, Q)/GL(n, A_Q)/K$ with a rank n vector bundle modular space $Bun_{n/C}$;
5. Replacing self-defensive forms with anomalous layers;
6. Replace the Hecke eigenform with the Hecke eigenlayer.

9.7 *Dark Matter Metabolism*

Modern astronomical observations show that [19], the visible matter in the universe is only 1/5 of the total matter, while the dark matter accounts for 4/5.

$$\frac{\text{visible matter}}{\text{dark matter}} = \frac{1/5}{4/5} = \frac{1}{4} \tag{87}$$

In modern astrophysics, dark energy is a hypothetical form of energy that fills the entire space and exhibits the opposite effect of gravity, acting as a repulsive force between matter. Because this energy is not in the well-known category of matter energy and cannot be directly measured by instruments, it is called dark energy.

9.8 *Several Past and Future Quantum Theoretical Explanations*

In deterministic classical physical concepts [2], the future is uniquely determined by the past: **past → future**

For cutting-edge physical concepts, Penrose [21]-186 gives more detailed patterns for various time processes.

In classical quantum theory assumptions, a stochastic measurement process is required to form a split between the past and the future: collapse mainstream explanation Bohr, von Neumann, etc.

$$\text{Past} \rightarrow \begin{array}{l} \cdot \rightarrow \text{Future 1} \\ \updownarrow \quad \text{Stochastic Measurement Process - Collapse - (Mainstream View)} \\ \cdot \rightarrow \text{Future 2} \end{array}$$

However, in Einstein's hypothesis, there is an unknown hidden variable mechanism in the intermediate process. The general view is that work on tests such as Bell's inequality from the 1960s onward refuted Einstein's hypothesis.

$$\text{Past} \rightarrow \begin{array}{l} \cdot \rightarrow \text{Future 1} \\ \updownarrow \quad \text{hidden variable? (Einstein assumption)} \\ \cdot \rightarrow \text{future 2} \end{array}$$

In the assumption of many worlds theory, different split states are automatically divided into different worlds.

$$\left\{ \begin{array}{lll} \text{Past 1} \rightarrow \text{World 1} \rightarrow & \text{Future 1} \\ \quad\quad \cdots & \text{(Multiple Worlds Hypothesis)} \\ \text{Past n} \rightarrow \text{World n} \rightarrow & \text{future n} \end{array} \right.$$

In the face of the latest experimental evidence, most cutting-edge theoretical researchers [2, 21, 19, 20] view modern quantum mechanics theory as classical quantum theory: {particle + wave function}, for general applications, it is a class of relatively crude models.

It is hoped that the next generation of quantum field theory, string theory, superstring theory, etc., can form a more powerful and unified quantum interaction model and provide a harmonious and unified basis for quantum theory from a holistic perspective.

10. Basic Representation of Quantum States and Detection of Quantum Entangled States

10.1 Origin and Development of Quantum Entangled State Detection

In the development process of quantum mechanics, the presentation of entanglement theory is the most absurd, wildest, most dramatic, most incredible, most legendary, and most ridiculous event in the quantum mechanical system [32].

In 1935, Einstein proposed the ERP paradox [33] to describe the entanglement effect. As one of the most difficult concepts in quantum mechanics to understand and accept, quantum entanglement reflects the nonlocal characteristics of quantum states, which Einstein calls spooky action at a distance. Characterizing such interactions as **entanglement** is Schrodinger's own contribution.

In the development history of quantum mechanics, Bohr and Einstein engaged in a long-term debate on related fundamental issues until Einstein's death in 1955; he still insisted that the fundamental theoretical system of quantum mechanics was incomplete.

10.2 Quantum State Complex Functions

Electrons have 1/2 spin$\uparrow, \downarrow$, and paired spin electrons are represented as $\{o\uparrow, o\downarrow\}$. The superposition state is expressed as

$$\phi = |\downarrow\rangle + |\uparrow\rangle$$

$$\phi = A|\downarrow\rangle + B|\uparrow\rangle, \quad (A^2 + B^2 = 1)$$

$$\phi = \begin{pmatrix} A \\ B \end{pmatrix}; |\downarrow\rangle = \begin{pmatrix} 1 \\ 0 \end{pmatrix}; |\uparrow\rangle = \begin{pmatrix} 0 \\ 1 \end{pmatrix}$$

$$\text{Rotate 90 Degrees}: \left\{ \begin{pmatrix} \frac{\sqrt{2}}{2} \\ -\frac{\sqrt{2}}{2} \end{pmatrix}, \begin{pmatrix} \frac{\sqrt{2}}{2} \\ \frac{\sqrt{2}}{2} \end{pmatrix} \right\}$$

$$\text{Parallel to x-Axis}: \left\{ \begin{pmatrix} \frac{\sqrt{2}}{2} \\ -i\frac{\sqrt{2}}{2} \end{pmatrix}, \begin{pmatrix} \frac{\sqrt{2}}{2} \\ i\frac{\sqrt{2}}{2} \end{pmatrix} \right\}$$

$$\text{Superposition}: (o\downarrow, o\uparrow) \begin{pmatrix} \frac{\sqrt{2}}{2} \\ i\frac{\sqrt{2}}{2} \end{pmatrix} = \frac{\sqrt{2}}{2} o\downarrow + i\frac{\sqrt{2}}{2} o\uparrow$$

$$\text{Superposition Probability}: |A|^2 = \left| \frac{\sqrt{2}}{2} \right|^2 = \frac{1}{2}; |B|^2 = \left| i\frac{\sqrt{2}}{2} \right|^2 = \frac{1}{2}$$

In the quantum state formulation, the typical pairwise wave function is expressed as

$$\Psi = \frac{1}{\sqrt{2}}(\psi_L + i\psi_R)$$
$$\psi_0 = \frac{1}{\sqrt{2}}(\psi_{1L} + i\psi_{1R})$$
$$\psi_\pi = \frac{1}{\sqrt{2}}(\psi_{1L} - i\psi_{1R})$$

Pairwise electron entanglement can be described in the form of

$$\text{Two Macrostates}: \{0, 1\} \to \{00, 01, 10, 11\}$$
$$\text{Two Electrons}: \{|o\downarrow\rangle, |o\uparrow\rangle\} \to \{|o\downarrow o\downarrow\rangle, |o\downarrow o\uparrow\rangle, |o\uparrow o\downarrow\rangle, |o\uparrow o\uparrow\rangle\}$$
$$\text{Quantum State}: \phi(|o\downarrow\rangle, |o\uparrow\rangle) = A|o\downarrow o\downarrow\rangle + B|o\downarrow o\uparrow\rangle + C|o\uparrow o\downarrow\rangle + D|o\uparrow o\uparrow\rangle$$
$$A^2 + B^2 + C^2 + D^2 = 1; A, B, C, D \in \mathbf{C}$$
$$\text{Entangled State}: \begin{pmatrix} A \\ B \\ C \\ D \end{pmatrix} \to \begin{pmatrix} 0 \\ N \\ M \\ 0 \end{pmatrix} = N|o\downarrow o\uparrow\rangle + M|o\uparrow o\downarrow\rangle$$

In the entangled state representation, two pairs of particles with opposite characteristics are annihilated into photons, and the characteristic measure is zero. The remaining two pairs of states $\{N, M\}$ are opposite to each other, forming a quantum entangled state representation.

However, at present, neither classical quantum theory, hidden variable theory nor many-world theory can properly describe the quantum entangled state system, and a more powerful description system is expected.

Lemma 15 (Complex Function Quantum State Representation). *The quantum state complex function representation expresses different rotation angles, superposition states, superposition state probabilities, and paired entangled state representations for the relevant state through the complex variable function formula.*

10.3 Two-Body Entanglement and Time-Reversal Symmetry

In [40], Wootters discussed the measurement representation of an entangled state formed by any two qubits. From the perspective of formal representation, if the two quantum states $\{\phi_1, \phi_2\}$ are entangled, the interaction of $\{\phi_1, \phi_2\}$ cannot be expressed as a direct product

$$|\psi\rangle \neq |\phi_1\rangle \otimes |\phi_2\rangle$$

Conversely, if it can be written in the form of a direct product, it means that the two quantum states are entangled. Let X, Y be a nonzero measure operator, and for unentangled quantum states, the reciprocity condition satisfies

$$[X, Y] = XY - YX = 0$$

For entangled state calculations, The time reversal operator $T = i\sigma_y K$, K is a complex conjugate operator. For a single quantum state

$$\langle\psi|T|\psi\rangle = 0$$

This operator converts any state to its orthogonal state whose inner product is zero.

By multiplying the vector by the imaginary number, under the action of the time operator, the original vector K is rotated 90 degrees counter-clockwise to become the vector T, and the corresponding result is shown in Fig. 6.

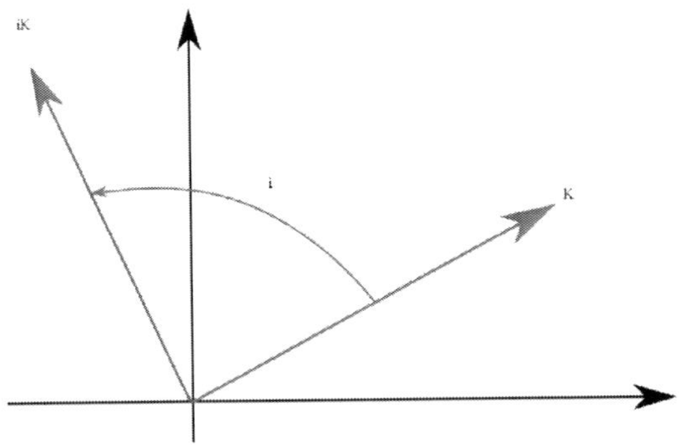

Figure 6. Time convert operator $T = iK$.

10.3.1 Entanglement Measurements

The complete set of orthogonal eigenvectors $\{|v_i\rangle\}$ corresponds to the nonzero eigenvalue ρ, and the i eigenvalue is $\langle v_i|v_i\rangle$ Each eigenvalue decomposing $|w_i\rangle$ is determined by the following relationship. U is an $m \times m$ unitary matrix

$$|w_i\rangle = \sum_{j=1}^{n} U_{ij}^{*}|v_j\rangle, m \geq n$$
$$\rho = \sum_{i} |w_i\rangle\langle w_i|$$

The first level of decomposition is represented by the state $\{|x_i\rangle\}, i = 1, \ldots, n$

$$\langle x_i|\tilde{x}_j\rangle = \lambda_i \delta_{ij}$$
$$\langle x_i|\tilde{x}_j\rangle = (U\tau U^T)_{ij}$$
$$\tau_{ij} \equiv \langle v_i|\tilde{v}_j\rangle$$
$$U\tau U = U\tau\tau^{*}U^{\dagger}$$

The second level of decomposition is represented by the state $\{|y_i\rangle\}, i = 1, \ldots, n$

$$|y_1\rangle| = |x_1\rangle$$
$$|y_j\rangle| = i|x_j\rangle \text{ for } j \neq 1$$

Use $|y_i\rangle$ to express V as an $n \times n$ unitary matrix for any decomposition

$$|z_j\rangle| = \sum_{j=1}^{n} V_{ij}^{*}|y_j\rangle$$

$$\langle c\rangle = \sum_i |z_i\rangle\langle z_i| = \sum_i (VYV^T)_{ii} = \mathrm{Tr}(VYV^T)$$

$$Y_{ij} = \langle y_i|\tilde{y}_j\rangle,\ V^T = V^{-1}$$

To average

$$\begin{aligned}\langle C\rangle &= \sum_i |(VYV^T)_{ii}| \\ &= \sum_i |\sum_j (V_{ij})^2 Y_{jj}|\end{aligned}$$

For the pure quantum state $|\psi\rangle$, the preconcurrence measure preconcurrence c is

$$c(\psi) = \frac{\langle\psi|\tilde{\psi}\rangle}{\langle\psi|\psi\rangle}$$

$$\begin{aligned}\langle c\rangle &= \sum_i \langle y_i|y_i\rangle \frac{\langle y_i|\tilde{y}_i\rangle}{\langle y_i|y_i\rangle} = \sum_i \langle y_i|\tilde{y}_i\rangle \\ &= C(\rho)\end{aligned}$$

10.3.2 Typical Example

The ground state of a single qubit can be expressed as

$$|\tilde{\psi}\rangle = \sigma_y|\psi\rangle;\ \psi, \tilde{\psi} \in \{|0\rangle, |1\rangle\}$$

$$\sigma_y = \sigma_2 = \begin{pmatrix} 0 & -i \\ i & 0 \end{pmatrix}$$

The ground state of two qubits is expressed as $\{|00\rangle, |01\rangle, |10\rangle, |11\rangle\}$, Pauli matrix direct product: $\sigma_y \otimes \sigma_y = \sigma_2 \otimes \sigma_2$

$$\tilde{\rho} = (\sigma_2 \otimes \sigma_2)\rho^{*}(\sigma_2 \otimes \sigma_2)$$

$$|\tilde{\psi}\rangle = \sigma_2 \otimes \sigma_2 K|\psi\rangle = \sigma_2 \otimes \sigma_2|\psi^{*}\rangle$$

The concurrency measure C is

$$E(\psi) = \mathscr{E}(C(\psi))$$

$$C(\psi) = |\langle\psi|\tilde{\psi}\rangle| = |\langle\psi|T|\psi\rangle| \neq 0$$

The entangled state is $|\psi\rangle = (|00\rangle + |11\rangle)/\sqrt{2}$

$$|\langle\psi|T|\psi\rangle| = 1$$

In general, the two quantum states satisfy

$$|\psi\rangle = a_1|00\rangle + a_2|01\rangle + a_3|10\rangle + a_4|11\rangle$$
$$C = 2|a_1a_4 - a_2a_3|$$

For the paired quantum system A, B single quantum density matrix σ, the related entanglement entropy is

$$E(\rho) = \mathcal{E}(C(\rho))$$
$$\rho = \sum_i p_i|\psi_i\rangle\langle\psi_i|$$
$$\mathcal{E}(\psi) = -\mathrm{Tr}(\rho_A \log_2 \rho_A) = -\mathrm{Tr}(\rho_B \log_2 \rho_B)$$
$$\mathcal{E}(\psi) = min \sum_i p_i E(\psi_i)$$
$$\rho_1 = \mathrm{Tr}|\psi\rangle\langle\psi| = \begin{pmatrix} |a_0|^2 + |a_1|^2 & a_0a_2^* + a_1a_3^* \\ a_2a_0^* + a_3a_1^* & |a_2|^2 + |a_3|^2 \end{pmatrix}$$

When the eigenvalues of the mixed state ρ density matrix are p_1, p_2, the entanglement entropy is

$$\mathcal{E}(\rho) = -\mathrm{Tr}\ \rho_1 \log_2 \rho_1 = -p_1 \log_2 p_1 - p_2 \log_2 p_2$$
$$p_1 = \frac{1 + \sqrt{1 - C^2}}{2}$$
$$p_2 = \frac{1 - \sqrt{1 - C^2}}{2}$$
$$p_1 + p_2 = 1$$

10.3.3 Measurement Basis

The relationship between entanglement entropy and the concurrency measure established by Wootters is

$$R = \sqrt{\sqrt{\rho}\tilde{\rho}\sqrt{\rho}}$$
$$\tilde{\rho} = \sum \rho^* \sum$$
$$\sum = \sigma_y \otimes \sigma_y$$
$$\lambda_1 \geq \lambda_2 \geq \lambda_3 \geq \lambda_4 \geq 0$$
$$C = \max(0, \lambda_1 - \lambda_2 - \lambda_3 - \lambda_4)$$
$$C(\rho) \geq C = \max(0, \lambda_1 - \lambda_2 - \lambda_3 - \lambda_4)$$

Further demonstration

$$\mathcal{E}(\rho) = h\left(\frac{1 + \sqrt{1 - C^2}}{2}\right)$$
$$h(x) = -x \log_2 x - (1 - x) \log_2(1 - x)$$

Lemma 16 (Two-Body Entangled State Discrimination). *The two-body entangled state is symmetric with time reversal, and the following properties can be used to determine: 1) if there are two entangled quantum states, they cannot be expressed as a direct product; 2) it can be written in the form of a direct product as entanglement-free, and the nonzero measured state satisfies the reciprocity condition; and 3) the entangled state time inversion operator converts any state into an orthogonal state.*

10.4 Symmetry+Supersymmetry

For the arbitrary wave function ψ, Pauli's principle is expressed as $\psi^2 = 0$, that is, $\psi\psi = 0$. Such a feature is just a special case of anti-commutation

$$\psi\phi = -\phi\psi \tag{88}$$

However, for the same type of boson operator Θ, Ξ, the commutation relation is

$$\Theta\Xi = \Xi\Theta \tag{89}$$

The n-dimensional complex vector space can be regarded as a $2n$-dimensional real vector space. The operation J satisfies $J^2 = -1$, and the effect on the real space is equivalent to multiplying i on the complex space.

The solution space F of the free Maxwell equations is an infinite-dimensional real vector space, and each solution can be expressed as the sum of the positive frequency solution F^+ and the negative frequency solution F^-.

$$F = F^+ + F^- \tag{90}$$

When operator J transforms real space to complex space

$$J(F) = iF^+ - iF^- \tag{91}$$

When the equation satisfies $ab \pm ba = c$, $\psi^*\phi \pm \phi\psi^* = i^k\langle\psi|\phi\rangle I, k \in \{0, 1, 2, 3\}$.

Using the basic Poisson operator $[a, b] = ab - ba$, the expansion operator $[a, b]_+ = ab + ba$. The supersymmetric operator $[]^s$ is expressed as

$$[a, b]^s = ab \pm ba \tag{92}$$

Lemma 17 (Symmetry + Supersymmetry). *Symmetry + Supersymmetry Criterion Poisson Operator Forming Extension Operator $[a, b]_+ = ab + ba$ Supersymmetry Operator $[]^s$ is represented as $[a, b]^s = ab \pm ba$.*

10.5 Vector Representation of Quantum State

The two quantum state representations of a single primitive [42] are defined as

$$|0\rangle \equiv \begin{pmatrix} 1 \\ 0 \end{pmatrix} \tag{93}$$

$$|1\rangle \equiv \begin{pmatrix} 0 \\ 1 \end{pmatrix} = \begin{pmatrix} 1 \\ 1 \end{pmatrix} - \begin{pmatrix} 1 \\ 0 \end{pmatrix} \equiv \neg \begin{pmatrix} 1 \\ 0 \end{pmatrix} \tag{94}$$

Through the tensor product $\otimes$ operator, the four quantum states of the two primitives after expansion are expressed as

$$|00\rangle \equiv |0\rangle \otimes |0\rangle = \begin{pmatrix} 1 \\ 0 \end{pmatrix} \otimes \begin{pmatrix} 1 \\ 0 \end{pmatrix} = \begin{bmatrix} 1 \times \begin{pmatrix} 1 \\ 0 \end{pmatrix} \\ 0 \times \begin{pmatrix} 1 \\ 0 \end{pmatrix} \end{bmatrix} = \begin{bmatrix} 1 \\ 0 \\ 0 \\ 0 \end{bmatrix}$$

$$|01\rangle \equiv |0\rangle \otimes |1\rangle = \begin{pmatrix} 1 \\ 0 \end{pmatrix} \otimes \begin{pmatrix} 0 \\ 1 \end{pmatrix} = \begin{bmatrix} 1 \times \begin{pmatrix} 0 \\ 1 \end{pmatrix} \\ 0 \times \begin{pmatrix} 0 \\ 1 \end{pmatrix} \end{bmatrix} = \begin{bmatrix} 0 \\ 1 \\ 0 \\ 0 \end{bmatrix}$$

$$|10\rangle \equiv |1\rangle \otimes |0\rangle = \begin{pmatrix} 0 \\ 1 \end{pmatrix} \otimes \begin{pmatrix} 1 \\ 0 \end{pmatrix} = \begin{bmatrix} 0 \times \begin{pmatrix} 1 \\ 0 \end{pmatrix} \\ 1 \times \begin{pmatrix} 1 \\ 0 \end{pmatrix} \end{bmatrix} = \begin{bmatrix} 0 \\ 0 \\ 1 \\ 0 \end{bmatrix}$$

$$|11\rangle \equiv |1\rangle \otimes |1\rangle = \begin{pmatrix} 0 \\ 1 \end{pmatrix} \otimes \begin{pmatrix} 0 \\ 1 \end{pmatrix} = \begin{bmatrix} 0 \times \begin{pmatrix} 0 \\ 1 \end{pmatrix} \\ 1 \times \begin{pmatrix} 0 \\ 1 \end{pmatrix} \end{bmatrix} = \begin{bmatrix} 0 \\ 0 \\ 0 \\ 1 \end{bmatrix}$$

For any n primitives, 2^n quantum states can be iteratively generated by the $\otimes^n$ compound operator, forming a $0 \leq |j\rangle < 2^n$ state representation, specified at the j position in the 2^n long vector.

Lemma 18 (Tensor Product Expansion). *The tensor product operator forms an expansion vector representation, and n primitives are expanded into 2^n quantum state vector representations after tensor product expansion.*

10.5.1 Quantum Entangled Theory and Implicit Assumptions

The mathematical representation of quantum entanglement is the superposition of several eigenstates [43], and the Bell state associated with two qubits is expressed as:

$$|\varphi^{\pm}> = \frac{1}{\sqrt{2}}(|0\rangle_A \otimes |0\rangle_B \pm |1\rangle_A \otimes |1\rangle_B) = \frac{1}{\sqrt{2}}(|00\rangle \pm |11\rangle) = \frac{1}{\sqrt{2}} \begin{bmatrix} 1 \\ 0 \\ 0 \\ \pm 1 \end{bmatrix}$$

$$|\phi^{\pm}> = \frac{1}{\sqrt{2}}(|0\rangle_A \otimes |1\rangle_B \pm |1\rangle_A \otimes |0\rangle_B) = \frac{1}{\sqrt{2}}(|01\rangle \pm |10\rangle) = \frac{1}{\sqrt{2}} \begin{bmatrix} 0 \\ 1 \\ \pm 1 \\ 0 \end{bmatrix}$$

For $m > 2$, the GHZ state is expressed as

$$|GHZ\rangle = \frac{1}{\sqrt{2}}(|0\rangle^{\otimes m} + |1\rangle^{\otimes m})$$

For $m = 2$, it degenerates to the Bell state $|\varphi^+\rangle$. The classical GHZ state is $m = 3$, and the system goes beyond a two-dimensional representation structure.

For two bosons, the $N00N$ state: N photons in one mode and N photons in the other mode are expressed as

$$|\psi_{N00N}\rangle = \frac{1}{\sqrt{2}}(|N\rangle_A|0\rangle_B + |0\rangle_A|N\rangle_B)$$

Bell, GHZ, and N00N state representations are capable of exhibiting extremely entangled state properties when entanglement measurements are properly chosen.

Lemma 19 (Two-Quantum Entangled State). *The quantum entangled superposition state vector related to two qubits can be expressed as: {Bell, GHZ, NOON} and other pairwise description forms.*

10.6 Bell Quantum State

In the classical quantum entangled state formulation, Bell-type entangled states are represented [42, p. 25].

Input	Output
$\|00\rangle$	$(\|00\rangle + \|11\rangle)/\sqrt{2}$
$\|01\rangle$	$(\|01\rangle + \|10\rangle)/\sqrt{2}$
$\|10\rangle$	$(\|00\rangle - \|11\rangle)/\sqrt{2}$
$\|11\rangle$	$(\|01\rangle - \|10\rangle)/\sqrt{2}$

Question 1: The following four formulas with negative signs, which are not placed in the table, can be properly defined in the system?

$$\{-(|00\rangle + |11\rangle)/\sqrt{2}, -(|01\rangle + |10\rangle)/\sqrt{2}, -(|00\rangle - |11\rangle)/\sqrt{2}, -(|01\rangle - |10\rangle)/\sqrt{2}\}$$

Answer: The quantum states $-|\psi\rangle$ and $+|\psi\rangle$ are considered to be the same final state, with a negligible overall phase $e^{i\psi}$ ($e^{i0} = +1, e^{i\pi} = -1$). Despite the apparent sign difference, the overall phase is not observed in actual measurements.

Question 2: In what way are linearly correlated quantum states handled in the representation of quantum entangled states? Answer: For linear correlation expressions, the

$$|\psi_1\rangle = c|\psi_2\rangle$$

correlation constant can be expressed as an overall phase $c = e^{i\psi}$. Due to the normalization effect of the quantum state,

$$||\psi_1\rangle|^2 = |c|^2||\psi_2\rangle|^2$$

$$||\psi_1\rangle|^2 = 1; ||\psi_2\rangle|^2 = 1; |c|^2 = 1 \Rightarrow c = e^{i\psi}$$

The states of $|\psi_1\rangle$ and $|\psi_2\rangle$ can be considered to be the same final state representation.

From these two questions and the answers obtained, the following can be seen. The sign characteristic of the overall phase difference between quantum states does not need to be distinguished in a classical quantum system. However, the conversion factor has such a stark difference between -1 and $+1$.

Lemma 20 (Global Phase Difference). *In a Bell quantum state representation, the sign feature of the global phase difference between quantum states does not need to be distinguished in classical quantum systems.*

10.6.1 Entangled State of Different Clusters

In these representations, the specific meaning of each symbol is not important. Importantly, each symbol is a constant. That is, starting from the definition of entanglement, the object under study does not change with time and is a constant. In response to such characteristics, Lei Yi'an emphasized in his blog [44] that this is an implicit assumption, but no one has strictly demonstrated the rationality of such an assumption.

Entanglement quantification can be used for simple quantum systems [48]:

$$\psi_A = \frac{|0000> + |1111>}{\sqrt{2}}$$

$$\psi_B = \frac{|0011> + |0101> + |0110> + |1001> + |1010> + |1100>}{\sqrt{6}}$$

On the surface, it appears that ψ_A has a stronger correlation between spins than ψ_B, but the entropy of state B is slightly larger when tracing the two spins. From the perspective of quantitative analysis, ψ_B is more entangled than ψ_A.

From the perspective of distinguishing states and clusters, ψ_B as a cluster represents six different states that can be distinguished.

10.6.2 Quantum Entangled State Theory

Described in Nielson & Chuang's monograph on quantum computing and quantum information [42, p. 12]. Although the current frontier research on quantum information has not established a complete theory of entanglement, there has been a series of progress in this direction to understand such strange properties. It is expected that subsequent basic research can effectively promote the development of quantum computing and quantum information and open up new frontier applications after an in-depth understanding of quantum entanglement effects.

10.7 Hadamard Matrix

Hadamard matrices play a fundamental role in describing complex quantum systems by using orthogonal function techniques and methods. Hadamard matrix representations of iterative patterns can be seen in a variety of quantum gate applications, from the Deutsch–Jozsa algorithm to the Grover retrieval mechanism.

Through the iterative operation of the $H^{\otimes^n}$ operator, similar to the $\otimes^n$ operator acting on the fundamental quantum state to form a vector, the expanded matrix maintains the $2^n \times 2^n = 2^{2n}$ dimension, and each row/column vector is orthogonal to each other. From the perspective of structure generation, it can be regarded as the corresponding generation $0 \le k < 2^{2n}$ state vector representation.

Lemma 21 (Hadamard Matrix Tensor Product). *The Hadamard matrix tensor product operator forms an extended matrix representation, and n tensor product operators become 2^{2n} element quantum state matrix representation after expansion.*

10.7.1 Basic Operator

$$H_2 = \begin{bmatrix} 1 & 1 \\ 1 & -1 \end{bmatrix} \tag{95}$$

$$-H_2 = \begin{bmatrix} -1 & -1 \\ -1 & 1 \end{bmatrix} \tag{96}$$

$$H_{2^m} = H_2 \otimes H_{2^{m-1}} = \begin{bmatrix} H_{2^{m-1}} & H_{2^{m-1}} \\ H_{2^{m-1}} & -H_{2^{m-1}} \end{bmatrix} \tag{97}$$

Generated after iteration,

$$H_4 = H_2 \otimes H_2 = \begin{bmatrix} H_2 & H_2 \\ H_2 & -H_2 \end{bmatrix} = \begin{bmatrix} 1 & 1 & 1 & 1 \\ 1 & -1 & 1 & -1 \\ 1 & 1 & -1 & -1 \\ 1 & -1 & -1 & 1 \end{bmatrix}$$

$$H_8 = H_4 \otimes H_2 = \begin{bmatrix} H_4 & H_4 \\ H_4 & -H_4 \end{bmatrix} = \begin{bmatrix} 1 & 1 & 1 & 1 & 1 & 1 & 1 & 1 \\ 1 & -1 & 1 & -1 & 1 & -1 & 1 & -1 \\ 1 & 1 & -1 & -1 & 1 & 1 & -1 & -1 \\ 1 & -1 & -1 & 1 & 1 & -1 & -1 & 1 \\ 1 & 1 & 1 & 1 & -1 & -1 & -1 & -1 \\ 1 & -1 & 1 & -1 & -1 & 1 & -1 & 1 \\ 1 & 1 & -1 & -1 & -1 & -1 & 1 & 1 \\ 1 & -1 & -1 & 1 & -1 & 1 & 1 & -1 \end{bmatrix}$$

10.7.2 Number of Positive and Negative Symbols

The numbers of $\{+, -\}$ are not equal in the 2×2 Hadamard matrix H_2 construction.

$$\begin{aligned} |H_2 : +| &= 3,\ |H_2 : -| = 1 \\ |H_2 : +| + |H_2 : -| &= 4 = 2^2 \\ |H_{2^2} : +| &= 10,\ |H_{2^2} : -| = 6 \\ |H_{2^2} : +| + |H_{2^2} : -| &= 16 = 2^4 \\ &\cdots \end{aligned}$$

$$|H_{2^n}:+| = 2^n + \sum_{j=0}^{2^n-2} 2^{n-1} = 2^n + (2^n-1)2^{n-1} = 2^{2n-1} + 2^{n-1}$$

$$|H_{2^n}:-| = \sum_{j=0}^{2^n-2} 2^{n-1} = 2^{2n-1} - 2^{n-1}$$

$$|H_{2^n}:+| + |H_{2^n}:-| = 2^{2n}$$

10.7.3 Number of Entire Balance Positive and Negative Signs

To balance the overall number of symbols, the paired negative and reflection representations are used to perform the following operator expansion, converting the H_2 matrix representation of unbalanced positive and negative signs into $\boxed{H_2}$ expansion of balanced positive and negative signs matrix structure. It can be verified that the $\boxed{H_{2^m}}$ matrix structure after any m iterations also maintains sign balance.

$$\overrightarrow{H}_2 = \begin{bmatrix} 1 & 1 \\ -1 & 1 \end{bmatrix} \tag{98}$$

$$-\overrightarrow{H}_2 = \begin{bmatrix} -1 & -1 \\ 1 & -1 \end{bmatrix} \tag{99}$$

$$\boxed{H_2} = \begin{bmatrix} -H_2 & H_2 \\ -\overrightarrow{H}_2 & \overrightarrow{H}_2 \end{bmatrix} = \left[\begin{array}{cc|cc} -1 & -1 & 1 & 1 \\ -1 & 1 & 1 & -1 \\ \hline -1 & -1 & 1 & 1 \\ 1 & -1 & -1 & 1 \end{array}\right] \tag{100}$$

$$\boxed{H_{2^m}} = \begin{bmatrix} -H_{2^m} & H_{2^m} \\ -\overrightarrow{H}_{2^m} & \overrightarrow{H}_{2^m} \end{bmatrix} \tag{101}$$

Corollary 13 (Hadamard Matrix Balance Positive and Negative Sign). *The positive and negative quantities formed by the Hadamard matrix tensor product are unbalanced in the single block representation, and the balanced state distribution of positive and negative quantities can be formed by four region expansion. The total number of elements involved is* $2^{2n+2} = 4 \times 2^{2n}$.

10.8 Matrix of Complex Numbers

Using the matrix representation [45] extension structure, it is possible to simulate the quantized representation of complex ordinal pairs and the results of algebraic operations. When $a = b$, 2×2 indicates that the matrix is a Hadamard matrix, and the sign imbalance problem discussed in the previous section will obviously appear.

10.8.1 Base Representation

$$(a,b) \to a + bi \to \begin{pmatrix} a & -b \\ b & a \end{pmatrix} \tag{102}$$

$$(a,-b) \to a - bi \to \begin{pmatrix} a & b \\ -b & a \end{pmatrix} \tag{103}$$

$$(-a,-b) \to -a - bi \to \begin{pmatrix} -a & -b \\ b & -a \end{pmatrix} \tag{104}$$

$$(-a,b) \to -a + bi \to \begin{pmatrix} -a & b \\ -b & -a \end{pmatrix} \tag{105}$$

For example: $(a,b):(-a,-b)$. The number of positive and negative symbols is $+2:-2$, $\begin{pmatrix} a & -b \\ b & a \end{pmatrix}:\begin{pmatrix} -a & -b \\ b & -a \end{pmatrix}$. The number of positive and negative symbols is $+3-1:+1-3$. In addition, $(a,-b):(-a,b)$. The number of positive and negative symbols is $+1-1:+1-1$, $\begin{pmatrix} a & b \\ -b & a \end{pmatrix}:\begin{pmatrix} -a & b \\ -b & -a \end{pmatrix}$. The number of positive and negative symbols is $+3-1:+1-3$.

From a sign extension point of view, the sign swap of the real part $-2 \leftrightarrow +2$ relative change is 4, while the sign of the imaginary part $-1+1 \leftrightarrow +1-1$ relative change is 0. From this point of view, the asymmetric characteristics of the quantization change after matrix expansion can be seen.

10.8.2 Rotation Matrix

For regular representation structures,

$$e^{i\theta} = (\cos\theta, \sin\theta) = \cos\theta + i\sin\theta \to \begin{pmatrix} \cos\theta & -\sin\theta \\ \sin\theta & \cos\theta \end{pmatrix} \tag{106}$$

The matrix representation is reasonable for 2D rotation models that only focus on quantized features. However, expecting to use the matrix mode to properly simulate the complex exponential representation, the defects caused by dimension expansion need to be studied in depth.

Due to the classical sequence pair representation, various quantized symbol features are rarely properly represented by the extended-dimensional analog matrix representation. Any model based on a balanced description needs to pay attention to the difference after such dimensional changes. The simple equivalence of completely unequal structural features would be a disaster for the extended system of formal representation.

Introduce insurmountable dimensional complexity and inappropriate coverage extension.

10.9 Pan Complex Variable Function

In the Pan complex variable function structure [46], the extended complex number is expressed as three imaginary unit forms $\{i,j,k\}$, which satisfy

Ellipse number $a = \alpha + \beta i$, satisfying $i^2 + 1 = 0, a \in C$
Parabolic number $a = \alpha + \beta k$, satisfying $k^2 = 0, a \in P$
Hyperbolic numbers $a = \alpha + \beta j$, satisfying $j^2 - 1 = 0, a \in H$

Under such an extension, the corresponding addition, subtraction, multiplication, and division rules and their complex conjugate operations can be defined. Connect corresponding models and methods to address fundamental topics in mathematical physics.

Similar to the four-element Hamilton extension of matrices, generic complex functions constitute ternary operator extensions. Since each primitive operator satisfies different operation rules, three sets of independent operation mechanisms can be used to provide consistent operation results. Hybrid operations often require more complex structural supports.

Can a more natural expansion model be used to support all the basic operator expansion requirements of generic complex functions? This topic deserves further attention.

11. Formal Measurement of Various Quantum Paradoxes

11.1 Quantum State Measurement and Reflection Representation

Using semitransmissive mirrors (semitransparent mirrors) and total reflection mirrors (total reflective mirrors), the symbolic formal representation of transforming different photons to form quantum states [32, 42] is shown in Fig. 7.

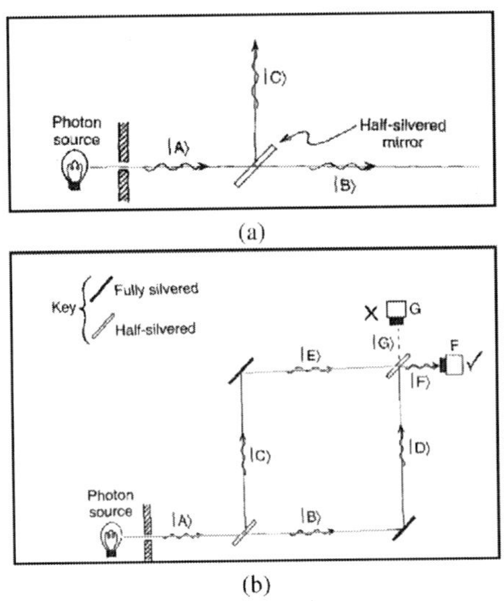

Figure 7. Representation of the interaction of different quantum states: (a) transflective mode; (b) mach–zehnder mode.

In graph 7(a), the interaction representation of the relevant input/output quantum states is expressed in complex function form as $|A\rangle \rightarrow |B\rangle + i|C\rangle$ or $A = B + iC$. The complex interaction of two-way interference in Fig. 7(b) is expressed as

$$A \rightarrow B + iC \text{ Semitransparent}$$
$$B \rightarrow iD \text{ Total Reflection}$$
$$C \rightarrow iE \text{ Total Reflection}$$

$$
\begin{aligned}
D &\to G + iF \text{ Conjugate Semitransparent} \\
E &\to F + iG \text{ Conjugate Semitransparent} \\
B + iC &\to iD + i(iE) = iD - E \text{ Intermediate Results} \\
iD - E &\to i(G + iF) - (F + iG) \\
&= iG - F - F - iG \\
&= -2F
\end{aligned}
$$

Through such transformation mapping, the measurement result is finally obtained from the measurement port $\{F, G\}$.

Lemma 22 (Modern Optical Measurement Conversion Equipment). *Modern optical equipment based on semitransparent and total reflection mirrors, providing a variety of complex variable function transformation modes in the measurement path channel: Basic complex functions such as total reflection and conjugate semitransparent transform the measurement mode.*

11.2 EPR Paradox

For paired spin particles $\{\phi_1, \phi_2\}$ [2, 32], the corresponding pairwise direction vector is $\{\phi_1 : \hat{n}_1, \phi_2 : \hat{n}_2\}$, entanglement measurement output: $\{+1 \to \text{Up}, -1 \to \text{Down}\}$ The simplified effect is shown in Fig.8(a), and the associated probability is: $\{P^{\phi}_{++}(\hat{n}_1, \hat{n}_2), P^{\phi}_{+-}(\hat{n}_1, \hat{n}_2), P^{\phi}_{-+}(\hat{n}_1, \hat{n}_2), P^{\phi}_{--}(\hat{n}_1, \hat{n}_2)\}$

$$E^{\phi} = P^{\phi}_{++}(\hat{n}_1, \hat{n}_2) - P^{\phi}_{+-}(\hat{n}_1, \hat{n}_2) - P^{\phi}_{-+}(\hat{n}_1, \hat{n}_2) + P^{\phi}_{--}(\hat{n}_1, \hat{n}_2)$$

The expected value of the measured output is

$$E^{\phi}(\hat{n}_1, \hat{n}_2) = -\hat{n}_1 \cdot \hat{n}_2, \ \ \hat{n}_1 = \hat{n}_2 \text{ Full relevant}$$

The four basic conditions (i)–(iv) corresponding to the measurement are

i) Full correlation: Measure outputs are inverse of each other
ii) Locality: No real change in the second system
iii) Reality: An element of physical reality corresponds to this physical quantity
iv) Completeness: Every element in physical reality has a counterpart in a complete theory

The corresponding entangled state is expressed as

$$|\phi\rangle = \frac{1}{\sqrt{2}}[|+\rangle_1|-\rangle_2 - |-\rangle_1|+\rangle_2] \tag{107}$$

where $|+\rangle_1, |-\rangle_1, \hat{n}_1$ represents the particle ϕ_1, and $|+\rangle_2, |-\rangle_2, \hat{n}_2$ represents the particle ϕ_2 related state.

Corollary 14 (EPR Paradox Measurement). *The EPR paradox can be described under the mechanism of measuring output using a complex function formal representation, distinguished from four aspects: full correlation, locality, reality, and completeness.*

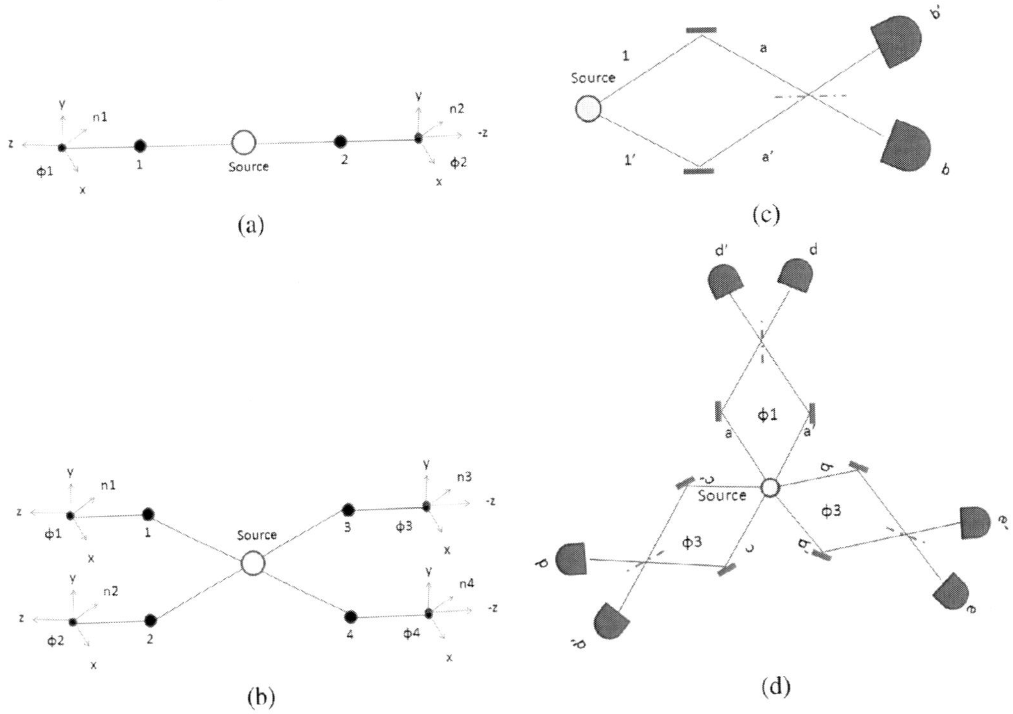

Figure 8. Paired spin particle interaction diagram: (a) paired spins; (b) four spins; (c) paired particles; (d) three paired particles.

11.3 Bell State

In 1964, Bell proposed the measurement and detection inequality [49] for local features, and for the system with eight states [50] marked as $\{0,1,2,3,4,5,6,7\}$ represents the eight zones of the three-dimensional space. Describe the spin states of eight zone entangled particles $\{A,B\}$. Due to the entangled nature of the two paired particles, the spin directions are opposite. That is, the direction of one particle (white vector) is determined, and the paired particle (black vector) will point in the other direction. Assuming that the probability of white vector direction particles appearing in the eight zones is $\{p_0,\ldots,p_j,\ldots,p_7\}$, $0 \leq j < 8$, each pointing to a zone, the probability of the group and for

$$1 = \sum_{j=0}^{7} p_j$$

Different correlation functions can be defined $\{Pxx(I), Pxz(I), Pzy(I), Pxy(I)\}$, $0 \leq I < 8$

$$Pxx(I) = \begin{cases} +1, & \text{If } Ax(I) \equiv Bx(I) \\ -1, & \text{Then} \end{cases}$$

$$Pxz(I) = \begin{cases} +1, & \text{If } Ax(I) \equiv Bz(I) \\ -1, & \text{Then} \end{cases}$$

$$Pzy(I) = \begin{cases} +1, & \text{If } Az(I) \equiv By(I) \\ -1, & \text{Then} \end{cases}$$

$$Pxy(I) = \begin{cases} +1, & \text{If } Ax(I) \equiv By(I) \\ -1, & \text{Then} \end{cases}$$

The eight entangled state spin vectors and the four correlation function values are

I	$Ax\ Ay\ Az$	$Bx\ By\ Bz$	P	Pxx	Pxz	Pzy	Pxy
0	+ + +	- - -	p_0	-1	-1	-1	-1
1	- + +	+ - -	p_1	-1	+1	-1	+1
2	+ - +	- + -	p_2	-1	-1	+1	+1
3	- - +	+ + -	p_3	-1	+1	+1	-1
4	+ + -	- - +	p_4	-1	+1	+1	-1
5	- + -	+ - +	p_5	-1	-1	+1	+1
6	+ - -	- + +	p_6	-1	+1	-1	+1
7	- - -	+ + +	p_7	-1	-1	-1	-1

If $p_I = 1/8, 0 \leq I < 8$, accumulate the relevant probability functions to obtain

$$Pxx = \sum_{I=0}^{7} Pxx(I) \times p_I = -1$$

$$Pxz = \sum_{I=0}^{7} Pxz(I) \times p_I = 0$$

$$Pzy = \sum_{I=0}^{7} Pzy(I) \times p_I = 0$$

$$Pxy = \sum_{I=0}^{7} Pxy(I) \times p_I = 0$$

It can be observed from these measures that when the probability measures of the occurrence of each state are equal, the spins of A and B measured in the same direction should be anticorrelated, while the spins of A and B measured in different directions after the accumulation are irrelevant.

From the measurement point of view, various quantities can be taken as absolute values after the operation. For example:

$$\begin{aligned} |Pxz - Pzy| &= 2|p_1 - p_2 - p_5 + p_6| \\ &= 2|(p_1 + p_6) - (p_2 + p_5)| \quad \text{Because } |x - y| \leq |x| + |y| \\ &\leq 2(p_1 + p_2 + p_5 + p_6) \\ &= p_0 + p_1 + p_2 + p_3 + p_4 + p_5 + p_6 + p_7 \\ &\quad -p_0 + p_1 + p_2 - p_3 - p_4 + p_5 + p_6 - p_7 \\ &= 1 + Pxy \end{aligned}$$

Bell's inequality is derived as follows:

$$|Pxz - Pzy| \leq 1 + Pxy \tag{108}$$

Using the fundamental formula expressed by quantum measurements,

$$\begin{aligned} E^{\phi}(x, y) &= -cos(x - y) \\ x - y &: 0, \pi/3, 2\pi/3 \\ E^{\phi}(x, y) &= E^{\phi}(y, z) = -1/2 \\ E^{\phi}(x, z) &= +1/2 \\ |E^{\phi}(x, y) - E^{\phi}(x, z)| - E^{\phi}(y, z) - 1 &= 1/2 \end{aligned}$$

The numerical value shows the result obtained by the quantum measurement formula, which does not satisfy Bell's inequality.

Corollary 15 (Bell Inequality). *Bell's inequality provides multiple quantum state correlation combinations based on state probability-based quantitative measurement conditions.*

11.4 CHSH–Bell Inequality

In 1969, four physicists, CHSH [51], published a paper in the *Physics Review Letter*, improving Bell's inequality. The obtained CHSH–Bell inequality is:

$$|P(a_1, b_1) + P(a_1, b_2) + P(a_2, b_1) - P(a_2, b_2)| \leq 2 \tag{109}$$

where $P(a_i, b_j)$ is the correlation function, which is the statistical mean of a_i, b_j in the experiment. For related experiments, a pair of measurement direction variables (a_1, a_2) can be done on one subsystem, while another pair of measurement variables (b_1, b_2) can be done on another subsystem. In the CHSH inequality, assuming that the total spin is zero, under special conditions $a_1 = b_1, P(b_1, b_1) = -1$, the CHSH inequality is reduced to Bell's inequality.

In 1972, Clauser and Freedman published their measurements of entangled states [52]. Interestingly, the measurement turned out to be the opposite of what the measurer expected. Clauser hopes that Einstein's statement can be confirmed, but his experimental measurement results determine that the value far exceeds the inequality constraint, with a clear deviation of five times the error, which strongly proves the correctness of the quantum mechanical measurement formula.

In 1982, Aspect et al. began to conduct quantum entanglement experiments [53], and in 2015, the follow-up work of Hensen et al. [54] excluded the theory based on local realism from the interactive experiment and proved the quantitative measurement expression of quantum mechanics on correctness of the formula. See Fig. 9 for the underlying entanglement representation.

The accepted experimental judgment of the nonlocality of quantum mechanics was completed by Aspect in 1982 through three experiments. Quantum mechanics is delocalized, and it is basically accepted in the physics community. Whether superluminal speed is hidden behind this conclusion, the scientific community is still uncertain.

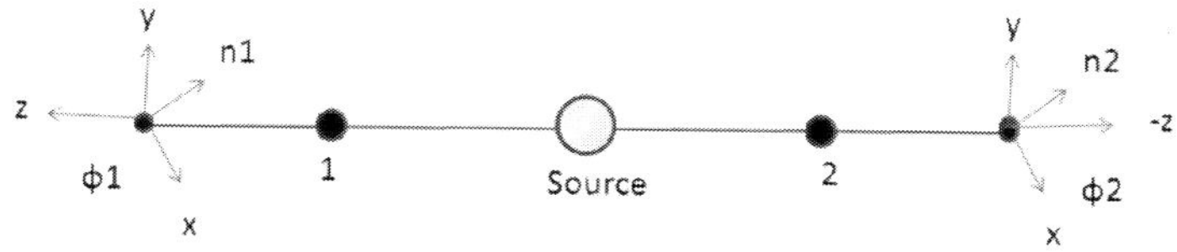

Figure 9. Pairwise spin-particle interaction.

Corollary 16 (CHSH–Bell Inequality). *The CHSH–Bell inequality provides multiple quantum state correlation combinations based on state probability-based quantitative measurement conditions.*

11.5 Single Qubit Entanglement

The superposition qubit of a single particle can be expressed as $|\text{qubit}\rangle = a|0\rangle + b|1\rangle$, $|a|^2 + |b|^2 = 1, a, b \in \mathbf{C}$ is plural. $a = \text{Re}(a) + i\,\text{Im}(a); a = \text{Re}(a) + i\,\text{Im}(a), i = \sqrt{-1}$, the square root of -1.

At first glance, a qubit has four arbitrary constants $(\text{Re}(a), \text{Im}(a), \text{Re}(b), \text{Im}(b))$. Due to the following constraints, a qubit has only two degrees of freedom. Between the four constants, there are two constraints: 1) a, b satisfy the probability normalization condition $(|a|^2+|b|^2 = 1)$; 2) for two complex numbers, only the relative phase difference has a physical meaning, and the absolute phase of the quantum superposition state cannot be observed.

For this reason, a is simplified to a real number $cos(q/2)$, the phase difference between a, b is f, and the formula is $a = \cos(q/2), b = \exp(i \cdot f)\sin(q/2)$.

$$|\text{qubit}\rangle = |y\rangle = a|1\rangle + b|0\rangle \tag{110}$$

Represent the qubit $|y\rangle$ in three-dimensional polar coordinates as a point on the Bloch sphere. A qubit has an infinite number of states spread over the entire sphere. Each state corresponds to a point on the unit sphere. Applying operations on qubits to change from one state to another, that is, the transformation corresponding to one point on the sphere to another point, is the unitary transformation. Performing a series of transformation operations is equivalent to applying a series of unitary transformations.

Unlike classical bit representations, with qubits, when multiple qubits are linked together, they can correlate to form an entangled state. Studying the detailed composition of various entangled state combination modes has become a frontier topic in international advanced quantum information.

Corollary 17 (Single Quantum Entanglement). *For special value conditions, single qubits can be used to represent entanglement.*

11.6 Bell's Theorem of Noninequalities

For four particles $\{1, 2, 3, 4\}$ in two directions, 1, 2 is forward, and 3, 4 is reverse; see Fig. 10. The spin representations of the four particles $\hat{n}_1, \hat{n}_2, \hat{n}_3, \hat{n}_4$ are polar coordinates (θ, ϕ), which are projected in the $x - y$ plane.

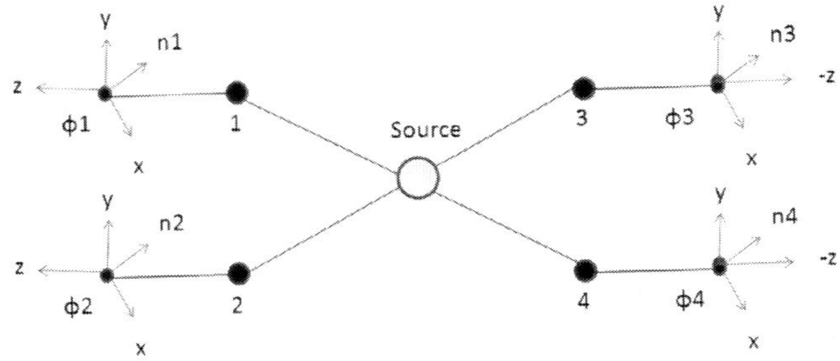

Figure 10. Four spin-particle interaction.

$$\begin{aligned}
|\phi\rangle &= \frac{1}{\sqrt{2}}[|+\rangle_1|+\rangle_2|-\rangle_3|-\rangle_4 - |-\rangle_1|-\rangle_2|+\rangle_3|+\rangle_4] \\
E^{\phi}(\hat{n}_1,\hat{n}_2,\hat{n}_3,\hat{n}_4) &= \cos\theta_1\cos\theta_2\cos\theta_3\cos\theta_4 \\
&\quad - \sin\theta_1\sin\theta_2\sin\theta_3\sin\theta_4 \times \cos(\phi_1+\phi_2-\phi_3-\phi_4) \\
&\rightarrow \quad \text{Project in } x-y \text{ plane:} \\
&\rightarrow \quad \text{projected on the } x-y \text{ plane:} \\
E^{\phi}(\hat{n}_1,\hat{n}_2,\hat{n}_3,\hat{n}_4) &= -\cos(\phi_1+\phi_2-\phi_3-\phi_4)
\end{aligned}$$

Under the condition of full correlation, if $\phi_1+\phi_2-\phi_3-\phi_4=0$, then $E^{\phi}(\hat{n}_1,\hat{n}_2,\hat{n}_3,\hat{n}_4) = -1$; If $\phi_1+\phi_2-\phi_3-\phi_4=\pi$, then $E^{\phi}(\hat{n}_1,\hat{n}_2,\hat{n}_3,\hat{n}_4) = +1$.

The four basic conditions (i)–(iv) corresponding to the measurement are

(i) Full Correlation: The output of any three particles can predict the condition of the fourth particle
(ii) Locality: Acting on any particle does not actually change the other three particles
(iii) Reality: An element of physical reality corresponds to this physical quantity
(iv) Completeness: Every element in physical reality has a counterpart in a complete theory

There are four sets of complete state functions: $A_\lambda(\phi_1), B_\lambda(\phi_2), C_\lambda(\phi_3), D_\lambda(\phi_4) \rightarrow pm1$.

For functions A, B, C, D, satisfy

If $\phi_1+\phi_2-\phi_3-\phi_4=0$, then $A_\lambda(\phi_1)B_\lambda(\phi_2)C_\lambda(\phi_3)D_\lambda(\phi_4) = -1$;
If $\phi_1+\phi_2-\phi_3-\phi_4=\pi$, then $A_\lambda(\phi_1)B_\lambda(\phi_2)C_\lambda(\phi_3)D_\lambda(\phi_4) = +1$.

$$A_\lambda(0)B_\lambda(0)C_\lambda(0)D_\lambda(0) = -1 \tag{111}$$
$$A_\lambda(\phi)B_\lambda(0)C_\lambda(\phi)D_\lambda(0) = -1 \tag{112}$$
$$A_\lambda(\phi)B_\lambda(0)C_\lambda(0)D_\lambda(\phi) = -1 \tag{113}$$
$$A_\lambda(2\phi)B_\lambda(0)C_\lambda(0)D_\lambda(\phi) = -1 \tag{114}$$

Equations (111) & (112)

$$A_\lambda(0))C_\lambda(0) = A_\lambda(\phi)C_\lambda(\phi)$$

Equations (111) & (113)

$$\begin{aligned}
A_\lambda(\phi)D_\lambda(\phi) &= A_\lambda(0)D_\lambda(0) \rightarrow C_\lambda(\phi)/D_\lambda(\phi) = C_\lambda(0)/D_\lambda(0) \\
C_\lambda(\phi)D_\lambda(\phi) &= C_\lambda(0)D_\lambda(0) \rightarrow D_\lambda(\phi) = \pm 1 \\
A_\lambda(2\phi)B_\lambda(0)C_\lambda(0)D_\lambda(\phi) &= -1 \rightarrow A_\lambda(2\phi) = A_\lambda(\phi) = \text{ Constant}
\end{aligned}$$

When $\phi = \pi/2, \theta = 0$,

$$A_\lambda(\theta + \pi)B_\lambda(0)C_\lambda(\theta)D_\lambda(0) = 1 \rightarrow A_\lambda(\theta + \pi) = -A_\lambda(\theta)$$

this result contradicts that $A_\lambda(2\phi) = A_\lambda(\phi)$ is a constant. Such quantitative properties reveal the inherent inconsistency underlying the measurement of the four underlying conditions (i)–(iv). Inherent inconsistency is contained in the degrees of freedom of combinatorial growth.

11.7 Multiphoton Entangled State

In 1995, Shih first realized ghost imaging of two-photon entangled sources; see Fig. 11. This area of research is currently known as quantum imaging.

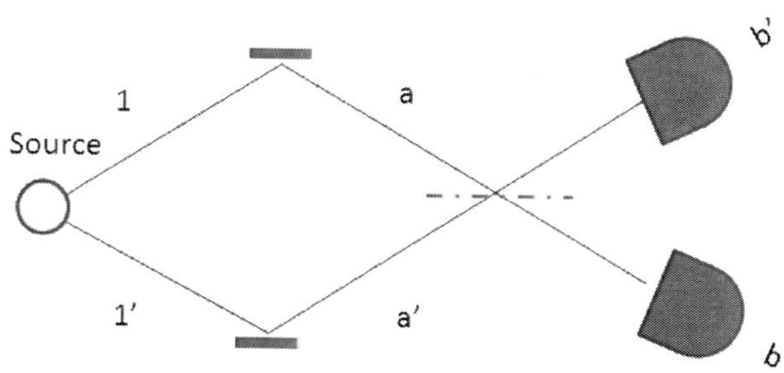

Figure 11. Pairwise particle interaction.

For more complex detection modes without spins, the most famous is the GHZ theorem [55, 56]; see Fig. 12, where all particles carry the same energy and maintain momentum conservation, and the space position directions are separated by 120 degrees.

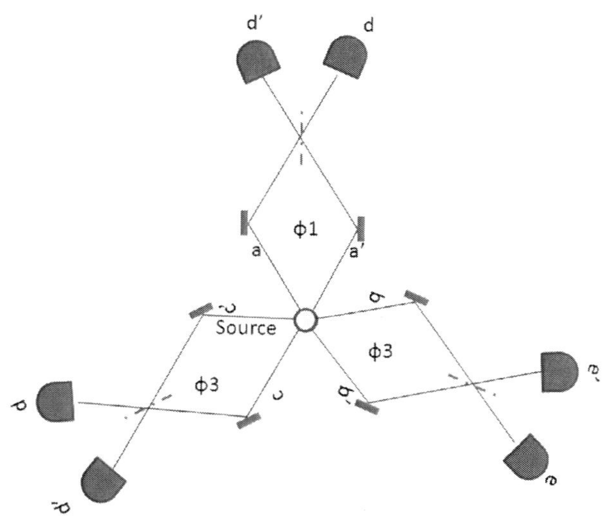

Figure 12. Three-pair particle interactions.

There are six groups of channels distributed around the central particle source: $\{a, b, c\}$ are separated by 120 degrees, and $\{a', b', c'\}$ are also separated by 120 degrees. The superposition formula of the three groups of particle entanglement states is:

$$|\psi\rangle = \frac{1}{\sqrt{2}}[|a\rangle_1|b\rangle_2|c\rangle_3 - |a'\rangle_1|b'\rangle_2|c'\rangle_3]$$

The particle beam a & a' arrives at the semitransparent mirror with a transmission and reflectivity of 50–50 in the mode of total reflection. After passing through the detector, each output stream is converted to form d & d' signal output.

$$|a\rangle = \frac{1}{\sqrt{2}}[|d\rangle_1| + i|d'\rangle_1]$$

$$|a'\rangle = \frac{1}{\sqrt{2}}e^{i\phi_1}[|d'\rangle_1| + i|d\rangle_1]$$

The triplet detector can output $(d, e, f)\cdots(d', e, f)\cdots$ with a combined variation of $d|d', e|e', f|f'$, etc. The corresponding detection probability is

$$P^{\psi}_{def} = \frac{1}{8}[1 + \sin(\phi_1 + \phi_2 + \phi_3)] \tag{115}$$

$$P^{\psi}_{d'ef} = \frac{1}{8}[1 - \sin(\phi_1 + \phi_2 + \phi_3)] \tag{116}$$

Has even : +, odd : − characteristics, then

$$E^{\psi}_{def}(\phi_1, \phi_2, \phi_3) = \sin(\phi_1 + \phi_2 + \phi_3)$$

Under the condition of full correlation,
if $\phi_1 + \phi_2 + \phi_3 = \pi/2$, then $E^{\phi}(\phi_1, \phi_2, \phi_3) = -1$.

Formula (115) & (116) sums to 1/4 independent of phase. Choose from four sets of parameters: $(\phi_1, \phi_2, \phi_3) : \{(\pi/2, 0, 0), (0, \pi/2, 0), (0, 0, \pi/2), (\pi/2, \pi/2, \pi/2)\}$

$$A_\lambda(\pi/2)B_\lambda(0)C_\lambda(0) = +1$$

$$A_\lambda(0)B_\lambda(\pi/2)C_\lambda(0) = +1$$

$$A_\lambda(0)B_\lambda(0)C_\lambda(\pi/2) = +1$$

$$\rightarrow A_\lambda(\pi/2)B_\lambda(\pi/2)C_\lambda(\pi/2) = +1$$

However, the formula $A_\lambda(\pi/2)B_\lambda(\pi/2)C_\lambda(\pi/2) = -1$, where in $a \sim c$, at least one of the conditions is untrue.

Three particles $\{a, b, c\}$ have two stationary states $0, 1(a, a', b, b', c, c')$ through a single particle stationary state and can form $2^3 = 8$ three-particle stationary states:

$$\{|111\rangle, |110\rangle, |101\rangle, |100\rangle, |011\rangle, |010\rangle, |001\rangle, |000\rangle\}$$

The ternary entangled quantum state of GHZ is expressed as

$$|GHZ\rangle = |111\rangle + |000\rangle \tag{117}$$

$|GHZ\rangle$ is the superposition state of two three-particle eigenquantum stationary states $|111\rangle$, $|000\rangle$. This three-particle system is both $|111\rangle$ and $|000\rangle$.

The GHZ entangled state is proven to be the most entangled state among the three particle quantum states:

$$|W\rangle = |001\rangle + |010\rangle + |100\rangle \tag{118}$$

From the GHZ state and the W-state of the three-particle system, it can be generalized from the three-qubits, three-element qubit system to the n-qubits, n-bit qubit system.

$$|GHZ\rangle_n = |11\cdots1\rangle + |00\cdots0\rangle \tag{119}$$

$$|W\rangle_n = |100\cdots00\rangle + |010\cdots00\rangle + \cdots + +|000\cdots01\rangle \tag{120}$$

Corollary 18 (Noninequalities Express Multiquantum Entanglement). *For multiquantum state detection mechanisms, such as GHZ ternary entangled state detection, the noninequalities discriminant condition can be used to effectively detect the composite superposition feature of multientangled states.*

11.8 Hardy Theorem

For the two-particle entanglement interaction model, the paradox [57] derived by Hardy in 1992 is a landmark for the description of quantum entanglement measurements. The experimental principle model is illustrated in Fig. 13.

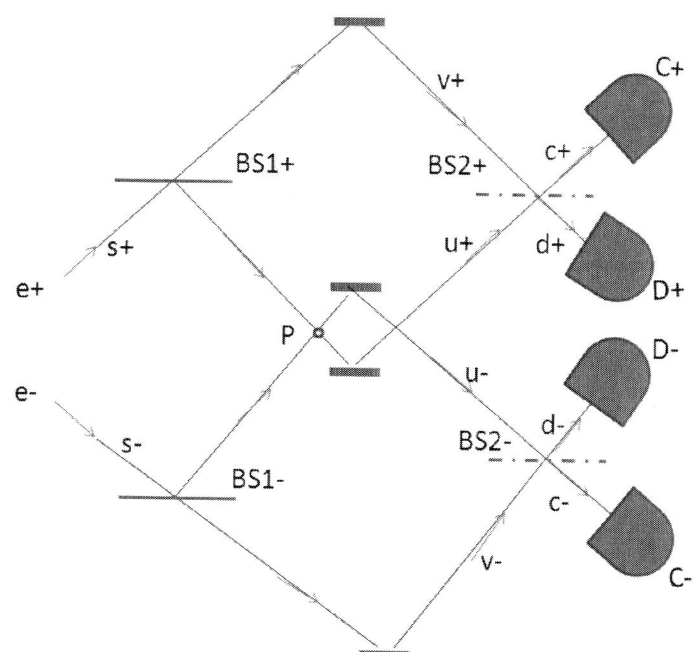

Figure 13. Hardy two-particle interaction.

11.8.1 Principle of Experiment

The ideal experimental setup consists of two sets of Mach-Zehnder interferometers: $MZ^{\pm}$, one for detecting positrons MZ^{+}, and the other for detecting electrons MZ^{-}. Each group of interferometers $MZ^{\pm}$ has two input signals: $s^{\pm}$, two pairs of internal signals: $v^{\pm}$ and $u^{\pm}$, and two pairs of output signals: $c^{\pm}$ and $d^{\pm}$, and then correspondingly output two pairs of signals through the detector: $c^{\pm} \to C^{\pm}$ and $d^{\pm} \to D^{\pm}$. The conjugate semitransparent mirror $BS2^{\pm}$ is a removable part of the system.

When a set of electron-positron pairs are simultaneously generated, corresponding pairs are injected into the associated interferometer. That is, the positrons follow the interferometer MZ^{+} path u^{+}, and the electrons follow the interferometer MZ^{-} path u^{-}. Then, the two particles will be at P, and the probability of each annihilation is equal to 1. Such interactions can be expressed as

$$|u^{+}\rangle|u^{-}\rangle \to |\gamma\rangle \text{ (paired particle annihilation)} \tag{121}$$

where the state $|u^{\pm}\rangle$ is the path $u^{\pm}$ of positrons and negatives, and $|\gamma\rangle$ is the radiation due to annihilation. As a possible interaction between two particles, positrons and electrons can be detected from the detector $D^{\pm}$.

11.8.2 Local Reality Check

The role of the conjugate semitransparent mirror $BS1^{\pm}$ can be expressed as

$$|s^{\pm}\rangle \to \frac{1}{\sqrt{2}}(i|u^{\pm}\rangle + |v^{\pm}\rangle) \text{ (conjugate semitransparent mirror } BS1^{\pm}) \tag{122}$$

The role of the conjugate semitransparent mirror $BS2^{\pm}$ can be expressed as

$$|u^{\pm}\rangle \to \frac{1}{\sqrt{2}}(|c^{\pm}\rangle + i|d^{\pm}\rangle) \text{ (conjugate semitransparent mirror } BS2^{\pm}) \tag{123}$$

$$|v^{\pm}\rangle \to \frac{1}{\sqrt{2}}(i|c^{\pm}\rangle + |d^{\pm}\rangle) \tag{124}$$

The effect of removing the conjugate semitransparent mirror $BS2^{\pm}$ is expressed as

$$|u^{\pm}\rangle \to |c^{\pm}\rangle \tag{125}$$

$$|v^{\pm}\rangle \to |d^{\pm}\rangle \text{ (Remove } BS2^{\pm}) \tag{126}$$

The initial state of the system is

$$|s^{+}\rangle|s^{-}\rangle \text{ (Initial state)} \tag{127}$$

The state after the action of the conjugate semitransparent mirror $BS1^{\pm}$ is expressed as

$$\frac{1}{2}(i|u^{+}\rangle + |v^{+}\rangle)(i|u^{-}\rangle + |v^{-}\rangle) \text{ (}BS1^{\pm}\text{ After interaction)} \tag{128}$$

After another P point, the state becomes

$$|e^+\rangle|e^-\rangle \to \frac{1}{2}(-|\gamma\rangle + i|u^+\rangle|v^-\rangle + i|v^+\rangle|u^-\rangle + |v^+\rangle|v^-\rangle) \text{ (Status after } P \text{ points)} \quad (129)$$

After removing the two sets of conjugate semitransparent mirrors $BS2^+$ and $BS2^-$, the state is expressed as

$$\frac{1}{2}(-|\gamma\rangle + i|c^+\rangle|d^-\rangle + i|d^+\rangle|c^-\rangle + |d^+\rangle|d^-\rangle) \text{ (remove } BS2^{\pm} \text{ state)} \quad (130)$$

Retain the conjugate semitransparent mirror $BS2^+$; after removing $BS2^-$, the state is expressed as

$$\frac{1}{2\sqrt{2}}(-\sqrt{2}|\gamma\rangle - |c^+\rangle|c^-\rangle + 2i|c^+\rangle|d^-\rangle + i|d^+\rangle|c^-\rangle) \text{ (Retain } BS2^+ \text{ Remove } BS2^-) \quad (131)$$

Removing $BS2^+$ and retaining the conjugate semitransparent mirror $BS2^-$, the state is expressed as

$$\frac{1}{2\sqrt{2}}(-\sqrt{2}|\gamma\rangle - |c^+\rangle|c^-\rangle + i|c^+\rangle|d^-\rangle + 2i|d^+\rangle|c^-\rangle) \text{ (Remove } BS2^+ \text{ Retain } BS2^-) \quad (132)$$

Place the conjugate semitransparent mirror

Put $BS2^{\pm}$, use the formula (123), (124) to bring the formula (129) to express the final state (123), (124) in equation (129) for final state

$$\frac{1}{4}(-2|\gamma\rangle - 3|c^+\rangle|c^-\rangle + i|c^+\rangle|d^-\rangle + i|d^+\rangle|c^-\rangle - |d^+\rangle|d^-\rangle) \text{ (Final State)} \quad (133)$$

Let λ be a hidden variable, 0 for particles passing through the conjugate semitransparent mirror, and ∞ for removing the conjugate semitransparent mirror. If positive and negative electrons are detected from $D^{\pm}$ via $BS2^{\pm}$: $D^{\pm}(0,\lambda) = 1$, not detected: $D^{\pm}(0,\lambda) = 0$. If positive and negative electrons are detected from $C^{\pm}$ after removing $BS2^{\pm}$: $C^{\pm}(\infty,\lambda) = 1$, not detected: $C^{\pm}(\infty,\lambda) = 0$

$$C^-(\infty,\lambda)C^-(\infty,\lambda) = 0 \text{ (Equation 130)} \quad (134)$$

There is no $|c^-\rangle|c^+\rangle$ detection in formula (130).

If the positron is detected by D^+,

$$\text{if } D^+(0,\lambda) = 1 \text{ then } C^-(\infty,\lambda) = 1 \text{ (Equation 131)} \quad (135)$$

If the positron is detected by D^-,

$$\text{if } D^-(0,\lambda) = 1 \text{ then } C^+(\infty,\lambda) = 1 \text{ (Equation 132)} \quad (136)$$

$$\to D^+(0,\lambda)D^-(0,\lambda) = 1 \text{ (Experiment 1/16)} \quad (137)$$

$$\to C^-(\infty,\lambda)C^-(\infty,\lambda) = 1 \text{ (The result contradicts the formula 134)} \quad (138)$$

The contradiction formed by the derivation shows the conflict between local realism and quantum mechanics. In terms of the form of comparison, the intrinsic structure is similar to the GHZ theorem and does not require inequality expression and only needs to use 1/16 of the experimental results in the state space to obtain the comparison effect.

11.8.3 Lorentz Invariance Detection

Whether the realistic theories have Lorentz invariance is another question that needs to be explored. EPR [33] is used to describe a sufficient condition for an element of physical reality: if we can deterministically predict (with probability 1) a quantitative measurement in the system, then there is an element in the real system that corresponds to the physical quantity and can provide a measurement with a numerical value equal to the predicted quantity.

In quantum mechanical representation, this kind of sufficient condition can be described as: If the characteristic state of a system transformation operator $\hat{A}$ is $|a\rangle$, that is, $\hat{A}|a\rangle = a|a\rangle$, then $[A] = a$ does not even need to be measured, where $[A]$ is the existing element value corresponding to the observable A.

Some observations satisfy Lorentz invariance when the measurement framework is independent of quantum mechanics and classical simulations. For example, the results of the measurement are used to confirm that the determination of whether a particle is within a given area does not depend on the observer's reference position. From this point of view, it is possible to make a judgment based on whether the particles are contained in the specified area, even without any quantitative measurement, using the relevant environmental conditions.

Asserts that the real elements are not expected to depend on the observer's frame of reference, are satisfying the Lorentz invariance in classical physics. Expect the same conclusion to hold for quantum mechanics. This gives the necessary condition of Lorentz invariance: the observable Lorentz invariant itself corresponding to reality also needs to have the Lorentz invariance feature.

In the following, according to the two sets of Lorentz invariance conditions, starting from model shown in Fig. 13, the contradiction is deduced.

11.8.4 Lorentz Invariance Derivation Process

In Fig. 13, two sets of $BS2^{\pm}$ conjugate semitransparent mirrors are placed, and the observable quantity can be expressed as

$$\hat{U}^{\pm} = |u^{\pm}\rangle\langle u^{\pm}| \text{ (forms the unitary transformation operator } U) \quad (139)$$

$$\hat{U}^{\pm}|u^{\pm}\rangle = |u^{\pm}\rangle \rightarrow [U^{\pm}] = 1 \text{ (reality condition)} \quad (140)$$

$$\hat{U}^{+}\hat{U}^{-}|u^{+}\rangle|u^{-}\rangle = |u^{+}\rangle|u^{-}\rangle \rightarrow [\hat{U}^{+}\hat{U}^{-}] = 1 \quad (141)$$

$$\hat{U}^{+}\hat{U}^{-}|u^{+},u^{-}\rangle_{\perp} = 0 \rightarrow [\hat{U}^{+}\hat{U}^{-}] = 0 \text{ (orthogonality condition)} \quad (142)$$

Here, $|u^{+},u^{-}\rangle_{\perp}$ is any state vector orthogonal to $|u^{+}\rangle|u^{-}\rangle$. According to reality condition,

$$\text{if } [U^{+}][U^{-}] = 1 \rightarrow [U^{+}U^{-}] = 1 \text{ (if electron and positron arrive at the same time)} \quad (143)$$

However, considering a class of reference coordinates F^+, the positron arrives at $BS2^-$ before the electron, and the positron passes through $BS2^+$, but the electron fails to pass through $BS2^+$, and the system state is

$$\frac{1}{2\sqrt{2}}(-\sqrt{2}|\gamma\rangle - |c^+\rangle|u^-\rangle + 2i|c^+\rangle|v^-\rangle + i|d^+\rangle|u^-\rangle) \tag{144}$$

If the positron at D^+ also precedes the electron to reach $BS2^-$, the electron state is $|u^-\rangle$.

$$\text{If an electron is detected at } D^+\text{, then } [U^-] = 1 \tag{145}$$

On the other hand, considering the case where F^- electrons precede positrons, using the principle of symmetry, we can obtain

$$\text{If a positron is detected at } D^-\text{, then } [U^+] = 1 \tag{146}$$

When two measurements take place simultaneously, the state before the particle reaches $BS2^\pm$ is determined by the formula (133). However, this state is orthogonal to $|u^+\rangle|u^-\rangle$, so

$$\text{For all experiments } [U^+U^-] = 0 \tag{147}$$

The observed quantities under investigation all satisfy the Lorentz invariance, and formulas (145)–(147) are independent of the coordinate reference system, so they can be compared with each other. Combined with the previous analysis, consider an experiment where D^+ and D^- detect states satisfying the formula (133) appearing in the 1/16th experiment. The results are obtained for these experimental formulas (145)–(147):

$$[U^+U^-] = 1 \tag{148}$$

Easy to check: the formula (147) contradicts (148)!

It can be asserted that based on the constraint that the observed quantity itself has Lorentz invariance, it is impossible to form quantum mechanics using reality theory.

Such assertions deny all arbitrary quantum explanations that expect microscopic particles to have actual measured orbits, including de Broglie–Bohm quantum interaction explanations.

Corollary 19 (Hardy Theorem—Double Quantum Entanglement). *Hardy's theorem strictly describes the interaction of two-particle entanglement. Using modern optical equipment, such as semitransparent mirrors and fully transparent mirrors, under the transformation of complex functions, the series may result in an impossible procedure analysis.*

11.9 Cluster State Formula

From the perspective of frontier exploration, [58, 59] proposed a cluster state in 2001. Cluster states represent a more general entangled state structure as a scalable resource for multiqubit entangled states (multiquantum GHZ states). From an experimental point of view, these states can be prepared in optical lattices or similar systems, which not only include the entangled state properties of the GHZ state and the W state but also have maximum connectivity and entanglement stability, which are rarely destroyed by local operations, which is convenient for experiments and widely used in quantum communication channels. The cluster state can be written as

$$|\psi_N\rangle = \frac{1}{2^{N/2}} \otimes_{a=1}^{N} (|0\rangle_a \rho_z^{(a+1)} + |1\rangle_a) \tag{149}$$

$\rho_z^{(N+1)} \equiv 1$. For $N = 2$,

$$\begin{aligned} |\psi_2\rangle &= \frac{1}{2}(|0\rangle_1 \rho_z^{(2)} + |1\rangle_1)(|0\rangle_2 + |1\rangle_2) \\ &= \frac{1}{2}(|0\rangle_1(|0\rangle_2 - |1\rangle_2) + |1\rangle_1(|0\rangle_2 + |1\rangle_2)) \end{aligned}$$

is a maximally entangled state.

In [60]–[63], some schemes are provided to separate the four-particle and five-particle cluster states, and the concept, properties, preparation, and unitary transformation of the entangled cluster state are introduced in detail.

Corollary 20 (Cluster State Quantum Entanglement). *For more complex entanglement state representations, the cluster state representation formula provides further extended modes of many types of quantum entanglement effects.*

11.10 Classic Model and Method of Quantum Information

The fundamental models and methods of quantum information are illustrated in Table 3.

12. Comprehensive Analysis

Multivariate complex functions provide convenient expressions for the representation of vectors and matrices. Different matrix groups are represented as frontier physical equation descriptions, which lay the foundation for mathematical descriptions. The Jones vector and matrix representations, which are commonly used in physical optical analysis and design, use complex variable functions to perform optical polarization measurements and provide measurement tools. In the basic theories of various modern physical measurements, overall invariance, including gauge invariance, occupies the core position.

Various complex geometric integral invariants represent structures, and different analytical frameworks have been developed for complex dynamic systems: lattice theory, system stability, topological gravity research, etc. From the level of principles, models, and methods, symmetries and conserved quantities are identified between transformation groups and domain disciplines, which support modern scientific theories and cutting-edge technologies and assist in the in-depth exploration of unknown problems. These tools and methods will provide the necessary technical and theoretical support for solving the mystery of quantum entanglement.

The discriminative mode provided by Wootters is worth paying attention to for choosing a suitable quantum entangled state for a complex function. First, it is described that two quantum entangled states cannot be expressed as a direct product form, and the two components are expressed as a direct product mode that satisfies the reciprocity condition or satisfies the complex conjugate operator under 90-degree rotation. In other words, quantum entangled states must be clusters of nonorthogonal relations, closely related to the selected feature. This type of discrimination mode plays an auxiliary discriminative role in properly checking whether the newly constructed quantum entanglement theory and technology are reasonable.

Through quantum state tensor product expansion, various forms of entangled state structures, such as multiquantum entangled states, Bell quantum states, and cluster

Table 3. Typical representation of quantum state/quantum entanglement

Representation	Characterization	Formulas	Quantitative Features
Bell State	Combination Probability	$\lvert Pxz - Pzy\rvert \leq 1 + Pxy$	Inequality
CHSH–Bell State	Combination Probabilities	$\lvert P(a_1, b_1) + P(a_1, b_2)+$ $P(a_2, b_1) - P(a_2, b_2)\rvert \leq 2$	Inequality
Reflection State	Semitransmission	$A = C + iD$	Equation
Input: $\{A, B\}$ Output: $\{C, D\}$	2-way Semitransmission	$\begin{cases} A = C + iD \\ B = D + iC \end{cases}$	Conjugate Pair
Complex Function	Wave Function	$\Psi = \frac{1}{\sqrt{2}}(\psi_L + i\psi_R)$	Wave Function
Single Entanglement	Complex Function Unit Circle	$\lvert\text{qubit}\rangle = a\lvert 0\rangle + b\lvert 1\rangle$ $\lvert a\rvert^2 + \lvert b\rvert^2 = 1;\ a, b \in \mathbf{C}$	Complex Function Complex Plane
Double Entanglement	Complex Functions Time Reversal	$\lvert\psi\rangle \neq \lvert\psi_1\rangle \otimes \lvert\psi_2\rangle$ $[X, Y] \neq 0$	Non-Direct Product Nonorthogonal
Noninequality Bell's Theorem	Spatial Angle	$E^{\phi}(\hat{n}_1, \hat{n}_2, \hat{n}_3, \hat{n}_4) =$ $-\cos(\phi_1 + \phi_2 - \phi_3 - \phi_4)$	Projection Combination
Multiphoton State	Multiphoton Entangled	$E^{\psi}_{def}(\phi_1, \phi_2, \phi_3) =$ $\sin(\phi_1 + \phi_2 + \phi_3)$ $\lvert GHZ\rangle_n = \lvert 11\cdots 1\rangle + \lvert 00\cdots 0\rangle$	Pairs of Combinations
Cluster Entanglement	Cluster Entanglement	$\lvert\psi_N\rangle = \frac{1}{2^{N/2}} \otimes_{a=1}^{N} (\lvert 0\rangle_a \rho_z^{(a+1)} + \lvert 1\rangle_a)$	Grouping Combination
Hardy's Theorem	Paired Interferometers	$\{e^+, e^-\} \to \{C^-, C^+, D^-, D^+\}$	2 Input 4 Output
Quantum State Vector	Qubit State	$\lvert 0\rangle = \begin{bmatrix} 1 \\ 0 \end{bmatrix},\ \lvert 1\rangle = \begin{bmatrix} 0 \\ 1 \end{bmatrix}$	Pairwise Vectors
	Qubit Tensor Product	$\lvert 0\cdots 0\rangle_n \equiv \lvert 0\rangle \otimes \lvert 0\rangle_{n-1} = \begin{bmatrix} 1 \\ 0 \\ \vdots \\ 0 \end{bmatrix}$	$n \to 2^n$ Phase Space
Hadamard Matrix	Fundamental Matrix	$H_2 = \begin{bmatrix} 1 & 1 \\ 1 & -1 \end{bmatrix},$	2×2 Matrix
	Matrix Tensor Product	$H_{2^m} = H_2 \otimes H_{2^{m-1}}$	$m \to 2^m$ Matrix Space

entangled states, can be properly analyzed. In the Hadamard matrix representation of tensor operator expansion, the number of $\{+,-\}$ involved in the matrix can be balanced under the condition of four blocks.

The main content of this chapter is to bring together the analysis, detection, and representation mechanisms of various quantum measurement paradoxes. Based on laser technology and various mirror reflection technologies, including semitransparent and total reflection films, the complex function measurement expressions are transformed through different optical paths, and the measurement process can be expressed through a series of standard transformations and their complex interactions. Regardless of the EPR paradox, Bell quantum state, CHSH–Bell inequality, single quantum entanglement, noninequality theorem, multiphoton entanglement state, Hardy's theorem, etc., from the principle of action to the main part of the detailed proof, the complex variable function transformation derived by using mirror relationships plays a central role in the analysis process of measurement. The cluster state representation of composite states takes the multiquantum entangled state representation structure to a new level of measurement.

The typical representations of the various quantum states and quantum entangled states involved in the chapter are summarized in Table 3.

Various basic complex function transformation operators and typical classic conclusions are summarized in the text from **Lemmas 1–22 and Corollaries 1–20** to facilitate the orderly reference for subsequent strict proofs.

References

[1] 莫里斯 克莱因,古今数学思想,第3,4册,上海科学技术出版社 2002 M. Kline. *Mathematical Thought from Ancient Tomodern Times*, Oxford University Press, 1972.

[2] R. Penrose. *The Road to Reality - A Complete Guide to the Laws of the Universe*, Vintage Book, 2005.

[3] P. J. Nahin. *An Imaginary Tale: The Story of $\sqrt{-1}$*, Preston University Press, 1998.

[4] M. A. Nielsen and I. L. Chuang. *Quantum Computation and Quantum Information*, Cambridge University Press & Tsing Hua University Press, 2015.

[5] L. H. Kauffman. Majarana Fermions and representations of the braid group. *Int. J. Mod. Phys. A* 33(23), 2018, 1830023. http://doi.org/10.1142/S0217751X18300235.

[6] C. N. Yang. *Elementary Particles*, Princeton University Press, 1962.

[7] Yang-Mills Equations. http://en.wikipedia.org/wiki/Yang-Mills_equations.

[8] BSPT Instanton. http://en.wikipedia.org/wiki/BSPT_instanton.

[9] H. Weyl. *Space Time Matter* (Translated by H. L. Brose), Dover, 1950.

[10] 华罗庚,华罗庚文集 - 多复变函数论卷 *II*, 科学出版社 2013 Hua Luogeng. *Hua Luogeng's Collected Works - Theory of Multiple Complex Variables, Vol. II*, Science Press, 2013.

[11] 张奠宙 王善平,陈省身文集,华东师范大学出版社 2002 Zhang Dianzhou and Wang Shanping. *Chen Shishen's Essays*, East China Normal University Press, 2002.

[12] R. S. Varga. *Matrix Iterative Analysis*, Prentice-Hall Inc., 1962.

[13] V. I. Arnold. *Mathematical Methods of Classical Mechanics*, Springer-Verlag, 1978.

[14] R. Penrose. *The Emperor's Mind: Concerning Computers, Minds and the Law of Physics*, Penguin, 1989.

[15] T. Siegfried. *The Bit and the Pendulum: From Quantum Computing to M Theory - The New Physics of Information*, John Wiley & Sons, 2000.

[16] C. Seife. *Decoding the Universe: How the Science of Information is Explaining Everything in the Cosmos, from our Brains to Black Holes*, Penguin, 2006.

[17] 吉姆 巴戈特 著;徐彬 于秀秀译, 量子通史 - 量子物理史上的40个重大时刻,中信出版社 2021。J. Baggott. *The Quantum Story in 40 Moments*, Oxford University Press, 2011.
[18] 许天周 李炳照, 线性正则变换及其应用,科学出版社 2013 Xu Tianzhou and Li Bingzhao. *Linear Regular Transformation and Its Application*, Science Press, 2013.
[19] 马天, 从数学观点看物理世界 - 基本粒子与统一场理论, 科学出版社 2015 Ma Tian. *Viewing the Physical World from a Mathematical Perspective - Elementary Particles and Unified Field Theory*, Science Press, 2015.
[20] G.'t Hooft. *The Cellular Automaton Interpretation of Quantum Mechanics*, Springer, 2016.
[21] 杨振宁,葛墨林,何杨辉编,拓扑与物理,江西科学技术出版社 2021 Yang Chen-ning, Ge Molin and He Yanghui. *Topology and Physics*, Jiangxi Science and Technology Press, 2021.
[22] 艾小文,数论问题 - 指数和的均值分析与研究,电子工业出版社 2021。Ai Xiaowen, *Problems in Number Theory - Mean Analysis and Research of Exponential Sums*, Electronic Industry Press, 2021.
[23] 艾小文, 数论问题 - 指数和的均值分析与研究,电子工业出版社 2021。
[24] *Roger Penrose. The Emperor's Mind: Concerning Computers, Minds and the Law of Physics*, Penguin, 1989.
[25] T. Siegfried. *The Bit and the Pendulum: From Quantum Computing to M Theory - The New Physics of Information*, John Wiley & Sons, 2000.
[26] L. Smolin. *Time Reborn: From the Crisis in Physics to He Future of the Universe*, Penguin, 2004.
[27] H. Weyl. *Space Time Matter* (Translated by H. L. Brose), Dover, 1950.
[28] L. Z. Xu, M. S. Jiang and Z. Q. Zhu. *Combinatorial Mathematics of Computation*, Shanghai Science & Technology Press, 1983 (in Chinese).
[29] J. H. van Lint and R. M. Wilson. *A Course in Combinatorics*, 2nd edition, Cambridge University Press, 2001.
[30] L. X. Wang. *An Elementary Treatise on Combinations*, Harbin Institute of Technology Press, 2012 (in Chinese).
[31] B. Riemann. *Foundations for a General Theory of Functions of a Complex Variable*. Collected papers of Bernhard Riemann (translated by R. C. Baker, C. Christensen and H. Orde), Kendrick Press, 2004, 1–41.
[32] 张永德,量子信息物理原理,科学出版社 2005.
[33] A. Einstein, B. Podolsky and N. Rosen. Can quantum-mechanical description of physical reality be considered complete? *Phys. Rev.* 47, 1935, 777–780.
[34] G. Birkhoof. *Lattice Theory*, The American Mathematical Society, 1948.
[35] G. Birkhoof. Von Neumann and lattice theory, momories of the American mathematical society, no. 10, 1955.
[36] R. J. Simpson. Convex lattice polygons on mimimum area. *Bull. Austral. Math. Soc.* 42, 1990, 353–367.
[37] A. Day. Doubling constructions in lattice theory. *Can. J. Math.* 44(2), 1992, 252–269.
[38] M. Stern. *Semimodular Lattices Theory and Applications*, Cambridge University Press, 1999.
[39] 许光午,王小云, 格的计算和密码学应用, 中国科学:数学 2020年 第50卷第10期 1417–1436。Xu Guangwu and Wang Xiaoyun. Computation and cryptography applications of lattice. *Sci. China Math.* 50(10), 2020, 1417–1436.
[40] W. K. Wootters. Entenglement of formation of an arbitrary state of two qubits. *Phys. Rev. Lett.* 80, 1998, 2245.
[41] J. von Neumann. Allgemeine Eigenwerttheorie Hermitescher Funktionaloperatoren. *Math. Ann.* 102, 1930, 49–131.
[42] M. A. Nielson and I. L. Chuang. *Quantum Computation and Quantum Information*, 10th Anniversary Edition, Cambridge University Press and Tsinghua University Press, 2015.
[43] Entanglement. https://quantiki.org/wiki/entanglement.
[44] 雷奕安, 量子纠缠究竟是怎么回事? http://blog.sciencenet.cn/blog-268546-1102443.html.
[45] Complex number. https://enwikipedia.org/wiki/Complex_number.
[46] 熊锡金, 泛复变函数及其在数学与物理中的应用,东北师范大学出版社 1988.
[47] General linear group. https://en.wilipedia.org/wiki/General_linear_group.

[48] 克里斯多佛门罗,关于量子力学,量子计算机能告诉我们什么,实在终极之问第8章,200-216, 湖南科学技术出版社 2018.
[49] J. S. Bell. On the Einstein-Podolsky-Rosen paradox. *Phys.* 1, 1964, 195–200.
[50] 张天蓉,走进量子纠缠-7- 贝尔不等式. http://blog.sciencenet.cn/blog-677221-537543.html.
[51] J. F. Clauser, M. A. Horne, M. A. Shimony, R. A. Holt, D. M. Greenberger, M. A. Horne, A. Shimony and A. Zeilinger. Proposed experiment to test local hidden-variable theories. *Phys. Rev. Lett.* 23, 880. http://doi.org/10.1103/physRevLett.23.880.
[52] S. J. Freedman and J. S. Clauser. Experimental test of local hidden-variable theories. *Phys. Rev. Lett.* 28, 1972, 938–941.
[53] A. Aspect, J. Dalibard and G. Roger. Experimental test of Bell's inequalities using time-varying analyzers. *Phys. Rev. Lett.* 47, 1982, 1804–1807.
[54] M. Redhead. *Incompleteness, Nonlocality and Realism*, Clarendon, 1987, 107–113.
[55] D. M. Greenberger, M.A. Horne and A. Zeilinger. Going beyond Bell's theorem. In *Bell's Theorem, Quantum Theory and Conceptions of the Universe*, edited by M. Kafatos, Kluwer Academic, 1989, 73–76.
[56] D. M. Greenberger, M. A. Horne, A. Shimony and A. Zeilinger. Bell's theorem without inequalities. *Am. J. Phys.* 58, 1990, 113. http://doi.org/10.1119/1.16243.
[57] L. Hardy. Quantum mechanics, local realistic theories and Lorentz-invariant realistic theories. *Phys. Rev. Lett.* 68(20), 1992, 2981–2984.
[58] H. J. Briegel and R. Raussendorf. Persistent entanglement in arrays of interacting particles. *Phys. Rev. Lett.* 86(5), 2001, 910.
[59] R. Raussendorf and H. J. Briegel. A one-way quantum computer. *Phys. Rev. Lett.* 86(22), 2001, 5188.
[60] S. Muralidharan and P. K. Panigrahi. Quantum-information splitting using multipartite cluster states. *Phys. Rev. A* 78(6), 2008, 062333.
[61] 张小龙. 多粒子纠缠态的制备[D].中国科学院研究生院(武汉物理与数学研究所), 2007.
[62] 连士曼. 四粒子Cluster态的制备和离子阱中二粒子一般态的概率隐形传输[D].河北师范大学, 2008.
[63] 樊仁杰. 通过腔Input-Output过程制备Cluster态及其相关应用[D].国防科学技术大学, 2009.
[64] Kaspar Wessel: On the Analytical Representation of Direction. An attempt Applied Chiefly to Solving Plane and Spherical Polygons. 1797. Translated by Flemming Damhus. Introductory chapters by Bodil Branner, Nils Voje Johansen, and Kirsti Andersen. Edited by Bodil Branner and Jesper Liitzen. Matematiskfysiske Meddelelser 46:1. Det Kongelige Danske Videnskabernes Selskab, Copenhagen 1999.

Chapter 6

Entanglement Foundation of Quantum Optics on Conjugate Phase Space

Three Measuring Models of Quantum Interactions on Conjugate Complex Vector Functions

Jeffrey Zheng[1]*, Fan Peng[2], Rui Wang[2], and Youping Li[3]
[1]Conjugate System Research, Melbourne, Australia
[2]Yunnan University, Kunming, China
[3]Chinese Academy of Engineering Physics, Beijing, China

Abstract

In this chapter, the four-region attribute property of the conjugate complex function is used to expand the inner product operation to cover the Gauss quaternary complex number unit and systematically solve the core problem of the basic theory of quantum information to provide n cobit phase spaces that satisfy the Wigner phase space condition. The sufficient and necessary conditions are 2^{2n} (full space) instead of the recognized 2^n. In quantum interaction, the state transition of the synchronized time structure $\{\{1, i, -1, -i\}, \{1, -i, -1, i\}\}$ is the conjugate basic time order to provide iterative expression formulas for the twelve entangled states of the eight basic states that interact synchronously in the quantum measurement process, to determine the conjugate orthogonal transform operation and complex function exchange group qualified effective scope, and to establish geometric transformation rules between the complex number plane and display plane. Using the analysis results of quantum optics and photonic theory, the experimental support evidence predicts that the output results of the observable fine strong interaction correspond to three types of topological clustering structures 25, 73, and 81, and the hyperfine measurement mechanism further improves the ability of the 25 and 73 groups to distinguish between two different distributions. Lemmas 1–5, Assumptions 1–6, Predictions 1–6, Theorems 1–16, and Corollaries 1–16 are provided.

Keywords: Quantum phase space, wigner phase space, conjugate complex vector function, complete phase space, measurement process, four attributes, conjugate algebra, strong interaction, quantum optics

---

*Corresponding Author's Email: conjugatelogic@yahoo.com.

In: Conjugate Construction of Quantum Optics
Editors: Jeffrey Zheng
ISBN: 979-8-88697-651-9

Truth is ever to be found in the simplicity, and
not in the multiplicity and confusion of things.
– Isaac Newton

Analysis does not owe its truly significant successes of the last century
to any mysterious use of $\sqrt{-1}$, but to the quite natural circumstances
that one has infinitely more freedom of mathematical movement
if he lets quantities very in a plan instead of only on a line.
– Leopold Kronecker

It will serve as a prototype enabling us to write down corresponding quantities
involving the exchange of two or more quanta between two electrons or
the interaction of an electron with itself.
– Richard Feynman

1. Research Purposes

The 2022 Nobel Prize in Physics is awarded to three scientists, Aspect, Clauser, and Zeilinger, who have made fundamental contributions to the fundamental theory of quantum entanglement and quantum optical measurement technology, but the basic quantum optical measurement theory and the principle are still foggy.

In this study, using conjugate complex vector functions and the four-element attribute domain division system for the quantum interaction and entanglement measurement process, a basic description system for quantum optics and light quantization is established.

The chapter is divided into three parts. The first part discusses the core matrix structure of quantum interaction. The second part briefly summarizes a series of the most important results published by the international quantum entanglement research team based on modern quantum optics and photon quantum cutting-edge technology during 2002–2022. Pictures and text show the basic representation of photon quanta, the description of quantum entanglement, and the experimental model and measurement mechanism, as well as typical measurement results.

In the third part, based on the conjugate complex vector function, aiming at the basic representation of quantum optics and light quanta, the description principle of quantum entanglement measurement, the measurement process and the result display system, the property complex plane structure, and quantum basic state corresponding to the complete complex vector function are used to support the necessary and sufficient conditions for establishing a self-consistent description of space and time in the quanta optical measurement system, and a formal description of the measurement paradox-free for the basis and application of quantum entanglement.

2. Quantum Interaction of Core Matrix Structure Representation

2.1 *Wigner's Complete Phase Space Puzzle*

In the field of basic quantum information research, Wootters et al. regard n qubits as variables, and Wigner expresses the 2^{2n} phase space [1] for continuous quantum systems.

Compared with the classical Turing machine, the quantum Turing machine adopts the vector description. A single qubit needs to choose $|0\rangle, |1\rangle$ as a pair of complex vectors to

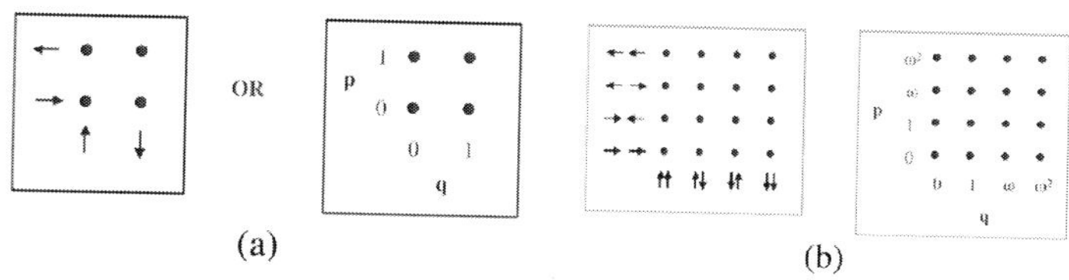

Figure 1. 2004 qubit structure: (a) 2×2, (b) 4×4.

form the basic representation of quantum information. The selected vector is constrained in the upper half-plane. Fig. 1(a) shows the 2×2 matrix formed by two complex conjugate variables $\{p, q\}$, and in Fig. 1(b), the complex conjugate itself is expressed as a 4×4 matrix representation after pairwise combination.

Lemma 1 (n qubit phase space 2^n). *The discrete form partially satisfies the Wigner condition to be a classical quantum information phase space, and the combined state of n qubits is a 2^n phase space.*

Can the 2^{2n} dimension of the Wigner phase space be fully expressible in a discrete geometric environment?

This question has become a mystery in basic research on quantum information.

However, this type of basic topic has almost nothing to do with the frontier experiments of quantum entanglement and is far from the core topics that frontier researchers generally pay attention to.

In the mainstream frontier research field of quantum information, the conclusion of Wootters is generally accepted, and the phase space corresponding to n qubits is 2^n.

2.2 Basic Transformation Group

The Klein group provides the smallest set for the noncommutative structure. Based on the symmetry and anti-symmetry properties, Hermite used the matrix to express the complex number pair to provide the matrix representation basis. The Pauli 2×2 matrices and the Dirac 4×4 matrices are expressed as the core part of quantum mechanics formulas. From the perspective of structural generation, the internal connection of the Klein, Hermite, Pauli, and Dirac series of studies is distinct, and the relationship between extended structures such as conjugate algebra is represented.

2.3 Conjugate Klein Groups

The conjugated Klein group is divided into two parts $\{K, \tilde{K}\}$, the fundamental group K, the conjugate group $\tilde{K}$,

$$K = \left\{ \begin{pmatrix} 1 & 0 \\ 0 & 1 \end{pmatrix}, \begin{pmatrix} -1 & 0 \\ 0 & 1 \end{pmatrix}, \begin{pmatrix} -1 & 0 \\ 0 & -1 \end{pmatrix}, \begin{pmatrix} 1 & 0 \\ 0 & -1 \end{pmatrix} \right\} \tag{1}$$

$$\tilde{K} = \left\{ \begin{pmatrix} 0 & 1 \\ 1 & 0 \end{pmatrix}, \begin{pmatrix} 0 & -1 \\ 1 & 0 \end{pmatrix}, \begin{pmatrix} 0 & -1 \\ -1 & 0 \end{pmatrix}, \begin{pmatrix} 0 & 1 \\ -1 & 0 \end{pmatrix} \right\} \tag{2}$$

The cluster of complex functions is $K(x), x = (p,q)$, and the cluster of conjugate complex functions is $\tilde{K}(x) = K(\tilde{x}), \tilde{x} = (q,p)$. Each cluster consists of four elements,

$$K(x) = \{(p,q),(-p,q),(-p,-q),(p,-q)\} \tag{3}$$

$$K(\tilde{x}) = \{\widetilde{(p,q)},\widetilde{(-p,q)},\widetilde{(-p,-q)},\widetilde{(p,-q)}\} \tag{4}$$

Lemma 2 (Conjugate Klein Complex Group). *For any complex number $x = (p,q)$, the conjugate Klein complex group is expressed as two clusters $\{K(x), K(\tilde{x})\}$, each containing four elements, forming a rectangle spanning four quadrants in the complex plane.*

2.4 Hermite Matrices

2.4.1 Matrix Representation of Complex Numbers

Let $x = (p,q) \to x = \begin{pmatrix} p & -q \\ q & p \end{pmatrix}$, $\tilde{x} = \widetilde{(p,q)} \to \widetilde{\begin{pmatrix} p & -q \\ q & p \end{pmatrix}}$

$$K(x) = \{(p,q),(-p,q),(-p,-q),(p,-q)\}$$

$$(p,q) = \begin{pmatrix} p & -q \\ q & p \end{pmatrix}_{p=q=1} \equiv \left[(1,0) = \begin{pmatrix} 1 & 0 \\ 0 & 1 \end{pmatrix} + (0,1) = \begin{pmatrix} 0 & -i \\ i & 0 \end{pmatrix}\right] = \begin{pmatrix} 1 & -i \\ i & 1 \end{pmatrix}$$

$$(-p,q) = \begin{pmatrix} -p & -q \\ q & -p \end{pmatrix}_{p=q=1} \equiv \left[(-1,0) = \begin{pmatrix} -1 & 0 \\ 0 & -1 \end{pmatrix} + (0,1) = \begin{pmatrix} 0 & -i \\ i & 0 \end{pmatrix}\right] = \begin{pmatrix} -1 & -i \\ i & -1 \end{pmatrix}$$

$$(-p,-q) = \begin{pmatrix} -p & q \\ -q & -p \end{pmatrix}_{p=q=1} \equiv \left[(-1,0) = \begin{pmatrix} -1 & 0 \\ 0 & -1 \end{pmatrix} + (0,-1) = \begin{pmatrix} 0 & i \\ -i & 0 \end{pmatrix}\right] = \begin{pmatrix} -1 & i \\ -i & -1 \end{pmatrix}$$

$$(p,-q) = \begin{pmatrix} p & q \\ -q & p \end{pmatrix}_{p=q=1} \equiv \left[(1,0) = \begin{pmatrix} 1 & 0 \\ 0 & 1 \end{pmatrix} + (0,-1) = \begin{pmatrix} 0 & i \\ -i & 0 \end{pmatrix}\right] = \begin{pmatrix} 1 & i \\ -i & 1 \end{pmatrix}$$

$$K(\tilde{x}) = \{\widetilde{(p,q)},\widetilde{(-p,q)},\widetilde{(-p,-q)},\widetilde{(p,-q)}\}$$

$$\widetilde{(p,q)} = \widetilde{\begin{pmatrix} p & -q \\ q & p \end{pmatrix}}_{p=q=1} \equiv \left[\widetilde{(1,0)} = \begin{pmatrix} i & 0 \\ 0 & i \end{pmatrix} + \widetilde{(0,1)} = \begin{pmatrix} 0 & -1 \\ 1 & 0 \end{pmatrix}\right] = \begin{pmatrix} i & -1 \\ 1 & i \end{pmatrix}$$

$$\widetilde{(-p,q)} = \widetilde{\begin{pmatrix} -p & -q \\ q & -p \end{pmatrix}}_{p=q=1} \equiv \left[\widetilde{(-1,0)} = \begin{pmatrix} -i & 0 \\ 0 & -i \end{pmatrix} + \widetilde{(0,1)} = \begin{pmatrix} 0 & -1 \\ 1 & 0 \end{pmatrix}\right] = \begin{pmatrix} -i & -1 \\ 1 & -i \end{pmatrix}$$

$$\widetilde{(-p,-q)} = \widetilde{\begin{pmatrix} -p & q \\ -q & -p \end{pmatrix}}_{p=q=1} \equiv \left[\widetilde{(-1,0)} = \begin{pmatrix} -i & 0 \\ 0 & -i \end{pmatrix} + \widetilde{(0,-1)} = \begin{pmatrix} 0 & 1 \\ -1 & 0 \end{pmatrix}\right] = \begin{pmatrix} -i & 1 \\ -1 & -i \end{pmatrix}$$

$$\widetilde{(p,-q)} = \widetilde{\begin{pmatrix} p & q \\ -q & p \end{pmatrix}}_{p=q=1} \equiv \left[\widetilde{(1,0)} = \begin{pmatrix} i & 0 \\ 0 & i \end{pmatrix} + \widetilde{(0,-1)} = \begin{pmatrix} 0 & 1 \\ -1 & 0 \end{pmatrix}\right] = \begin{pmatrix} i & 1 \\ -1 & i \end{pmatrix}$$

Corollary 1 (Conjugated Hermite Matrix Eigenvalues). *For a complex number $x = (p,q)$ Hermite 2×2 matrix representation, $K(x)$ contains the four matrices corresponding to the complex cluster. These matrices provide real eigenvalue solutions. The $K(\tilde{x})$ conjugate complex cluster of four matrices provides the imaginary eigenvalue solutions.*

2.4.2 Feature Matrix Clustering

Relevant feature matrices can be organized into four sets.

$$K(E) \equiv \left\{ \left\{ \begin{pmatrix} 1 & 0 \\ 0 & 1 \end{pmatrix}, \begin{pmatrix} -1 & 0 \\ 0 & -1 \end{pmatrix} \right\}, \left\{ \begin{pmatrix} 0 & -i \\ i & 0 \end{pmatrix}, \begin{pmatrix} 0 & i \\ -i & 0 \end{pmatrix} \right\} \right\} \tag{5}$$

$$K(\overline{E}) \equiv \left\{ \left\{ \begin{pmatrix} 0 & 1 \\ 1 & 0 \end{pmatrix}, \begin{pmatrix} 0 & -1 \\ -1 & 0 \end{pmatrix} \right\}, \left\{ \begin{pmatrix} -i & 0 \\ 0 & i \end{pmatrix}, \begin{pmatrix} i & 0 \\ 0 & -i \end{pmatrix} \right\} \right\} \tag{6}$$

$$K(\tilde{E}) \equiv \left\{ \left\{ \begin{pmatrix} i & 0 \\ 0 & i \end{pmatrix}, \begin{pmatrix} -i & 0 \\ 0 & -i \end{pmatrix} \right\}, \left\{ \begin{pmatrix} 0 & -1 \\ 1 & 0 \end{pmatrix}, \begin{pmatrix} 0 & 1 \\ -1 & 0 \end{pmatrix} \right\} \right\} \tag{7}$$

$$K(\overline{\tilde{E}}) \equiv \left\{ \left\{ \begin{pmatrix} 0 & i \\ i & 0 \end{pmatrix}, \begin{pmatrix} 0 & -i \\ -i & 0 \end{pmatrix} \right\}, \left\{ \begin{pmatrix} -1 & 0 \\ 0 & 1 \end{pmatrix}, \begin{pmatrix} 1 & 0 \\ 0 & -1 \end{pmatrix} \right\} \right\} \tag{8}$$

The collection of matrices contains eight different representations, adding reflection operators to obtain a complete set of sixteen representations of matrices.

Theorem 1 (Conjugate Hermite Matrix Quadruple on Primitive Eigenmatrix Representation). *Utilizing the* $\{0, \{1, i, -1, -i\}\}$ *primitive representation, the conjugate extension structure of the Hermite form is four groups of* $\{1, i, -1, -i\}$. *The feature matrix represents* $\{K(E), K(\overline{E}), K(\tilde{E}), K(\overline{\tilde{E}})\}$, *and each group of feature clusters contains two real matrices and two imaginary matrices.*

2.5 *Pauli Matrices*

2.5.1 Basic representation

The three generators of the Pauli matrix are

$$\sigma_1 = \begin{pmatrix} 0 & 1 \\ 1 & 0 \end{pmatrix}, \sigma_2 = \begin{pmatrix} 0 & -i \\ i & 0 \end{pmatrix}, \sigma_3 = \begin{pmatrix} 1 & 0 \\ 0 & -1 \end{pmatrix}$$

Comparing Pauli representations and Hermite eigengroups, the four matrices $\{I, \sigma_1, \sigma_2, \sigma_3\}$ are contained in three Klein eigengroups,

$$I, \sigma_2 \in K(E); \sigma_1 \in K(\overline{E}); \sigma_3 \in K(\overline{\tilde{E}})$$

The $K(\tilde{E})$ feature group does not contain the associated Pauli matrices.

2.5.2 Matrix Eigenvalues

Lemma 3 (Four Real Eigenvalues of Pauli Matrices). *The eigenvalues of the four Pauli matrices* $\{I, \sigma_1, \sigma_2, \sigma_3\}$ *are* $\{I : \{\lambda = ++\}, \sigma_1 : \{\lambda = +-\}, \sigma_2 = \{\lambda = -+\}, \sigma_3 : \{\lambda = +-\}\}$. *All eigenvalues are* $\{+1, -1\}$ *combinations of unit real numbers.*

Corollary 2 (Four Groups of Imaginary Matrix Eigenvalues). *The eigenvalues of the four* $K(\tilde{E})$ *matrices* $K(\tilde{E}) \equiv \left\{ \left\{ \begin{pmatrix} i & 0 \\ 0 & i \end{pmatrix}, \begin{pmatrix} -i & 0 \\ 0 & -i \end{pmatrix} \right\}, \left\{ \begin{pmatrix} 0 & -1 \\ 1 & 0 \end{pmatrix}, \begin{pmatrix} 0 & 1 \\ -1 & 0 \end{pmatrix} \right\} \right\}$. *The main diagonal* *is an imaginary number, the eigenvalues of the four matrices,* $\{\{\lambda_1 = +i, \lambda_2 = +i\}, \{\lambda_1 = -i, \lambda_2 = -i\}, \{\lambda_1 = +i, \lambda_2 = -i\}, \{\lambda_1 = -i, \lambda_2 = +i\}\}$. *All are unit imaginary numbers.*

From the perspective of matrix eigenvalue representation, this type of eigensolution structure is $\{+i, -i\}$, which is not directly related to the three Pauli matrices.

2.5.3 Product Expansion Mode

Lemma 4 (16 Matrix Product Expansions). *Using matrix product expansion, the* 2×2 *matrix group contains eight real matrices and eight imaginary matrices, for a total of sixteen matrices.*

Proof. The sixteen matrices are

$$\begin{pmatrix}1&0\\0&1\end{pmatrix}, \begin{pmatrix}-1&0\\0&1\end{pmatrix}, \begin{pmatrix}-1&0\\0&-1\end{pmatrix}, \begin{pmatrix}1&0\\0&-1\end{pmatrix}, \begin{pmatrix}0&1\\1&0\end{pmatrix}, \begin{pmatrix}0&-1\\1&0\end{pmatrix}, \begin{pmatrix}0&-1\\-1&0\end{pmatrix}, \begin{pmatrix}0&1\\-1&0\end{pmatrix}$$

$$\begin{pmatrix}i&0\\0&i\end{pmatrix}, \begin{pmatrix}-i&0\\0&i\end{pmatrix}, \begin{pmatrix}-i&0\\0&-i\end{pmatrix}, \begin{pmatrix}i&0\\0&-i\end{pmatrix}, \begin{pmatrix}0&i\\i&0\end{pmatrix}, \begin{pmatrix}0&-i\\i&0\end{pmatrix}, \begin{pmatrix}0&-i\\-i&0\end{pmatrix}, \begin{pmatrix}0&i\\-i&0\end{pmatrix}$$

2.6 *Dirac Matrices*

From the perspective of matrix generation, there is a correspondence between the Dirac matrix and Pauli matrix.

2.6.1 Matrix Eigenvalue Representation

Let $\gamma' = \begin{pmatrix}1&0\\0&-1\end{pmatrix}, \gamma = \begin{pmatrix}0&1\\-1&0\end{pmatrix}$. The four generators of the Dirac matrix are

$$\gamma_0 = \gamma' \otimes I_2 = \begin{pmatrix}I_2&0\\0&-I_2\end{pmatrix}, \gamma_1 = \gamma \otimes \sigma_1 \begin{pmatrix}0&\sigma_1\\-\sigma_1&0\end{pmatrix}$$

$$\gamma_2 = \gamma \otimes \sigma_2 \begin{pmatrix}0&-\sigma_2\\\sigma_2&0\end{pmatrix}, \gamma_3 = \gamma \otimes \sigma_3 \begin{pmatrix}0&\sigma_3\\-\sigma_3&0\end{pmatrix}$$

$$\gamma^0 = \begin{pmatrix}1&0&0&0\\0&1&0&0\\0&0&-1&0\\0&0&0&-1\end{pmatrix}, \gamma^1 = \begin{pmatrix}0&0&0&1\\0&0&1&0\\0&-1&0&0\\-1&0&0&0\end{pmatrix}$$

$$\gamma^2 = \begin{pmatrix}0&0&0&-i\\0&0&i&0\\0&i&0&0\\-i&0&0&0\end{pmatrix}, \gamma^3 = \begin{pmatrix}0&0&1&0\\0&0&0&-1\\-1&0&0&0\\0&1&0&0\end{pmatrix}$$

Lemma 5 (Six Dirac Matrices). *Six Dirac matrices* $\{I_4, -I_4, \gamma_0, \gamma_1, \gamma_2, \gamma_3\}$ *have eigenvalues*

$$I_4 : \{\lambda = + + ++\}, \gamma_0 : \{\lambda = + + --\}, \gamma_2 : \{\lambda = + - -+\}$$

$$-I_4 : \{\lambda = - - --\}, \gamma_1 : \{\lambda = - - ++\}, \gamma_3 : \{\lambda = - + +-\}$$

The eigenvalues of both Dirac and Pauli matrices are represented by the combination of unit real numbers of $\{+1, -1\}$.

2.6.2 Matrix Product Expansion

Corollary 3 (Dirac Matrix Product Produces 64 Matrices). *Using matrix product expansion, the* 4×4 *matrix group contains corresponding* $4 \times 8 = 32$ *real matrices and the same number of imaginary matrices, including a total of sixty-four matrices.*

Taking the real number matrix as an example, it is expressed as four shape matrices and eight symbol matrices:

$$\begin{pmatrix} 1&0&0&0 \\ 0&1&0&0 \\ 0&0&1&0 \\ 0&0&0&1 \end{pmatrix}, \begin{pmatrix} 0&0&0&1 \\ 0&0&1&0 \\ 0&1&0&0 \\ 1&0&0&0 \end{pmatrix}, \begin{pmatrix} 0&1&0&0 \\ 1&0&0&0 \\ 0&0&0&1 \\ 0&0&1&0 \end{pmatrix}, \begin{pmatrix} 0&0&1&0 \\ 0&0&0&1 \\ 1&0&0&0 \\ 0&1&0&0 \end{pmatrix}$$

$$\begin{pmatrix} 1&0&0&0 \\ 0&1&0&0 \\ 0&0&1&0 \\ 0&0&0&1 \end{pmatrix}, \begin{pmatrix} -1&0&0&0 \\ 0&-1&0&0 \\ 0&0&1&0 \\ 0&0&0&1 \end{pmatrix}, \begin{pmatrix} -1&0&0&0 \\ 0&-1&0&0 \\ 0&0&-1&0 \\ 0&0&0&-1 \end{pmatrix}, \begin{pmatrix} 1&0&0&0 \\ 0&1&0&0 \\ 0&0&-1&0 \\ 0&0&0&-1 \end{pmatrix}$$

$$\begin{pmatrix} -1&0&0&0 \\ 0&1&0&0 \\ 0&0&1&0 \\ 0&0&0&-1 \end{pmatrix}, \begin{pmatrix} -1&0&0&0 \\ 0&1&0&0 \\ 0&0&-1&0 \\ 0&0&0&1 \end{pmatrix}, \begin{pmatrix} 1&0&0&0 \\ 0&-1&0&0 \\ 0&0&-1&0 \\ 0&0&0&1 \end{pmatrix}, \begin{pmatrix} 1&0&0&0 \\ 0&-1&0&0 \\ 0&0&1&0 \\ 0&0&0&-1 \end{pmatrix}$$

2.6.3 Product Transformation of Nonequilibrium Effect

The Dirac matrix is decomposed into four components to record the number of four elements in the matrix $\{1(4), i(4), -1(4), -i(4)\}$ on the matrix subscript as the initial value sequence. Taking the step process to multiply -1 on the first column, the number of matrices is transformed into $\{1(4), i(6), -1(4), -i(2)\}$. Select the first and fourth columns multiplied by -1, the measure is transformed into $\{1(4), i(8), -1(4), -i(0)\}$, and then the first and fourth rows are multiplied by -1, the measure is $\{1(4), i(4), -1(4), -i(4)\}$.

Compare the original matrix with the transformed matrix, which is the transpose of the former. It can be observed from the symbol ratio in the conversion process that although each 2×2 substructure locally maintains the forms of the Hermite matrix, the numbers of values of the overall components of the matrix are unbalanced.

$$\left(\begin{array}{cc|cc} 1 & i & i & -1 \\ -i & 1 & -1 & -i \\ \hline -i & -1 & 1 & -i \\ -1 & i & i & 1 \end{array}\right)_{\begin{Bmatrix} 1(4) & i(4) \\ -i(4) & -1(4) \end{Bmatrix}}$$

$$= \begin{pmatrix} 1&0&0&0 \\ 0&1&0&0 \\ 0&0&1&0 \\ 0&0&0&1 \end{pmatrix} + \begin{pmatrix} 0&0&0&-1 \\ 0&0&-1&0 \\ 0&-1&0&0 \\ -1&0&0&0 \end{pmatrix} + \begin{pmatrix} 0&i&0&0 \\ -i&0&0&0 \\ 0&0&0&-i \\ 0&0&i&0 \end{pmatrix} + \begin{pmatrix} 0&0&i&0 \\ 0&0&0&-i \\ -i&0&0&0 \\ 0&i&0&0 \end{pmatrix}$$

$$\left(\begin{array}{cc|cc} 1 & i & i & -1 \\ -i & 1 & -1 & -i \\ \hline -i & -1 & 1 & -i \\ -1 & i & i & 1 \end{array}\right) \begin{pmatrix} -1&0&0&0 \\ 0&1&0&0 \\ 0&0&1&0 \\ 0&0&0&1 \end{pmatrix} = \left(\begin{array}{cc|cc} -1 & i & i & -1 \\ i & 1 & -1 & -i \\ \hline i & -1 & 1 & -i \\ 1 & i & i & 1 \end{array}\right)_{\begin{Bmatrix} 1(4) & i(6) \\ -i(2) & -1(4) \end{Bmatrix}}$$

$$= \begin{pmatrix} -1&0&0&0 \\ 0&1&0&0 \\ 0&0&1&0 \\ 0&0&0&1 \end{pmatrix} + \begin{pmatrix} 0&0&0&-1 \\ 0&0&-1&0 \\ 0&-1&0&0 \\ 1&0&0&0 \end{pmatrix} + \begin{pmatrix} 0&i&0&0 \\ i&0&0&0 \\ 0&0&0&-i \\ 0&0&i&0 \end{pmatrix} + \begin{pmatrix} 0&0&i&0 \\ 0&0&0&-i \\ i&0&0&0 \\ 0&i&0&0 \end{pmatrix}$$

$$\left(\begin{array}{cc|cc} 1 & i & i & -1 \\ -i & 1 & -1 & -i \\ \hline -i & -1 & 1 & -i \\ -1 & i & i & 1 \end{array}\right) \begin{pmatrix} -1&0&0&0 \\ 0&1&0&0 \\ 0&0&1&0 \\ 0&0&0&-1 \end{pmatrix} = \left(\begin{array}{cc|cc} -1 & i & i & 1 \\ i & 1 & -1 & i \\ \hline i & -1 & 1 & i \\ 1 & i & i & -1 \end{array}\right)_{\begin{Bmatrix} 1(4) & i(8) \\ -i(0) & -1(4) \end{Bmatrix}}$$

$$= \begin{pmatrix} -1&0&0&0 \\ 0&1&0&0 \\ 0&0&1&0 \\ 0&0&0&-1 \end{pmatrix} + \begin{pmatrix} 0&0&0&1 \\ 0&0&-1&0 \\ 0&-1&0&0 \\ 1&0&0&0 \end{pmatrix} + \begin{pmatrix} 0&i&0&0 \\ i&0&0&0 \\ 0&0&0&i \\ 0&0&i&0 \end{pmatrix} + \begin{pmatrix} 0&0&i&0 \\ 0&0&0&i \\ i&0&0&0 \\ 0&i&0&0 \end{pmatrix}$$

$$\begin{pmatrix} -1&0&0&0 \\ 0&1&0&0 \\ 0&0&1&0 \\ 0&0&0&-1 \end{pmatrix} \left(\begin{array}{cc|cc} 1 & i & i & -1 \\ -i & 1 & -1 & -i \\ \hline -i & -1 & 1 & -i \\ -1 & i & i & 1 \end{array}\right) \begin{pmatrix} -1&0&0&0 \\ 0&1&0&0 \\ 0&0&1&0 \\ 0&0&0&-1 \end{pmatrix}$$

$$= \begin{pmatrix} -1&0&0&0 \\ 0&1&0&0 \\ 0&0&1&0 \\ 0&0&0&-1 \end{pmatrix} \left(\begin{array}{cc|cc} -1 & i & i & 1 \\ i & 1 & -1 & i \\ \hline i & -1 & 1 & i \\ 1 & i & i & -1 \end{array}\right) = \left(\begin{array}{cc|cc} 1 & -i & -i & -1 \\ i & 1 & -1 & i \\ \hline i & -1 & 1 & i \\ -1 & -i & -i & 1 \end{array}\right)_{\begin{Bmatrix} 1(4) & i(4) \\ -i(4) & -1(4) \end{Bmatrix}}$$

$$= \begin{pmatrix} 1&0&0&0 \\ 0&1&0&0 \\ 0&0&1&0 \\ 0&0&0&1 \end{pmatrix} + \begin{pmatrix} 0&0&0&-1 \\ 0&0&-1&0 \\ 0&-1&0&0 \\ -1&0&0&0 \end{pmatrix} + \begin{pmatrix} 0&-i&0&0 \\ i&0&0&0 \\ 0&0&0&i \\ 0&0&-i&0 \end{pmatrix} + \begin{pmatrix} 0&0&-i&0 \\ 0&0&0&i \\ i&0&0&0 \\ 0&-i&0&0 \end{pmatrix}$$

$$\begin{pmatrix} -1 & 0 & 0 & 0 \\ 0 & -1 & 0 & 0 \\ 0 & 0 & 1 & 0 \\ 0 & 0 & 0 & 1 \end{pmatrix} \left(\begin{array}{cc|cc} 1 & i & i & -1 \\ -i & 1 & -1 & -i \\ \hline -i & -1 & 1 & -i \\ -1 & i & i & 1 \end{array}\right) \begin{pmatrix} -1 & 0 & 0 & 0 \\ 0 & -1 & 0 & 0 \\ 0 & 0 & 1 & 0 \\ 0 & 0 & 0 & 1 \end{pmatrix}$$

$$= \left(\begin{array}{cc|cc} 1 & i & -i & 1 \\ -i & 1 & 1 & i \\ \hline i & 1 & 1 & -i \\ 1 & -i & i & 1 \end{array}\right) \begin{Bmatrix} 1(8) & i(4) \\ -i(4) & -1(0) \end{Bmatrix}$$

From the above examples of symbolic matrix products, the dynamic effect of the operation process is extremely complex. How to balance the overall number of symbols of the four primitives after the operation of operators is a difficult problem in finite group theory.

In 1928, Dirac used the Dirac matrix to construct the basis for the successful expression of nonrelativistic quantum mechanics formulas, seeking the construction of high-dimensional special real number solutions and strictly ensuring that the eigenvalue solutions of the equations are represented by quaternary real number vectors.

2.7 *Intrinsic Symmetry of Dirac Equation*

The Dirac equation structure forms a 4×4 representation matrix. From the perspective of eigenvalue measurement, classical quantum mechanics, and quantum information formulas, the complex conjugate operator makes the measured eigenvalue $xx^* = x^*x = 1$, which satisfies the Hermite symmetry condition through the inner product of the orthogonal operation. The equation solution is the eigenvector determined by the eigenvalues of the main diagonal. Considering the complete solution space $\{1, i, -1, -i\}$, this level of the measure dimension is $1/4$.

Using the quaternary extension operator, the conjugate complex function passes through eight sets of vectors in four attribute regions, two sets of conjugate cross structures, and two sets of conjugate cross tensor products after orthogonal transformation. Based on $\{1, i, -1, -i\}, \{1, -i, -1, i\}$ being sixteen groups of counter-clockwise and clockwise matrices, the main diagonal or auxiliary diagonal forms a quaternary complex unit vector $\{1, i, -1, -i\}, \{1, -i, -1, i\}$.

In the conjugate complex vector function, it is 4×4. The characteristic matrix exists in all involved in the YY^{-1}, $\tilde{Y}\tilde{Y}^{-1}$ computing environment, and there is a constant matrix associated with sixteen sets of eigenvectors as the conjugate quadrature output part.

Eight matrices of main diagonal eigenvalues,

$$\overline{Y}^{0}\overleftarrow{\tilde{Y}} = \begin{pmatrix} 1 & i & -1 & -i \\ -i & 1 & i & -1 \\ -1 & -i & 1 & i \\ i & -1 & -i & 1 \end{pmatrix}, \overline{Y}^{2}\overleftarrow{\tilde{Y}} = \begin{pmatrix} i & -1 & -i & 1 \\ 1 & i & -1 & -i \\ -i & 1 & i & -1 \\ -1 & -i & 1 & i \end{pmatrix}$$

$$S(\overline{Y}^{0}\overleftarrow{\tilde{Y}}) = \{R_sI_a|R_sI_s\} \ : \ S(\overline{Y}^{2}\overleftarrow{\tilde{Y}}) = \{R_aI_s|R_sI_s\}$$

$$\overline{Y}^{3}\overleftarrow{\tilde{Y}} = \begin{pmatrix} -1 & -i & 1 & i \\ i & -1 & -i & 1 \\ 1 & i & -1 & -i \\ -i & 1 & i & -1 \end{pmatrix}, \overline{Y}^{1}\overleftarrow{\tilde{Y}} = \begin{pmatrix} -i & 1 & i & -1 \\ -1 & -i & 1 & i \\ i & -1 & -i & 1 \\ 1 & i & -1 & -i \end{pmatrix}$$

$$S(\overline{Y}^{3}\overleftarrow{\tilde{Y}}) = \{R_sI_a|R_sI_s\} \ : \ S(\overline{Y}^{1}\overleftarrow{\tilde{Y}}) = \{R_aI_s|R_sI_s\}$$

$$\tilde{Y}^{0}\overleftarrow{\overline{Y}} = \begin{pmatrix} 1 & -i & -1 & i \\ i & 1 & -i & -1 \\ -1 & i & 1 & -i \\ -i & -1 & i & 1 \end{pmatrix}, \tilde{Y}^{2}\overleftarrow{\overline{Y}} = \begin{pmatrix} -i & -1 & i & 1 \\ 1 & -i & -1 & i \\ i & 1 & -i & -1 \\ -1 & i & 1 & -i \end{pmatrix}$$

$$S(\tilde{Y}^{0}\overleftarrow{\overline{Y}}) = \{R_sI_a|R_sI_s\} \ : \ S(\tilde{Y}^{2}\overleftarrow{\overline{Y}}) = \{R_aI_s|R_sI_s\}$$

$$\tilde{Y}^{3}\overleftarrow{\overline{Y}} = \begin{pmatrix} -1 & i & 1 & -i \\ -i & -1 & i & 1 \\ 1 & -i & -1 & i \\ i & 1 & -i & -1 \end{pmatrix}, \tilde{Y}^{1}\overleftarrow{\overline{Y}} = \begin{pmatrix} i & 1 & -i & -1 \\ -1 & i & 1 & -i \\ -i & -1 & i & 1 \\ 1 & -i & -1 & i \end{pmatrix}$$

$$S(\tilde{Y}^{3}\overleftarrow{\overline{Y}}) = \{R_sI_a|R_sI_s\} \ : \ S(\tilde{Y}^{1}\overleftarrow{\overline{Y}}) = \{R_aI_s|R_sI_s\}$$

$$S(d) = \{R_s(4)R_a(4)I_s(4)I_a(4)|R_s(8)I_s(8)\}$$

Eight matrices of auxiliary diagonal eigenvalues,

$$\tilde{Y}^{0}\overleftarrow{Y} = \begin{pmatrix} i & -1 & -i & 1 \\ -1 & -i & 1 & i \\ -i & 1 & i & -1 \\ 1 & i & -1 & -i \end{pmatrix}, \tilde{Y}^{2}\overleftarrow{Y} = \begin{pmatrix} -1 & -i & 1 & i \\ -i & 1 & i & -1 \\ 1 & i & -1 & -i \\ i & -1 & -i & 1 \end{pmatrix}$$

$$S(\tilde{Y}^{0}\overleftarrow{Y}) = \{R_sI_s|R_sI_a\} \ : \ S(\tilde{Y}^{2}\overleftarrow{Y}) = \{R_sI_s|R_aI_s\}$$

$$\tilde{Y}^{3}\overleftarrow{Y} = \begin{pmatrix} -i & 1 & i & -1 \\ 1 & i & -1 & -i \\ i & -1 & -i & 1 \\ -1 & -i & 1 & i \end{pmatrix}, \tilde{Y}^{1}\overleftarrow{Y} = \begin{pmatrix} 1 & i & -1 & -i \\ i & -1 & -i & 1 \\ -1 & -i & 1 & i \\ -i & 1 & i & -1 \end{pmatrix}$$

$$S(\tilde{Y}^{3}\overleftarrow{Y}) = \{R_sI_s|R_sI_a\} \ : \ S(\tilde{Y}^{1}\overleftarrow{Y}) = \{R_sI_s|R_aI_s\}$$

$$\overline{\tilde{Y}}^{0}\overleftarrow{\overline{Y}} = \begin{pmatrix} -i & -1 & i & 1 \\ -1 & i & 1 & -i \\ i & 1 & -i & -1 \\ 1 & -i & -1 & i \end{pmatrix}, \overline{\tilde{Y}}^{2}\overleftarrow{\overline{Y}} = \begin{pmatrix} -1 & i & 1 & -i \\ i & 1 & -i & -1 \\ 1 & -i & -1 & i \\ -i & -1 & i & 1 \end{pmatrix}$$

$$S(\overline{\tilde{Y}}^{0}\overleftarrow{\overline{Y}}) = \{R_sI_s|R_sI_a\} \ : \ S(\overline{\tilde{Y}}^{2}\overleftarrow{\overline{Y}}) = \{R_sI_s|R_aI_s\}$$

$$\overline{\tilde{Y}}^{3}\overleftarrow{\overline{Y}} = \begin{pmatrix} i & 1 & -i & -1 \\ 1 & -i & -1 & i \\ -i & -1 & i & 1 \\ -1 & i & 1 & -i \end{pmatrix}, \overline{\tilde{Y}}^{1}\overleftarrow{\overline{Y}} = \begin{pmatrix} 1 & -i & -1 & i \\ -i & -1 & i & 1 \\ -1 & i & 1 & -i \\ i & 1 & -i & -1 \end{pmatrix}$$

$$S(\overline{\tilde{Y}}^{3}\overleftarrow{\overline{Y}}) = \{R_sI_s|R_sI_a\} \ : \ S(\overline{\tilde{Y}}^{1}\overleftarrow{\overline{Y}}) = \{R_sI_s|R_aI_s\}$$

$$S(u) = \{R_s(8)I_s(8)|R_s(4)R_a(4)I_s(4)I_a(4)\}$$

$$S(d) = \{R_s(4)R_a(4)I_s(4)I_a(4)|R_s(8)I_s(8)\}$$

Observing that the $\{S(d), S(u)\}$ overall measure exhibits the polarization distinguishability of the main and auxiliary diagonal symmetry properties, and ignoring the auxiliary diagonal features in the strict quantum information solution space leads to the system being one of the reasons for incompleteness.

Corollary 4 (Quantum Mechanics and Quantum Information/Conjugated Structure Distinguishable Eigenvectors). *Compared with a set of quaternary unit eigenvectors in classical quantum mechanics and quantum information, the conjugate complex vector function provides sixteen sets of quaternary distinguishable output eigenvectors, and the dimension ratio of the extended structure is* $1/16$.

In the conjugate structure solution space, the $15/16$ part is beyond the classical quantum information. This kind of structural failure makes classical quantum mechanics and quantum information architectures unable to accommodate quantum entanglement, and it is difficult to resolve many paradoxes in modern quantum entanglement experiments.

2.8 *Complete Phase Space Dimensions for Binary and Quaternary Qubit Systems*

Corollary 5 (Complete Solution Space of Pauli Matrices). *The complete solution space of the Pauli matrices needs to expand the auxiliary diagonal elements to include the adjoint solution space* $\{+i, -i\}$. *The dimension of the complete phase space is* $2^2 \times 2^2 = 2^4 = 4^2 = 16$ *to completely cover the* $\{K(E), K(\tilde{E})\}$ *cluster. Each set of matrices is guaranteed to provide a pair of eigenvalue solutions.*

Proof. Using the feature matrix cluster of the conjugated Klein group, the Pauli matrix represents that the real number solution space of the two-qubit adjoint is $\{+1, -1\}$, four matrices $\{I, \sigma_1, \sigma_2, \sigma_3\}$, and the phase space dimension is equal to $2^2 = 4$.

Corollary 6 (Complete Solution Space of Dirac Matrices). *The complete solution space of the Dirac matrices needs to expand the auxiliary diagonal elements to include the quaternion solution structure represented by the imaginary number of the adjoint solution space* $\{+i, -i\}$, *and the dimension of the phase space is* $2^4 \times 2^4 = 2^8 = 256$.

Proof. Using the Dirac matrices, quaternary qubit representation, each feature matrix provides a quaternion solution structure based on $\{+1, -1\}$ real number representation, the number of intrinsic features is the same as the four shape matrices closely related, and the phase space dimension is $2^4 = 16$. The additional eight groups of $\{+1, -1\}$ sign matrices cannot provide a benign balanced structure, which ensures that after each local operation transformation, the matrix as a whole maintains the quantitative balance of the quaternary features. Usually, in addition to using the original 4×4 Dirac matrices, other transformations cannot be added at will.

2.9 *Conjugate Complex Vector Representation*

In conjugate complex vector algebra, sixteen matrices can be generated by two matrices $\{\overline{Y}\ {}^{0}\overleftarrow{\tilde{Y}}, \tilde{Y}\ {}^{0}\overleftarrow{\overline{Y}}\}$, obtained after the row and column rotation and its reflection operator.

The elements of the middle partition matrix are four blocks, and each block is a combined representation of a 2×2 Pauli matrix. Each block contains $3(+)1(-)|1(+)3(-)$, which is a Hermite matrix expression mode.

$$\overline{Y}^{\,0}\overleftarrow{\tilde{Y}} = \left(\begin{array}{cc|cc} 1 & i & -1 & -i \\ -i & 1 & i & -1 \\ \hline -1 & -i & 1 & i \\ i & -1 & -i & 1 \end{array}\right), \quad \tilde{Y}^{\,0}\overleftarrow{\overline{Y}} = \left(\begin{array}{cc|cc} i & -1 & -i & 1 \\ -1 & -i & 1 & i \\ \hline -i & 1 & i & -1 \\ 1 & i & -1 & -i \end{array}\right)$$

$$\overline{Y}^{\,0}\overleftarrow{\tilde{Y}} = \begin{pmatrix} 1&0&0&0 \\ 0&1&0&0 \\ 0&0&1&0 \\ 0&0&0&1 \end{pmatrix} + \begin{pmatrix} 0&i&0&0 \\ 0&0&i&0 \\ 0&0&0&i \\ i&0&0&0 \end{pmatrix} + \begin{pmatrix} 0&0&-1&0 \\ 0&0&0&-1 \\ -1&0&0&0 \\ 0&-1&0& \end{pmatrix} + \begin{pmatrix} 0&0&0&-i \\ -i&0&0&0 \\ 0&-i&0&0 \\ 0&0&-i&0 \end{pmatrix}$$

The corresponding sixteen groups of feature matrices are

$$\{1, i, -1, -i\}_d \times \left\{ \begin{pmatrix} 1&0&0&0 \\ 0&1&0&0 \\ 0&0&1&0 \\ 0&0&0&1 \end{pmatrix} + \begin{pmatrix} 0&1&0&0 \\ 0&0&1&0 \\ 0&0&0&1 \\ 1&0&0&0 \end{pmatrix} + \begin{pmatrix} 0&0&1&0 \\ 0&0&0&1 \\ 1&0&0&0 \\ 0&1&0& \end{pmatrix} + \begin{pmatrix} 0&0&0&1 \\ 1&0&0&0 \\ 0&1&0&0 \\ 0&0&1&0 \end{pmatrix} \right\}_d$$

$$\tilde{Y}^{\,0}\overleftarrow{\overline{Y}} = \begin{pmatrix} 0&0&0&1 \\ 0&0&1&0 \\ 0&1&0&0 \\ 1&0&0& \end{pmatrix} + \begin{pmatrix} 0&0&-i&0 \\ 0&-i&0&0 \\ -i&0&0&0 \\ 0&0&0&-i \end{pmatrix} + \begin{pmatrix} 0&-1&0&0 \\ -1&0&0&0 \\ 0&0&0&-1 \\ 0&0&-1&0 \end{pmatrix} + \begin{pmatrix} i&0&0&0 \\ 0&0&0&i \\ 0&0&i&0 \\ 0&i&0&0 \end{pmatrix}$$

The corresponding sixteen groups of feature matrices are

$$\{1, i, -1, -i\}^u \times \left\{ \begin{pmatrix} 0&0&0&1 \\ 0&0&1&0 \\ 0&1&0&0 \\ 1&0&0& \end{pmatrix} + \begin{pmatrix} 0&0&1&0 \\ 0&1&0&0 \\ 1&0&0&0 \\ 0&0&0&1 \end{pmatrix} + \begin{pmatrix} 0&1&0&0 \\ 1&0&0&0 \\ 0&0&0&1 \\ 0&0&1&0 \end{pmatrix} + \begin{pmatrix} 1&0&0&0 \\ 0&0&0&1 \\ 0&0&1&0 \\ 0&1&0&0 \end{pmatrix} \right\}^u$$

Theorem 2 (Complete Phase Space of Conjugated Structures). *The conjugated structure realizes the Wigner phase space based on discrete geometry, which ensures that the discrete quantum information architecture can provide complete support for four qubit interaction modes.*

Proof. There are thirty-two pure state basic modes $\{d, u\}$ to provide sixteen modes, the two are independent of each other, and the corresponding phase space dimension $16 \times 16 = 2^4 \times 2^4 = 2^8 = 256$.

3. Frontier Quantum Entanglement Analysis and Experiments

3.1 Quantum Entanglement Measurement Process

In Fig. 2(a),(b), the paper [2] published in 2002 describes the quantum entanglement interaction model, explores the possible combination modes of quantum entangled states, and shows the projected image and entanglement from polarized light in which the level of interaction classification is studied.

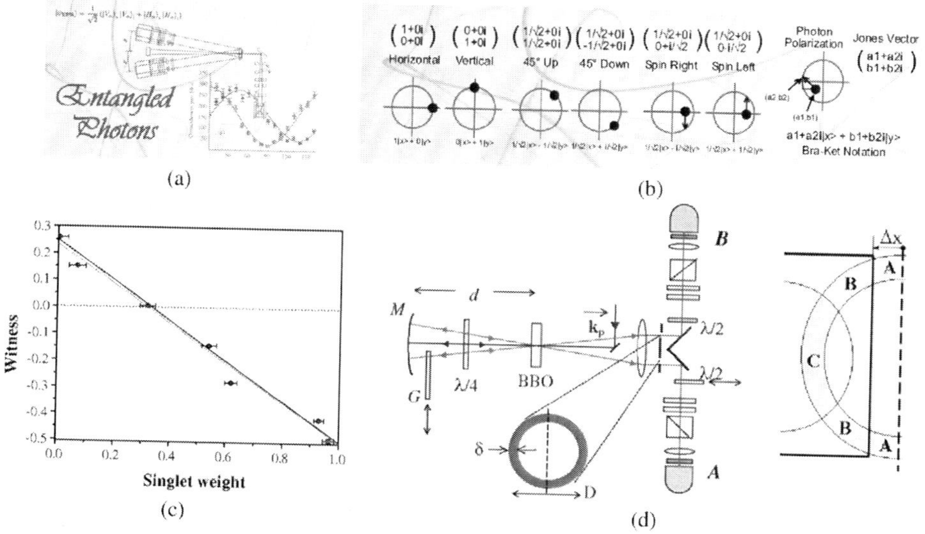

Figure 2. Entanglement model: (a) action mode in 2002; (b) interaction formula of entangled states; (c) polarization measurement results in 2003; (d) polarization optical path design.

The optical path design and the detected signal of the entanglement experiment are shown in Fig. 2(a). In Fig. 2(b), the widely used Jones vector and matrix express the possibility of quantum entanglement by using modern optical instruments in a typical state of occurrence.

The paper [3] demonstrates the design and results in Fig. 2(c),(d) that quantum entanglement researchers have used polarized light quanta filtering modes for the detection of entangled states and typical assay results shown in Fig. 2(c). In Fig. 2(d), the coherent light passes through the BBO crystal to form a frequency-multiplied signal. Using the dual-channel control mechanism and polarization angle filtering, the signal passes through the BBO crystal again through the mirror, and the multifrequency pole is transformed by the optical lens. The polarized signal is projected and imaged through the circular slit, and by extracting specific coherent frequency characteristics, the fine distribution characteristics of the equal energy surface of the polarized signal sequence are displayed.

This kind of polarization detection mechanism has been gradually accepted by quantum information researchers and has become a typical detection mode of the entanglement effect [5].

3.2 *Typical Quantum Entanglement Theory and Experimental Results*

Through years of cutting-edge research, quantum entanglement theory and experiments have been developed in depth, and typical achievements are shown in Figs. 3–5 in a series of diagrams.

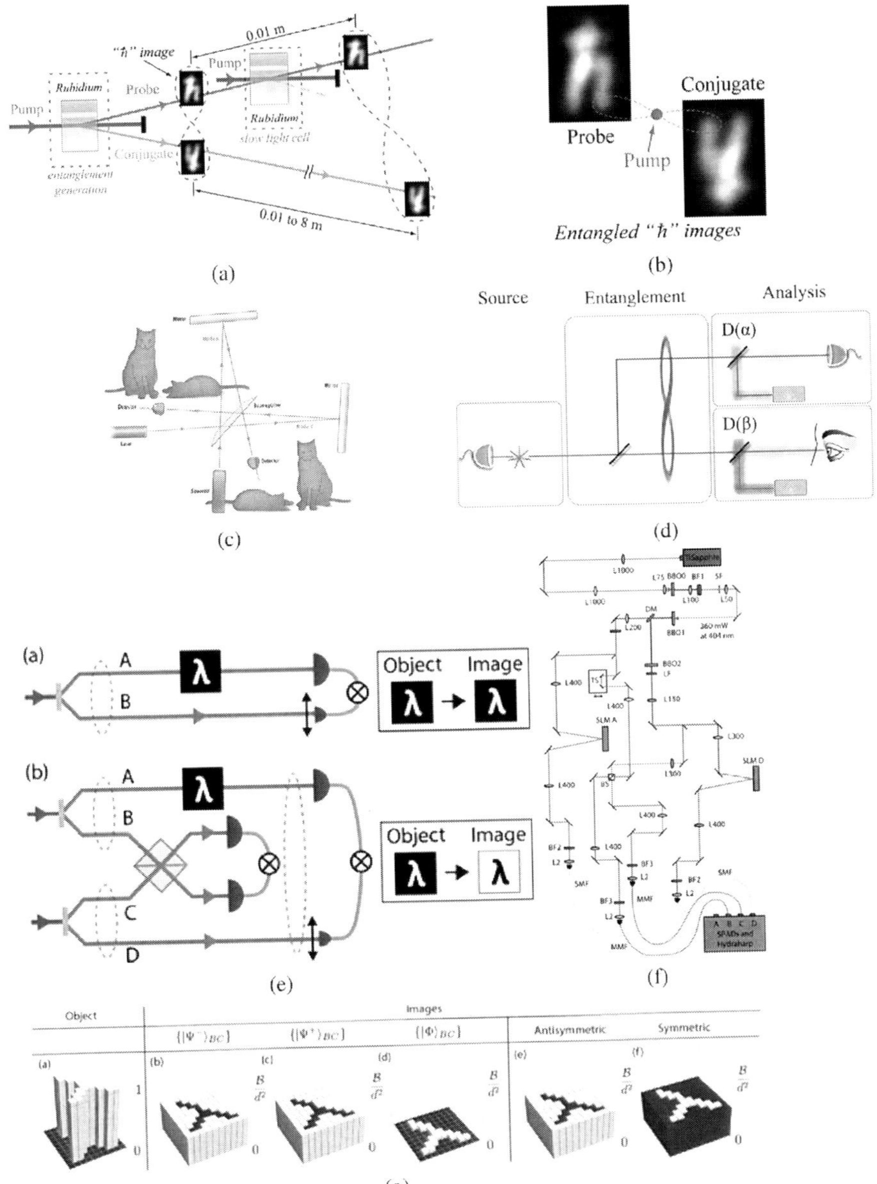

Figure 3. Measurement mode for entanglement detection: (a) 2009 ghost image system design; (b) ghost image conjugate results; (c) 2010 detection schrodinger cat state diagram; (d) 2016 observation entanglement state design. conjugate ghost image detection measurement mode in 2019: (e) correlation system design; (f) ghost image generation mechanism; (g) group projection illustration.

In Fig. 3(a),(b) of the 2009 quantum entanglement detection experiment [5], the signal is passed through a rubidium gas filter to form a conjugate ghost projection distribution of ghost images.

In the field of quantum information popularization, Schrodinger's cat paradox is deeply rooted in [7]. In Fig. 3(c), the formal expression of quantum entanglement using lasers and various reflective optical devices has become the focus of popular science education. In the illustration, the laser source interacts with the reference light source in the squeezed state, and the combined signal is detected by two detectors after decomposition through transmission and reflection.

In Fig. 3(d) observing the quantum entanglement effect for human-machine interaction [8], the laser signal forms upspin and downspin after the transmission and reflection polarization decomposition after the first pass through the half-mirror polarization signal output, and the two-way polarization signal sequence obtains two-way measurement output after passing through the semimirror for the second time.

Comparing and observing the schemas corresponding to Schrodinger cat detection and entanglement observation, it is not difficult to see the commonality of the corresponding detection models.

In Fig. 3(e)–(g), the quantum entangled state technology in 2019 is used to form the photoelectron inversion of the conjugate ghost image detection result [6], (e) shows two sets of conjugate images on the generation path, (f) gives symmetric and antisymmetric conjugate image distributions for the optical path detection process and four sets of output interfaces, and (g) provides symmetric and antisymmetric conjugate images.

In 2016, the first observation of quantum entanglements was performed on four photons [9]. Typical entanglement models and methods, measuring path, projection, and other information are shown in Fig. 4(a)–(d). From the level of the angular trace, four meta states are illustrated in entangled distributions.

Research on high-dimensional quantum entanglement has been rich and colorful in recent years, and three sets of models and measurements are presented in Fig. 5(a)–(g).

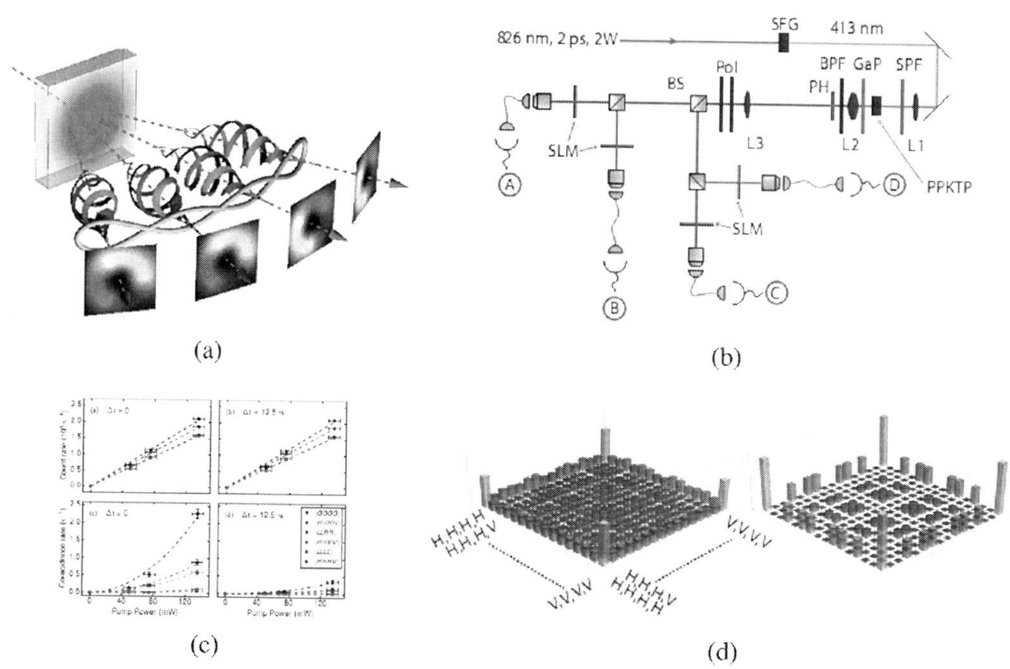

Figure 4. The first four-qubit entanglement detection measurement mode in 2016: (a) effect display; (b) entanglement system design; (c) two-dimensional diagram; (d) three-dimensional diagram.

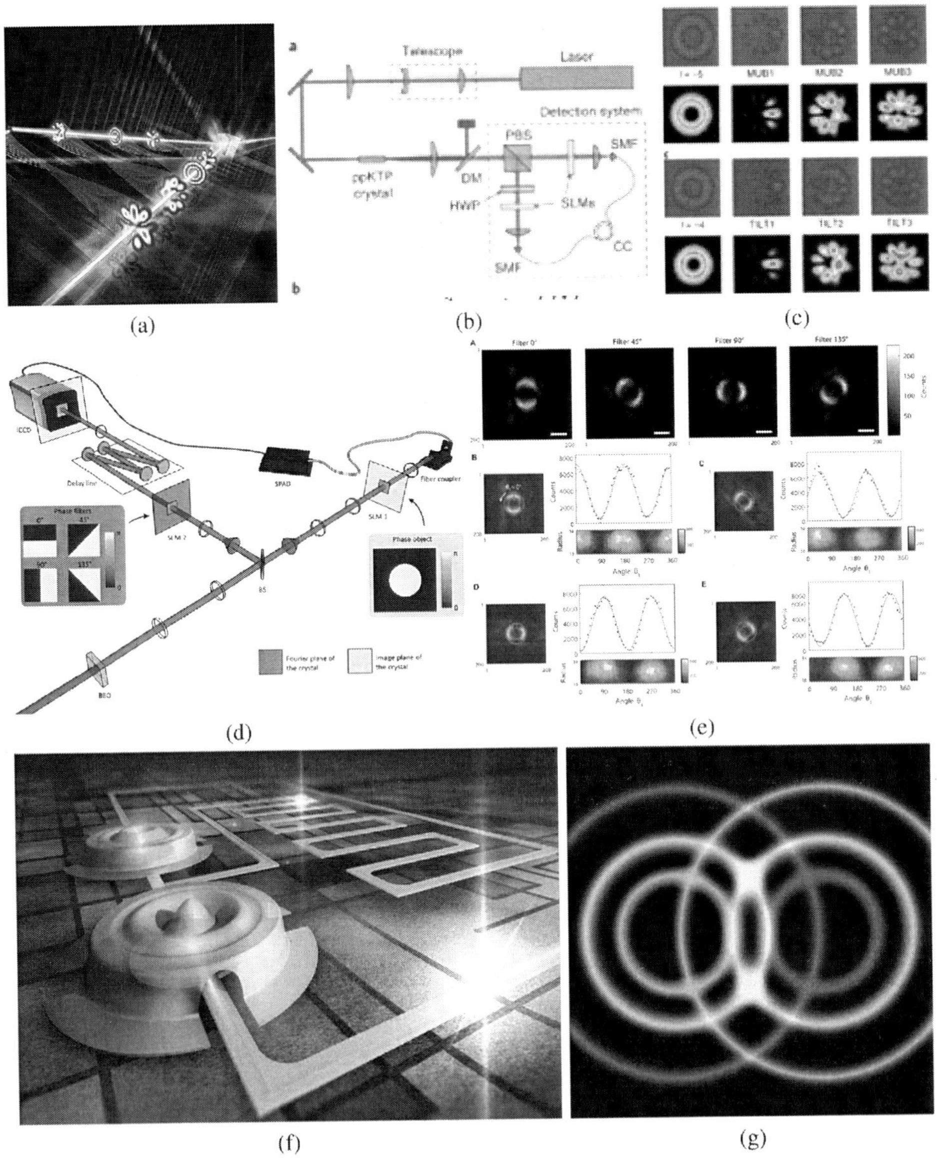

Figure 5. High-dimensional quantum entanglement in 2018: (a) results schematically; (b) measurement model; (c) output result. Illustration of the first observed entanglement measurement of qubits with four polarization directions: (d) system design; (e) 1D+2D illustration. The 2018 intense laser frequency doubling entangled state generation mechanism: (f)(g) (f) model schematic; (g) measurement results.

Fig. 5(a)–(c) explores high-dimensional entangled state patterns based on two-state patterns, briefly illustrating (a) path-dependent image changes, (b) measurement system design, and (c) series of projection results [11].

In Fig. 5(d)–(e), it is the first time that human beings have used the polarization signal to control the output mode and to observe the distribution image of the quantum entanglement on polarization arrangement [12]. using $\{0, \pi/4, \pi/2, 3\pi/2\}$ radian polarized filter control, four sets of sinusoidal quantization curves and the corresponding relationship between the series of shifts are successfully observed.

In Fig. 5(f)–(g), the powerful quantum entanglement state generation mechanism [10] is shown through the comprehensive utilization of strong lasers and planar decomposition of BBO crystals to form multifrequency visualized entanglement state diagrams. From the corresponding output distributions of different frequencies, the entanglement state interaction is formed at the intersection of equal energy rings. The output frequency changes along with the successful polarization effect, and the interaction area exhibits different color distributions (Fig. 5(g)).

3.3 Interaction Results under Strong Interactions

Two series of figures are shown in Figs. 6–7.

In Fig. 6(a')–(b'), the superposition characteristics of entanglement interactions of two photons (a') and multiphoton (b') are comprehensively applied to visualize high-dimensional output images [13].

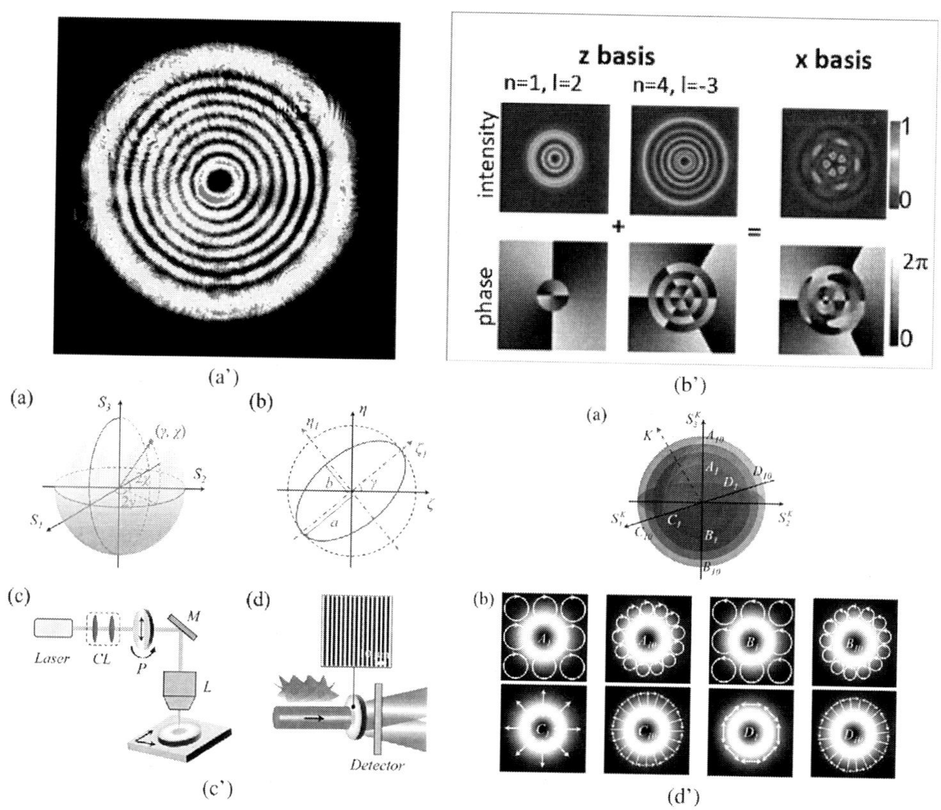

Figure 6. Illustration of two-photon and multiphoton strong correlation measurements in 2014: (a') 103-dimensional results; (b') diagram of multigroup parameter control. 2022 geometric phase adjustment: (c') system design mode; (d') multiprojection vacant diagram.

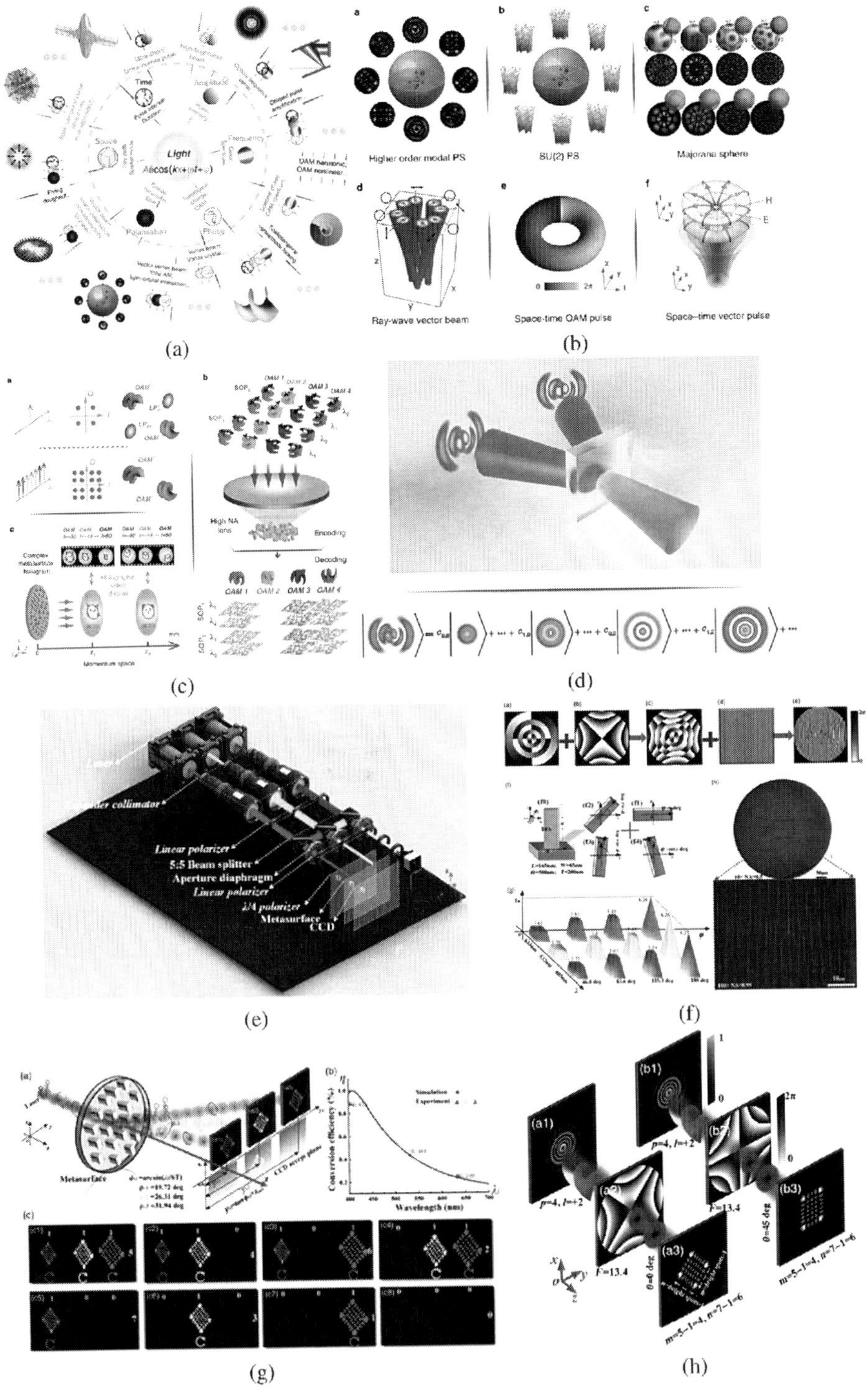

Figure 7. Illustration of high-dimensional structured light quanta measurement in 2022: (a) correlation design; (b) eight-element correlation; (c) projection mode; (d) multispectral pooling diagram. Diagram of optical rotation interaction measurement in 2022: (e) experimental system; (f) filtering mode; (g) 2^3 combination encoding mode; (h) rhombus and rectangle point group characteristic spectrum diagram.

Geometric phase controls are shown in Fig. 6(c')–(d'), through grating modulation (c'), forming a wide-spectrum spinor state encoding (d'), opening the way for the next generation of entangled devices [14].

From the control mode accompanied by geometric regulation, the control mechanism of the eight vector orientations plays a central role. Four sets of images in Fig. 7(a)–(d) show the latest research results: (a) the different interaction forms of the comprehensive analysis model and the accompanying observable effects, (b) eight typical models described in three directions, based on the octet pattern and its combinatorial characteristics to form a rich and observable entanglement state effect, (c) adding more quantum states from the basic level, and (d) showing supercomplex combination entanglement decomposition and synthesis structure. The corresponding system provides a higher-dimensional controllable structured light source [15] to describe the decomposition and combination and interaction effects.

The four sets of images in Fig. 7(e)–(h) show the results of the latest strong-interaction optical rotation signature measurements. (e) The experimental model of the system. (f) The polarization filtering mode and the visualization model used distribution. (g) Three-primitive eight-combination lattice-coded quantum characteristic group aggregation characteristic distribution. (h) Two sets of typical rhombus and rectangular point group characteristic distribution patterns. The system provides a controllable combination mechanism with strong interaction, a series of filtering modes described in [16], decomposition, and combination lattice feature projection characteristics.

Seven images in Fig. 8(a)–(g) show the results of three sets of strong interaction optical rotation measurements reported at the 2022 International Quantum Conference (Quantum 2022). (a) Quantum error-correcting logic state dynamics [21]. Diagrams (b)–(d) show the series of results of structured light under the action of a nonorthogonal state [22]. (b) System processing modulus. (c) Corresponding filter encoding modes and output distribution. (d) Encoding and transforming point group eigenspectrum effects. In illustrations (e)–(g), (e) provides the entangled state simultaneous multichannel characteristic spectrum measurement model. (f) The four color measurement curves correspond to $\{0, \pi/2, \pi, 3\pi/2\}$ concurrent eigenvector projections of four sets of measured angles. (g) Two sets of concurrent quaternary eigenvectors, arranged in sequence and reverse order [23].

3.4 Typical Measurement Models and Application Methods

From the perspective of high-tech promotion and application, two series of images are shown in Figs. 9–10. A series of studies and developments have been carried out for application scenarios such as quantum key distribution, network security assurance, optoelectronic circuits, photon angular momentum detection, and femtosecond laser technology.

In Fig. 9(a)–(b), results of communication and quantum cryptography sequence analysis are shown: (a) Modal transformation intrinsic relations. (b) Different antisymmetric quantum networks and their multiparticipant quantum control coupling-detection effects of entangled modes [17].

In Fig. 9(c)–(d), the frontier exploration mode of the optical quantum circuit is shown: (a) shows the entangled state optical circuit design and (b) the control switching mechanism and detection results [18].

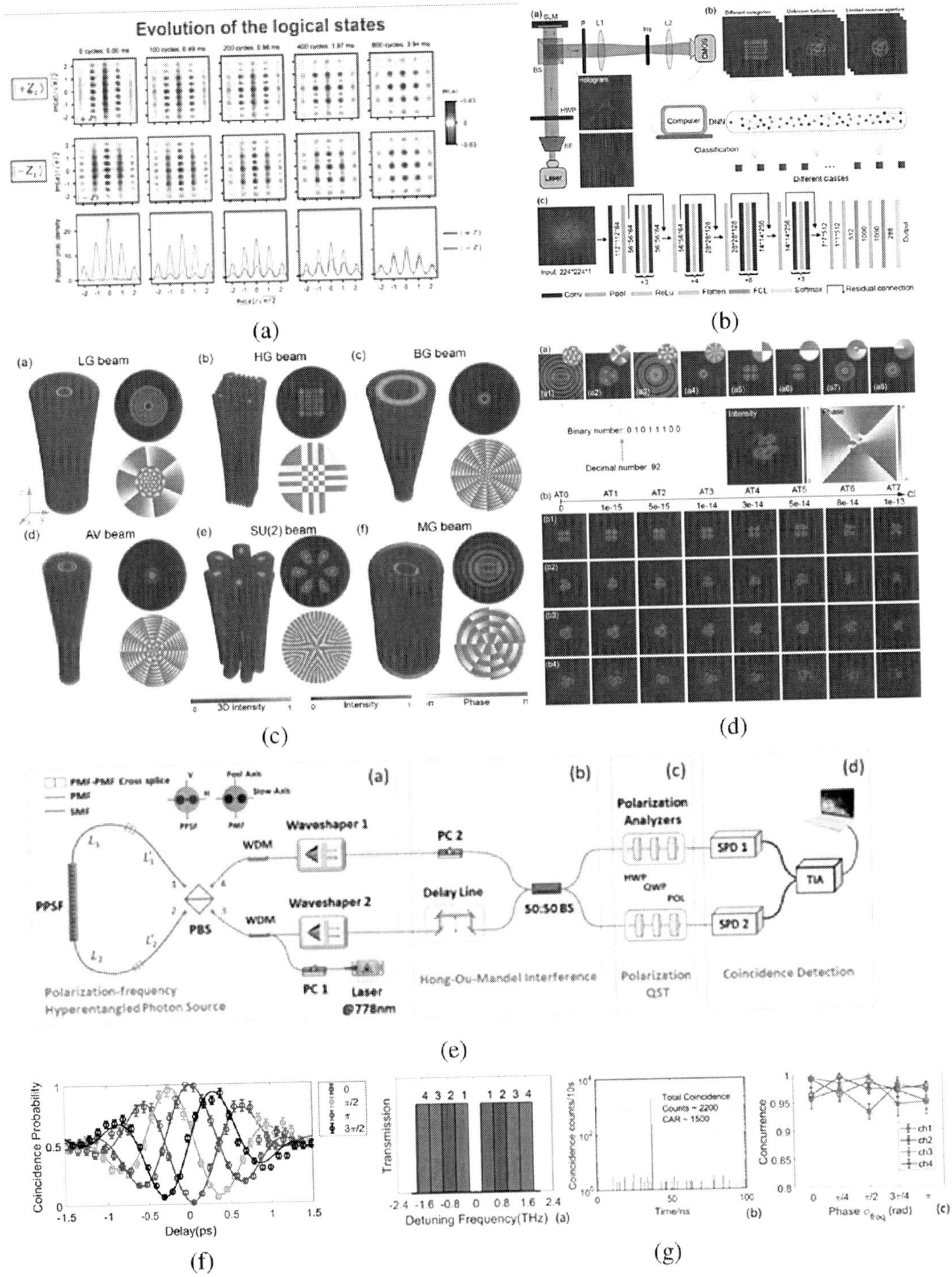

Figure 8. 2022 international quantum conference (quantum 2022) interaction measurement diagram: (a) logic state change; (b) processing mode; (c) coding mode and output distribution; (d) transformation point group characteristics spectrum diagram; (e) simultaneous multichannel characteristic spectrum measurement model; (f) dynamic projection of quaternary eigenvectors; (g) two sets of quaternary eigenvectors are arranged in sequence and reverse order.

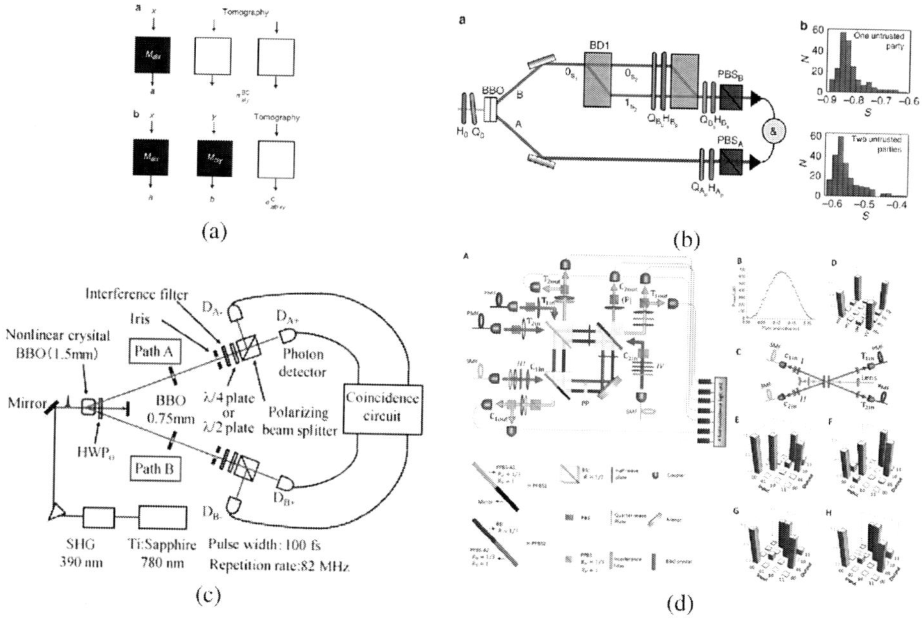

Figure 9. 2015 optical quantum communication and quantum cryptography channel design: (a) association design; (b) processing model. 2019 optical quantum entanglement circuit detection mode: (c) optical circuit design; (d) schematic results.

From the point of view of quantum optical detection, a two-way interactive system is configured in Fig. 10(a)–(b): (a) suitable for interacting with molecules through coherent light sources; (b) observing from a single molecule measured effects of reciprocal synthesis and decomposition of paired polarized photons [19].

Various ultrafast laser technologies play an important role in the field of analysis and measurement, providing a reliable source of generation for visible entangled photons. In Fig. 10(c)–(d), high-tech equipment support [20] for fine quantitative analysis is shown: (c) The polarization signal source providing various quantum interaction effects. (d) Detection signal distribution.

3.5 *Optical Measurement Component*

Various devices are distributed in modern optical measurement systems, and special grating devices are shown in Fig. 10(e) to provide support for the orderly display of the fine distribution characteristics of composite optical flow signals [24].

In Fig. 10(f), the light-splitting effect of the prism forms a filter control mechanism related to the position. Through the enhanced interaction mode, the enhanced optical signal is collected into a higher frequency and shorter period super signal sequence to achieve a shorter pulse period and stronger output [25].

The special microstructure graphene off-angle mode shows the characteristics of the formed lattice change in Fig. 10(g). Around the characteristics of this special material, the potential of the material as a new detection device is discussed [26].

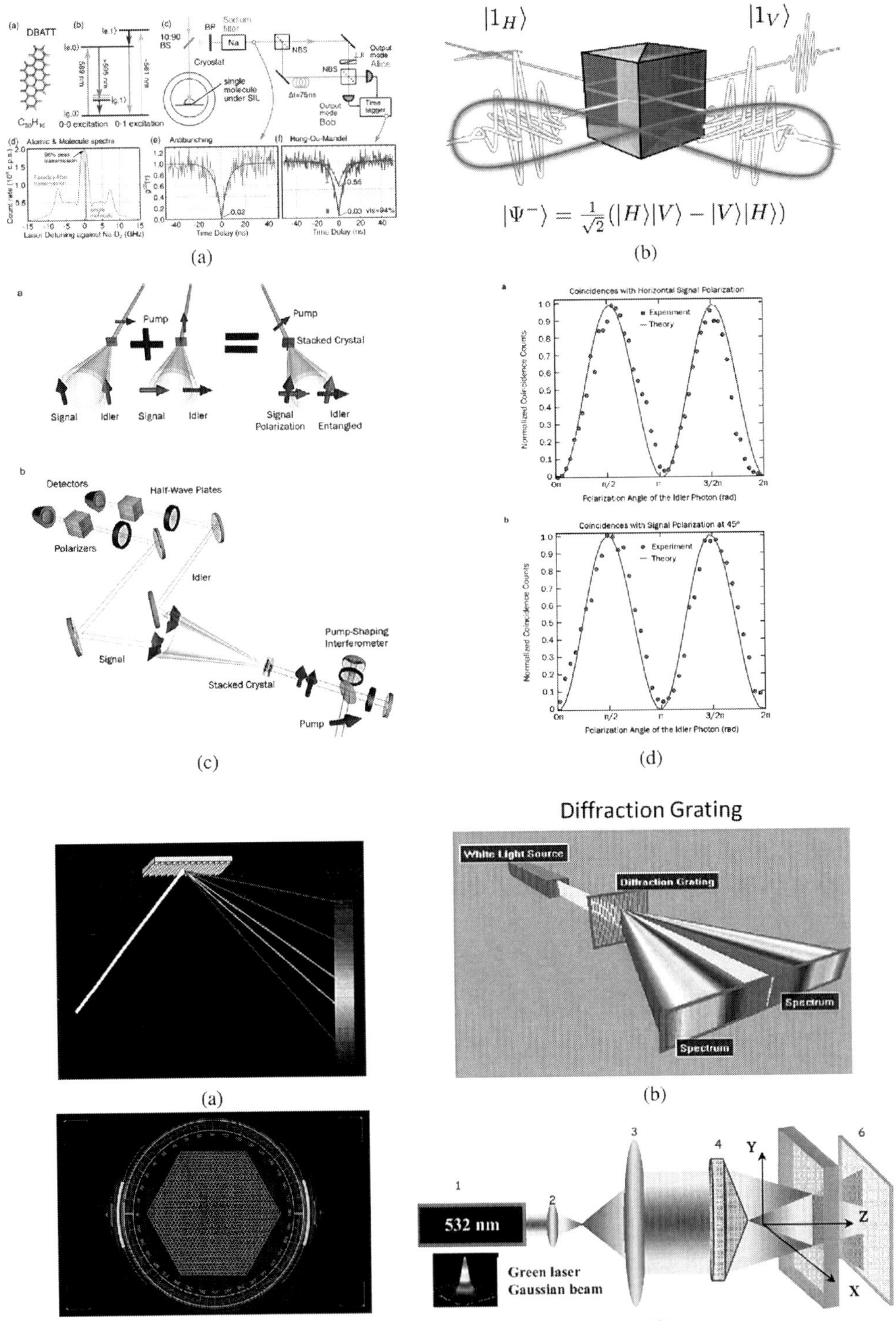

Figure 10. Molecular detection with quantum optics in 2019: (a) correlation model; (b) entanglement core representation. Ultrafast laser measurements in 2019: (c) correlation design; (d) schematic results. Optical measurement components: (e) grating components; (f) prism light splitting; (g) small corner graphene; (h) holographic image engineering.

In Fig. 10(h), a comprehensive integrated quantum optics technology is demonstrated for 2D and 3D holographic image engineering to support complex frontier technology applications [27].

3.6 Topology Surface Applications

All kinds of Dirac cone surfaces [28] are used in the forefront of condensed matter physics [33], topological quantum materials [30], nanosemiconductor materials [29], nonlinear optics [31], super cold atom simulation of graphene [32], quantum transport [34], and other research fields playing an important role. The series of visualized topological surfaces are given in Fig. 11(a)–(g).

In Fig. 11(a), the key time nodes and research topics of topological quantum materials from 1980 to 2019 are represented. In Fig. (b), different types of Dirac cone surfaces are shown, and in Fig. (c), complex bonding surfaces show connection surfaces. Fig. (d) shows the characteristic distribution of graphene simulated by ultracold atoms. Figs. (e) and (f) show that special topological quantum materials have photon emission spectra. Fig. (g) shows the topological surface structure of graphene under the condition of small angle deflection.

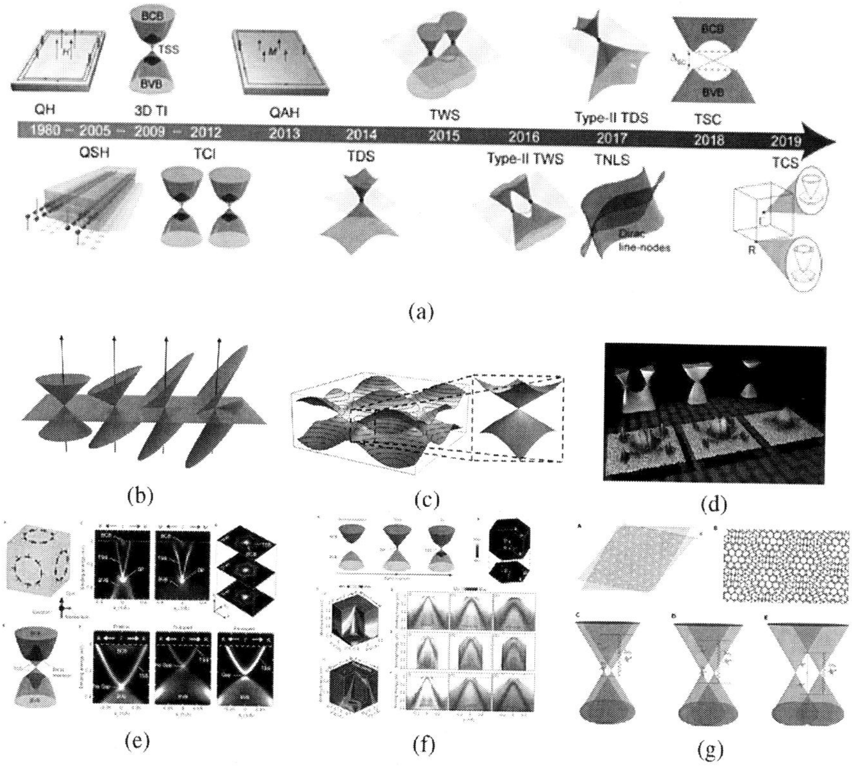

Figure 11. Topological surfaces: (a) historical development of topological quantum materials; (b) dirac cones; (c) complex junction surface; (d) ultracold atom simulation graphene; (e) topological quantum materials; (f) light quantum emission spectrum; (g) graphene small angle deflection.

4. Basics of Quantum Optics Measurements

To ensure the rationality of the quantum measurement model, the appendix of this book is used to collect batch lemmas, theorems, corollaries, and other forms of each chapter to provide basic support for the principle, aiming at the special structural characteristics of the measurement basis, discrete geometric transformation, multilevel information entanglement measurement, global analysis, matrix representation, key steps, etc., to establish adaptive theorems and inferences.

4.1 Discrete Transformation of Complex Vector Primitives

Under the action of different operators, complex vector primitives exhibit extremely rich transformation characteristics. In the case of various power exponent calculations, it is necessary to strictly distinguish the order of action at different levels to ensure the correctness of the calculation of primitive coordinates.

Theorem 3 (Square Root Product of Conjugate Vectors in First Quadrant of Complex Plane). *The vector* $z = re^{i\theta} = (x, y), \theta = \angle z \in [0, \pi/2)$; $\tilde{z} = re^{i\tilde{\theta}} = (y, x), \tilde{\theta} = \angle \tilde{z} \in (0, \pi/2]$ *in the first quadrant of the complex plane corresponds to the square root vectors* $\{\sqrt{z}, \sqrt{\tilde{z}}\}$ *in the first quadrant of the complex plane. The product of vectors is* $\sqrt{z}\sqrt{\tilde{z}} = ir$.

Proof.

$$\begin{aligned}
\sqrt{z} &= z^{1/2} = r^{1/2}e^{i\theta/2} \\
\sqrt{\tilde{z}} &= \tilde{z}^{1/2} = r^{1/2}e^{i\tilde{\theta}/2} \\
\sqrt{z}\sqrt{\tilde{z}} &= r^{1/2}e^{i\theta/2}r^{1/2}e^{i\tilde{\theta}/2} \\
&= re^{i(\theta/2+\pi/2-\theta/2)} = re^{i\pi/2} \\
&= r(0, 1) = ir
\end{aligned}$$

Theorem 4 (Product of Squares of Conjugate Vectors in First Quadrant of Complex Plane). *The vector* $z = re^{i\theta}, \theta, 2\theta \in [0, \pi/2)$; $\tilde{z} = re^{i\tilde{\theta}}, \tilde{\theta}, 2\tilde{\theta} \in (0, \pi/2]$ *are in the first quadrant of the square complex plane, the corresponding vector* $\{z^2, \tilde{z}^2\}$. *The product of paired vectors is* $z^2\tilde{z}^2 = ir^4$.

Proof.

$$\begin{aligned}
z^2 &= (z)^2 = r^2e^{i2\theta} \\
\tilde{z}^2 &= (\tilde{z})^2 = r^2e^{i2\tilde{\theta}} \\
z^2\tilde{z}^2 &= r^2e^{i2\theta}r^2e^{i2\tilde{\theta}} \\
&= r^4e^{i(2\theta+\pi/2-2\theta)} = r^4e^{i\pi/2} \\
&= r^4(0, 1) = ir^4
\end{aligned}$$

Theorem 5 (Complex Plane First Quadrant Conjugate Vector on *n* Power Root Product). *Vector* $z = re^{i\theta}\tilde{z} = re^{i\tilde{\theta}}$ *in the first quadrant of the complex plane corresponds to the vector* $\{\sqrt[n]{z}, \sqrt[n]{\tilde{z}}\}$. *The product of paired vectors is* $\sqrt[n]{z}\sqrt[n]{\tilde{z}} = ir^{2/n}$.

Proof.

$$\begin{aligned}
\sqrt[n]{z} &= z^{1/n} = r^{1/n}e^{i\theta/n} \\
\sqrt[n]{\tilde{z}} &= \tilde{z}^{1/n} = r^{1/n}e^{i\tilde{\theta}/n} \\
\sqrt[n]{z}\sqrt[n]{\tilde{z}} &= r^{1/n}e^{i\theta/n}r^{1/n}e^{i\tilde{\theta}/n} \\
&= r^{2/n}e^{i(\theta/n+\pi/2-\theta/n)} = r^{2/n}e^{i\pi/2} \\
&= r^{2/n}(0,1) = ir^{2/n}
\end{aligned}$$

Theorem 6 (Product of First Quadrant Conjugate Vector *n* Power of Complex Plane). *Vector $z = re^{i\theta}, \theta \in [0, \pi/2)$; $\tilde{z} = re^{i\tilde{\theta}}, \tilde{\theta}, n\tilde{\theta} \in (0, \pi/2]$ in the first quadrant of the n power complex plane, the corresponding vector $\{z^n, \tilde{z}^n\}$. The product of paired vectors is $z^n\tilde{z}^n = ir^{2n}$.*

Proof.

$$\begin{aligned}
z^n &= (z)^n = r^n e^{in\theta} \\
\tilde{z}^n &= (\tilde{z})^n = r^n e^{in\tilde{\theta}} \\
z^n\tilde{z}^n &= r^n e^{in\theta}r^n e^{in\tilde{\theta}} \\
&= r^{2n}e^{i(n\theta+\pi/2-n\theta)} = r^{2n}e^{i\pi/2} \\
&= r^{2n}(0,1) = ir^{2n}
\end{aligned}$$

Theorem 7 (Noncommutative Cross-level Operator). *Complex vector product operation with nonlinear projection, the operation involves power exponent function n power root | n power operation. The order of operation of operators is not commutative $n > 1, \sqrt{z}\sqrt{\tilde{z}} = ir \neq i^{1/2}r = \sqrt{z\tilde{z}}$; $\sqrt[n]{z}\sqrt[n]{\tilde{z}} = ir^{2/n} \neq i^{1/n}r^{2/n} = \sqrt[n]{z\tilde{z}}$; $z^2\tilde{z}^2 = ir^4 \neq i^2r^4 = (z\tilde{z})^2$; $z^n\tilde{z}^n = ir^{2n} \neq i^n r^{2n} = (z\tilde{z})^n$. After opening the n power root | n power transformation, the vector product of the n power root | n power is not equal to the cross-level multiplication of conjugate vectors in the original space and then n power root | n power. In the results of a power operation, the order of operations follows the transformation rules of the same level to avoid cross-level operations.*

Proof. Square root operation,

$$\begin{aligned}
\sqrt{z}\sqrt{\tilde{z}} &= re^{i\pi/2} = ir \\
&\neq \\
\sqrt{z\tilde{z}} &= \sqrt{re^{i\theta}re^{i\tilde{\theta}}} \\
&= \sqrt{r^2e^{i(\theta+\pi/2-\theta)}} = r\sqrt{e^{i\pi/2}} \\
&= re^{i\pi/4} = i^{1/2}r
\end{aligned}$$

n power root operation,

$$\begin{aligned}
\sqrt[n]{z}\sqrt[n]{\tilde{z}} &= r^{2/n}e^{i\pi/2} = ir^{2/n} \\
&\neq \\
\sqrt[n]{z\tilde{z}} &= \sqrt[n]{re^{i\theta}re^{i\tilde{\theta}}}
\end{aligned}$$

$$
\begin{aligned}
&= \sqrt[n]{r^2 e^{i(\theta+\pi/2-\theta)}} = r^{2/n} \sqrt[n]{e^{i\pi/2}} \\
&= r^{2/n} e^{i\pi/(2n)} = i^{1/n} r^{2/n}
\end{aligned}
$$

Square operation,

$$
\begin{aligned}
z^2 \tilde{z}^2 &= r^4 e^{i\pi/2} = ir^4 \\
&\neq \\
(z\tilde{z})^2 &= (re^{i\theta} re^{i\tilde{\theta}})^2 \\
&= (r^2 e^{i(\theta+\pi/2-\theta)})^2 = r^4 (e^{i\pi/2})^2 \\
&= r^4 e^{i\pi} = i^2 r^4
\end{aligned}
$$

n power operation,

$$
\begin{aligned}
z^n \tilde{z}^n &= r^{2n} e^{i\pi/2} = ir^{2n} \\
&\neq \\
(z\tilde{z})^n &= (re^{i\theta} re^{i\tilde{\theta}})^n \\
&= (r^2 e^{i(\theta+\pi/2-\theta)})^n = r^{2n} (e^{i\pi/2})^n \\
&= r^{2n} e^{in\pi/2} = i^n r^{2n}
\end{aligned}
$$

Corollary 7 (Product of Pairs of Conjugate Vectors in Other Quadrants). *The product of conjugate vectors* $\{\sqrt[n]{z}, \sqrt[n]{\tilde{z}}\}$ *vectors is* $\sqrt[n]{z}\sqrt[n]{\tilde{z}} = \pm ir^{2/n}$.

Proof. For the first and third quadrants, the product is $+ir^{2/n}$, and for the second and fourth quadrants, the product is $-ir^{2/n}$.

4.2 Visualization of Discrete Geometric Transform

The discrete transformation of complex function primitives provides various transformation modes to facilitate compound operator mapping functions such as rotation and reflection. However, from the perspective of image visualization, human visual understanding relies on the support of geometric optics. In the discrete geometric configuration conversion, the optical measurement assists the complex number plane in providing an enhanced coordinate transformation function for the flexible display of conjugate complex vector functions to maintain the obtained measurement parameters in each attribute region. Let $|.|$ be the arithmetic absolute value operator $S(.)$ sign function, n power root operator maps the complex plane to be

$$
\sqrt[n]{(,)} : (x, y) \to (S(x)\sqrt[n]{|x|}, S(y)\sqrt[n]{|y|})
$$

multiplied by n power mapping

$$
(,)^n : (x, y) \to (S(x)|x|^n, S(y)|y|^n)
$$

The corresponding power exponent mapping geometric transformation is called the power function cobweb (spider web) projection.

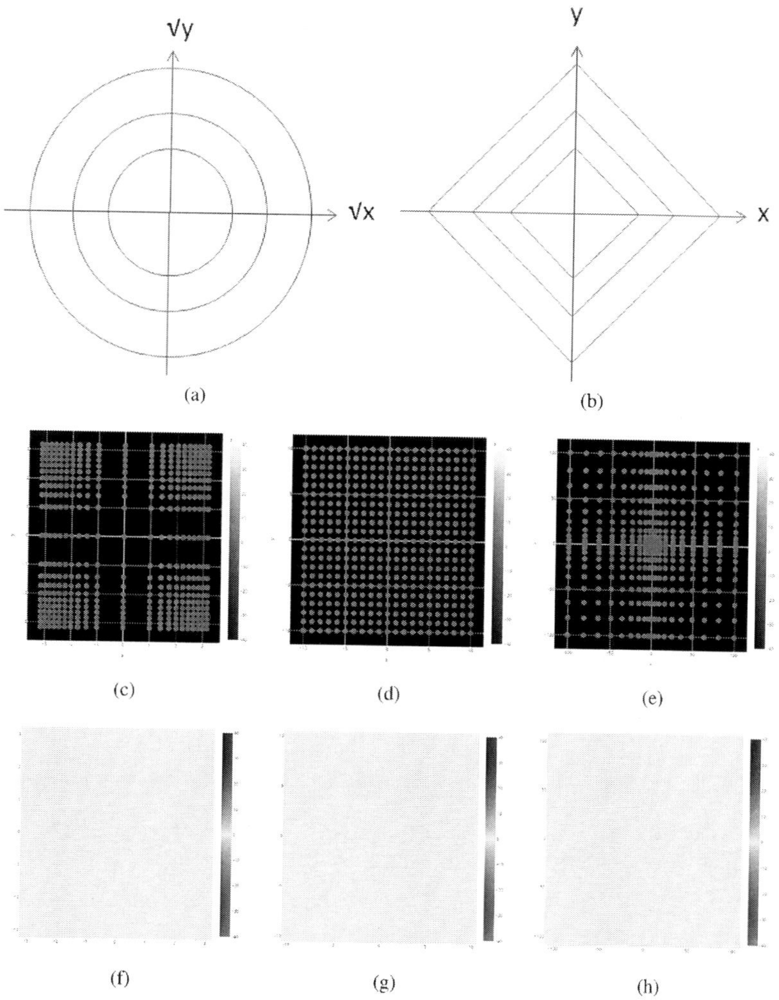

Figure 12. Circle-to-rhombus power transformation: (a) square root complex plane $n = 1/2$; (b) normal complex plane $n = 1$. Complex base coordinate geometric transformation with zero point value: (c)–(e) cobweb basis and (f)–(h) heatmaps. (c) square-rooted complex plane $n = 1/2$; (d) linear complex plane $n = 1$; (e) square complex plane $n = 2$; (f) square heatmap $n = 1/2$; (g) linear heatmap $n = 1$; (h) square heatmap $n = 2$.

Theorem 8 (Cobweb Projection Operator Eight-direction Projection Invariance). *The cobweb projection operator can keep the projection direction invariant in eight directions. They are* $\{0, \pi/4, \pi/2, 3\pi/4, \pi, 5\pi/4, 3\pi/2, 7\pi/4\}$, *respectively.*

Proof. For the four coordinate directions, any power function will not change the original direction. When $|p| = |q|$, different power exponents will also maintain the original direction. However, for other angles, the direction changes slightly. For example, the cobweb square transformation: $(p, q) = (2, 1), \cos\theta = 2/\sqrt{5} \rightarrow (p^2, q^2) = (4, 1), \cos\rho = 4/\sqrt{17} \rightarrow \theta \neq \rho$.

The mapping effect from the circular coordinates of the square root operator to the diamond coordinates of the linear operator is illustrated in Fig. 12(a)–(b). The three-point

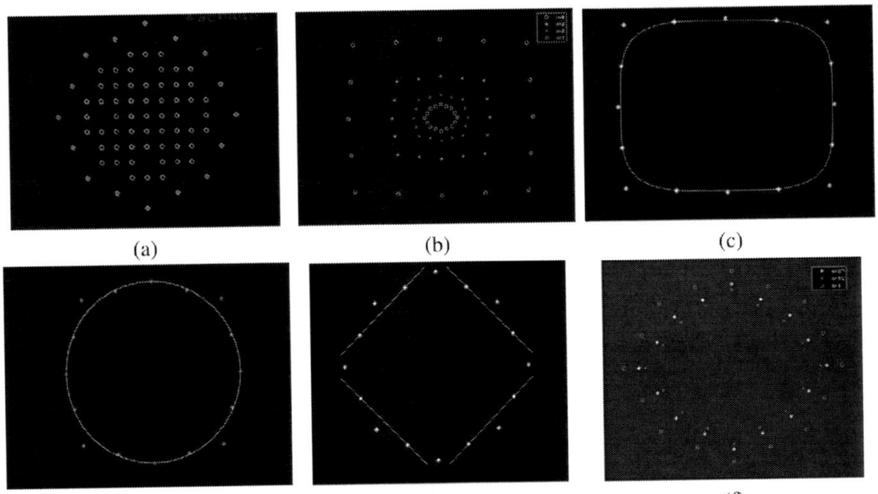

Figure 13. Two-dimensional coordinate projection supported by conjugate complex variable vector primitives: (a) $p = 2q$ conjugate group qualified basic complex coordinate family; (b) power exponent expansion $n = \{1, 2, 3, 4\}$; (c) $n = 4$ loop; (d) $n = 3$ loop and circle; (e) basic coordinate periphery and rhombus; (f) power exponent contraction $n = \{1, 2/3, 1/2\}$.

group coordinate plane display modes are shown in Fig. 12(c)–(e), and the complex number primitive coordinates are visually displayed from the distribution structure of all points with zero value in $n = \{1/2, 1, 2\}$. Various distribution grids show grid point change modes. In Fig. 12(f)–(h), the information carried by (c)–(e) point primitives is displayed in the heatmap mode, and the equivalent distribution effect of the three graphs is displayed as a whole region.

The ring configuration corresponding to each concentric circle in the square root complex plane constitutes the equivalent feature of the complex vector function. Using the pairwise conjugation properties of complex pairs, the product and projection properties described in Chapter 1 are maintained. Using local area smoothing, the global characteristics are difficult to judge in point coordinates, and it is relatively easier to make the mode judgment of the heatmap to observe the corresponding change characteristics from the coordinate values of each map.

In Fig. 13(a)–(f), there is a list of the outer contour links on the $p = 2q$ conjugate group combinatorial lattice of the complex coordinate family, and the outer contour is shown by different modes of exponential expansion and contraction changes in the link. In Fig. (a), the 16 points of the outer contour selected from the 73 grid points are marked as red *, and the 16 initial points of Fig. (b) are placed in the middle as green * $n = 1$, and the three expanded parts are in $n = \{2, 3, 4\}$. Fig. (c) shows that the $n = 4$ link distribution is close to a rectangle, Fig. (d) shows that the $n = 2$ link distribution is close to a circle, and Fig. (e) shows that the $n = 1$ link distribution is close to a rhombus. The 16 initial points in Fig. (f) are placed in the outermost red * $n = 1$, and the index of the two shrink parts in turn is $n = \{2/3, 1/2\}$. Observing these different parameter coordinate points, although the lengths of the vectors involved in the corresponding loops are different, all the associated exponential coordinates lie on nearly straight lines with briefly equal angles. It embodies the characteristic cobweb transformation.

Theorem 9 (Cobweb Projection Operator Quantitative Measurement Real Function Output Mode). *In the complex plane $[-N, N]^2$ region, it involves the quantitative measurement value carried by the complex primitive coordinate point, though the correlation functions are represented as real measurements at z coordinates perpendicular to the complex plane.*

Proof. The complex function vector structure based on the power exponent can be used in various modes: probability, absolute value, modulo function, and other real function output operator modes to meet this type of condition.

In Figs. 14–15(a)–(o) five sets of functions $1.(p, q)$, $2.\{(p, q), (p, -q)$, $3.\{(p, q), (-p, q)\}$, $4.\{(p, q), (-p, -q)\}$, $5.\{(p, q), (q, p)\}, n = \{1/2, 1, 2\}$ are displayed in two modes: spider web coordinates and heatmap.

In the information displayed by the cobweb coordinates in Fig. 14, the visual judgment is easily affected by the position, and in the heatmap of Fig. 15, the visual effect of local and global correlation is clearly displayed. From the horizontal polarization formed by complex conjugate operators in heatmap mode, the vertical polarization formed by real conjugate operators, the zero-value plane formed by negative conjugate operators, and the block effect of paired conjugate operators from the output diagrams of the three different parameters are observed to obtain corresponding differences.

The nine groups of functions selected in Figs. 16 (a)–(i) are: (a) $\{V(x)\}$ four-element rotation cross; (b) $\{K(x)\}$ Klein group; (c) $\{K(\tilde{x})\}$ conjugate Klein group; (d) $\{V(x) + V(\tilde{x})\}$ conjugate eight elements; (e) $\{|x + \tilde{x}|\}$ absolute value of the paired conjugate group; (f) $\{x + \tilde{x}\}$ paired conjugate group; (g) $\{(p, q), (p, q) + (p, -q)\}$ fundamental surface + complex conjugate surface; (h) $\{(p, q), (p, q) + (q, p)\}$ fundamental plane + conjugate plane; (i) $\{(p, q), (p, q) + (-p, q)\}$ fundamental plane + real conjugate plane.

With the selected projection function in Fig. 16, a three-dimensional diagram is formed to reflect the intrinsic properties of the conjugate vector function. In the distribution surface shown in Fig. (e), all sample points are greater than or equal to zero, which is suitable for expressing the quantum change form expressed using the counting statistics mode to comprehensively utilize the quantitative features associated with the geometric shape and coordinate points to display the associated measurement content.

Corollary 8. *Start from any $(\sqrt[n]{m}, 0), m \geq n$ position to $(0, \sqrt[n]{m})$ end, there are $m + 1$ vectors $\{(\sqrt[n]{j}, \sqrt[n]{k})\}_{j+k=m}, j, k \in [0, m]$ associated with the ring chain.*

Proof. From the relationship between corresponding points on the n-th power root complex plane, $m + 1$ coordinate positions that satisfy the $j + k = m$ constraint condition change monotonically. The chain $n = 1$ is connected by a straight line, and $n = 2$ is connected by an arc.

Theorem 10 (Number of Equipotential Loops in $[-N, N]^2$ Region of Complex Plane). *The number of points involved in the $[-N, N]^2$ region of the complex plane. The number of cycles more than 2 is $2N - 1$.*

Proof. Considering the first quadrant, remove the shared zero point, each coordinate point supports a loop, and the length of the link in $\{(N, 0), (0, N)\}$ is the $N + 1$ range. Beyond this position it is also possible to define complementary N links of decreasing length from $N - 1$ to 1.

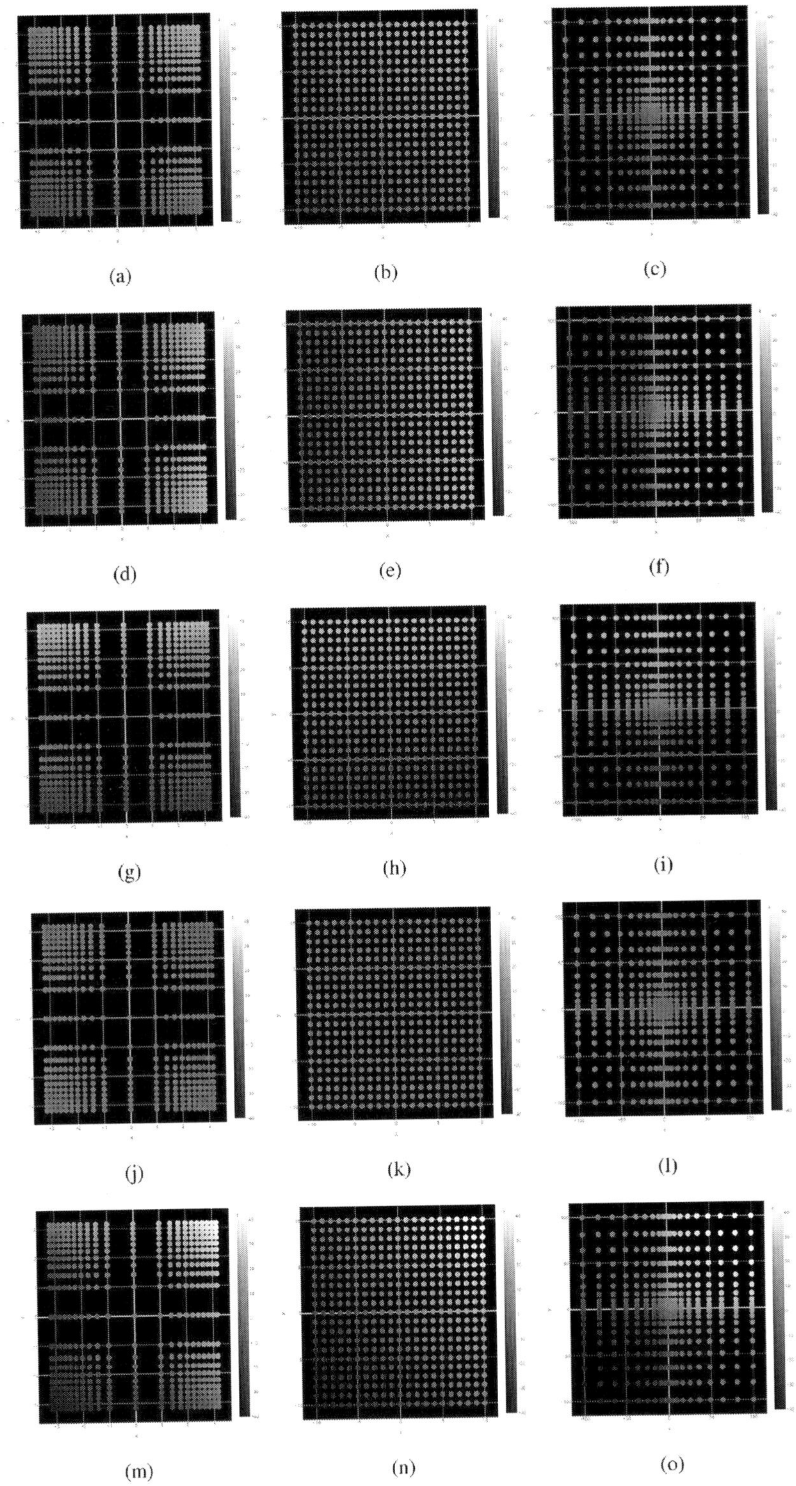

Figure 14. Five groups of two-dimensional coordinate function projections: (a) $(p,q)^{1/2}$; (b) (p,q); (c) $(p,q)^2$; (d) $[(p,q)+(p,-q)]^{1/2}$; (e) $[(p,q)+(p,-q)]$; (f) $[(p,q)+(p,-q)]^2$; (g) $[(p,q)+(-p,q)]^{1/2}$; (h) $[(p,q)+(-p,q)]$; (i) $[(p,q)+(-p,q)]^2$; (j) $[(p,q)+(-p,-q)]^{1/2}$; (k) $[(p,q)+(-p,-q)]$; (l) $[(p,q)+(-p,-q)]^2$; (m) $[(p,q)+(q,p)]^{1/2}$; (n) $[(p,q)+(q,p)]$; (o) $[(p,q)+(q,p)]^2$.

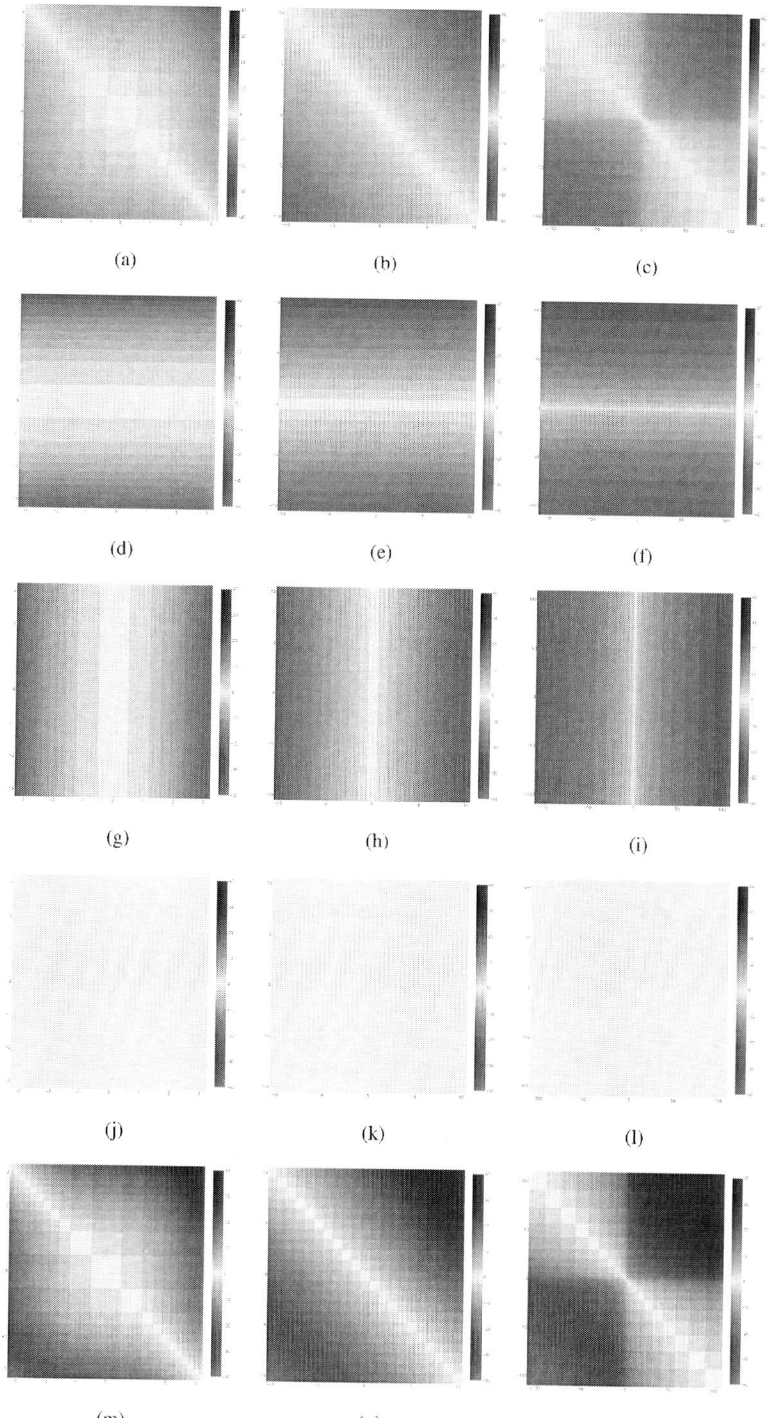

Figure 15. Five Sets of two-dimensional coordinate function area smoothing heatmap: (a) $(p,q)^{1/2}$; (b) (p,q); (c) $(p,q)^2$; (d) $[(p,q)+(p,-q)]^{1/2}$; (e) $[(p,q)+(p,-q)]$; (f) $[(p,q)+(p,-q)]^2$; (g) $[(p,q)+(-p,q)]^{1/2}$; (h) $[(p,q)+(-p,q)]$; (i) $[(p,q)+(-p,q)]^2$; (j) $[(p,q)+(-p,-q)]^{1/2}$; (k) $[(p,q)+(-p,-q)]$; (l) $[(p,q)+(-p,-q)]^2$; (m) $[(p,q)+(q,p)]^{1/2}$; (n) $[(p,q)+(q,p)]$; (o) $[(p,q)+(q,p)]^2$.

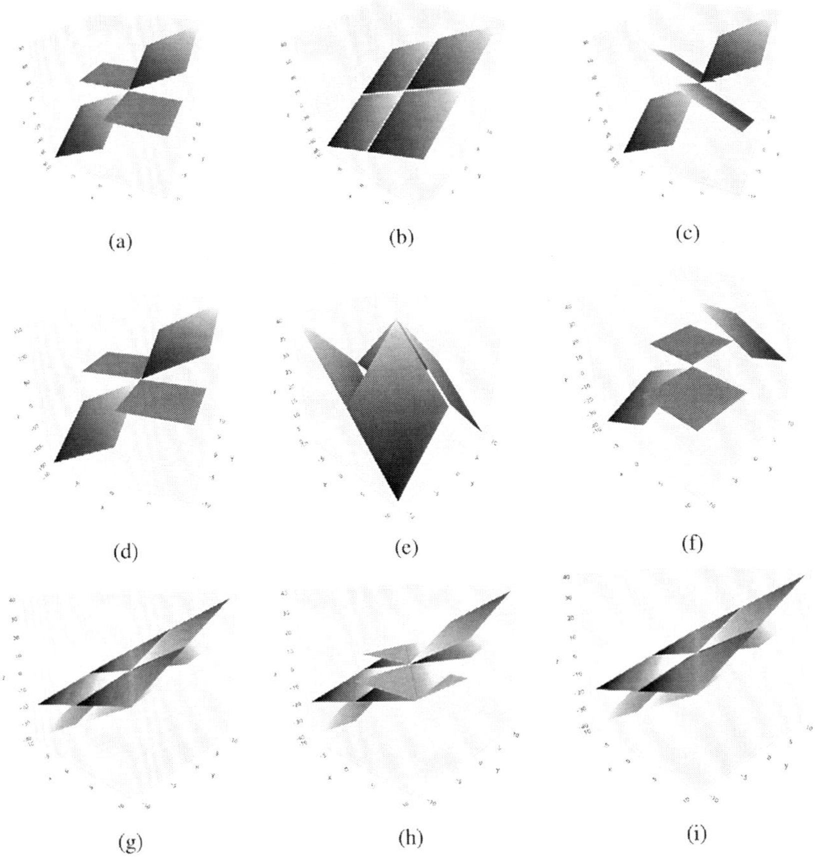

Figure 16. Nine groups of three-dimensional coordinate function projections: (a)–(i) $x = (p, q)$ (a) $\{V(x)\}$; (b) $\{K(x)\}$; (c) $\{K(\tilde{x})\}$; (d) $\{V(x) + V(\tilde{x})\}$; (e) $\{|x + \tilde{x}|\}$; (f) $\{x + \tilde{x}\}$; (g) $\{(p, q), (p, q) + (p, -q)\}$; (h) $\{(p, q), (p, q) + (q, p)\}$; (i) $\{(p, q), (p, q) + (-p, q)\}$.

Theorem 11 (Three-dimensional Rhomboid Equipotential Torus in Complex Plane $[-N, N]^2$). *From $(0, 0)$ position in $[-N, N]^2$ complex plane, it starts to reach $(N, 0), (0, N)$ link diamond area before stepping up and expanding outward, reaching $\{(N, 0), (0, N)\}$ link diamond area then step and shrink until (N, N) contains a step number of $2N + 1$ to form a rhomboid three-dimensional structure.*

Proof. The $2N - 1$ loops formed in the plane, plus two zero points, form a $2N + 1$ layered rhomboid three-dimensional structure.

For example, 2D and 3D rhomboid structures are shown in Fig. 17(a)–(d).

The nine groups of functions selected in Figs. 18 (a)–(i) constituting the visualization surfaces are (a) double diamond cone surface-Type II Dirac cone, (b) double paraboloid-potential function distribution, (c) multiple compound saddle surface - complex coupling surface, (d) complex spherical connection coupling surface—complex potential function, (e) single saddle surface, (f) single high power surface, and (g)–(i) strong interaction space potential surface from different viewing angles—special conjugate vector functions of complex variables.

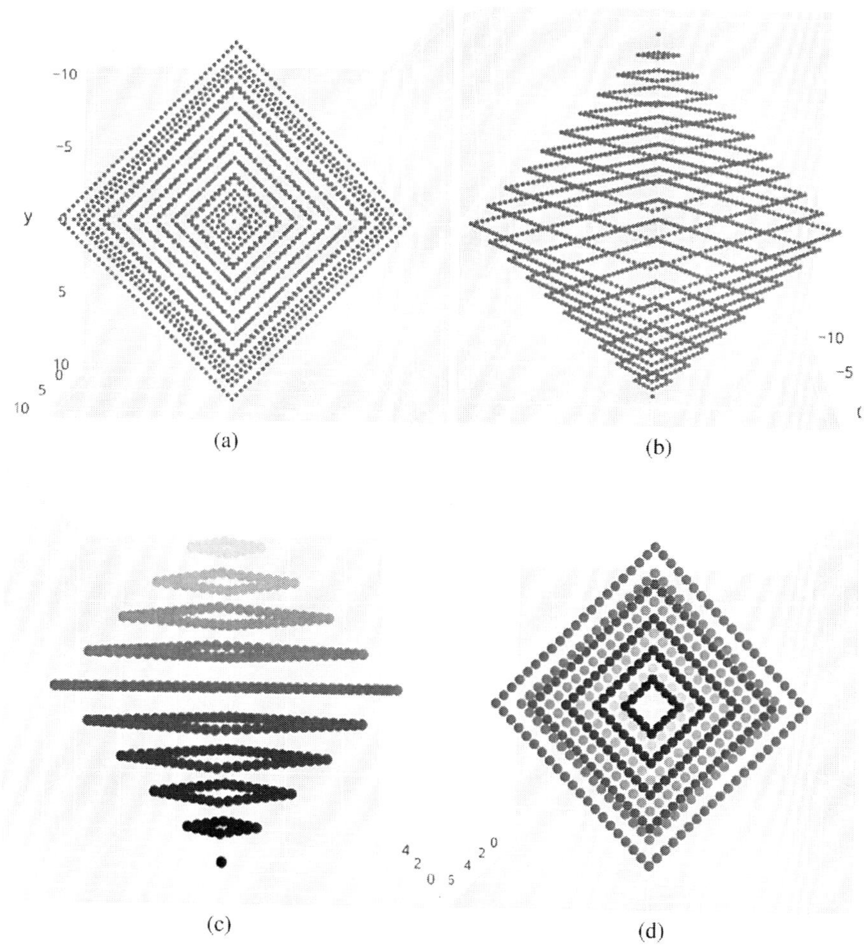

Figure 17. Rhombic equipotential surface: (a), (d) planar rhomboid ring, (b), (c) three-dimensional rhomboid ring.

These macroscopic non-Euclidean geometric shape distributions show the flexible and precise basic geometric measurement support mode that conjugate complex vector functions may provide in different application scenarios from the intrinsic levels.

5. Quantum Optics Determine Prerequisites for Detection of Quantum Entanglement

5.1 Measurement Interaction of Information Entangled States

Based on the linear algebra of operation rules on the matrix and eigenvectors, the mutual product of a large number of eigenvectors with different eigenvalues is zero in the conjugate orthogonal operation mode, and those nonzero eigenvectors need to be further distinguished as entangled eigenvectors. With the rapid development of cutting-edge quantum optics and photon quantum technology, a variety of rich optical experimental models and

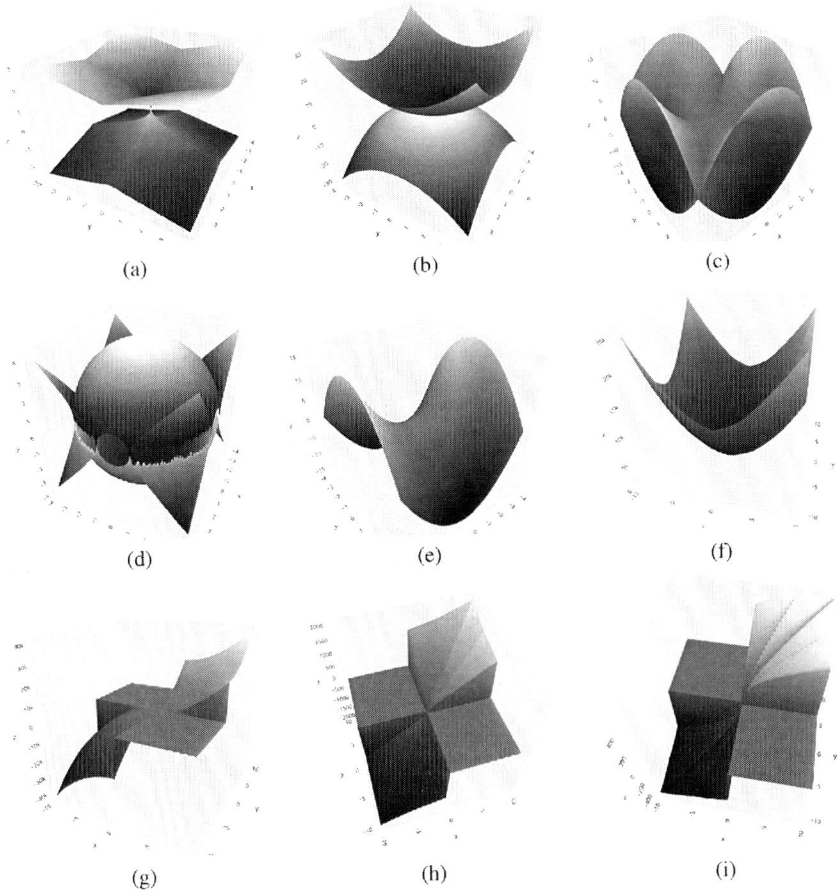

Figure 18. Topological surface—cobweb projection mode: (a) double diamond cone surface; (b) double paraboloid; (c) multiple compound saddle surface; (d) complex spherical connection coupling surface; (e) single saddle surface; (f) single high-power surface; (g)–(i) strong interaction space potential surface from different viewing angles.

photon quantum detection equipment constitute an operable basic condition for the precise measurement of the quantitative characteristics of the cluster state of complex photon quantum interactions.

From the perspective of the measurement operation, normalization conditions are the basis of the information description. For the field of complex signal sending and receiving systems of remote communication of arbitrary information transmission type, it is usually effective to normalize complex interaction modes. Entangled information may be formed when the normalized information vectors have a matching argument angle relationship. However, from the perspective of the whole analysis of complex systems, due to the indiscriminate integrated information, the various local information contained in the normalized system itself conflicts with each other, and it is easy to introduce the overall logical paradox of the intrinsic series, including Mobius rings, Klein bottles, geometric paradox model, etc.

For special application fields such as atomic physics, photochemistry, biological genes, nanomaterials, quantum optics, and optical quantum where entangled information has obvious characteristics, it is necessary to add a distinct series of constraints on normalization.

A complex number $x = (p, q), \theta = \angle x, r = \sqrt{p^2 + q^2}$ in the first quadrant of the conjugate complex vector function has the same radius. The ring elements of eight coherent vector states are intrinsically associated with $\{\theta, r\}$ from the level of invariance measurement. It is essential to determine the coherence radius and accurately describe the characteristics of related entangled states.

5.2 *Multiple Levels of Quantum Optics and Optical Entanglement Measurement*

From the comprehensive level of the detection environment, there are quantitative measurements, single state measurements, coherent clustering, multiple iteration effects, and other optical quantum measurement specifications, description models, and quantitative projections. It is necessary to take the appropriate quantitative analysis of optical quantum entanglement, visual measurement, and analysis system to meet six sets of assumptions and prerequisites.

Assumption 1 (Quantum Optics and Optical Quantum Measuring—Application Constraints). *There are correlation measurement, analysis model method and its system, belonging to the field of quantum optics and optical quantum science, involving constraints on the interaction characteristics of optical quantum, multiqubit feature description and analysis, entangled state interaction, quantum information, and comprehensive measurement result as well as distribution visualization scope.*

Assumption 2 (Geometric Optics+Cobweb Projection+Coincidence Counting—Measurement Visualization). *Based on light quanta detector and geometric optics transformation mechanism, the measurement satisfies geometric optics, cobweb projection, coincidence counting, and other modes, as discrete metric information interaction on visualization support of the quantitative feature.*

Assumption 3 (Paired Conjugate Complex Function Eigenvalue—Conjugate Measurement). *In the configuration space of the optical quantum information measurement system, the eigenvalue measure of the paired optical quantum is the conjugate complex pair* $\{x, \tilde{x}\}$, *the measured parameter is* $\{\theta, \tilde{\theta}\}$, *and the intrinsic vector radius r is the region invariant eigenvalue.*

Assumption 4 (Photon Quark Conjugate Discrete Cluster—Smallest Complete Set). *In the configuration space, paired light quanta conjugate complex eigenvalues, the four attribute regions represented by* $\{0, \pi/2, \pi, 3\pi/2\}$ *are the starting point to distribute in each region* $\pi/2$ *containing eight light quanta, associated with the region property code* $\{0231, 0132\}$ *conversion to determine the inverse of the interaction in the clock/clockwise rotations on the order of time. There are at most four groups of coherent light quanta pairs at the same time, forming the conjugate Klein group-quark discrete cluster.*

Assumption 5 (*n* Iterative Photons Form Single Energy Level Continuous Ring—Single Ring). *In the n iterative processing mode, four groups of coherent photon sequence pairs form complex entangled particles in the measurement. A closed loop is formed in the phase space, which satisfies the conjugate complex vector function, and the quantum eigenvalues after n iterations remain unchanged.*

Assumption 6 (n Iterative Quanta Form Multiple Energy Level Continuous Rings—Multiple Rings). *Under the strong interaction, the light quanta n iterative processing under the condition of Assumption 3, the complex entanglement composite states are distributed in the measurement phase space to form multiple closed loops, which satisfy the conjugate group qualifying conditions, and the light quanta grouping is constrained in the specific eigenvalue region.*

6. Analysis Model

6.1 Complete Quantum Information Phase Space

6.1.1 Basic Logical Primitives

Compared with the classical logical Turing machine, which is a serial operation mechanism, the unit information amount of $\{0, 1\}$ is described as bits, and the quantum logic information is formed by complex vector expansion, using two sets of primitives $\{|0\rangle, |1\rangle\}$ as the initial generation element. Since this kind of extended incomplete description [1] cannot cover the Wigner phase space, it is necessary to explore a complete description mechanism. Conjugate logic is based on vector functions of complex variables, and four primitive representations $\{(1, 0), (0, 1), (-1, 0), (0, -1)\}$ are the initial generating elements.

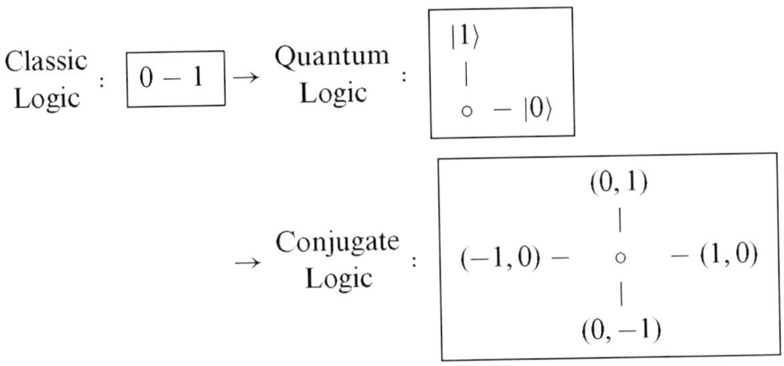

Corollary 9 (Classical Logic, Quantum Logic, and Conjugate Logic). *Classical logic is based on a single primitive $\{0, 1\}^n$ linear structure, quantum logic is based on two primitive vectors based on $\{|0\rangle, |1\rangle\}^n$ planar structure, and conjugate logic is based on four primitive vectors $\{(1, 0), (0, 1), (-1, 0), (0, -1)\}^n$ complex plane structure.*

Proof. Classical logic and quantum logic refer to Lemmas 1–7 in Chapter 5 of this book. For the conjugate logic part, refer to Corollaries 1–2 in Chapter 1 of this book.

6.1.2 Number of Phase Space Eigensolutions

From the perspective of primitive distribution, classical logic is a linear structure, quantum logic is generalized to two vectors, and conjugate logic is generalized to four elements of vectors on a balanced distribution structure. Let a conjugate or complete quantum bit / a complete qubit be a cobit. The corresponding phase space dimensions are as follows:

Classical Phase Space : $\{0,1\}$/bit → n bits → 2^n States Linear Combination Solutions

Quantum Phase Space : $\{|0\rangle, |1\rangle\}$/qubit → n qubits → 2^n States $\{1,-1\}$ Real Number Eigensolutions

Conjugate Phase Space : $\{0,1,\tilde{0},\tilde{1}\}$/cobit → n cobits → 2^{2n} States $\{1,i,-1,-i\}$ Complex Eigensolutions

Corollary 10 (Number of Classical, Quantum, and Conjugate Phase Space Eigensolutions). *From the perspective of primitive distribution, classical logic is a linear structure, quantum logic is generalized to be two vectors, and conjugate logic is generalized to be four meta-vectors on balanced distribution structure, the numbers of phase space states involved in the n meta-variable are* $\{2^n, 2^n, 2^{2n}\}$.

Proof. For a proof of the phase space of classical logic and quantum information logic, check [1], Lemma 1, and Chapter 5. For a demonstration of conjugate logical phase spaces, see Chapter 2, Theorem 10.

6.1.3 Fine Algebraic Structure

From the perspective of algebraic analysis, the similarities and differences between the classical quantum description and the conjugate algebraic transformation are investigated using the interaction mechanism, such as the subgroup ring/field and symmetry operator, accompanied by the fine description formula.

Quantum Transformation Group : $xx^* = 1,\ U_T^{-1} = U_T^*$, $tr(U_T U_T^*) = tr(U_T^* U_T) = I$ → $SU(n)$ Transformation Group Lie Algebra Real Eigensolutions

Conjugate Transformation Group : $y\tilde{y} = i,\ Y^{-1} = \overleftarrow{\tilde{Y}}$, $tr(Y\tilde{Y}) = tr(\tilde{Y}Y) = iI$ → $CTS(n)$ Transformation Group Conjugate Algebra Complex Eigensolutions

Corollary 11 (Quantum and Conjugate Transformation Groups). *$SU(n)$ is the core of the classical quantum transformation group, and the inner product sum of pairs of complex conjugate vectors is the unit vector on the global transformation structure supported by Lie algebras.*

CTS(n) is the core of the conjugate transformation group, and the product of pairwise conjugate vectors is an imaginary unit vector.

Proof. For the quantum transformation, see Chapter 5, Lemmas 1–2. For the conjugate transformation part, refer to Theorems 10–12 in Chapter 1.

The basic theory part of this book consists of four chapters, showing the accompanying complex number plane on four quadrant regions with conjugate algebraic support.

Quantum information requires complete phase space support, and then it is partially demonstrated that n cobits (cobit : complete or conjugate qubit) have a conjugate phase space dimension of $4^n = 2^n \times 2^n$, rather than the currently recognized 2^n. Because of this huge gap in phase space, both classical quantum mechanics and modern quantum information are faced with many problems as a series of paradoxes presented through the analysis and application of quantum entanglement theory.

6.2 *From Lie Algebra to Conjugate Algebra*

Starting from the given variable x, add $\{*, -1, -, \leftarrow, [,], \{,\}\}$ sequentially to form group/ring/field, reflection, rotation, and symmetry tensor-like extensions such as symmetric/antisymmetric extraction to obtain reciprocal symmetry information associated with Poisson, Lagrange, Lie, Dirac brackets. The core processing flow is described as

$x(1:1) \rightarrow$ | $\{*\}(2)$ Vector Group, Ring 2:4 | $\rightarrow$ | $\{-1\}(2)$ Vector Field 4:16 |

$\rightarrow$ | $\{-\}(2)$ Vector Reflection 8:64 | $\rightarrow$ | $\{\leftarrow\}(4)$ Rotation 32:1024 |

$\rightarrow$ | $\{[,]_d, \{,\}_d\}(2)$ Symmetry/Antisymmetry 64:4096 |

$\rightarrow \{X(64), H(4096)\}$

$y(1:1) \rightarrow$ | $\{\sim, \sharp, /, *\}(8)$ Conjugate Vector Group, Ring 8:64 |

$\rightarrow$ | $\{-1\}(2)$ Vector Field 16:256 | $\rightarrow$ | $\{-\}(2)$ Vector Reflection 32:512 |

$\rightarrow$ | $\{\leftarrow\}(4)$ Rotation $128:2^{14}$ | $\rightarrow$ | $\{[,]_d^u, \{,\}_d^u\}(4)$ Symmetry/Antisymmetry $512:2^{18}$ |

$\rightarrow \{Y(512), C(2^{18})\}$

Corollary 12 (Lie and Conjugate Algebras). *From basic variables to symmetric operators, both Lie algebra and conjugate algebra need to go through six levels of extended structure, where on* $\{1,3,4,5\}$ *layers, the content of the level expansion is the same, and the essential difference lies in the second and the sixth levels. In the second layer, the classical quantum involves pairs of complex conjugate vectors* $\{x, x^*\}$ *and pairs of complex conjugate angles* $\{\theta, -\theta\}$*, while conjugate structures involve four conjugate extension operators* $\{\sim, \sharp, /, *\}$*, paired conjugate angles* $\{\theta, \tilde{\theta}\}$*, and eight conjugate vectors with an expansion factor of 1:4. In the sixth layer, the symmetric and antisymmetric operators are expanded from the main diagonal to the main (real number solution)/auxiliary (imaginary number solution) double diagonals, indicating that the expansion factor is 1:2. The composite structure expansion factors are 1/8 and 1/64. A unit cobit different from a qubit requires 64 bits of additional information for support.*

Table 1. The six levels of quantum information representation and the conjugate complex vector function represent the number of vectors/tensors and ratio

L	Op	x	xD	H	xD^2	y	yD	C	yD^2	xD/yD	xD^2/yD^2
1		$\{x\}$	1	$\{x\}^2$	1	$\{y\}$	1	$\{y\}^2$	1	1	1
2	$\{\sim,\sharp,\prime,*\}$	$\{x,x^*\}$	2	$\{x,x^*\}^2$	4	$V=\{Y,\tilde{Y}\}$	8	V^2	64	1/4	1/16
3	-1	$\{X\}$	4	X^2	16	$E=\{V,V^{-1}\}$	16	E^2	256	1/4	1/16
4	$-$	$W=\{X,\overline{X}\}$	8	W^2	64	$G=\{E,\overline{E}\}$	32	G^2	1024	1/4	1/16
5	$\leftarrow$	$U=\{^{0231}\overleftarrow{X},\ ^{0231}\overleftarrow{\overline{X}}\}$	32	U^2	1024	$K=\{^{0231}\overleftarrow{G}\}$	128	K^2	1024×16	1/4	1/16
6	$[,],\{,\}$	$[U,U],\{U,U\}$	64	$\{[,]\{,\}\}^2$	4096	$[,]^u_d\{,\}^u_d$	512	$\{[,]^u_d\{,\}^u_d\}^2$	2^{18}	1/8	1/64

Proof. The expansion of each level refers to the first half of the book Lemmas, Theorems, and Corollaries and the second half of the second chapter. Parameter calculation reference Table 1 estimates expansion variables and operators for each expansion level.

Corollary 13 (Classical Quantum Structure Is Simplified Representation of Conjugate Structure). *From Lie algebra to conjugate algebra representation, it can be observed that the classical quantum structure is a conjugate algebra that simplifies the eight-variable vector cluster into two argument vectors on complex conjugate pair representation. The complete quantum solution space structure needs to be equipped with principal and auxiliary double diagonal complex eigenvalues, and the n cobit phase space representation is extended from* 2^n *to* 2^{2n}.

Proof. See the discussion in the second half of Chapter 2. A complete description requires the addition of auxiliary diagonal eigenvalues.

6.3 *Comparison of Classical Quantum Information Variable and Conjugate Vector Function on Description Space*

Using the support provided in Chapters 1–5 of this book, the two formulas describe the extension operators into six levels and compare the number of vectors/tensors involved and the quantization ratio of each level.

The abbreviated symbols in the table are the L layer number, Op operator, x initial vector, xD vector dimension, H Hilbert tensor, xD^2 tensor dimension, y initial vector, yD vector dimension, C Conjugate tensor, yD^2 tensor dimension, xD/yD quantum/conjugate vector dimension ratio, and xD^2/yD^2 Hilbert/Conjugate tensor dimension ratio. $*$ complex conjugate operator, $\{\sim,\sharp,\prime,*\}$ quaternary conjugate extension operator, group/ring/Galois field -1 inverse operator, $-$ vector reflection operator, $\leftarrow$ rotation operator, $\{[,],\{,\}\},\{[,]^u_d\{,\}^u_d\}$ Lie/Poisson/Lagrange/Dirac/Conjugate bracket operator, $\{X=\{x,x^{-1},x^*,x^{*-1}\},W,U\}$, $Y=\{y,y^\sharp,y',y^*\}$, $\tilde{Y}=\{\tilde{y},\tilde{y}^\sharp,\tilde{y}',\tilde{y}^*\}$, $\{V,E,G,K\}$ auxiliary intermediate variables.

Observing Table 1 associated with the increase in the dimension of the action operator, the quantitative indicators of the dimensions of each level of the classical quantum information or the conjugate algorithm expand according to the superpower exponential growth mode, but what is interesting is that from the 2nd to the 5th layer, the two-dimensional index quantization ratio remains unchanged and fixed at 1/4 and 1/16 until the 6th layer becomes 1/8 and 1/64. The second layer introduces two variables and eight variable associations each. After adding three kinds of inverse, reflection, and rotation operators in layers 3–5, the related dimensions are obviously multiplied, but the change rate of the dimension ratio is maintained. The difference in the 6th layer is the introduction of auxiliary pairs in conjugate algebra on symmetry and anti-symmetry determination mechanisms.

From the perspective of the architecture, adding these operators to the algebraic expansion of conjugate complex vectors, most of which correspond to the structural modes of classical quantum theory, ensures that the classical quantum information system is a simplified mode of the algebraic structure of conjugate complex vector functions to be a self-consistent complex vector function subsystem. From the single variable in the first layer to the sixth layer, the dimension of classical quantum information increases to 64:4096, the expression of the conjugate complex vector function reaches $512 : 2^{18}$, and the dimension parameter indicates that under the complete phase space with a larger presentation space. It is equivalent to classical quantum information to support the solution of a single qubit. Each extension level provides 1–2 bits of combined control information. The entire architecture requires an auxiliary description of $64 = 2^6$ dimensions. However, it is necessary to support the solution of a single cobit in the conjugate algebraic system, and auxiliary information of 64 dimensions is needed. The classical quantum information architecture needs to distinguish between symmetric and antisymmetric levels, supplement $\{[,]^u\{,\}^u\}$ information, and complete the solution mechanism.

7. Key Steps of Measurement Process

The detection process is divided into three levels: 1) pairwise entanglement state detection, 2) single/multiring of n iteration mode, 3) strong interaction of conjugate group qualified mode.

7.1 Pairwise Entanglement Detection

Type A Measurement Description
Input: Photon flow
Output: Pairwise conjugate features
Processing: Two-stage half-lens separation of spin and complex pairs

The light quanta flow enters the detection module from the input channel and is separated by two stages of semimirror BS. The first-stage half-mirror uses reflection and transmission to distinguish the spin into two polarized light quanta channels, and the light quanta of the separated polarization channels pass through the second-stage half-mirror to determine the real part and imaginary part of the light quanta eigenvalue.

Processing flow,

Input: $\rightarrow$ | I. BS1 : $\{x, \tilde{x}\}$ | $\rightarrow$ | II. BS2: $\{x = (p, q), \tilde{x} = (q, p)\}$ | $\rightarrow$ Output

In the measurement examples collected from the first part, Figs. 3(d) and 4(b) describe the detection model of the paired light quanta entanglement state and illustrate the process of paired light quanta detection.

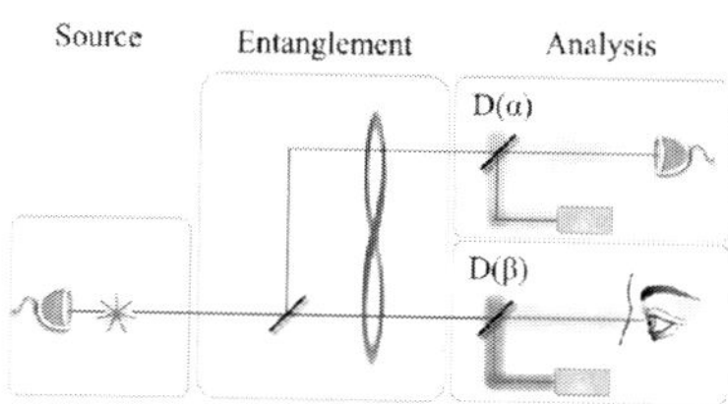

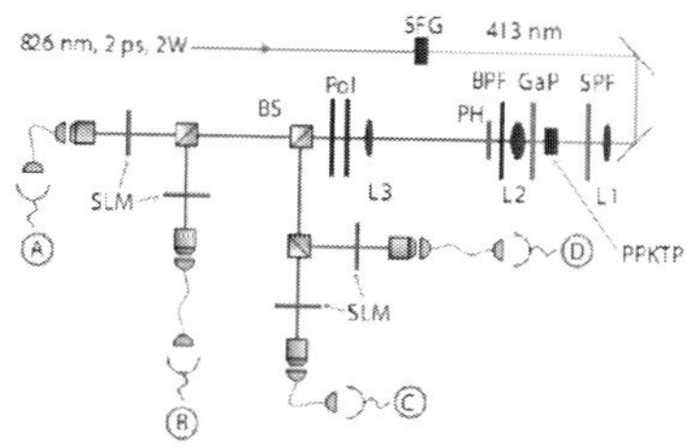

The light quanta are separated by two polarizations of transmission and reflection through the beam splitter. The first component of the optical mirror is divided into two paths according to the entanglement pair spin characteristic up/down, and each path is polarized and separated by the second component of the optical mirror to determine the two complex quantitative characteristics of the real and imaginary parts of an optical quanta state.

7.2 Single/Multiple Ring of n Iteration Detection

Type B Measurement Description	
Input:	Photon flow
Output:	Single/multiple concentric rings
Processing:	Multilevel interference, polarization, filtering, projection

Processing flow,

Input: $\rightarrow$ | I. DM Interference | $\rightarrow$ | II. Filters | $\rightarrow$ | III. PBS Splitter | $\rightarrow$ Output

In Figs. 2(d) and 5(b), the single/multiring optical quantum entanglement detection models are shown.

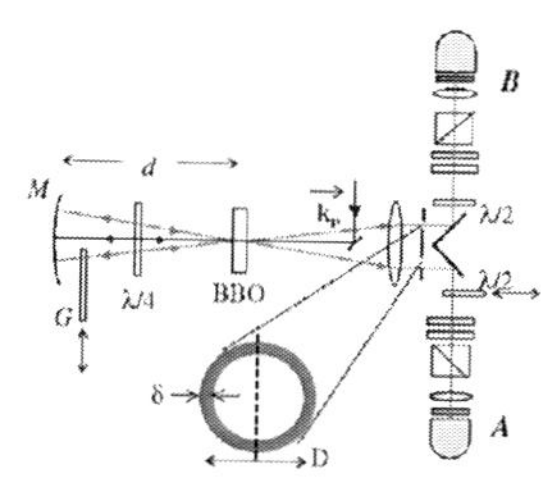

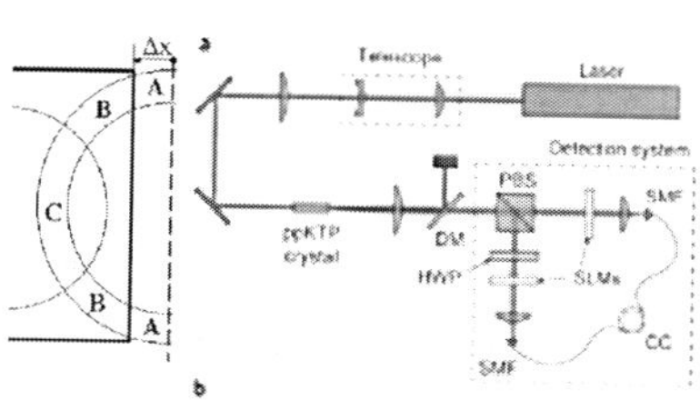

7.3 *Strong Interaction Detection*

	Type C Measurement Description
Input:	Photon flow
Output:	Diamond/rectangle and other point-like feature clusters, point group feature distribution
Processing:	Multilevel interference, polarization, filtering, strong interaction, diffractive projection

Processing flow,

Input: → [I. Interference → II. Interaction Filters → III. Grating] → Output

The strong interaction feature point cluster detection entanglement cluster visualization results are shown in Figs. 7(b) and 5(e).

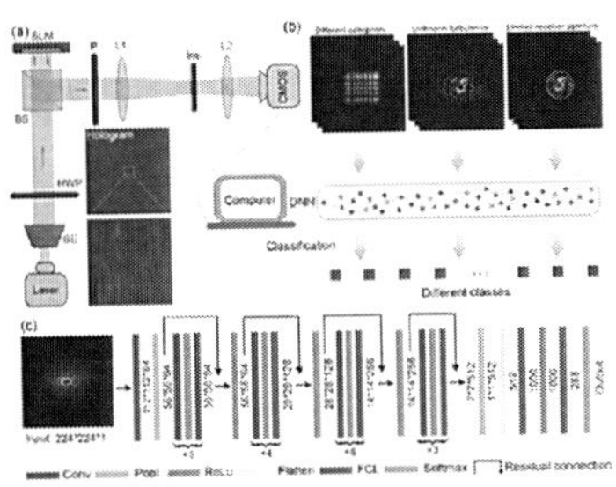

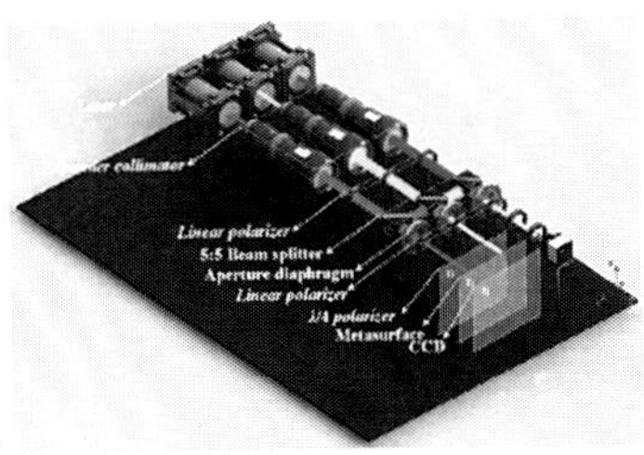

7.4 *Type A Pairwise Entangled State Detection*

We check the quantum entangled state detection model and method in Chapter 5 and combine the series of theorems and inferences in Section 4 to establish the equivalence theorem for paired conjugate vectors and paired entangled state detection.

Theorem 12 (Paired Conjugate Vector Detection ≡ Paired Quantum Entangled State Detection). *Eight vectors of a conjugate quark cluster are located in the four attribute regions; in the same region, paired conjugated states provide measurable characteristic information carried in the representation of paired entangled states in the description of quantum information interaction.*

Proof. From the perspective of paired conjugate state detection, eight vectors of a conjugate quark cluster guarantee that there is a pair of conjugate state vectors in each attribute region satisfying $\theta/2 + \tilde{\theta}/2 = \{\pi/2, \pi, 3\pi/2, 2\pi\}$. In the first quadrant, $\{z = r^{1/2}e^{i\theta/2}, \tilde{z} = r^{1/2}e^{i\tilde{\theta}/2}\}$ is a pair of conjugate vectors, and the two vectors $\{z, \tilde{z}\}$ themselves represent complex numbers, $z : \theta/2 \in [0, \pi/2)$ represents the anticlockwise spin feature, $\tilde{z} : \tilde{\theta}/2 = \pi/2 - \theta/2 \in (0, \pi/2]$ represents the clockwise spin on the integrated spin feature $\theta/2 + \tilde{\theta}/2 = \pi/2$. The pairwise conjugate vector product is $z\tilde{z} = \tilde{z}z = re^{i\,pi/2} = ir$ on the imaginary axis of the complex plane. On other quadrants $z\tilde{z} = \tilde{z}z = \pm ir$.

Corollary 14. *The Type A measurement pipeline is for detecting and separating pairs of conjugate eigenvectors $\{z, \tilde{z}\}$ satisfying Theorem 12.*

Proof. In the two-stage separation process, $\{\theta, \tilde{\theta}\}$ is used to distinguish the spin direction in the first stage and distinguish the real part and imaginary part of the complex sequence pair in the second stage after spin separation of quantitative features.

7.5 Type B Multiple Iteration Interaction Detection

From the perspective of the potential function isosurface distribution, rings with different radii indicate that the measurement system has converged to a spatially continuous distribution that satisfies certain characteristics. The paired conjugate vector structure described by Theorem 12, under the condition of a sufficiently large number of samples, the loop formed is represented by Theorem 10 for multiple sets of rhombus loops on the complex plane. Combined with the cobweb projection, the Type B detection mode can be completed.

Theorem 13 (Single Loop Detection). *Select a single diamond-shaped loop on the complex plane and use the spider web transformation of the square root mode to determine the density matrix of each coordinate point on the complex plane of the accompanying projected density values, forming a circular feature visualization distribution.*

Proof. The number of samples gathered on the loop is determined by the density matrix, and the square root operation can transform the diamond-shaped loop on the complex plane into a circular ring.

Corollary 15 (Multiple Loop Detection). *Select multiple diamond-shaped loops on the complex plane. Through the square root mode spider web transformation, the density matrix of each coordinate point on the complex plane determines the accompanying projected density values to form multiple circular features on visualization distributions.*

Proof. Selects multiple diamond-shaped loops and naturally obtains multiple circular distributions through the root spider web projection.

Theorem 14 (Maximum Value of Complete Distribution Located on Diagonal of Four Attribute Regions). *On the complex plane satisfying the complete distribution condition, the accompanying density matrix has maximum values near the four diagonals passing through the zero point.*

Proof. In the third chapter, the state distribution based on the transformation of the precision time series into the complex plane vector representation is described. The density distribution formed is the product of two sets of normal distributions of the biquadratic binomial distribution, and the diagonal has a maximum in continuous conditions.

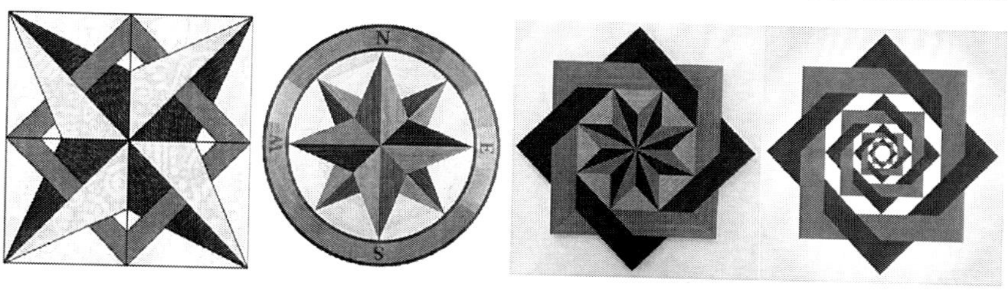

Theorem 15 (Maximum Value of Product Distribution Located on Four Coordinates). *The distribution is formed by the product of conjugate vectors in the complex plane. The matrix has a maximum value of the accompanying density on the four coordinate lines of the complex plane.*

Proof. The vector products of mutual conjugates in the same loop are accumulated on the four coordinate lines, and the remaining vector products form the relevant density distribution as a whole to form the maximum value of the density distribution on the coordinate line. From the aspect of transformation distribution continuity, this type of distribution is equal to taking the product of the Hadamard matrix to rotate the original distribution $\pi/4$ in a rectangle, with the radius unchanged, the overall rotation becomes a rhombus, and the corresponding distribution along the coordinate. The axis direction has the maximum value.

Corollary 16 (Distribution of Maximum Value in Eight Directions). *Both distribution and product distribution are superimposed to form a distribution of maximum value in eight directions, located at* $\{0, \pi/4, \pi/2, 3\pi/4, \pi, 5\pi/4, 3\pi/2, 7\pi/4\}$, *decomposed into two sets of conjugate cross structures.*

Proof. Theorems 14–15, each set provides a cross-structure distribution, which is combined into eight-directional maxima.

7.6 Type C Strong Interaction Detection

When the visual projection does not have circular spectral line distribution constraints, the point group cluster distribution provided by the complex plane itself can be displayed through the $n = 1$ spider web distribution. In the fourth chapter, it is discussed that starting from the given eigenvector to be measured, it is further combined and converted into nine eigenvectors through Klein group combination, and further expansion is formed through conjugate addition to form a special lattice cluster structure qualified for the conjugate structure.

Theorem 16 (Strong Interaction Detection). *The density matrix formed on the coordinates of the complex primitives in the complex plane for the conjugate group qualification, through* $n = 1$ *cobweb transformation, shows from diamond to rectangular distribution features, fine distribution details are determined by the real and imaginary parts of the input vector quantization scale and the number of iterations of the interaction.*

Proof. In the fourth chapter, several examples are provided to show the special distribution under different conditions. The macroscopic distribution of complete clusters can converge to rhombus, octagon, and rectangle, and fractal stratification can be observed in the recursive distributions of iteration effects.

8. Illustration of Measurement Results

8.1 Four-quadrant Statistical Distribution and Conjugate Quark Clustering

The phase distribution of seven quantum states is illustrated in Fig. 19(a).

The two groups of conjugated quark clusters generated by the conjugated structure are shown in Fig. 19(b)–(e). Two octary quark clusters are shown in (b) and (c) and two sets of multinomial distributions in (d) and (e) show four quadrants on complex planes.

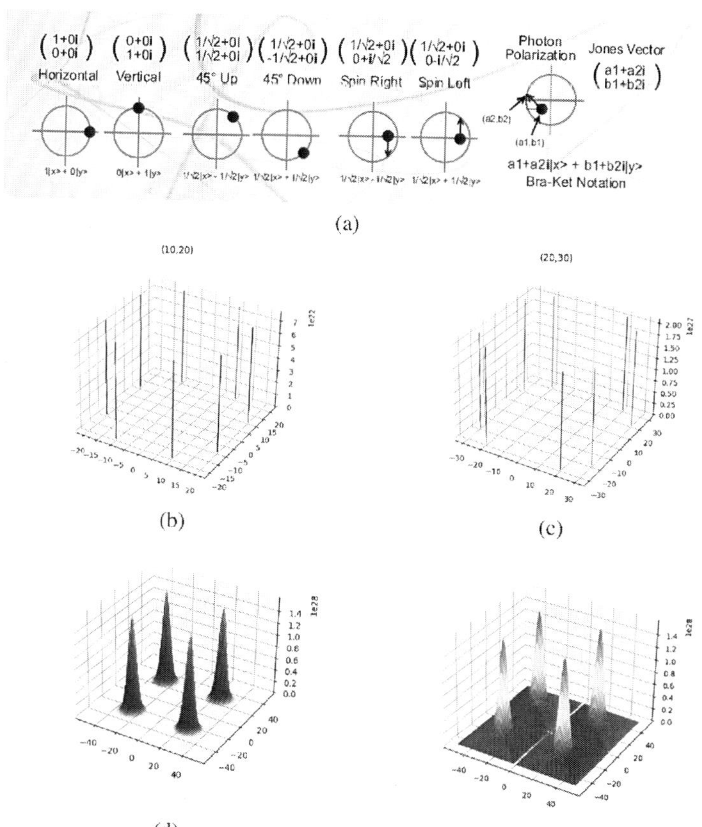

Figure 19. Two-level measurement results: (a) quantum state detection. Conjugate clusters: (b)–(c) octary quark clusters; (d)–(e) two sets of 3D multinomial distributions on four-quadrant regions of complex planes $m = 40$.

8.2 Single-ring and Multiring Distributions

It is a natural way to illustrate the entangled measurements and to retain entangled photons in a circular distribution with a constant radius. This fact makes Type B measurement mode play a key role in general measuring applications.

Five sets of Type B samples are shown in Figs. 20–24.

In Fig. 20(a)–(i), nine maps are originally generated by the multinomial equations $m = 18$ proposed in Chapter 3 to form 2D heatmap distributions in four quadrants. Three cobweb projections are applied in three vertical groups {(a), (d), (g): $n = 1/2$, (b), (c), (h): $n = 1$, (c), (f), (i): $n = 2$}.

Three maps of the first row (a)–(c) correspond to original distributions on four quadrants, three maps of the second row (d)–(f) are generated results from complex products under $\pi/4$ rotation distributions on the original distributions of the four quadrants, and three maps of the third row (g)–(i) are integrated from the first and second maps; i.e., each original map is superposited with its rotational $\pi/4$ map to generate the third map.

It is interesting to note that map (g) shows typical classic accumulated results. However, wave interferences are significantly shown in maps (h) and (i), and there are much stronger interferences on map (h) than on map (h).

Three sets of sample images are shown in Fig. 20(j)–(l) selected from Fig. 5(e) and (c) and Fig. 6(a'). There are eight significant cluster patterns and special visual effects very similar to the conjugate in Fig. 20(a)–(i). It is convenient to make visual comparisons.

Three sets of simulating samples are generated from three sets of random sample sequences on lengths = {1K,5K,50K}, $m = 20$. Each set provides two symmetric/asymmetric groups. The nine maps (a)–(i) show symmetric distributions, and another nine maps (j)–(r) illustrate asymmetric patterns. The interferences are contained in (g)–(i) and (p)–(r).

Two sets of six enlarged maps are shown in Fig. 24 to select each superposition group of three maps from multinomial distribution on Fig. 20(g)–(i) and Fig. 23(g)–(i). Two groups are shown similar visual distributions, especially on wave interferences on $n \geq 1$ parameters.

Observing visual effects from Fig. 20 to Fig. 24, various differences can be identified.

In addition to mathematical derivations on lemmas and theorems and corollaries, these sets of sample images provide multiple pieces of evidence to show conjugate schemes useful in various simulations of entangled interactions satisfying comprehensive applications.

8.3 Conjugation Group Combinatorial Lattice

In Fig 25(a)–(e), five typical output clusters generated by conjugate group combinatorial lattices are shown: 25-point rhombus, 25-point rectangle, 73-point octagon, 73-point octagon, and 81-point octagon.

Fig. 25(f)–(g)(5(f,g)) shows the visualization results of strong interaction feature point cluster detection and entanglement cluster detection. (f) Eight types of system formations combined mode. (g) Diamond and rectangular typical output feature distributions.

Comparing these visualized distributions of entangled states comprehensively, the graphical distributions of the series of measurements can all be explained under the three measurement models. From the comparison results of these measurement diagrams, the diversity of the measurement results is fully supported by the conjugate complex vector function and the cobweb projection visualization.

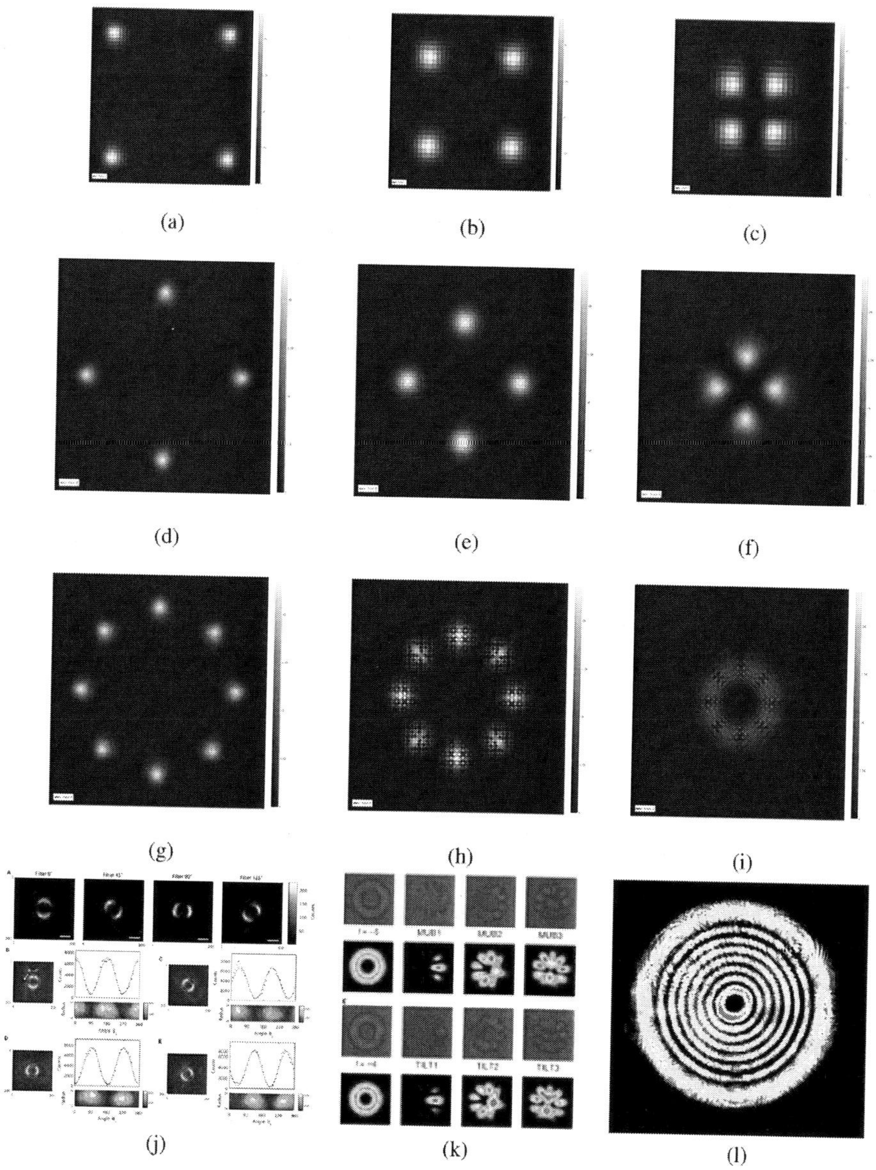

Figure 20. Three sets of nine cobweb projections on conjugate binomial products $m = 18$: (a) $n = 1/2$; (b) $n = 1$; (c) $n = 2$; (d) $n = 1/2, \theta = \pi/4$; (e) $n = 1, \theta = \pi/4$; (f) $n = 2, \theta = \pi/4$; (g) $n = 1/2$: (a)+(d); (h) $n = 1$: (b)+(e); (h) $n = 2$: (c)+(f). 2014–2018 three cases of advanced entangled measurements (j)–(l): (j) the first observed entangled measurement of qubits in 2016, (k) high-D quantum entanglement in 2018, (l) two-photon correlation 103-D results in 2014.

9. Forecast

To facilitate experimental scientists in the field of quantum optics and quantum information to test the rationality and reliability of the solution on the proposed conjugate eigenvalue system, six achievable predictions are collected to test the provided conjugate logic and conjugate complex vector functions on structural models and methods, along with the core conclusions of the derivation and systematic correctness of the visualization system.

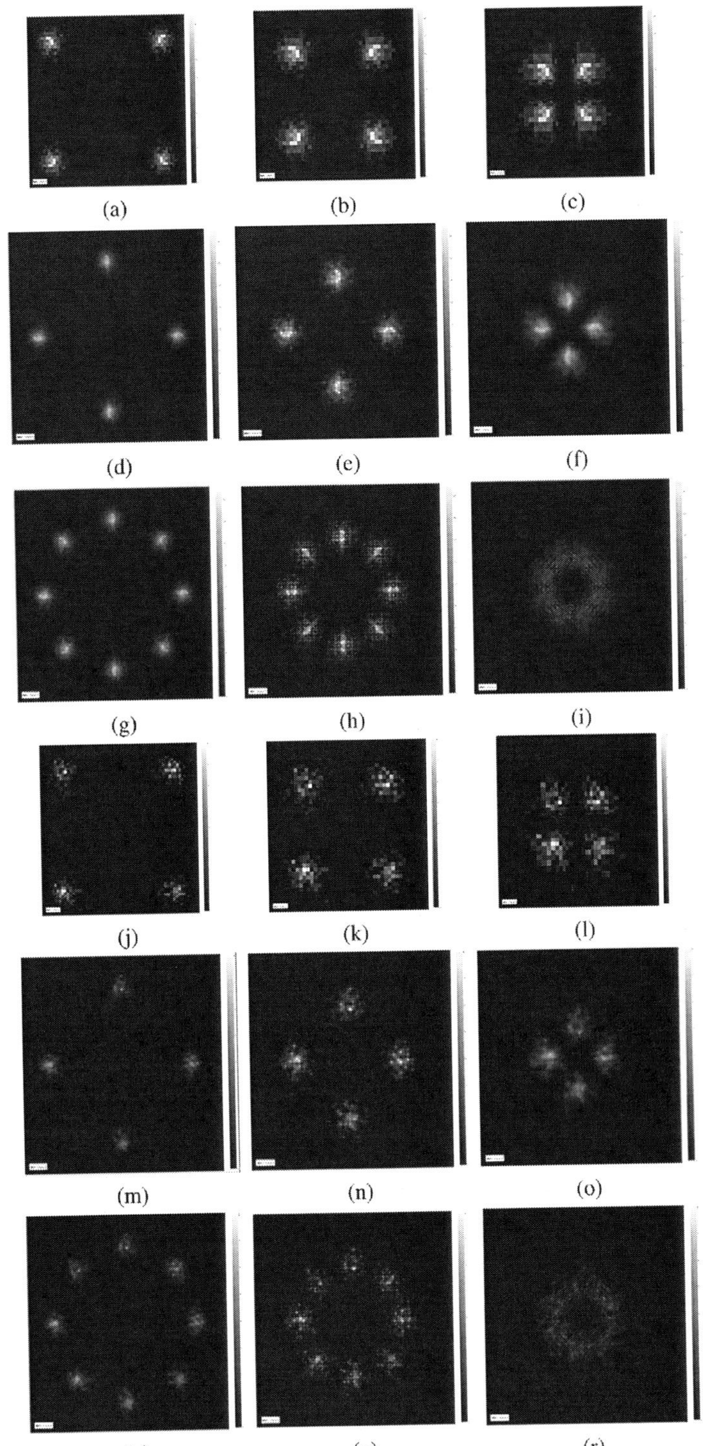

Figure 21. Two sets of eighteen cobweb projections on random data set length 1K $m = 20$, (a)–(i) symmetric mode; (j)–(r) asymmetric mode: (a), (j) $n = 1/2$; (b), (k) $n = 1$; (c), (l) $n = 2$; (d), (m) $n = 1/2, \theta = \pi/4$; (e), (n) $n = 1, \theta = \pi/4$; (f), (o) $n = 2, \theta = \pi/4$; (g), (p) $n = 1/2$: (a)+(d); (h), (q) $n = 1$: (b)+(e); (i), (r) $n = 2$: (c)+(f).

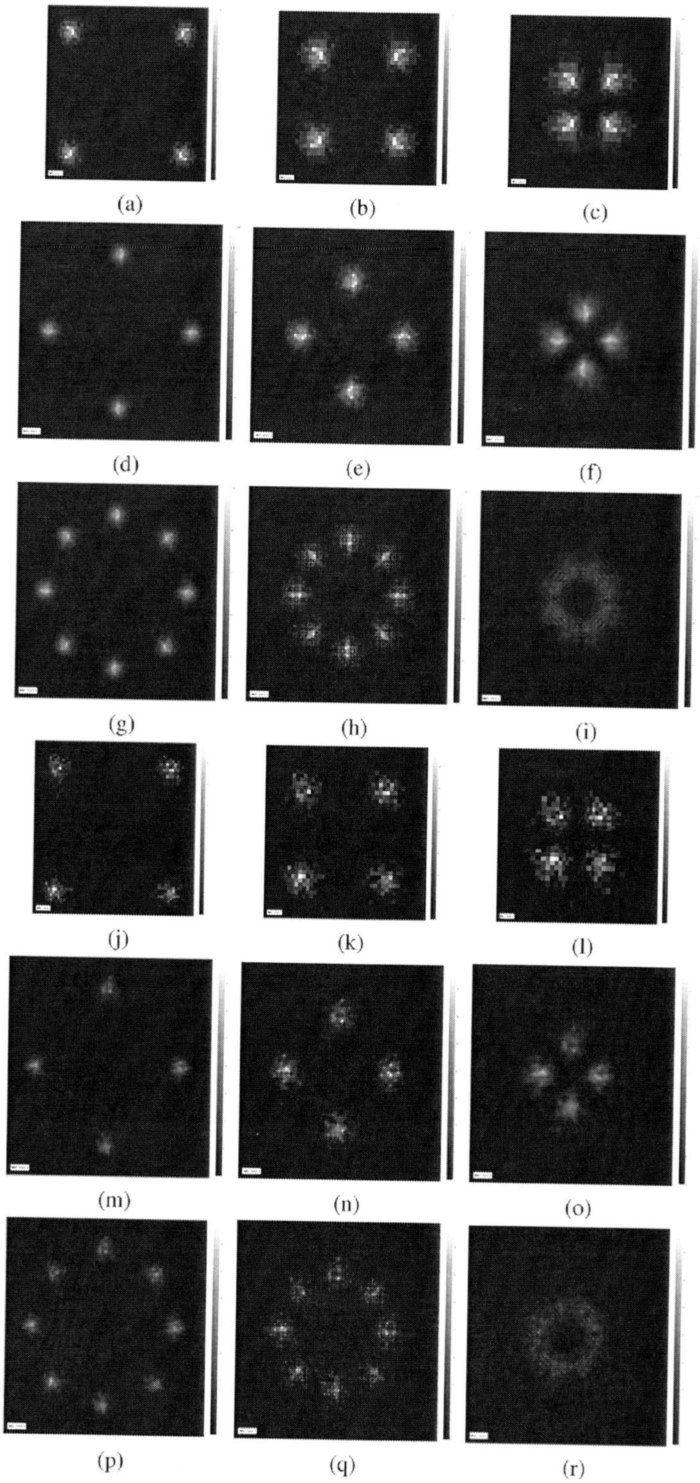

Figure 22. Two sets of eighteen cobweb projections on random data set length 5K $m = 20$ (a)–(i) symmetric mode; (j)–(r) asymmetric mode: (a), (j) $n = 1/2$; (b), (k) $n = 1$; (c), (l) $n = 2$; (d), (m) $n = 1/2, \theta = \pi/4$; (e), (n) $n = 1, \theta = \pi/4$; (f), (o) $n = 2, \theta = \pi/4$; (g), (p) $n = 1/2$: (a)+(d); (h), (q) $n = 1$: (b)+(e); (i), (r) $n = 2$: (c)+(f).

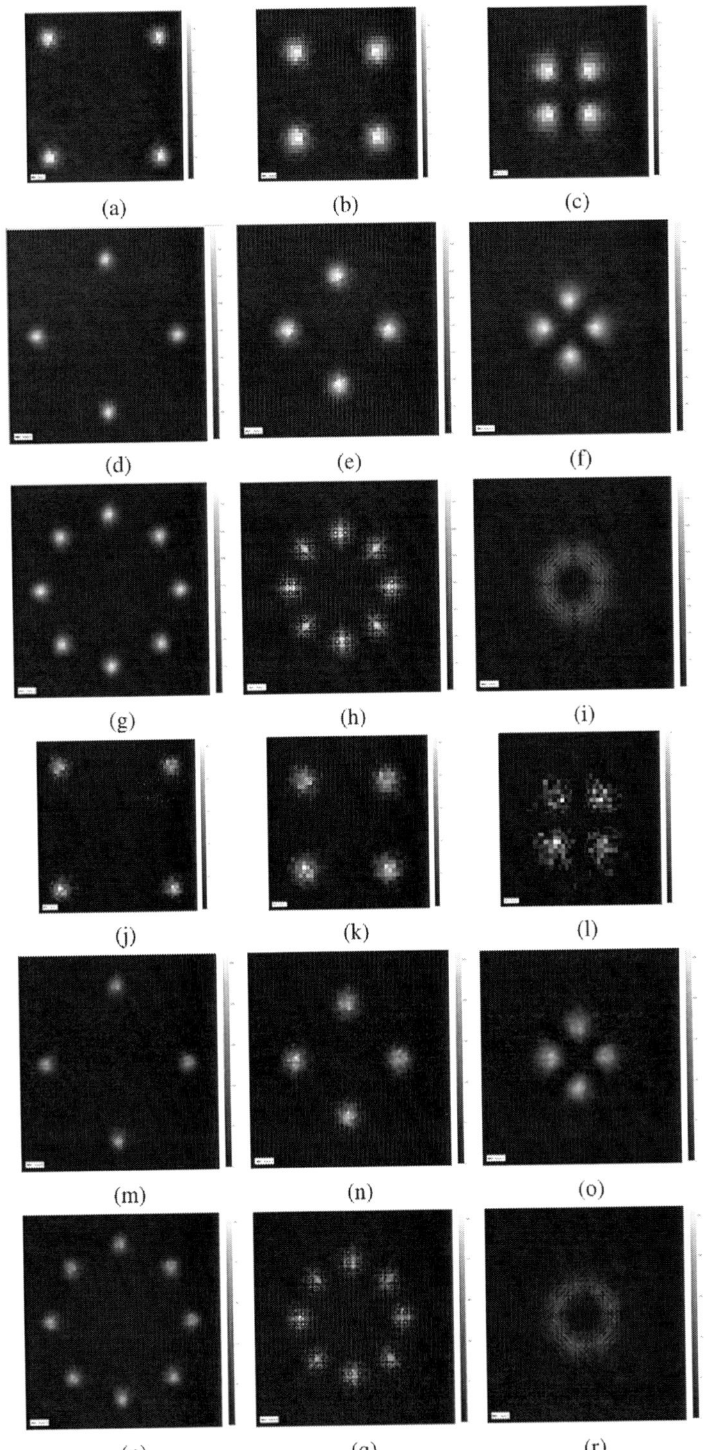

Figure 23. Two sets of eighteen cobweb projections on random data set length 50K $m = 20$, (a)–(i) symmetric mode; (j)–(r) asymmetric mode: (a), (j) $n = 1/2$; (b), (k) $n = 1$; (c), (l) $n = 2$; (d), (m) $n = 1/2, \theta = \pi/4$; (e), (n) $n = 1, \theta = \pi/4$; (f), (o) $n = 2, \theta = \pi/4$; (g), (p) $n = 1/2$: (a)+(d); (h), (q) $n = 1$: (b)+(e); (i), (r) $n = 2$: (c)+(f).

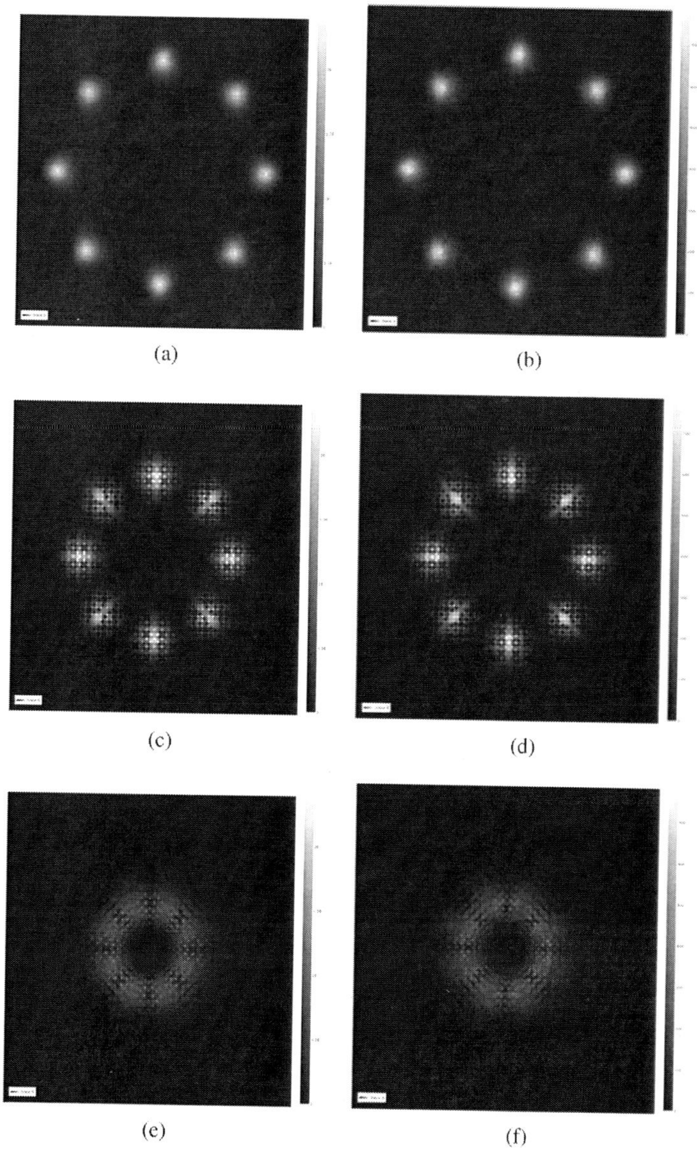

Figure 24. Two Sets of six cobweb projections of interferences on conjugate binomial products $m = 18$ (a), (c), (e) and data sequence 50K $m = 20$ (b), (d), (f): (a), (b) $n = 1/2$; (c), (d) $n = 1$; (e), (f) $n = 2$

Prediction 1 (Complete Quantum Phase Space). *The dimension of the complete phase space of n cobits is 2^{2n}, the imaginary eigenvalue quantifies the solution space of the system, and the statistical distribution of the phase space is transformed by the conjugate orthogonal transformation into a visualized image distribution.*

Prediction 2 (Conjugate Phase Space: Complete Phase Space 1:3). *Under the condition that the conjugate operator satisfies the time transformation order, the conjugate eigenvalue solution space covers one-third of the complete phase space and two-thirds of the complete phase space transcends the conjugate systems.*

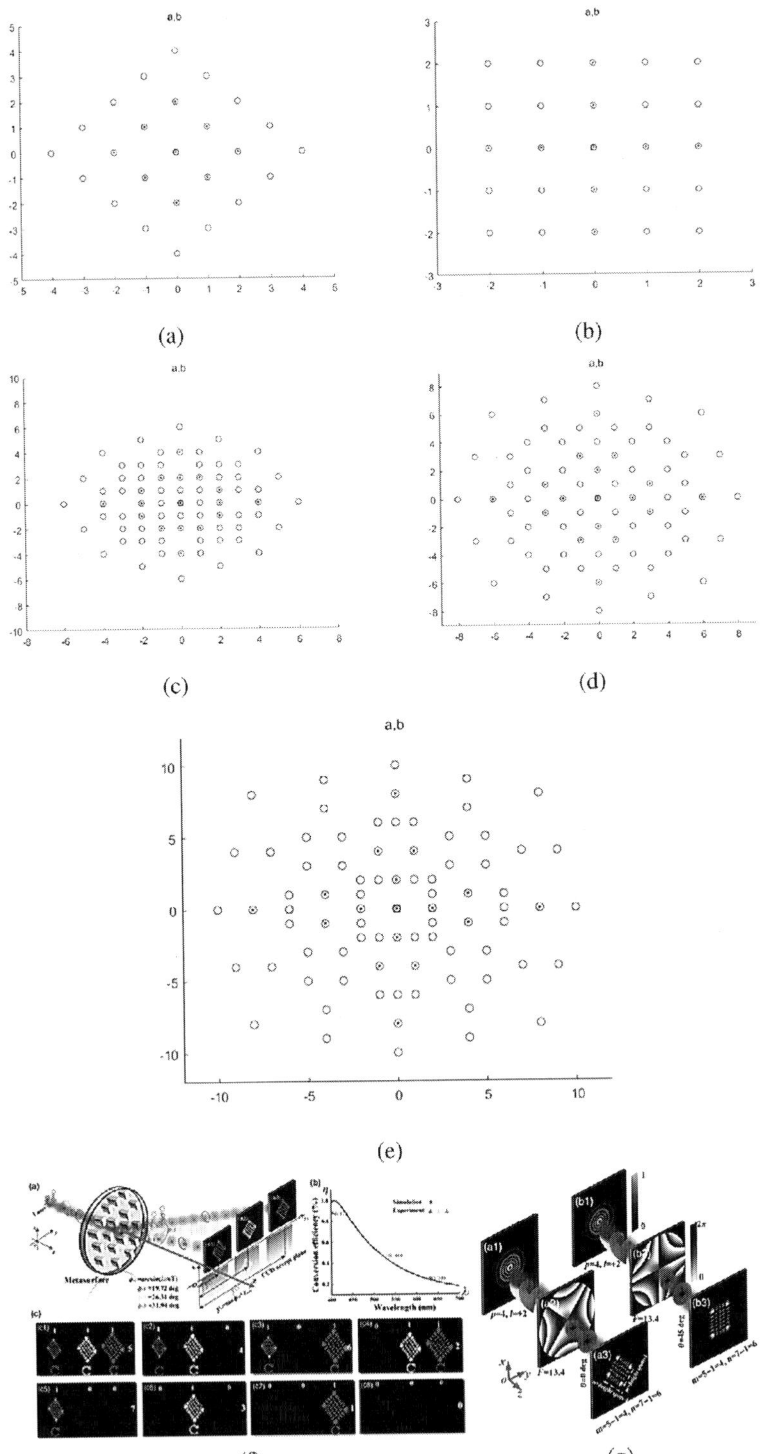

Figure 25. Five sets of conjugate group combinatorial lattices: (a) $p = q$ diamond 25 points; (b) $(p, 0)|(0, q)$ rectangle 25 points; (c) $p = 2q$ octagon 73 points; (d) $p = 3q$ octagon 73 points; (e) other octagon 81 points. Aggregation detection of two sets of strong interactions on feature point groups: (f) eight combination modes; (g) typical output diamond and rectangle.

Prediction 3 (Determination of Dirac and Conjugate Eigenvalue Solutions Based on Rotational Symmetry). *Using the matrix diagonal and auxiliary eigenvalues of a complex plane, the rotational symmetry distribution accurately determines the Dirac equation of the system eigenvalue solution and common differences between solutions of eigenvalue systems. The solution space of the conjugate structure eigenvalue system has complete global rotational symmetry, but the solution space of the Dirac eigenvalue system does not provide rotational symmetry.*

Prediction 4 (Complex Plane Rhombus Link Rotational Symmetry). *Satisfying* $|j| + |k| = n, n > 0, j, k \in \mathbf{N}$ *on the complex plane, the rhombus link corresponds to the conjugate orthogonal normalized solution space of n quantum states, the dimension of the solution space is* $4n$*, and the independent measure of each projection space dimension* $n+1$ *is* $n-1$*. The four coordinates* $(n, 0), (0, n), (-n, 0), (0, -n)$ *are shared as the start and end of the link.*

Prediction 5 (Precise Conjugate Expansion Operators). *Four conjugate operators* $\{\sim, \sharp, \prime, *\}$ *paired with conjugate variables* $\{y, \tilde{y}\}, \theta, \tilde{\theta}$ *is the smallest cluster of quantum entanglement structures. The enigma of entangled states, which has plagued quantum optics and quantum information for a long time, can be naturally eliminated by accurately measuring the extended structure of complete conjugate vector state clusters.*

Prediction 6 (Strong Interaction Characteristic Cluster Distribution). *In addition to zero points, five nonzero distribution patterns in the quantum optics of a strong interaction environment can be rigorously tested by quantitative precision measurement mechanisms. Six sets of conjugated groups qualified a typical distribution.*

(0).	$p = q = 0$	*Single zero point on the* $(0, 0)$ *position*
(1).	$p = q$	*Diamond 25 points*
(2).	$(p, 0)\|(0, q)$	*Rectangular 25 points*
(3).	$p = 2q$	*Octagon 73 points*
(4).	$p = 3q$	*Octagon 73 points*
(5).	*Others*	*Octagon 81 points*

10. Summary

How to solve the mystery of quantum entanglement is the core topic faced by modern quantum information and quantum computing. Based on the theoretical and experimental models and results of conjugate complex vector functions, conjugate transformation structures, complex planes formed by precise clock pulses, conjugate group combinatorial lattices, and a series of modern quantum entanglement state detection models and methods. Six measurement assumptions, three precision measurement models, and six predictive targets are proposed. It is expected that the proposed quantum optical measurement model and method will gain the attention of subsequent theoretical and experimental experts worldwide and effectively promote the basic research on quantum mechanics, quantum information, quantum optics, other interdisciplinarity, and the development and application of cutting-edge high technology in future exploration.

References

[1] W. K. Wootters. Picturing qubits in phase space. *IBM J. Res. Dev.* 48(1), 2004, 99.

[2] Modelling quantum entanglement 2002. http://animatedphysics.com/insights/modelling-quantum-entanglement/.

[3] M. Barbieri, F. De Martini, G. Di Nepi, P. Mataloni, G. M. D'Ariano and C. Macchiavello. Detection of entanglement with polarized photons: Experimental realization of an entanglement witness. *Phys. Rev. Lett.* 91, 2003, 227901.

[4] 2009 entanglement detection Otfried Guhne. Geza Toth Axiv 0811.2803.pdf.

[5] Entangled images and delayed EPR entanglement 2009. https://jqi.umd.edu/news/entangled-images-and-delayed-epr-entanglement.

[6] N. Bornman, M. Agnew, F. Zhu et al. Ghost imaging using entanglement-swapped photons. *NPJ Quantum Inf.* 5, 2019, 63. https://doi.org/10.1038/s41534-019-0176-5.

[7] New method lets researchers entangle five photons for the first time, and that's just the start 2010. https://www.popsci.com/science/article/2010-05/researchers-entangle-five-photons-first-time/.

[8] 2016 an idea for allowing the human eye to observe an instance of entanglement by Bob Yirka, Phys.org. https://phys.org/news/2016-02-idea-human-eye-instance-entanglement.html

[9] B. C. Hiesmayr, M. J. A. de Dood and W. Löffler. Observation of four-photon orbital angular momentum entanglement. *Phys. Rev. Lett.* 2016. https://www.photonicsonline.com/doc/four-photons-entangled-in-orbital-angular-momentum-0001.

[10] 2018 massive quantum entanglement experiment could help solve a physics mystery 2018. https://nexusnewsfeed.com/article/science-futures/massive-quantum-entanglement-experiment-could-help-solve-a-physics-mystery/.

[11] J. Bavaresco, N. Herrera Valencia, C. Klöckl et al. Measurements in two bases are sufficient for certifying high-dimensional entanglement. *Nature Phys.* 14, 2018, 1032–1037. https://doi.org/10.1038/s41567-018-0203-z.

[12] P.-A. Moreau, E. Toninelli, T. Gregory, R. S. Aspden, P. A. Morris and M. J. Padgett. Imaging bell-type nonlocal behavior. *Sci. Adv.* 5(7), 2019, eaaw2563. https://doi.org/10.1126/sciadv.aaw2563.

[13] 2014 A record quantum entanglement: 103 dimensions. https://www.kurzweilai.net/a-record-quantum-entanglement-103-dimensions S. Ramelow and A. Zeilinger. Record quantum entanglement of multiple dimensions. 2014. https://www.uab.cat/web/newsroom/news-detail/record-quantum-entanglement-of-multiple-dimensions-1345668003610.html.

[14] Z. Lyu and C. Wang. Geometric-phase-based shearing interferometry for broadband vortex state decoding. *Sci. Rep.* 12, 2022, 3015. https://doi.org/10.1038/s41598-022-07083-w.

[15] C. He, Y. Shen and A. Forbes. Towards higher-dimensional structured light. *Light Sci. Appl.* 11, 2022, 205. https://doi.org/10.1038/s41377-022-00897-3.

[16] Dahai Yang, Jie Lin, Chen Chen, Chang Li, Junbo Hao, Baiying Lv, Keya Zhou, Yiqun Wang and Peng Jin. Multiwavelength high-order optical vortex detection and demultiplexing coding using a metasurface. *Adv. Photonics Nexus* 1(1), 2022, 016005. https://doi.org/10.1117/1.APN.1.1.016005.

[17] D. Cavalcanti, P. Skrzypczyk, G. Aguilar et al. Detection of entanglement in asymmetric quantum networks and multipartite quantum steering. *Nat. Commun.* 6, 2015, 7941. https://doi.org/10.1038/ncomms8941.

[18] T. Ono, R. Okamoto, M. Tanida, H. F. Hofmann, and S. Takeuchi. Implementation of a quantum controlled-SWAP gate with photonic circuits. *Sci. Rep.* 7, 2017, 45353.

[19] Mohammad Rezai, Jörg Wrachtrup and Ilja Gerhardt. Polarization-entangled photon pairs from a single molecule. *Optica* 6, 2019, 34–40.

[20] Visible entangled photons from an ultrafast fiber laser source 2019. https://www.photonics.com/Articles/Visible_Entangled_Photons_from_an_Ultrafast_Fiber/a64846.

[21] L. Hu, Y. Ma, W. Cai et al. Quantum error correction and universal gate set operation on a binomial bosonic logical qubit. *Nat. Phys.* 15, 2019, 503–508. https://doi.org/10.1038/s41567-018-0414-3.

[22] Hao Wang, Ziyu Zhan, Yijie Shen, Jianqi Hu, Xing Fu and Qiang Liu. Deep-learning-assisted communication capacity enhancement by non-orthogonal state recognition of structured light. *Opt. Express* 30, 2022, 29781–29795. https://doi.org/10.1364/OE.465318.

[23] Changjia Chen, Calvin Xu, Arash Riazi, Eric Y. Zhu, Alexander C. B. Greenwood, Alexey V. Gladyshev, Peter G. Kazansky, Brian T. Kirby and Li Qian. Telecom-band hyperentangled photon pairs from a fiber-based source. *Phys. Rev. A* 105, 2022, 043702.
[24] Bob Johnson - Jim Johnson March 30, 2012 - 11:30 pm essential guide to the EU. https://www.thunderbolts.info/wp/2012/03/30/essential-guide-to-the-eu-chapter-11/.
[25] 2022 achromatic second harmonic generation. http://wwwriedle/projects/UV_achromatic/UV_achromatic.php.
[26] 2019 with a simple twist, a 'magic' material is now the big thing in physics. https://www.quantamagazine.org/how-twisted-graphene-became-the-big-thing-in-physics-20190430/.
[27] A. Badalyan, T. Gevorgyan et al. Engineering of 2D and 3D holographic gratings in photorefractive media. In *Proceedings of SPIE, Photonics and Micro- and Nano-Structured Materials* 8414, 2011. (6) (PDF) Optical induction of 3D refractive lattices in doubly doped LiNbO3 photorefractive crystal. Available from: https://www.researchgate.net/publication/265250916_Optical_induction_of_3D_refractive_lattices_in_doubly_doped_LiNbO3_photorefractiv_crystal [accessed October 20, 2022].
[28] Dirac cone. https://en.wikipedia.org/wiki/Dirac_cone.
[29] Type-II Dirac fermions spotted in two different materials July 13, 2017. https://physicsworld.com/a/type-ii-dirac-fermions-spotted-in-two-different-materials/.
[30] Yujie Chen, Xu Gu, Yiwei Li, Xian Du, Lexian Yang and Yulin Chen. Recent advances in topological quantum materials by angle-resolved photoemission spectroscopy. *Matter* 3, 2020, 1114–1141.
[31] Luojun Du, Yunyun Dai and Zhipei Sun. Twisting for tunable nonlinear optics. *Matter* 3(4), 2020, P987–P988. https://doi.org/10.1016/j.matt.2020.09.013.
[32] Hamish Johnston. Graphene simulated using ultracold atoms. March 14, 2012. https://physicsworld.com/a/graphene-simulated-using-ultracold-atoms/.
[33] Hamish Johnston. 'Ring of exceptional points' appears on a Dirac cone. September 18, 2015. https://physicsworld.com/a/ring-of-exceptional-points-appears-on-a-dirac-cone/.
[34] Shuo Wang, Ben-Chuan Lin, An-Qi Wang, Da-Peng Yu, and Zhi-Min Liao. Quantum transport in Dirac and Weyl semimetals: A review. *Adv. Phy.* 2(3), 2017, 518–544. https://doi.org/10.1080/23746149.2017.1327329.
[35] Jeffrey Zheng, Chris Zheng and T. L. Kunii. A framework of variant logic construction for cellular automata. In *Cellular Automata – Innovative Modelling for Science & Technology*, Edited by Alejandro Salcido, InTech Press, 2011, 325–352.
[36] Jeffrey Zheng and Chris Zheng. A Framework to express variant and invariant functional spaces for binary logic. *Front. Electr. Electron. Eng. China* 5(2), 2010, 163–173.
[37] Jeffrey Zheng. *Variant Construction from Theoretical Foundation to Applications*, Springer Nature, 2019.
[38] 郑智捷,变值体系理论基础及其应用 第一卷 理论基础及其应用 科学出版社 2021 Jeffrey Zheng, *Variant Construction theoretical Foundation and Applications, Vol. 1, Theoretical Foundation and Applications*, Science Press, 2021 (in Chinese).

Chapter 7

Conjugate Construction in Future Explorations

Resolved Logical Paradoxes on Multivariate Complex Functions and Hierarchical Structures

Jeffrey Zheng*
Conjugate System Research, Melbourne, Australia

Abstract

Based on the conjugate vector logic construction, this chapter explores a series of problems related to conjugate vector algebra operations from several aspects, including classical paradox, modern logic, topological geometry, visual paradox, classical Tai Chi, resolving Mobius paradox, holistic conjugate representation and measurement of coherent states, geometry of complex numbers, modern representation of complex conjugate numbers, theory of complex functions, mathematical physics, complex function equations, variational methods, two-body entanglement and time-reversal symmetry, symmetry + supersymmetry criteria, and variant constructions + knowledge model. The 0-1 vector logic is established, leading to the unparalleled support of the multivariable complex function system and the basic module of the description system. Future challenges are predicted for further explorations.

Keywords: Logical paradox, multivariate complex function, mathematical physics equation, supersymmetry, variant construction, knowledge model, future exploration

God made the integers All the rest is the work of man.
– Leopold Kronecker

God exists because arithmetic is consistent. Devil exists because we cannot prove it!
– Hermann Weyl

Geometry is the most complete science We must know – We will know!
– David Hilbert

*Corresponding Author's Email: conjugatelogic@yahoo.com.

In: Conjugate Construction of Quantum Optics
Editors: Jeffrey Zheng
ISBN: 979-8-88697-651-9

1. Introduction

The frontier research of modern mathematics and logic had a very violent collision in the early 20th century. Russell's paradox proposed by Russell marked a feasible decision logic system compatibility mechanism. Along with Hilbert's series of achievements in geometric logic and basic mathematical theory, exploring whether logical paradoxes are included in the theoretical framework system is the core topic that a class of modern logic and mathematics systems has always been concerned about.

In 1930, in a series of exploratory studies on axiomatic model theory, Godel established the noncompleteness theorem of axiomatic systems. It sounded like the death knell for the ideals of Hilbert's school of formal theory of mathematical foundations. Along with Turing's proposed Turing machine model, recursive function theory, λ calculus, Shannon switch theory and information theory, and finite automata models and methods, the origin and development of modern computers, information science, and communication technology are comprehensively promoted.

The traditional logic system is based on a binary symbol system, starting from the Maxwell equation and the general theory of relativity. The mathematical and physical basis used has gradually transformed from the real variable function measurement system to the complex variable function model, such as the quantum mechanics system and Hilbert space. The kind of logical system that may be properly supported has become a fundamental issue that is difficult to determine.

For this reason, Bohr used the ancient oriental Tai Chi diagram to simulate the relevant laws of quantum mechanics. However, due to the inherent changing characteristics of such dynamic representation patterns, beyond the scope of application of the combinatorial logic framework, how to properly represent complex dynamic logic systems requires more powerful abstract models and method support.

Since ancient Greece, the development of logic has been closely related to the paradox of various changes and movements [1]–[6]. A brief examination of the historical development of paradox is an extremely interesting topic to explore.

1.1 Paradox (Fallacy, Contradiction)

Zeno's paradox in ancient Greece: Achilles, dichotomy, playing field, the paradox of flying arrows (everything is immutable, motion is an illusion, time does not exist), quantum measurement paradox, etc.

True Paradox: A statement that contradicts itself or is a circular argument, or some logically impossible situation Ex: This statement is false.

Entertainment Paradoxes: Bertrand Box Paradox, Birthday Paradox, Monty Hall Paradox, etc.

Scientific Paradoxes: Maxwell's Genie, Rod and Barn Paradox, Twins Paradox, Grandfather Paradox, Laplace Demon, Schrodinger's Cat, Fermi Paradox.

1.2 Ramsay Classification: Logical and Semantic Paradoxes

Logical paradox: If the axiom system can prove both the primitive A and the negative primitive $\neg A$.

Aristotle: The same thing cannot exist and does not exist at the same time, nor can it oppose itself in the same way.

Mo Jing (墨经): Others, neither can be neither, contradicting themselves. Two propositions that contradict each other cannot be true at the same time.

Gongsun Long (公孙龙): A white horse is not a horse (The connection paradox).

Substantial implication paradox: $p \to q \sim \neg p \vee q$ If the latter is an equivalent formula of the former, it will lead to a paradox!

Russell's Paradox: The Barber's Paradox, triggering the third math crisis.

Group paradox [7, 8]: $A = \{01, 10\}$, $\neg A = \{10, 01\} \to S = \{S_0, S_1, S_2, S_3\}$, $S_0 = 00, S_1 = 01, S_2 = 10, S_3 = 11; A = \{S_1, S_2\} = \{S_2, S_1\} = \neg A \to A = \neg A$.

Lemma 1 (Simple Paradoxes). *Paradox representations such as a white horse is not a horse, substantial implication, groups, etc., can use the local variables involved to form regional judgment conditions. However, to systematically describe Russell's paradox, it is necessary to establish the overall decision rules related to clusters.*

Theorem 1 (Group Paradox). *Group paradox provides the smallest and compact case of a state space as a pair of two states to satisfy the logic constraint $A = \neg A$.*

Corollary 1 (Gongsun Long Paradox). *The paradox "A white horse is not a horse" can be formally expressed as a logical statement.*

For example, let $A = \{\text{white}\}$, $B = \{\text{horse}\}$, $AB = A \cap B = \{\text{white horse}\}$, and let f be a logic function $f(A), f(B), f(AB) \in \{0, 1\}$.

If $f(A) \neq f(B)$, then $f(AB) = f(A) \cap f(B) = ?$

The equation $f(AB) = ?$ contains an intrinsic paradox.

1.3 Variant Construction and Applications

The comprehensive use of advanced theoretical systems for practical application is the only way for all abstract theories to be applied. This chapter makes use of basic theoretical models and methods in [7, 8] to describe: vector logic, meta-knowledge models, and hierarchical designs to form a comprehensive application fulcrum.

This chapter focuses on the analysis of [11], and the development of the architectural system must promote from classical logic to vector logic to resolve the inherent logical paradoxes. By analyzing the Mobius strip and inherent logical paradoxes, the conjugated ring can be used to systematically eliminate inherent logical paradoxes in architecture for knowledge systems.

To show this process clearly, the following paragraphs will gradually introduce the relevant historical stories, related models, and methods to study modern logic and paradoxes.

1.4 Conjugate Construction Solving Quantum Entangled Puzzles

From historical data collection, this chapter uses vector logic and conjugate construction to explore a series of problems related to vector algebra operations from several aspects, including the classical paradox, modern logic, topological geometry, visual paradox, classical Tai Chi, and digestion Mobius paradox, global conjugate representation and coherent state measurement, geometry of complex numbers, modern representation of complex conjugate numbers, theory of complex functions, mathematical physics complex function equations, variational methods, two-body entanglement and time-reversal symmetry, symmetry + supersymmetry criterion, and variant construction + knowledge model.

In Chapter 6, we list the basic modules of the unparalleled support and description leading to the multivariable complex function system based on 0-1 vector logic of the conjugate transformation to resolve quantum entanglement puzzles of quantum optics and quantum information as photonic application support.

2. Modern Logic Topological Geometry Visual Paradox Classic Tai Chi

Where logic fails, wisdom is born

– Kurt Godel

2.1 Godel's Paradox

Godel's incompleteness theorem published in 1931 was the fatal blow to Hilbert's formal system. The general idea of the theorem is that any sufficiently strong formal system that includes arithmetic has unprovable true propositions, and the system itself cannot prove that it has no contradictions. This study shows the essential limitations of formal methods. This theorem was the pinnacle of mathematical logic in the 20th century. The most important mathematical theorem of the 20th century.

Fig. 1(a) shows the effect of the human image after multiple reflections between two mirrors; Fig. 1(b) shows the reflection effect through the water surface.

Lemma 2 (Visual Representation of Godel's Paradox). *Based on the optical reflection effects of complex lenses, it is possible to show the limits of what is possible with objects through recursive transformations. This provides a visual paradox in the reflected world to contain an* $A = \neg A$ *property with self-contradiction.*

2.2 Topological Geometry Paradox: Mobius Ring and Klein Bottle

In 1858, two German mathematicians, Mobius and Listing, discovered [14], twisting 180 degrees of paper strips glued at both ends, which has a magical one-sided property, called the Mobius strip.

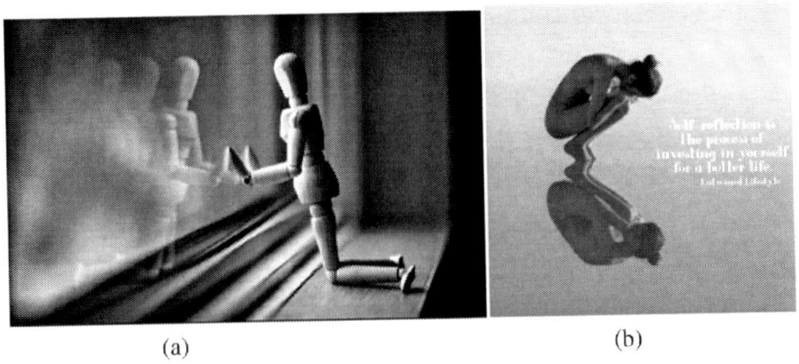

(a) (b)

Figure 1. Illustration of godel's paradox: (a) infinite reflection; (b) mirror reflection.

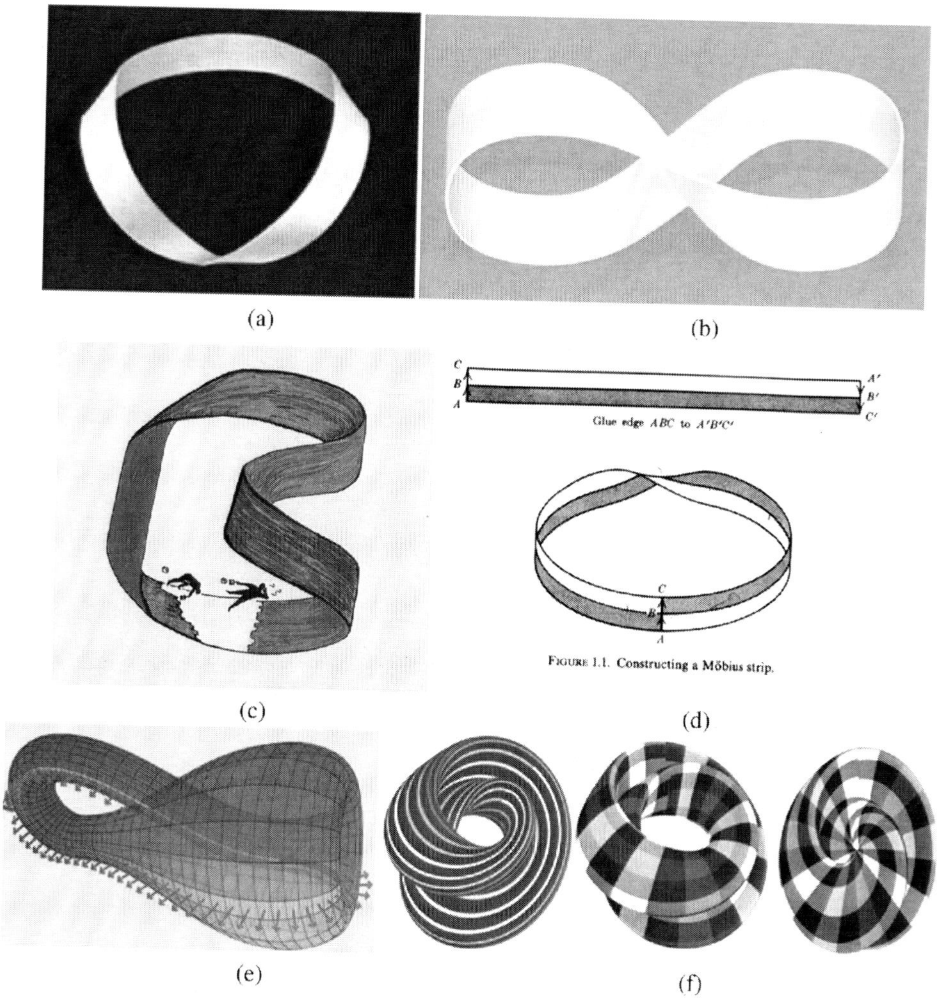

Figure 2. Mobius paradox and klein bottle representation: (a)–(b) mobius ring (c) mobius paradox; (d) construction of mobius ring; (e) klein bottle; (f) klein bottle equivalent representation.

From a logic viewpoint, this magic surface naturally provides an intrinsic logic paradox on any position and its opposite position satisfying the $A = \neg A$ property.

In Figs. 2(a) and (b), the Mobius strips are shown; in Figs. 2(c) and (d), typical Mobius strips are shown for the direct connection to illustrate the paradox effect.

In 1882 German mathematician Klein extended the Mobius structure to four-dimensional space, forming the Klein bottle [16]. Different Klein bottle topological geometries for dynamic Klein batch bottles and equivalent representations are shown in Figs. 2(e) and (f).

Lemma 3 (Mobius Ring and Klein Bottle). *The Mobius ring forms a 3D single leaf surface with twisted 2D strips; the Klein bottle uses the twisted 3D embedding hypersurface connections to show four-dimensional Mobius structures. On both constructions, there are intrinsic logic paradoxes on $A = \neg A$ naturally.*

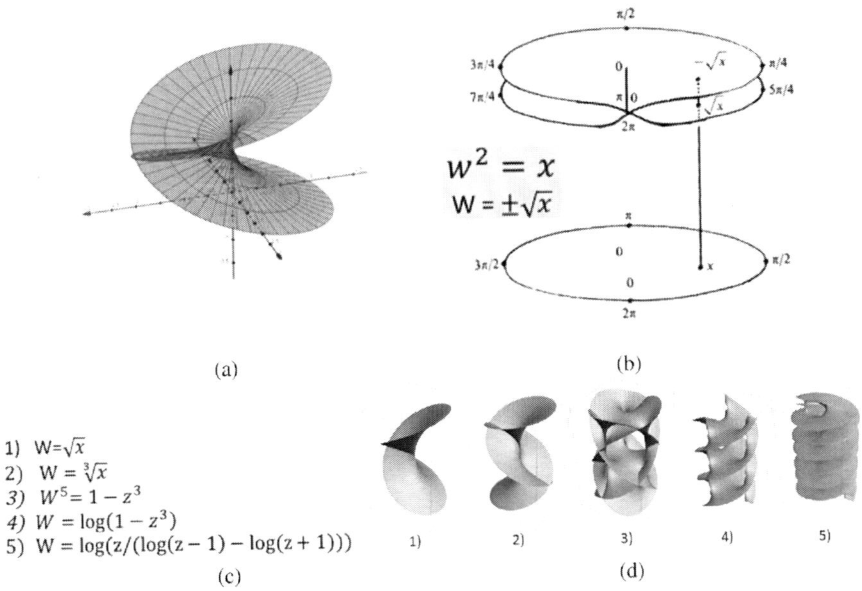

Figure 3. Riemann surface: (a) two-page riemann surface; (b) two-page decomposition mode; (c) five sets of riemann surface equations; (d) display surface with different parameters.

2.3 Riemann Surface—Double Leaf Surface

The complex plane C is probably the most trivial Riemann surface [17]. The mapping $f(z) = z$ (identity mapping) defines a set of graphs of C. The mapping $g(z) = z^*$ (conjugate) mapping also defines a graph of C as an atlas of C. Graphs f and g are incompatible, so they give each C a Riemann surface structure.

In fact, given a Riemann surface X and its atlas A, the conjugate atlas $B = \{f^* : f \in A\}$ is always incompatible with A, so X is assigned a different Riemann surface structure. $w^2 = x, w = \pm\sqrt{x}$. The manifold will be divided into two leaves of Riemann surfaces, corresponding to single-valued real variables. The separation of the Riemann surface into two pieces of corresponding points x has the Mobius property.

In Fig. 3(a), the two-leave projection of the Riemann surface diagram is shown. In Fig. 3(b), the corresponding points of the two pages and their projection positions are clearly visible. The surface itself is clearly Mobiusian; in Figs. 3(c) and (d), the corresponding Riemann surfaces are shown on five different algebraic equations.

Lemma 4 (Riemann Surface). *The Riemann surface based on the multivariate complex function formula from the overall distribution projection connection can observe the inherent Mobius ring paradox characteristics for the $A = \neg A$ property in the self-contradiction.*

2.4 Visual Paradox

Escher provides a rich paradigm [18] in the field of visual paradoxes. Whether it is from angels and devils, day and night, mirror reality, and the reincarnation of heaven and earth, all kinds of paradoxical icons are displayed among the rich themes. Since the 1930s, his

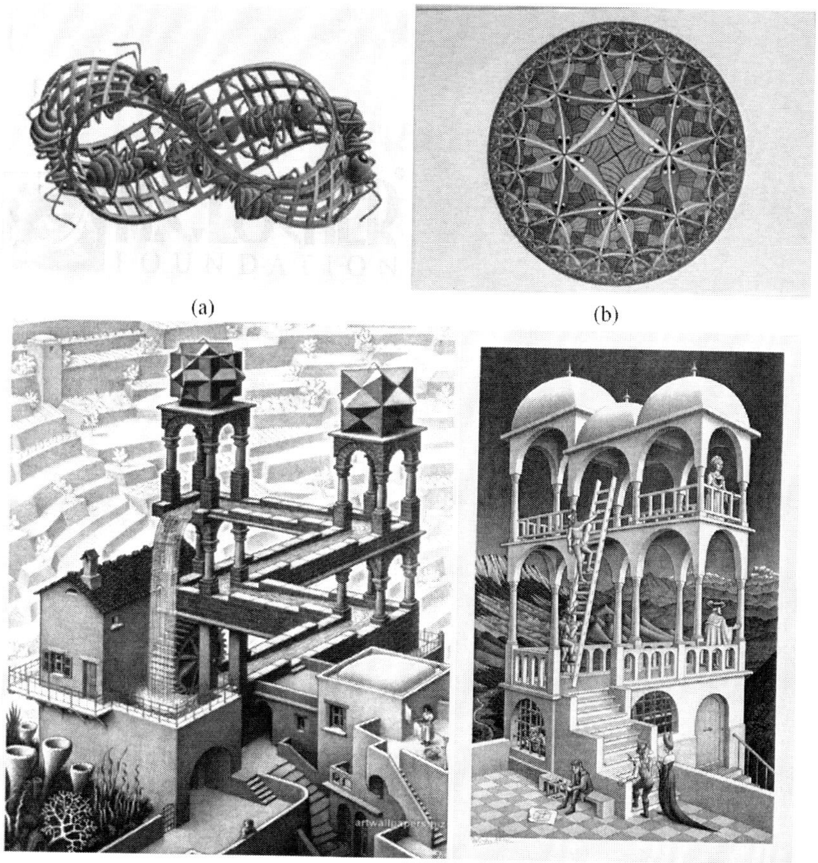

(a) (b) (c) (d)

Figure 4. Escher's diagram: (a) ants crawling through a mobius ring; (b) infinite finite projection disk; (c) impossible waterfall; (d) impossible building.

works have been circulated in the scientific community, showing a unique path for exploring the laws of symmetrical interaction of various categories.

In Fig. 4(a), several ants crawl on the Mobius torus in turn. In Fig. 4(b), the finite area of the disc can accommodate an infinite number of ants. Branches cover the entire board; in Figs. 4(c) and (d), the paradoxical circular pattern of the Impossible Falls and the beautiful appearance of an Impossible Building are shown.

Lemma 5 (Escher's Visual Paradox). *Based on the complex diagrams displayed by Escher's visual art, from the overall distribution projection connection and other levels, the inherent characteristics of the Mobius ring paradox can be observed for the* $A = \neg A$ *property in the self-contradiction.*

2.5 Classic Tai Chi Diagram Series

In the traditional oriental classics [19]–[21], "Book of Changes": "Easy to have Tai Chi," the first two line symbols Yin {- -} and Yang {—}, the two symbols give birth to the four states {⚌,⚍,⚎,⚏}, and the four states give birth to the eight states - eight trigrams {☰,☴,☲,☶,☱,☵,☳,☷}. The description starts from chaos to the law of the formation of all things in the world.

The ancient Tai Chi diagram originated from the observation of the sun and the shadow of the sun as natural clocks. Different eras have contributed. For example, Zhou Dunyi in the Song Dynasty initially formed the Tai Chi diagram, and the "Yin-Yang fish" Tai Chi diagram with overall characteristics. After being transformed by Zhao Weiqian in the early Ming Dynasty, it was finalized by Zhao Zhongquan in the late Ming Dynasty.

In Fig. 5(a), the sundial is used with the projection angle formed by the sunlight projection angles. In Fig. 5(b), the standard Tai Chi diagram is shown with the outer periphery of the Eight Trigrams of King Wen; Figs. 5(c) and (d) show the design steps of the Tai Chi diagram in the Pisces pattern, as well as the unique Tai Chi diagrams of Buddhism, Taoism, and Confucianism.

From the perspective of simplifying the visual representation, the Ming Dynasty Lai Zhide (1525–1604) [19] provided a series of creative graphic representations. In terms of simplified visual representation, the flag of modern Korea contains a simplified representation of Tai Chi in four states. Among the rich Tai Chi representations formed in history, the hierarchically structured Tai Chi change graph deserves attention.

Fig. 5(e) shows Lai's Tai Chi diagram, and Fig. 5(f) takes the Korean flag as an example to show another kind of simplified Tai Chi and the external representation of the four hexagrams. In Figs. 5(g) and (h), different Tai Chi diagrams are matched with the outer peripheral eight trigrams, as well as paired conjugated Tai Chi diagrams. Fig. 5(i) shows the circulation pattern that exists in Tai Chi.

Lemma 6 (Classical Tai Chi and Visual Paradox). *Based on the complex diagrams displayed by the classic Tai Chi diagram, the inherent characteristics of the Mobius ring paradox can be observed from the overall distribution projection connection and other levels. Escher's visual paradox is equivalent to the abstract dynamic change characteristic of traditional oriental Tai Chi visual expressions. Since chaotic states are located in the central position of the Tai Chi diagrams, they all contain the logic paradoxes $A = \neg A$ in dynamic activities of multiple integrated levels in their phase spaces.*

3. Dissolving Paradox of Classic Mobius Logic

3.1 Mobius's Classic Logic Paradox

For a torus, if the front side of the ring is A and the back side of A is $\neg A$, due to the unique one-sided property of the Mobius torus, the front side and the back side take the same values under arbitrary conditions. Both satisfy $A = \neg A$.

3.2 Conjugate Rings Resolve Paradox of Mobius Logic

Based on the model and method provided in [8] chapters 3 and 4 on the variant construction and chapter 2 on symmetric/antisymmetric brackets, the conjugate ring symmetrically

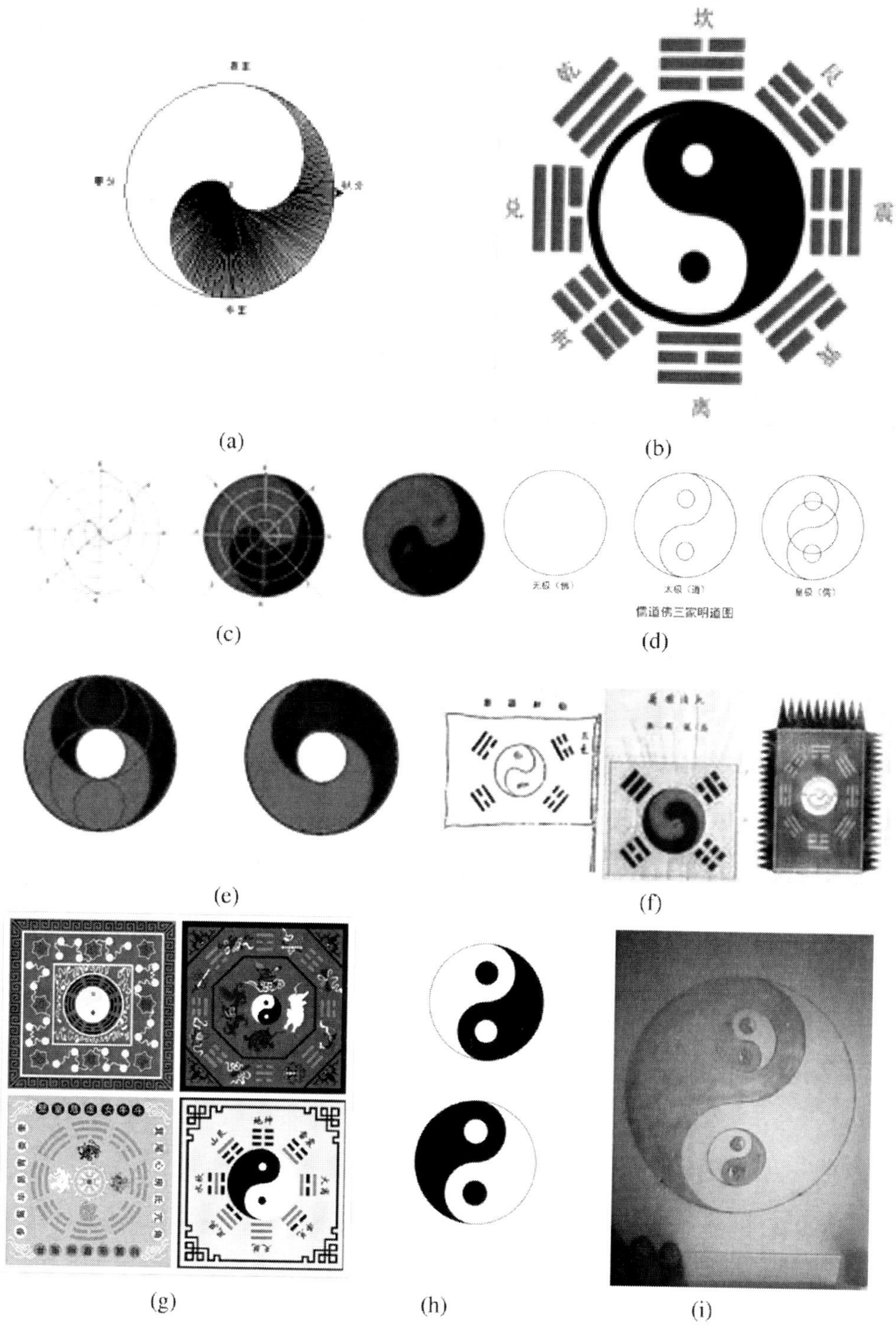

Figure 5. Tai Chi icons: (a) sundial projection diagram; (b) Tai Chi eight trigrams; (c) pisces diagram step-by-step design; (d) Buddhist, Taoist, and Confucian Tai Chi representations; (e) Lai's style of simplified Tai Chi; (f) Korean flag of simplified Tai Chi; (g) various eight trigrams of Tai Chi diagrams; (h) paired reflection symmetric Tai Chi diagrams; (i) Yin-Yang set of Yin-Yang Tai Chi diagrams.

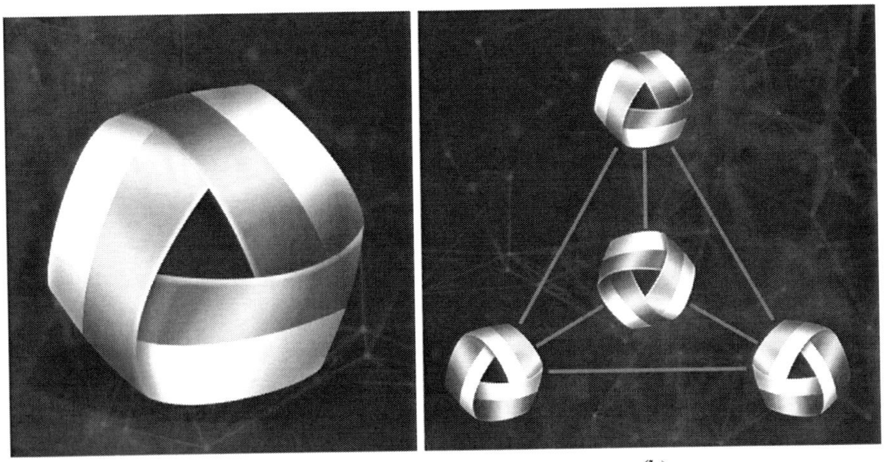

(a) (b)

Figure 6. Conjugate ring to resolve the mobius paradox: (a) a four-color conjugated ring; (b) four sixteen-color conjugated rings are located at the top of the tetrahedron.

divides the visible torus into two left and right, and one color is maintained all the way. Under conditions that form a bilayer ring, a set of conjugated rings exists with four torus faces.

Since the double-layer torus can form a set of close-fitting Mobius strip structures, after the torus rotates once, the left torus is staggered to become the right torus, which ensures that the values taken are different, that is, it satisfies $A \neq \neg A$. In other words, under the action of $\neg$ operators and four color bands, the system resolves the inherent paradox of the Mobius ring by using the pairing mode of the left and right exchange positions.

A single conjugated ring is shown in Fig. 6(a), and four conjugated rings on top of a tetrahedron are shown in Fig. 6(b), representing sixteen sets of distinguishable color bands.

Theorem 2 (Conjugate Ring Resolve Logic Paradoxes on Mobius Ring Types). *Since the double-layer torus can form a set of close Mobius strip structures, after one rotation of the ring, the left torus staggers into the right torus ensuring that the values taken are different to satisfy $A \neq \neg A$. Under the action of $\neg$ logic operators, the system resolves the inherent paradoxes of the Mobius ring types using the pairing mode of the left and right exchanges.*

4. Conjugate Representation and Entangled Measurement

In Chapters 1–6, multiple results are established. It is convenient to summarize each main contribution in compact representations.

4.1 Conjugate Complex Vector Systems

In Chapter 1, the description of the whole structure is composed of six lemmas, twenty-four theorems, and sixteen corollaries.

The foundation of the conjugate complex vector structure is established, and the key properties are listed.

For a complex number $x = (p, q), r = \sqrt{p^2 + q^2}$, the key results are summarized as follows.

Theorem 3 (Fundamental Theorem of Algebra Support Real Coefficients Only). *The fundamental theorem of algebra is a necessary and sufficient condition to correspond the solution space on the condition of* $\{$*real coefficients*$\} \leftrightarrow \{+\theta, -\theta\}$ *and* $\{x, x^*\}$ *a pair of measures, complex vectors, and complex conjugate operator* $*$.

Corollary 2 (Paradox in Gauss's Theorem). *Gauss's fundamental theorem of algebra is incomplete in covering the solution spaces under the coefficients on imaginary numbers to naturally contain the Mobius ring and intrinsic logic paradoxes* $A = \neg A$ *in the phase space.*

Theorem 4 (Conjugate Complex Vector Systems). *Four conjugate operators* $\{\sim, \sharp, \prime, *\}$ *are defined associated with a pair of conjugate angles* $\{\theta, \tilde{\theta}\}$ *and the four attribute regions in a complex plane on the synchronous parallel processing and recursive organization conditions. Eight feature vectors are composed of two crosses, and the whole construction provides six dynamic transformation modes associated with the basic recursive operations.*

Corollary 3 (Paradox Free on Complete Solution Spaces). *Based on the four distribution regions (four colors of the conjugate ring) and a pair of conjugate vectors (opposite rotations), and other comprehensive approaches, this extends the fundamental theorem of algebra to cover* $\{$*complex coefficients*$\}$ *in a complete solution space. The conjugate complex vector systems guarantee without logic paradoxes on the vector phase space in synchronous dynamic environments.*

4.2 Conjugate Transformation Structure and Conjugate Complex Vector Systems

In Chapter 2, the description of the whole structure is composed of three lemmas, twenty-two theorems, and seven corollaries in two parts.

The first part describes the eight complex vectors in a pair of conjugate feature classes generated by the conjugate transformation structure from $\{0, 1\}$ elements of an N length sequence.

Theorem 5 (Output from Conjugate Transformation Structure). *The conjugate transformation structure processes the initial 0–1 vector of N elements to be* $2n$ *feature classes and transforms feature vectors into* 2^{2n} *configurations. To output a pair of conjugate classes, eight elements are conjugate complex number vectors.*

The second part describes expanding operations to conjugate algebra with complete brackets identifying symmetric/antisymmetric measurements with the time order.

Corollary 4 (Restriction of Poisson, Lagrange, Lie, Dirac Brackets, and Lie Algebra). *Poisson, Lagrange, Lie, and Dirac Brackets support multiple real eigenvalue solutions* $\{-1, +1\}$ *of a dynamic system to identify symmetric/antisymmetric properties of the principal diagonal eigenvalues on the Hermite matrices, and those ignore multiple imaginary solutions of the auxiliary diagonal eigenvalues completely.*

Corollary 5 (Conjugate Brackets and Conjugate Algebra). *Using conjugate algebra, conjugate brackets provide the complex eigenvalue solutions* $\{+1, +i, -1, -i\}$, $\{+1, -i, -1, +i\}$ *beyond Poisson, Lagrange, Lie, and Dirac brackets to identify vector operators* $\{-1, \leftarrow\}$ *of symmetric/antisymmetric properties under {Group, Ring, Field} domains. The time order satisfies vector and tensor operational environments under a single circle of the eigenvalue matrix condition.*

4.3 Clock Reference Signals on Conjugate Complex Plane

In Chapter 3, clock reference signals sufficiently provide the controllable quantitative bases of the precise measuring foundation on a complex number plane.

The description of the whole structure contains one lemma, ten theorems, and eight corollaries.

Theorem 6 (Two Projects of Clock Reference Signals). *Conjugate measurements of clock reference signals provide the conjugate complementary measurement space, the narrow projection on the conjugate configuration, and the general projection on the complex number plane.*

Corollary 6 (Narrow and General Projections on Complex Plane). *A narrow projection provides* 2^{2m} *positions or* 4×2^{2m} *positions in a configuration space of one or four 2D uniform distributions without a metric order, and a general projection provides* $(m + 1)^2$ *positions or* $4(m + 1)^2$ *positions in 2D double-binormal distributions corresponding to a complex number plane on one quadrant or four quadrants.*

4.4 Conjugate Group Combinatorial Lattices

In Chapter 4, a conjugate group combinatorial lattice is proposed. The description of the whole structure is composed of one lemma, thirteen theorems, and twelve corollaries.

Theorem 7 (Conjugate Group Combinatorial Lattice and Abel Groups). *Three extension steps are required from a single vector to generate a conjugate group combinatorial lattice. Step 1. The conjugate Klein group expands one vector to eight vectors of two conjugate crosses in a conjugate cluster; Step 2. Four vectors of each cross take sixteen combinations to generate nine nontrivial vectors; Step 3. Two sets of eighteen nontrivel vectors create an integrated Abel group under complex addition to generate a conjugate group combinatorial lattice.*

Corollary 7 (Lattice and Strong Interaction). *Normally entangled interactions appear in the eight vectors of a conjugate cluster for the conjugate Klein group in Step 1, and extended entanglements are restricted in the same eigenvalue on a cycle with a constant radius r. However, the extensions in Steps 2 and 3 generate multiple cycles under combinatorial and Abel group operators to indicate that there are strong interactions involved beyond a single cycle condition.*

4.5 *Entangled Measurements of Quantum Optics*

In Chapter 5, the historic development of complex conjugate systems, complex geometry, multiple variable complex functions, and entangled measurements of quantum optics are reviewed.

The description of the whole structure is composed of twenty-two lemmas and twenty corollaries.

Lemma 7 (Modern Progress in Entangled Measurements of Quantum Optics). *The frontier quantum theory system and its corresponding various representations of the quantum entangled state are reviewed, and their typical complex geometric invariants satisfying the characteristics of global transformation are described.*

Lemma 8 (Non-Orthogonal Measurements in Quantum Entanglements). *Wootters has listed various measurement schemes to indicate entanglement vectors without orthogonal properties in quantitative measuring conditions.*

Corollary 8 (Reflective Mirrors ≡ Basic Complex Operators). *In modern quantum optics theory and experimental devices, various reflection mirrors play a key role in implementing the basic complex operators in quantum optical and photonic environments.*

For example, a full reflection mirror implements an imaginary operator i, and a semi-transparent mirror implements a conjugate operator $\sim$.

The proving procedure of the **Hardy Theorem** is a perfect case to illustrate a series of complex operators applied as the basic reasons in the derivation.

4.6 *Entanglement Foundation of Quantum Optics*

Chapter 6 is divided into three parts. The first part discusses the core matrix structure of quantum interaction. The second part summarizes the most important results of quantum entanglements during 2002–2022.

The third part describes the necessary and sufficient conditions of quantum entanglement to establish a self-consistent description of space and time in the quanta optical measurement system under a formal description of the measurement being paradox-free.

The description of the whole structure is composed of five lemmas, six assumptions, six predictions, sixteen theorems, and fifteen corollaries.

Lemma 9 (Real Eigenvalues of Quantum Optics and Quantum Information). *Quantum optics and quantum information under Dirac brackets provide a necessary and sufficient condition to support {Real Eigenvalue} solutions of* $\{+1, -1\} \leftrightarrow \{+\theta, -\theta\}$ *and* $\{x, x^*\}$ *a pair of measures, complex vectors, and complex conjugate operator* $*$.

Corollary 9 (Paradox in Real Eigenvalue Solutions). *The general solution of the Dirac bracket and Lie algebra schemes is a reverse version of the Gaussian fundamental theorem of algebra. It is incomplete to cover the complex solution spaces under the parameters of any imaginary numbers containing the Mobius ring and intrinsic logic paradoxes* $A = \neg A$ *in the phase space.*

Theorem 8 (Numbers of Logic, Quantum, and Conjugate Eigensolutions). *From the primitive distributions, classical logic is a linear structure, quantum logic is generalized to be two vectors, and conjugate logic is generalized to be four meta-vectors on balanced distributions. The numbers of state spaces involved in the n meta-variables are* $\{2^n, 2^n, 2^{2n}\}$.

Corollary 10 (Classical Quantum Formula ≡ Simplified Conjugate Structure). *From Lie algebra to conjugate algebra, the classical quantum formula simplifies the eight-variable vectors into two vectors on a pair of complex conjugate representations. The complete quantum solution needs to be equipped with principal and auxiliary double diagonal complex eigenvalues, and the n cobit phase space representation expands from* 2^n *to* 2^{2n}.

Theorem 9 (Paired Conjugate Vectors ≡ Paired Quantum Entangled States). *Eight vectors of a conjugate cluster are located in the four attribute regions; in the same region, paired conjugated states provide measurable characteristic information carried in the representation of paired entangled states in the description of quantum information interaction.*

Theorem 10 (Single Loop Detection). *Select a single diamond-shaped loop on the complex plane and use the spider web transformation of the square root mode to determine the density matrix of each coordinate point on the complex plane of the accompanying projected density values, forming a circular feature visualization distribution.*

Corollary 11 (Maximum Distributions in Eight Directions). *Both complete distribution and product distribution are superimposed to be a distribution with maximum values in eight directions* $\{0, \pi/4, \pi/2, 3\pi/4, \pi, 5\pi/4, 3\pi/2, 7\pi/4\}$ *that can be decomposed into two conjugate crosses.*

Theorem 11 (Strong Interaction Detection). *Through* $n = 1$ *cobweb transformation, the density matrix in the complex plane shows from diamond to rectangular feature distributions, fine eigenvalues are determined by the quantization scales of real and imaginary parts of the input vector, and the number of iterations of the interactions.*

5. Mysterious Ancient Vortex Theories

5.1 Mystery Number World

Different from modern number theory that divides integers into prime numbers and composite numbers, the eastern traditional I Ching map (易图) and Chinese I Ching numbers (术数) [22]–[24] and the arithmetic system [25] use symbols and positions to form corresponding mappings and conversion modes. Such structures appear repeatedly in the traditional river diagram (河图), Luo's diagram (洛书), and various I Ching diagrams.

The Tao originally from Tai Chi stands in one, one is divided into two as Yin and Yang, the {heaven, earth, self} is three, the coordinate directions are four, the five means star shape, and six elements indicate universal, seven up and eight down, and the maximal one is nine beyond the nine return to one [26]. Idioms such as the supreme number nine associated with heaven, earth and person, the nine heavens embracing the moon, and the ninety-five sole honor were combined with divination and arithmetic in the imperial power of all dynasty things.

Further explorations of history and further references are studied in [27] on multiple ancient mystery numbers on combinatorial, permutation, rolling dice, and set partition, etc., originally from China, India, Greece, Arabic, and European.

5.2 *Vortex Math*

Vortex mathematics [28], a system associated with decimal counting, is popularized by Tesla in China and India. Archaic number structure: $n \in [1,9], t \in N$, the decimal-expanded number sequence of n^t, add two digits in turn, and the number is one digit in $[1,9]$, continue to add to the next bit; more than 10 advanced bits become two, and then add; after the first step, the remaining number is still between $[1,9]$.

5.2.1 Arithmetic Process

Vortex arithmetic transformation conversion process.

Input: $n \in [0, 13], t \in [0, 13]$;
Output: $V(n, t)$;
Process:
  $n^t = x_m x_{m-1} \cdots x_j \cdots x_1$; Convert t exponent to m decimal sequence
  $y = x_m, j = m$; Calculation initialization
  while ($j > 1$) do From highest bit, modulo addition to end of lowest bit
  $z_2 z_1 = y + x_{j-1}$;
  if($z_2 == 0$) then $\{ y = z_1 \}$
  If there is no carry, assign a ones digit
  else then $\{ y = z_1 + 1 \}$
  add a carry if there is a carry
  $j = j - 1$; next loop
  Output: $V(n, t) = y$

5.2.2 Number Sequences

Generated sequences are shown in Table 1 on $t = 0 - 13$.

The exponential sequence is ignored, and the arithmetic transformation sequence is displayed.

To facilitate the comparison of related loop structures, it is cut down to ten lines and compared with a set of standard number theory mod 10 multiplicative group representation structures side by side.

Table 1. Vortex numbers from $t = 0 - 13$

t	0	1	2	3	4	5	6	7	8	9	10	11	12	13
$0^t =$	0	0	0	0	0	0	0	0	0	0	0	0	0	0
$V(0,t) =$	0	0	0	0	0	0	0	0	0	0	0	0	0	0
$1^t =$	1	1	1	1	1	1	1	1	1	1	1	1	1	1
$V(1,t) =$	1	1	1	1	1	1	1	1	1	1	1	1	1	1
$2^t =$	1	2	4	8	16	32	64	128	256	512	1024	2048	4096	8192
$V(2,t) =$	1	2	4	8	7	5	1	2	4	8	7	5	1	2
$3^t =$	1	3	9	27	81	243	729	2187	6561	19683	59049	177147	531441	1594323
$V(3,t) =$	1	3	9	9	9	9	9	9	9	9	9	9	9	9
$4^t =$	1	4	16	64	256	1024	4096	16384	65536	262144	1048576	4194304	16777216	67108864
$V(4,t) =$	1	4	7	1	4	7	1	4	7	1	4	7	1	4
$5^t =$	1	5	25	125	625	3125	15625	78125	390625	1953125	9765625	48828125	244140625	1220703125
$V(5,t) =$	1	5	7	8	4	2	1	5	7	8	4	2	1	5
6^t	1	6	36	216	648	3888	23328	139968	839808	5038848	30233088	181398528	1088391168	5630347008
$V(6,t) =$	1	6	9	9	9	9	9	9	9	9	9	9	9	9
$7^t =$	1	7	49	343	2401	16807	117649	823543	5764801	40353607	282475249	1977326743	13841287201	96889010407
$V(7,t) =$	1	7	4	1	7	4	1	7	4	1	7	4	1	7
$8^t =$	1	8	64	512	4096	32768	262144	2097152	16777216	134217728	1073741824	8589934592	68719476736	549755813888
$V(8,t) =$	1	8	1	8	1	8	1	8	1	8	1	8	1	8
$9^t =$	1	9	81	729	6561	59049	531441	4782969	43046721	387420489	3486784401	31381059609	282429536481	2541865828329
$V(9,t) =$	1	9	9	9	9	9	9	9	9	9	9	9	9	9
$10^t =$	1	10	100	1000	10000	100000	1000000	10..,0	10...0	10...0	10...0	10...0	10...0	10...0
$V(10,t) =$	1	1	1	1	1	1	1	1	1	1	1	1	1	1
$11^t =$	1	11	121	1331	14641	161051	1771561	10..,0	10...0	10...0	10...0	10...0	10...0	10...0
$V(11,t) =$	1	2	4	8	7	5	1	2	4	8	7	5	1	2
$12^t =$	1	12	144	1728	20736	248832	2985984	10..,0	10...0	10...0	10...0	10...0	10...0	10...0
$V(12,t) =$	1	3	9	9	9	9	9	9	9	9	9	9	9	9
$13^t =$	1	13	169	2197	28561	371293	4826809	10..,0	10...0	10...0	10...0	10...0	10...0	10...0
$V(13,t) =$	1	4	7	1	4	7	1	4	7	1	4	7	1	4

Table 2. Transformation number sequences $N = 0 - 13$

$V(n,t)$ t	0	1	2	3	4	5	6	7	8	9	10	11	12	13
0	0	0	0	0	0	0	0	0	0	0	0	0	0	0
1	1	1	1	1	1	1	1	1	1	1	1	1	1	1
2	1	2	4	8	7	5	1	2	4	8	7	5	1	2
3	1	3	9	9	9	9	9	9	9	9	9	9	9	9
4	1	4	7	1	4	7	1	4	7	1	4	7	1	4
5	1	5	7	8	4	2	1	5	7	8	4	2	1	5
6	1	6	9	9	9	9	9	9	9	9	9	9	9	9
7	1	7	4	1	7	4	1	7	4	1	7	4	1	7
8	1	8	1	8	1	8	1	8	1	8	1	8	1	8
9	1	9	9	9	9	9	9	9	9	9	9	9	9	9
10	1	1	1	1	1	1	1	1	1	1	1	1	1	1
11	1	2	4	8	7	5	1	2	4	8	7	5	1	2
12	1	3	9	9	9	9	9	9	9	9	9	9	9	9
13	1	4	7	1	4	7	1	4	7	1	4	7	1	4

$V(n,t)$ t	0	1	2	3	4	5	6	7	8	9	10	11	12	13
0	0	0	0	0	0	0	0	0	0	0	0	0	0	0
1	1	1	1	1	1	1	1	1	1	1	1	1	1	1
2	1	2	4	8	7	5	1	2	4	8	7	5	1	2
3	1	3	9	9	9	9	9	9	9	9	9	9	9	9
4	1	4	7	1	4	7	1	4	7	1	4	7	1	4
5	1	5	7	8	4	2	1	5	7	8	4	2	1	5
6	1	6	9	9	9	9	9	9	9	9	9	9	9	9
7	1	7	4	1	7	4	1	7	4	1	7	4	1	7
8	1	8	1	8	1	8	1	8	1	8	1	8	1	8
9	1	9	9	9	9	9	9	9	9	9	9	9	9	9

(nt mod 10) Cyclic Group t:	0	1	2	3	4	5	6	7	8	9
0	0	0	0	0	0	0	0	0	0	0
1	0	1	2	3	4	5	6	7	8	9
2	0	2	4	6	8	0	2	4	6	8
3	0	3	6	9	2	5	8	1	4	7
4	0	4	8	2	6	0	4	8	2	6
5	0	5	0	5	0	5	0	5	0	5
6	0	6	2	8	4	0	6	2	8	4
7	0	7	4	1	8	5	2	9	6	3
8	0	8	6	4	2	0	8	6	4	2
9	0	9	8	7	6	5	4	3	2	1

5.2.3 Recursive Structure

Compare the transformation tables for two different transformation modes, the arithmetic transformation modes associated with the start of the sequence. Counting mode starting from $n = 0$, the entire sequence will be zero, and the feature clusters are single cluster point $\{(0),(1)\}$, mapping cluster point $\{1 \rightarrow \{3,6,9\} \rightarrow (9)\}$, clockwise $\{(124875),(147)\}$, counter-clockwise $\{(157842),(174)\}$, swap structure $\{(18)\}$, and a total of five transformation structures.

n	0	$\{1, 10\}$	$\{2, 11\}$	$\{3, 12\}$	$\{4, 13\}$
$\{V(n,t)\}$ loop	(0)	(1)	(124875)	$1 \to 3 \to (9)$	(147)
Features	Single Points	Single Points	Six Synchronous	Injective Points	Triple Chronic
n	5	6	7	8	9
$\{V(n,t)\}$ loop	(157842)	$1 \to 6 \to (9)$	(174)	(18)	$1 \to (9)$
Features	Inverse quintuple	Injective assemblage	Triadic inverse	Inverse of binary	Injective assemblage

When $n > 0$, the vortex arithmetic transform output is greater than zero. Feature clusters are single point $\{(1), (9)\}$, clockwise $\{(248751), (471)\}$, counter-clockwise $\{(578421), (741)\}$, swap structure $\{(81)\}$, map cluster point $\{\{3, 6\} \to (9)\}$, and five transformation structures.

n		$\{1, 10\}$	$\{2, 11\}$	$\{3, 12\}$	$\{4, 13\}$
$\{V(n,t)\}$ loop		(1)	(248751)	$3 \to (9)$	(471)
Features		Single Points	Six Synchronous	Injective Points	Triple Chronic
n	5	6	7	8	9
$\{V(n,t)\}$ loop	(578421)	$6 \to (9)$	(741)	(81)	(9)
Features	Five-Element Inverse	Injective Convergence	Trinary Inverse	Binary Swap	Single Convergence

Under the condition of the multiplication modulo operation in number theory, when the input parameter is a prime factor of 10, zero will appear in the inner region of the matrix. List of different cyclic representations, including single aggregation point $\{(0)\}$, clockwise $\{(0123456789), (02468), (0369258147), (04826)\}$, swap structure $\{(05)\}$, counter-clockwise $\{(06284), (0741852963), (08642), (0987654321)\}$ Four kinds of transformation structures.

In the multiplicative group structure of cluster number theory, the transformation mode of number theory is more explicit and standardized than the arithmetic expression and does not include multistep mapping aggregation points. The action area is [0, 9], and the function of the zero element and its transformation is exerted as much as possible, which is required for the transformation of system characteristics.

n	0	1	2	3	4
$\{(nt \mod 10)\}$	(0)	(0123456789)	(02468)	(0369258147)	(04826)
Cycle Features	Single Convergence Point	Ten Single Clockwise	Five Double Clockwise	Ten Three Clockwise	Five Four Clockwise

n	5	6	7	8	9
$\{(nt \mod 10)\}$	(05)	(06284)	(0741852963)	(08642)	(0987654321)
Cycle Features	Two five exchange	Six four reverse	Ten three reverse	Five two reverse	Ten singl reverse

Lemma 10 (Recursive Structures of Vortex Math). *Vortex mathematics expresses a series of repeat loops as generated sequences to be four classes: 1. Single convergent points; 2. Fixed length anticlockwise loops; 3. Fixed length clockwise loops; and 4. Injective convergent points.*

6. Conjugate Structure of Vector Logic, Variant Construction, and Knowledge Model

6.1 Vector Logic Operations

In [7, 8], Chapters 1–3 use 0–1 logic vectors, variant vectors of switch algebra on the basis of the classical vector in the {AND, OR, NOT, XOR, NOR} operators, and extended permutation and complementary operators.

In the measurement space, the state clustering eigenvalues are formed, and the eigenvectors are constructed for the selected eigenshapes. The clustering effect and the conjugated representation structure express various changes in regular patterns. From the levels of the architecture, the complementary equilibrium structure, the conjugate invariant, and all partial/overall analyses can be judged based on quantitative metrics. A typical hierarchy of a cellular structure exhibits a classification structure of conjugate pairings, an arbitrarily complex system that may be decomposed into two modes: a local structure and a global organization.

A hierarchical representation of cellular structures is shown in Figure 7(a), and a complex system decomposed into local and global modes is shown in Figure 7(b).

The function space quantization properties related to the entire representation are shown in Table 3 under a small number of variables to list the number of variables, states, orthogonal functions, and logic functions.

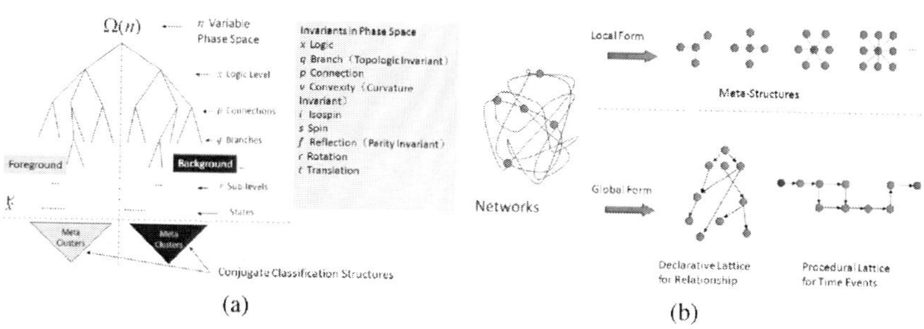

Figure 7. Vector logic operations: (a) hierarchy of cellular structure; (b) local and global decomposition modes.

Table 3. The n-ary logic variables, states, orthogonal functions, number of phase spaces for logic functions

Variables n	States 2^n	Orthogonal Functions 2^{2n}	Logic Functions 2^{2^n}
0	1	1	2
1	2	4	4
2	4	16	16
3	8	64	256
4	16	256	2^{16}
5	32	1024	2^{32}
6	64	4096	2^{64}

6.2 Complex Arithmetic Operation, Meta Structure Organization, and Variant Construction

In [7, 8], the fourth and fifth parts provide analytical models and methods for hierarchical structure designs. Historic studies link to complex computing topics: quantum Turing machines, quantum probability and statistics, quantum computing, quantum information, quantum entropy function, standard model, quantum gravity, dark matter, dark energy, black hole dynamics, supersymmetry, brain science, complex neural networks, models of consciousness, superstrings, integrated information theory, and nature Smart et al. [29]–[35].

Based on vector logic operations, it is necessary to extend integrated arithmetic operations, m-mer states, compound entropy projections, average/integrations, etc. In addition, various middleware substructures provide overall feature supports that are organized by meta-structure organization, including description + process knowledge model, local + whole analysis structure, and classification + retrieval mechanism. According to the relevant local characteristics of the system, the variant construction provides capacity to support a complex system analysis.

Fig. 8(a) shows the concept cell model of the meta-knowledge system, and Fig. 8(b) shows a real-world problem with a refined measurement procedure in solving diagrams of implementation of process analysis.

Theorem 12 (Variant Construction). *Variant construction is composed of variant logic, variant measurement, variant map, and hierarchical knowledge model, including description + process knowledge model, local + global analysis structure, and classification + retrieval mechanism. According to the relevant local characteristics of the system, variant construction forms a foundation to provide integrated holistic structural support on the multiple levels of a hierarchy on complex system analysis, design, implementation, and other engineering activities.*

7. Comprehensive Analysis

From the development of modern logic and mathematics, Godel's paradoxes are the most striking results in the study of logic in the 20th century. Although the visualization mode

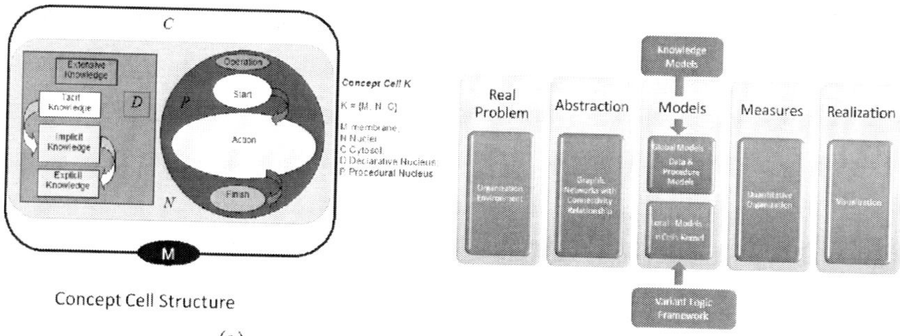

(a) (b)

Figure 8. Variant construction of complex arithmetic operations: (a) concept cell model; (b) partial and global analysis modes of complex systems.

Table 4. Basic n-ary logical variables, $\{A, B\}$ type state, corresponding manifold and computational model

n	Type	State	Manifold	Model
0	A	1	Monomorphic	Chaos/Tai Chi
0	B	2	1-D Real Manifolds	Turing Machines
1	{A,B}	4	1-D Complex Manifolds	Quantum Turing Machine
2	{A,B}	16	4-D Complex Manifolds	Natural Prototype
3	A	64	16-D Complex Manifolds	Standard Model
3	B	256	64-D Complex Manifolds	Supersymmetric Models
...	...	...	...	...
n	{A,B}	$\{2^{2n}, 2^{2^n}\}$	Super-D Complex Manifolds	Clustering Knowledge Machine

of complex variable function through the research of Klein and Poincare was noticeable in the late 19th century, more impact visual effects were created by Escher in a series of mathematically paradoxical paintings. Mobius belts and Klein bottles play a central role in modern mathematical theory in all geometrical visual paradoxes, from the solution of multi-variate field differential equations, Riemann surfaces, large-scale manifolds, fluid dynamics, vortices, von Karman vortex streets, vortex shedding, and surface integrals [36]–[39]. The potential shadows of paradoxes are there.

In ancient oriental traditional expressions [11], a series of Tai Chi illustrations have been used to express the complex dynamic characteristics of balance. From the simplified Tai Chi diagram, Lai's Tai Chi diagram (来氏太极图) and the four-phase pattern of the Korean flag are typical cases. The conjugated ring provides a systematic method to solve the Mobius paradoxes. The extended classification features are effective in naturally eliminating the intrinsic paradox of the one-sided surface using the complementary characteristics of four distinguishable spatial strips sequentially to support further complex structure representations.

Exploration of complex functional solutions includes cubic algebraic equations, complex conjugate representations, complex geometry, fundamental theorem of algebra, exponential representations of complex conjugated numbers, modern representations of complex conjugated numbers, theory of complex functions, Riemann surfaces, theory of elliptic functions, mathematical physics complex function equation, variational Hamilton-Jacobi,

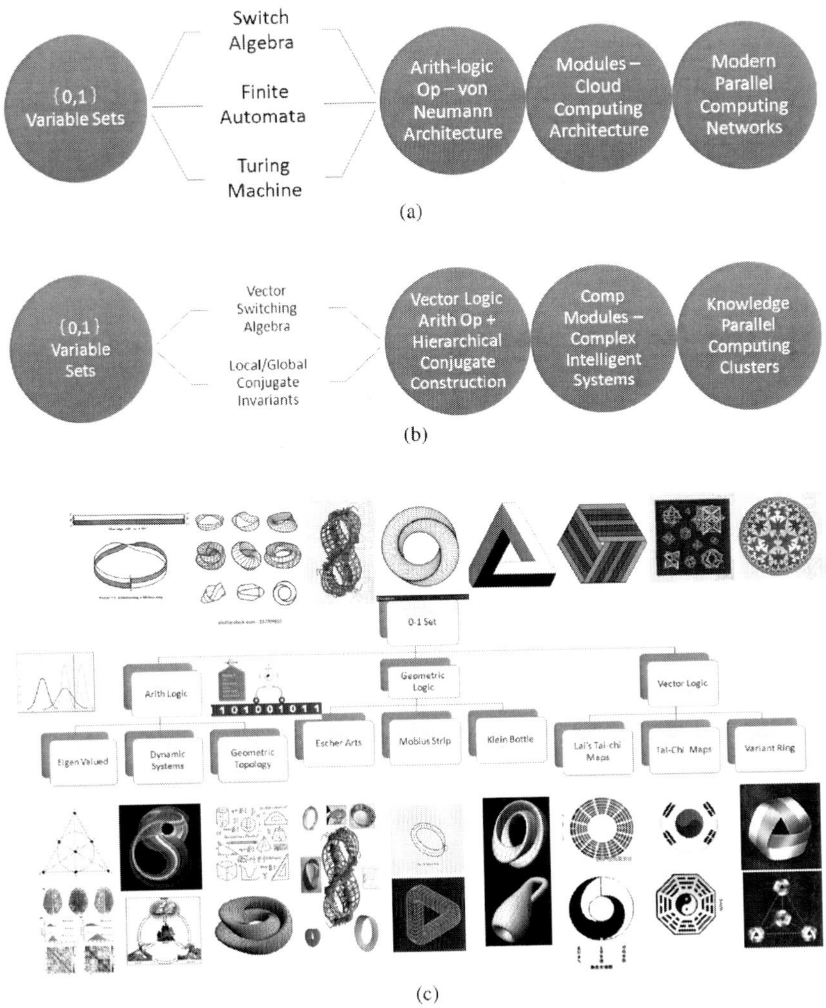

Figure 9. Core architecture: (a) classic logic + turing machine + von neumann architecture; (b) vector logic + conjugate structure + meta knowledge model; (c) three modes: classical logic, visual paradox, vector logic.

Hilbert space, two-body entanglement, time inversion symmetry, and symmetry + supersymmetry criterion.

Conjugate representation and photonic entangled measurements play a central role in frontier quantum optics and modern mathematical physics.

The multivariate complex functions provide the quantitative eigenvalue process to satisfy the Hamilton-Jacobi optimization. The integral equations of mathematical physics depend on the simplified conditions of differential equations under variational environments, in which the derivations are equal to zero, and partial integrals are constants associated with normalized initial values and special optimizations on various regions. The Hamilton-Jacobi optimization of the solution space must be fully guaranteed under $2n$ constant solutions.

The Hilbert space creates the basis for the formal representation of quantum mechanics. Two-body entangled states and time inversion provide a basis for quantitative measurement to determine entanglement properties. Symmetry and supersymmetry judgments are useful to check the sequential eigenvalues of the state sequences associated with a higher symmetry criterion and the locality measurement constraints.

Vector logic, variant construction, and knowledge models are constructed from basic $\{0,1\}$ components, effectively extending the basic model of modern computing Turing machines and the von Neumann architecture. Powerful hierarchical cellular structures support local and global decomposition models of complex dynamic systems and comprehensively formulate future knowledge and intelligent systems. Architectures and applications are explored on special strategies on typical cases on the concept cell models and local and global analysis of complex systems.

A classification scheme of conjugate complex vector systems listed in Table 4 toward expecting applications on ultrahigh-dimensional complex manifolds and clustering knowledge machines on future knowledge system architectures—computational models of hyper knowledge and intelligent systems. Among them, the number of A-type state spaces 2^{2n} and the number of B-type state spaces 2^{2^n}.

The important results are summarized in **Lemmas 1–10, Theorems 1–12, Corollaries 1–11, and Predictions 1–3**.

8. Forecast

Although a consistent foundation of quantum optics is proposed in Chapter 6, there are still many mysteries and hard scientific and technological problems in the world. It is better to make three predictions to complete the chapter. Wish the next generations have a better future.

Prediction 1 (Conjugate Complex Vector Solutions Useful in Vortex Mechanisms). *Conjugate complex vector functions provide powerful synchronous functions to recursively generate dynamic effects of increasing and decreasing radii in the complex plane to provide a foundation to support future exploration and applications on complicated vortex mechanisms.*

Prediction 2 (Coming General Computers). *Modern computers are restricted in Type B* $n = 0, State = 2$ *mode on* $\{0, 1\}$ *Turing machine in 1D real manifolds. The coming general computers are based on Type* $\{A, B\}$ $n = 1, State = 4$ *mode on full* $\{(1, 0), (0, 1), (-1, 0), (0, -1)\}$ *eigenvalue solutions on conjugate machines to support 1D complex manifolds for complete quantum computations.*

Prediction 3 (Extensive Applications on Supercomputation and Visualizations). *Resolved Mobius paradoxes, it will be useful to implement higher performance supercomputing and visualization targets on conjugate complex vector systems in hypersynchronous network environments for knowledge and intelligent content management for subsequent generations.*

References

[1] 吉姆 艾尔–哈利利著 戴凡惟译, 悖论:破解科学史上最复杂的9大谜团,中国青年出版社 2015 Jim Al-Khalili, translated by Dai Fanwei. *Paradox: Solving the 9 Most Complex Mysteries in the History of Science*, China Youth Press, 2015.

[2] 庞元正 李建华, 系统论 控制论 信息论 经典文献选编, 求实出版社 1989 Pang Yuanzheng and Li Jianhua. *Selected Classical Literature of System Theory, Cybernetics and Information Theory*, Qiushi Publishing House, 1989.

[3] 编辑委员会, 逻辑学词典,吉林人民出版社 1983 Editorial Board, *Dictionary of Logic*, Jilin People's Publishing House, 1983.

[4] 斯图尔特 夏皮罗 著 郝兆宽 杨睿之 译,数学哲学,复旦大学出版社 2009 Written by Stewart Shapiro, translated by Hao Zhaokuan and Yang Ruizhi. *Philosophy of Mathematics*, Fudan University Press, 2009.

[5] 罗伊 索伦森著 贾红雨译, 悖论简史 – 哲学和心灵的迷宫,北京大学出版社 2007 Roy Sorensen, translated by Jia Hongyu. *A Brief History of Paradox - The Labyrinth of Philosophy and Mind*, Peking University Press, 2007.

[6] 逻辑悖论 https://baike.baidu.com/item/逻辑悖论/126875 Logical paradox. https://baike.baidu.com/item/logical paradox/126875.

[7] Jeffrey Z. J. Zheng. *Variant Construction from Theoretical Foundation to Applications*, Springer Nature, 2019.

[8] 郑智捷,变值体系理论及其应用 第一卷 理论基础及其应用,科学出版社 2021. Jeffrey Z. J. Zheng. *Vairant Construction Theory and Applications, Vol. 1, Theoretical Foundation and Applications*, Science Press, 2021.

[9] Z. J. Zheng and A. Maeder. The conjugate classification of the kernel form of the hexagonal grid. In *Modern Geometric Computing for Visualization*, edited by T. L. Kunii and Y. Shinagawa, Springer-Verlag, 1992, 73–89.

[10] Z. J. Zheng. *Conjugate Transformation of Regular Plane Lattices for Binary Images*, PhD Thesis, Monash University, 1994.

[11] 郑智捷,消解逻辑悖论建立元知识智能化体系,计算机科学,49(1), 2022, 9–16. https://doi.org/10.118296/jsjkx.210700023 Zheng Zhijie. Dissolving logical paradoxes to establish a meta-knowledge intelligent system. *Comp. Sci.* 49(1), 2022, 9–16. https://doi.org/10.118296/jsjkx.210700023.

[12] Kleene, S.C.: Gödel's impression on students of logic in the 1931s. In: Weingartner, P., Schmetterer, L. (eds.) Gödel Remembered, pp. 49–64. Bibliopolis (1987)

[13] 莫比乌斯带 https://baike.baidu.com/item/莫比乌斯带/4457881 Mobius strip. https://baike.baidu.com/item/Mobius strip/4457881

[14] S. Barr. *Experiments in Topology*, Dover Publications Inc., 1964.

[15] 菲利克斯克莱因著舒湘芹等译高观点下的初等数学,复旦大学出版社 2008 Felix Klein, translated by Shu Xiangqin et al. *Elementary Mathematics from a High Perspective*, Fudan University Press, 2008.

[16] 克莱因瓶 https://baike.baidu.com/item/克莱因瓶/1187845 Klein bottle. <https://baike.baidu.com/item/Klein bottle/1187845>.

[17] B. Riemann. *Foundations for a General Theory of Functions of a Complex Variable*. Collected papers of Bernhard Riemann (translated by R. C. Baker, C. Christensen and H. Orde), Kendrick Press, 2004, 1–41.

[18] M. C. Escher. *The Graphic Work*, Taco, 1989.

[19] [明] 来知德, 来易经图解,中央编译出版社2010 (Ming Dynasty) Lai Zhide, *Lai Annotate the Book of Changes*, Central Compilation Press, 2010.

[20] 施维,周易八卦图解,四川出版集团巴蜀书社 2003 Shi Wei. *Zhouyi Bagua Diagrams*, Sichuan Publishing Group Bashu Publishing House, 2003.

[21] 雷泽,十分钟周易,中山大学出版社1993 Lei Ze. *Ten Minutes Zhouyi*, Sun Yat-Sen University Press, 1993.

[22] (宋)邵雍 著 李一圻 点校 郑同 增订,增广校正梅花易数,九州出版社 2009 (Song Dynasty) *Shao Yong Written by Li Yiqi Point Correction Zheng Tong Supplemented, augmented and corrected Plum Blossom Yishu*, Kyushu Publishing House, 2009.

[23] (清)黄宗,易学象数论,九州出版社 2007 (Qing Dynasty) Huang Zongyi. *Yi Xue Xiang Number Theory*, Kyushu Publishing House, 2007.

[24] 周学熹,易学教材6种,中医古籍出版社 2006 Zhou Xuexi. *6 Kinds of Teaching Materials for Easy Learning*, Traditional Chinese Medicine Ancient Books Publishing House, 2006.

[25] 王居恭,术数入门,华龄出版社 2009 Wang Jugong, *Introduction to Vortex Arithmetic*, Hualing Publishing House, 2009.

[26] 叶舒宪 田大宪,中国古代神秘数字,陕西出版集团陕西出版社 2011 Ye Shuxian and Tian Daxian. *Mysterious Numbers in Ancient China*, Shaanxi Publishing House of Shaanxi Publishing Group, 2011.

[27] D. E. Knuth. *The Art of Computer Programming, Vol. 4A: Combinatorial Algorithms*, Addison-Wesley, 2011.

[28] A. Vaish. Vortex mathematics: Source code of the univerrse. https://aayushivaish.medium.com/vortex-mathematics-source-code-of-the-universe-23749cc4ad6.

[29] J. von Neumann. Allgemeine Eigenwerttheorie Hermitescher Funktionaloperatoren. *Math. Ann.* 102, 1930, 49–131.

[30] A. Haun and G. Tononi. Why does space feel the way it does? Towards a principled account of spatial experience. *Entropy* 21(12), 2019, 1160.

[31] P. Kalita, J. A. Langa and F. Soler-Toscano. Informational structures and informational fields as a prototype for the description of postulates of the integrated information theory. *Entropy* 21, 2019, 493. https://doi.org/10.3390/e21050493.

[32] J. R. Hanson, S. I. Walker. Integrated information theory and isomorphic feed-forward philosophical zombies. *Entropy* 21, 2019, 1073. https://doi.org/10.3390/e21111073.

[33] L. Albantakis and G. Tononi. Causal composition: Structural differences among dynamically equivalent systems. *Entropy* 21, 2019, 989. https://doi.org/10.3390/e21100989.

[34] T. Bolognesi. Integrated information in process-algebraic compositions. *Entropy* 21, 2019, 805. https://doi.org/10.3390/e21080805.

[35] Harold Szu, Mike Wardlaw, Jeff Willey et al. Theory of glial cells & neurons emulating biological neural networks (BNN) for natural intelligence (NI) operated effortlessly at a minimum free energy (MFE). *MOJ Applied Bionics and Biomechanics* 1(1), 2017, 8–29. https://doi.org/10.15406/mojabb.2017.01.00002.

[36] Vortex. http://en.wikipedia.org/wiki/Vortex In addition, a list of interesting topics: Cyclonic Separation, Helmholtz's Theorems, History of Fluid Mechanics, Karman Vortex Street, Quantum Vortex, and Wormhole.

[37] Computational fluid dynamics. http://en.wikipedia.org/wiki/Computational_fluid_dynamics.

[38] Vortex theory of the atom. http://en.wikipedia.org/wiki/Vortex_theory_of_the_atom.

[39] K. Godel. An example of a new type of cosmological solutions of Einstein's field equations of gravitation. *Rev. Mod. Phys.* 21(3), 1949, 447–450.

Part III

Appendices

True is whole
but whole is nothing other than
essence consummating itself through its development.
– George Hegel

It is by logic that we prove
But by intuition that we discover
To know how to criticize is good
To know how to create is better.
– Henri Poincare

We adore chaos because we love to produce order
Simplicity and order are if not the principal then certainly
The most important guidelines for human beings in general.
– M.C. Escher

In this part, seven appendices are collected to show the most important results in each chapter as lemmas, assumptions, predictions, theorems, and corollaries.

Appendix 1: Lemmas 1–6, Theorems 1–24, and Corollaries 1–16 in Chapter 1

Appendix 2: Lemmas 1–3, Theorems 1–22, and Corollaries 1–7 in Chapter 2

Appendix 3: Lemma 1, Theorems 1–10, and Corollaries 1–8 in Chapter 3

Appendix 4: Lemma 1, Theorems 1–13, and Corollaries 1–12 in Chapter 4

Appendix 5: Lemmas 1–22 and Corollaries 1–20 in Chapter 5

Appendix 6: Lemmas 1–5, Assumptions 1–6, Predictions 1–6, Theorems 1–16, and Corollaries 1–16 in Chapter 6

Appendix 7: Lemmas 1–10, Theorems 1–12, Corollaries 1–11, and Predictions 1–3 in Chapter 7

Appendix 1

Lemmas, Theorems, and Corollaries in Chapter 1

Lemmas 1–6, Theorems 1–24, Corollaries 1–16:

1. Lemmas 1–6

Lemma 1 (Complex Conjugate Pairs of Elementary Algebraic Equations). *The elementary algebraic equations in n variables have n complex solutions, two related complex solutions, with pairwise complex conjugate properties.*

Lemma 2 (Length of Complex Conjugate Vector). *The modulo $|x|$ represents the inner product of pairwise complex conjugate vectors in Hilbert space, quantifying the general invariant of the vector length for the canonical projection of complex conjugate vectors.*

Lemma 3 (Noncompact Equivalent Representation). *The 2×2 real rotation matrix is an extended representation of complex ordinal pairs with more distinguishable symbols of combination states to be compared with the complex number states on sequence pairs.*

$$\boxed{x \in \mathrm{CP}} \to \begin{Bmatrix} x & +\theta \\ x^* & -\theta \end{Bmatrix} : \{x^*x = xx^* = 1\} \to \frac{+\theta \in \mathrm{UHP}}{-\theta \in \mathrm{LHP}} \to \frac{\{re^{i\theta}\}}{\{re^{-i\theta}\}}, \theta \in R$$

Lemma 4 (Klein Group). *The four states associated with $K(x)$ occupy the four vertex positions on the complex plane, forming a rectangle.*

Lemma 5 (Eight Vector Interaction). *In the $X = K(x)$ and $\tilde{X} = \tilde{K}(x)$ expression, pairs of products are multiplied to form 64 pairwise groups of output representations.*

Lemma 6 (Periodic Characteristic of Global Modulo Function). *For any complex function representation, the paired conjugate arguments $\{\theta, \tilde{\theta}\}$ rotate and transform the values, satisfying $\theta \bmod 2\pi \in [0, 2\pi), 0 \leq \theta < 2\pi$; $\tilde{\theta} \bmod \widetilde{2\pi} \in (0, 2\pi], 0 < \tilde{\theta} \leq 2\pi$ on periodic characteristics.*

2. Theorems 1–24

Theorem 1 (Conjugate Klein Groups). *$K(x)$ and $\tilde{K}(x)$ consist of two sets of Klein groups on eight states, and the cluster can generally be represented as a single rectangle or two rectangles.*

Theorem 2 (Unique Eight/Four Vectors of Conjugate Clusters beyond Gaussian Integers). *On the complex plane for any* $x = (p, q), p, q \in N, \theta = \angle(p, q), 0 \leq \theta < \pi/2$, *there is a unique conjugate state* $\tilde{x} = (q, p), p, q \in N, \tilde{\theta} = \angle(q, p), 0 < \tilde{\theta} \leq \pi/2$ *and* $\tilde{\theta} = \pi/2 - \theta$. *Beyond the four elements of Gaussian integers, it can simultaneously satisfy* $\{K(x), \tilde{K}(x)\}, \{V(x), \tilde{V}(x)\}$ *on the complex plane. Only two cases are satisfied: 1) eight distinguishable points; 2) four distinguishable points.*

Theorem 3 (Octary/Quaternary Vectors on Circle). *The octary vectors satisfying Theorem 2 are located on the circle with radius* $r = \sqrt{p^2 + q^2}$.

Theorem 4 (Number of Cocircular Clusters of Octary/Quaternary Vectors). *For the* $[0, N]^2$ *quadrant, the number of cocircular clusters* $C_I(N)$ *is* $C_I(N) = (N + 1)N/2$.

Theorem 5 (Twelve Output Vectors). *As output results of sixty-four products satisfying Lemma 6, a total number of twelve distinguishable vectors are included.*

Theorem 6 (Clustering Structure of Eight Vectors). *Two group representations have equivalent elements between* $\{K(x), \tilde{K}(x)\}$ *and* $\{V(x), \tilde{V}(x)\}$ *within different clustering partitions.*

Theorem 7 (Eight Vectors on Corrected Representations). *Using the* $\{K(x), \tilde{K}(x)\}$ *groups, the corrected vectors can be renamed from* $\{V(x), \tilde{V}(x)\} \rightarrow \{U(y), \tilde{U}(y)\}$. *The involved vectors represent the two equivalent clusters.*

Theorem 8 (Four Exponential Representations). *Combined with* $\{Q_j(y)\}_{j=0}^3, \{\tilde{Q}_j(y)\}_{j=0}^3$, *the exponent representations have arguments from* θ *and* $\tilde{\theta}$, *each region contains a pair of conjugate exponents, the adjacent numbering order of counter-clockwise is* $\{0, 2, 3, 1\}$, *and the adjacent angle increment is* $\pi/2$.

$$\boxed{x \in \text{CP}} \rightarrow \begin{Bmatrix} x & \theta \\ x^* & -\theta \end{Bmatrix} : \{x^*x = xx^* = 1\} \rightarrow \frac{+\theta \in \text{UHP}}{-\theta \in \text{LHP}} \rightarrow \frac{\{re^{i\theta}\}}{\{re^{-i\theta}\}}, \theta \in R$$

$$\boxed{x \in \text{CP}} \rightarrow \begin{Bmatrix} y\ \theta \\ \tilde{y}\ \tilde{\theta} \end{Bmatrix} \rightarrow \begin{Bmatrix} YY & Y\tilde{Y} \\ \tilde{Y}Y & \tilde{Y}\tilde{Y} \end{Bmatrix} \rightarrow \frac{\{2, \tilde{2}\} \| \{0, \tilde{0}\}}{\{3, \tilde{3}\} \| \{1, \tilde{1}\}} \rightarrow \frac{\{re_2^{i\theta}, re_2^{i\tilde{\theta}}\} \| \{re_0^{i\theta}, re_0^{i\tilde{\theta}}\}}{\{re_3^{i\theta}, re_3^{i\tilde{\theta}}\} \| \{re_1^{i\theta}, re_1^{i\tilde{\theta}}\}}$$

Theorem 9 (Four Coordinate Vectors). *Eight conjugate exponent functions are multiplied two by two to form a quaternary unit vector under the action of the formulas (46)–(49):* $\{1, i, -1, -i\}$.

Theorem 10 (Sixteen Product Vectors and Four Outputs). *In the* $\{e_j^{i\theta}\}_{j=0}^3 \times \{e_j^{i\theta}\}_{j=0}^3$ *action, the four transitions of output forms are the four types:* $\{A, B, C, D\}$.

Theorem 11 (Sixteen Conjugate Product Vectors and Four Outputs). *In the* $\{e_j^{i\tilde{\theta}}\}_{j=0}^3 \times \{e_j^{i\tilde{\theta}}\}_{j=0}^3$ *action, the four groups of output forms are:* $\{E, F, G, H\}$.

Theorem 12 (Three Vector Length Growths + Six Change Modes). *Use a pair of conjugate exponents* $\{e_0^{i\theta}, e_0^{i\tilde{\theta}}\}$ *iterates, from the starting point of the horizontal coordinate and the vertical coordinate, respectively forming three groups of counter-clockwise and clockwise. As the number of iterations increases, the argument increases synchronously, but the length of the vector is increased, unchanged, and shrinked. Three mechanisms showing six growth patterns are identified.*

Theorem 13 (Grow and Shrink Mode under Multiple Rotations). *Using four pairs of conjugate exponents* $\{\{r^k e_j^{ik\theta}\}_{j=0}^3, \{r^k e_j^{ik\tilde{\theta}}\}_{j=0}^3\}_{k=0}^n$ *to perform n iterations, eight groups of initial values are obtained from the corresponding coordinates as the starting points, and four groups of counter-clockwise and clockwise values are formed. With the increase in the number of iterations, the arguments grow synchronously, and the length of the vector is increased by* $r^k < r^{k+1}$ *and contracted by* $r^k > r^{k+1}$ *two mechanisms of action, the starting point and the end point are selected, the change patterns of the two trends are opposite, and the whole distribution characteristics are similar.*

Theorem 14 (Single-Cycle and Multi-Cycle Modes). *Using a pair of conjugate exponents* $\{\{r_j^k e_0^{ik\theta}\}_{j=1}^m, \{r_j^k e_0^{ik\tilde{\theta}}\}_{j=1}^m\}_{k=0}^n$ *to perform n iterations, paired initial values from the corresponding coordinate points, two groups of counter-clockwise and clockwise are formed. The number of iterations increases, the argument increases synchronously, and the j vector radius* $r_j^k = r_j^{k+1}$ *iteration remains unchanged. Single or multiple radii bicyclic structures are formed.*

Theorem 15 (Quarter Attribute Partition). *The complex plane is divided into eight regions* $\{Q_j(y)\}_{j=0}^3, \{\tilde{Q}_j(y)\}_{j=0}^3$ *mode, and four attribute partitions are formed.*

Theorem 16 (Modulo Representation in First Quadrant). *In the first quadrant, the paired conjugate arguments* $\{\theta, \tilde{\theta}\}$ *rotate the transformation value, satisfying* $\theta(\text{mod}\,\pi/2) \in [0, \pi/2), 0 \leq \theta < \pi/2$; $\tilde{\theta}(\text{mod}\,\tilde{\pi}/2) \in (0, \pi/2], 0 < \tilde{\theta} \leq \pi/2$ *on periodic characteristics.*

Theorem 17 (Interaction on Modular Parameters). *The algebraic equation transforms any m element partition condition on the solvable ring to form a regular polygon with m vertices through the cyclotomic fields. When m is a prime number, the polygon is expanded by the conjugate rotation group to form a total number of 4m vertex polygons, arguments* $\{n\theta, n\tilde{\theta}\}$ *constraint in* $n\theta(\text{mod}\,\pi/2) \in [0, \pi/2)$; $n\tilde{\theta}(\text{mod}\,\tilde{\pi}/2) \in (0, \pi/2]$.

Theorem 18 (Growing and Shrinking of Multiple Rotations in First Quadrant). *Using a pair of conjugate exponents* $\{\{r^k e_0^{ik\theta}\}, \{r^k e_0^{ik\tilde{\theta}}\}\}_{k=0}^n$ *to perform n iterations, the paired initial values start from the corresponding coordinate points, forming four groups of counter-clockwise and clockwise. The number of k iterations increases, the argument period changes, and the length of the vector is increased* $r^k < r^{k+1}$, *and the length of the conjugate vector is contracted* $r^k > r^{k+1}$ *on two actions. To select the starting point and the end point, the two change patterns are opposite, and the whole characteristics are similar to the two-color curl structure.*

Theorem 19 (Rotational Mode of Single Arc and Multi-Arc in First Quadrant). *Using a pair of conjugate exponents* $\{\{r_j^k e_0^{ik\theta}\}_{j=1}^m, \{r_j^k e_0^{ik\tilde{\theta}}\}_{j=1}^m\}_{k=0}^n$ *to perform n iterations, the initial value corresponds to the coordinate point, and two groups of counter-clockwise and clockwise are formed. As the number of iterations increases, the argument increases synchronously, and the jth vector radius* $r_j^k = r_j^{k+1}$ *iteratively remains unchanged, forming single or multiple two-color quarter arc structures.*

Theorem 20 (Number of Whole Divisions and Individual Projections on First Quadrant). *From parameters determined by formulas (84)–(88)* $\{F(m), P(m), P(\tilde{m}), P(m) \cup P(\tilde{m}), P(m) \cap P(\tilde{m}), P(m) + P(\tilde{m}\}$, *they provide the number of whole divisions and the number of individual projections for the selected quadrant.*

Theorem 21 (Common Part Equivalent). *For all common parts, whose values lie in* $0 < n < m$, *the placeholder count values* $T(n, s, m) = T(n, s, \tilde{m})$ *are equal.*

Theorem 22 (Total Number of Placeholders). *For all values in the common part of* $0 < n < m$, *placeholder count* $T(n, s, m) = T(n, s, \tilde{m})$ *value is equal. For all cases, the parameters satisfy* $T(m) = T(\tilde{m})$. *When* m *is a prime number,* $T(m) = T(\tilde{m}) = m^2 - m + 1$.

Theorem 23 (Global Conjugate Invariant). *The global conjugate invariant* $H(m) = H(\tilde{m}) = m^2 - m + 1$ *is a monotonically increasing sequence,* $\Delta(m) = \Delta(\tilde{m})$. *When* m *is prime* $\Delta(m) = \Delta(\tilde{m}) = 0$. $\{T(m) = T(\tilde{m})\}$ *is the wave growth sequence,* $\{\Delta(m) = \Delta(\tilde{m})\}$ *is the segmental wave of the pulse sequence, and* $\{T(m) - \Delta(m), T(\tilde{m}) - \Delta(\tilde{m})\}$ *is a fluctuating growth sequence.*

Theorem 24 (Extended Galois Field Algebraic Equivalent Representation). *The solution space of the* m *polynomial algebraic equation and the* m *conjugate complex functions are equivalent to increasing the solution space from* $\{x^m - 1\} \in G(m)$ *to* $\{x^m + 1, x^m + i, x^m - 1, x^m - i\}$, *expanded in the* $x^{4m} - 1 \in G(4m)$ *structure.*

3. Corollaries 1–16

Corollary 1 (Necessary and Sufficient Conditions of Octary/Quaternary Vectors on Circle). *When* $p, q > 0, p \neq q$, *each quadrant contains two vectors, and the four quadrants contain a total of eight vectors. When* $p, q > 0, p = q$ *Each quadrant contains one vector, and four quadrants contain four vectors. When* $\{p > 0, q = 0\}$ *or* $\{q > 0, p = 0\}$, *the four vectors lie on the axes.*

Corollary 2 (Number of Clusters in Four Quadrants Equivalent to Number of Circles). *Four quadrants* $\{[00], [10], [11], [01]\}$ *satisfy the condition of Theorem 4, and all quadrants are shared on the* $C_I(N)$ *distinct circles.*

Corollary 3 (Constraints on Number of Clusters of Eight Vectors on the Circle). *From any vector* $x = (p, q)$, *satisfying Theorem 4, there is a certain radius* $r = \sqrt{p^2 + q^2}$ *and paired conjugate angles* $\{\theta, \tilde{\theta}\}$, *related to four/eight vectors. The number of conjugate vectors in pairs satisfies* $4 \leq |\{K(x), \tilde{K}(x)\} = \{V(x), \tilde{V}(x)\}| \leq 8$.

Corollary 4 (Number of Vector Clusters in Four Quadrants). *When the complex plane region is* $[-N, N]^2$, *the distinct vector positions and the number of points on the complex plane are related. The total number of distinguishable vectors is* $F_I = 8N^2 + 1$.

Corollary 5 (Number of Distinct Vectors in $[-N, N]^2$**).** *There are a total number of* $(2N + 1)^2$ *sample points, and each point provides a pair of conjugate angles* $\{\theta, \tilde{\theta}\}$ *to act once. The total number of* $F_V(N)$ *is*

$$F_V(N) = 2(2N + 1)^2 = 8N^2 + 8N + 2$$

There is a number of $8N + 1$ *amounts that is larger than Corollary 4. Three excess parts are provided by the overlapped elements in the four axes, the four diagonals, and the zero position.*

Corollary 6 (Eight Corrected Vectors in Order). *The corrected* $\{U(y)\}$ *cluster contains four vectors associated with* θ *as a base cross, while the* $\{\tilde{U}(y)\}$ *cluster has four vectors associated with* $\tilde{\theta}$ *as a conjugate cross.*

Corollary 7 (Twelve Output Vectors after Single Product). *Satisfying Corollary 6, eight vectors take* 8×8 *times the number of products and then group the results as clusters. Twelve distinguishable vectors are generated.*

Corollary 8 (Twelve Output Vectors after *n* products). *Satisfying Corollary 6, eight vectors take multiple products and merging operations. A total of twelve distinguishable vectors are generated.*

Corollary 9 (Common Part of Two Conjugate Values in Same Region). *Pair-wise conjugate exponent representation of* $\{Q_j(y)\}_{j=0}^3, \{\tilde{Q}_j(y)\}_{j=0}^3$, *share arguments* $\theta, \tilde{\theta} \in \{(0, \frac{\pi}{2}), (\frac{\pi}{2}, \pi), (\pi, \frac{3\pi}{2}), (\frac{3\pi}{2}, 2\pi)\}$, *at four coordinates on the axis, the corresponding value ranges are different.*

Corollary 10 (Paired Conjugate Exponents Have Same Sign). *The conjugate exponent representation of paired* $\{Q_j(y)\}_{j=0}^3, \{\tilde{Q}_j(y)\}_{j=0}^3$ *has related signs*

$$S(e_j^{i\theta}) = S(e_j^{i\tilde{\theta}}) \in \{(+,+)_{j=0}, (-,+)_{j=2}, (-,-)_{j=3}, (+,-)_{j=1}\}$$

Corollary 11 (Products of Thirty-Two Conjugate Exponents). *Under the action of the formulas (46)–(49), the eight conjugate exponent vectors involve a total of thirty-two products of sequence pairs to form a quaternion unit vector:* $\{1, i, -1, -i\}$.

Corollary 12 (Four Transition Groups). *From the perspective of the transition attribute mode, the eight groups of pairs are clustered into four groups under the product transition:* $\{A, H\}$ *is equivalent to* $+1$*;* $\{B, E\}$ *counter-clockwise* $+i$*;* $\{C, F\}$ *Negative Conjugate* -1*;* $\{D, G\}$ *Clockwise* $-i$.

Corollary 13 (Eight Coloring Modes). *For the diagrams that satisfy Theorem 13–14, two different colors can be selected for each region, and the entire attribute distribution characteristics are distinguished by eight colors.*

Corollary 14 (Equivalent Category and Even Number Partition). *For any* $\{P(m), P(\tilde{m})\}$ *in the first quadrant provides the same type of equivalent distinction structure, and even partition numbers appear less frequently.*

Corollary 15 (Conjugate Invariant on Galois Field of Algebraic Biperiodic Function). *For* m *conjugate variables, it is natural to express the* m *order complex variables in the functional algebraic structure* $G(m)$*. The whole structure is intrinsically associated with the P-adic number theory of the modular structure and the Galois field of the algebraic representation.*

Corollary 16 (Invariant Property of $4m$ Dimensional Solution Space). *In the solution structure, the* $4m$ *variable algebraic equation* $\{x^{4m} - 1\} \equiv \{x^m + 1, x^m + i, x^m - 1, x^m - i\}$ *retains the whole cluster unchanged.*

Appendix 2

Lemmas, Theorems, and Corollaries in Chapter 2

Lemmas 1–3, Theorems 1–22, Corollaries 1–7:

1. Lemmas 1–3

Lemma 1. *For a given $K(\overrightarrow{x})$ and its phase space Ω, $|\Theta| = |\tilde{\Theta}| = 2^m$,*

$$|\Omega| = |\Theta| + |\tilde{\Theta}| = 2^{m+1}$$

Lemma 2. *For a $BVA(Y)$ with $2n$ distinct vector variables, there are 2^{2n} states and $2^{2^{2n}}$ logic functions.*

Lemma 3. *For an elementary equation, the conjugate operator and $\neg$ operator generate different results.*

$$\tilde{F}(\langle A, B\rangle X) \neq \neg F(\langle A, B\rangle X) \tag{1}$$

2. Theorems 1–22

Theorem 1. *For a given conjugate classification, $2n$ feature classes $\{\{\Gamma_j\}_{j=1}^{n}, \{\tilde{\Gamma}_j\}_{j=1}^{n}\}$ partition a pair of two conjugate phase spaces Θ and $\tilde{\Theta}$ respectively.*

Theorem 2. *For a given configuration X, two canonical feature vectors satisfy $\Gamma(X) = \tilde{\Gamma}(X)$.*

Theorem 3. *For two feature vectors $\{\Gamma_j(X), \tilde{\Gamma}_j(X)\}$, conjugate operator $\tilde{}$ reverse feature vector and original configuration.*

$$\tilde{\Gamma}_j(X) = \neg\Gamma_j(\neg X); \tag{2}$$

$$\neg\tilde{\Gamma}_j(\tilde{X}) = \Gamma_j(\neg X); \tag{3}$$

$$\Gamma_j(X) = \neg\tilde{\Gamma}_j(\neg X) = \tilde{\tilde{\Gamma}}_j(X); \tag{4}$$

$$\neg\Gamma_j(X) = \tilde{\Gamma}_j(\neg X) \tag{5}$$

Theorem 4. *For two distinguished feature vectors under $\{\neg, \cap, \cup, \tilde{}\}$ operators, there are only six groups of irreducible expressions.*

$$\neg\Gamma_j(X) \cap \neg\Gamma_k(X) = \tilde{\Gamma}_j(\neg X) \cap \tilde{\Gamma}_k(\neg X); \tag{6}$$

$$\Gamma_j(X) \cup \Gamma_k(X) = \neg\tilde{\Gamma}_j(\neg X) \cup \neg\tilde{\Gamma}_k(\neg X); \tag{7}$$

$$\neg\tilde{\Gamma}_j(X) \cup \neg\tilde{\Gamma}_k(X) = \Gamma_j(\neg X) \cup \Gamma_k(\neg X); \tag{8}$$

$$\tilde{\Gamma}_j(X) \cap \tilde{\Gamma}_k(X) = \neg\Gamma_j(\neg X) \cap \neg\Gamma_k(\neg X); \tag{9}$$

$$\neg\Gamma_j(X) \cap \tilde{\Gamma}_k(X) = \neg\Gamma_j(X) \cap \neg\Gamma_k(\neg X); \tag{10}$$

$$\Gamma_j(X) \cup \neg\tilde{\Gamma}_k(X) = \Gamma_j(X) \cup \Gamma_k(\neg X) \tag{11}$$

Theorem 5. *The elementary equation is a vector logic equation to reverse the selected sets of feature vectors and keep other unselected sets of feature vectors invariant.*

Theorem 6. *Under the $\neg$ operator, the elementary equation is a self-conjugate equation to exchange index sets of feature vectors and to be reversed on the configuration vector.*

$$\neg F(\langle A, B\rangle X) = F(\langle B, A\rangle \neg X) \tag{12}$$

Theorem 7. *For a given $\langle A, B\rangle$, if $A = \emptyset$ or $B = \emptyset$, then $F(\langle A, B\rangle X)$ for components $\langle A, B\rangle X$ is an equation of dilation or erosion.*

Theorem 8. *The elementary equation of conjugate transformation generates four extreme vectors $\{\Gamma(X), \vec{0}, \vec{1}, \neg\Gamma(X)\}$; four extreme vectors are $\langle A, B\rangle = \{\langle\emptyset, \emptyset\rangle, \langle\emptyset, I\rangle, \langle I, \emptyset\rangle, \langle I, I\rangle\}$.*

Theorem 9. *For a configuration X and its two sets of feature vectors $\{\Gamma_j(X)\}_{j=0}^{n}, \{\tilde{\Gamma}_k(X)\}_{k=0}^{n}$, the elementary equation of conjugate transformation can generate 2^{2n} vector functions.*

Theorem 10. *For an CTS(X) on $1 \le n \le 2^m$, there are 2^{2n} functions.*

Theorem 11. *For $2n$ vector variables in BVA and $2n$ feature vectors on $m + 1$ kernel form in CTS, if $n \le \lceil \frac{m+1}{2} \rceil$, then the function number of CTS is systematically much larger than the function number of BVA.*

Theorem 12. *For a configuration with N elements under a kernel form with $m + 1$ bits in $2n$ clusters, the total number of combinations $\Sigma(N, m, n)$ is determined by*

$$\Sigma(N, m, n) = 2^N \times 2^{2n}, 1 \le n \le 2^m \tag{13}$$

Theorem 13. *For a $\langle A, B\rangle X \in CTS(X)$, the elementary equation of CTS has four equivalent expressions.*

$$F(\langle A, B\rangle X) = (\Gamma(X) \cap \neg\Gamma_A(X)) \cup \neg\tilde{\Gamma}_B(X) \tag{14}$$

$$= \Gamma_{A'}(X)) \cup \neg\tilde{\Gamma}_B(X) \tag{15}$$

$$= \neg\Gamma_A(X)) \cap \tilde{\Gamma}_{B'}(X) \tag{16}$$

$$= (\Gamma(X) \cup \neg\tilde{\Gamma}_B(X)) \cap \neg\Gamma_A(X) \tag{17}$$

Theorem 14. *For an elementary equation, its ¬ description has two forms.*

$$\neg F(\langle A, B\rangle X) = F(\langle A', B'\rangle X) \qquad (18)$$
$$= F(\langle B, A\rangle \tilde{X})$$

Theorem 15. *For any $Y_1, Y_2 \in CTS(X)$ under logic operators $\{\cap, \cup\}$, two combined vectors are in $CTS(X)$.*

$$Y_1 \cap Y_2 = F(\langle A_1 \vee A_2, B_1 \wedge B_2\rangle X) \qquad (19)$$
$$Y_1 \cup Y_2 = F(\langle A_1 \wedge A_2, B_1 \vee B_2\rangle X) \qquad (20)$$

Theorem 16. *For an elementary equation, the conjugate operator exchanges two index sets of feature vectors.*

$$\tilde{F}(\langle A, B\rangle X) = F(\langle B, A\rangle X) \qquad (21)$$

Theorem 17. *For an elementary equation, a NOT operation can be performed by two twistors.*

$$\zeta^2 F(\langle A, B\rangle X) = F(\zeta^2\langle A, B\rangle X) = F(\langle A', B'\rangle X) = \neg F(\langle A, B\rangle X) \qquad (22)$$

Theorem 18. *For a selected $H(\langle A, B\rangle X)$, eight complex operators and eight conjugate measurement operators have 1-1 correspondence in equivalent relationships.*

Theorem 19 (Numbers of Configurations in Conjugate Phase Space). *Conjugate transformation structure 2n Group output variable cluster dimension $2^{2(n+1)}$, the corresponding complex vector function state occupies $(-2^n, 2^n]^2$ configuration space.*

Theorem 20 (Pairwise Conjugate [,], {, } Brackets). *Two types of interaction operators $\{[x,y]_d, \{x,y\}_d\}, \{[x,y]^u, \{x,y\}^u\}$ provide complementary symmetric and antisymmetric information for the selected product matrix cluster.*

Theorem 21 (4 x 4 Matrix Kernel). *If the vector Y is an exchange operator associated with a variety of extended vector operators $\{-1, \leftarrow\}$ matrix eigenvalues, the 4×4 eigenmatrix can be used as the kernel, $\{-1, \leftarrow\}$ $\{[,]_d, \{,\}_d\}, \{[,]^u, \{,\}^u\}$ matrix representation implementation.*

Theorem 22 (Time Order Restriction of Operators). *The 4×4 feature matrix that satisfies the time order in the conjugate algebra representation represents the overall invariant structure, and there are thirty-two groups of $\{1, i, -1, -i\} \times 4$ feature matrix representations, involving sixteen groups of different configurations, which can represent 1/3 of the possible substitutions.*

3. Corollaries 1–7

Corollary 1. *In general, CTS provides a hierarchy of flexible function representatives in wider regions from $2^2, \ldots, 2^{2n}$ to $2^{2^{m+1}}$ instead of $2^{2^{2n}}$ on BVA.*

Corollary 2. *Combining various configurations and feature functions, their controllable spaces have numeric regions from $\Sigma(N, m, 1) = 2^{N+2}$ to $\Sigma(N, m, 2^m) = 2^N \times 2^{2^{m+1}}$.*

Corollary 3. *For an elementary equation, four twistors are performed to be a unit operator; it is back to be itself.*

$$\zeta^4 F(\langle A, B\rangle X) = E(F(\langle A, B\rangle X)) = F(\langle A, B\rangle X) \tag{23}$$

Corollary 4. *Applying two operators* $\{\sim, \zeta\}$ *repeatedly, an elementary equation has eight expressions.*

$$\{\sim, \zeta\} : \langle A, B\rangle \Rightarrow \left\{ \begin{matrix} \langle A, B\rangle, \langle B', A\rangle, \langle A', B'\rangle, \langle B, A'\rangle \\ \langle B, A\rangle, \langle A, B'\rangle, \langle B', A'\rangle, \langle A', B\rangle \end{matrix} \right\} \tag{24}$$

Corollary 5 (1-1 Correspondences of Eight Eigenvectors). *Select a pair of conjugate vectors from the 2n group of variables and expand to four attribute regions to become a single-channel conjugate structure vector of output representation. The structure and the conjugate complex vector function are corrected, and the octuples of the complex cluster are transformed into a 1-1 correspondence.*

Corollary 6 (Pauli Matrix Solutions). *The exchange operator associated with the eigenvalues of the Pauli matrix, implemented by the* $[,]_d, \{,\}_d, 2 \times 2$ *eigenmatrix.*

Corollary 7 (Dirac Matrix Solutions). *The exchange operator associated with the eigenvalues of the Dirac matrix, implemented by the* $[,]_d, \{,\}_d, 4 \times 4$ *eigenmatrix.*

Appendix 3

Lemmas, Theorems, and Corollaries in Chapter 3

Lemma 1, Theorems 1–10, Corollaries 1–8:

1. Lemma 1

Lemma 1 (Decomposition of Classic Measurement Process). *The classical decomposition process involves four core parts, namely, the linear input variable m, the classical logic state space 2^m, the narrow projection measure space 2^m, and the general projection measure space represented by the binomial $m + 1$.*

2. Theorems 1–10

Theorem 1 (Conjugate Numbers in One Block of Narrow Distribution). *In the conjugate decomposition process, one block of the narrow distribution contains 2^{2m} pairs of complex numbers of a $[0, 2^m)^2$ region.*

Theorem 2 (Conjugate Numbers in Four Block of Narrow Distribution). *In the conjugate decomposition process, four blocks of narrow distribution contain $2^{2(m+1)}$ pairs of complex numbers of four $[0, 2^m)^2$ regions.*

Theorem 3 (Conjugate Numbers in One Block of General Distribution). *One block of the general distribution contains $(m + 1)^2$ pairs of complex numbers as complex basis positions in a partial complex plane of a $[0, m]^2$ region.*

Theorem 4 (Conjugate Numbers in Four Block of General Distribution). *Four blocks of general distribution contain $(m + 1)^2$ pairs of complex numbers as complex basis positions in a complete complex plane of a $[-m, m]^2$ region.*

Theorem 5 (Narrow Information Entropy Function). *For a system of m variables, 2^m states and 2^m measures, the narrow information entropy function satisfies*

$$E_a(m) = m \tag{1}$$

Theorem 6 (General Information Entropy Function). *For a system of m variables, 2^m states and 2^m measures, the general information entropy function satisfies*

$$E_b(m) \leq log_2(m+1) \tag{2}$$

Theorem 7 (Single Block of Narrow Information Entropy Function). *For $m+1$ variables, 2^{2m} measures and $(s, \tilde{s})$ conjugate function, the narrow system entropy function satisfies*

$$E_a((s, \tilde{s})(m)) = 2m \tag{3}$$

Theorem 8 (Single Block General Entropy Function). *For $m+1$ variables, 2^{2m} measures on $(s, \tilde{s})$ conjugate function, and the general entropy function satisfies*

$$E_b((s, \tilde{s})(m)) \leq 2log_2(m+1) \tag{4}$$

Theorem 9 (Difference between Two 0-Points Combined General Entropy Functions). *When $(s, \tilde{s})$ and $(s', \tilde{s}')$ or $(s', \tilde{s})$ and $(s, \tilde{s}')$ are selected, if the two 0 points of the two blocks are merged, then the difference of the general entropy function satisfies*

$$E(m,m) - E'(m,m) = \frac{1}{2^{2m}} \tag{5}$$

Theorem 10 (Difference between Two 0-Lines Combined General Entropy Functions). *When $(s, \tilde{s})$ and $(s', \tilde{s})$ or $(s', \tilde{s})$ and $(s', \tilde{s}')$ or $(s, \tilde{s}')$ and $(s', \tilde{s}')$ or $(s, \tilde{s})$ and $(s, \tilde{s}')$ are selected, if two blocks are combined, then the difference in the general entropy function satisfies*

$$E(m,m) - E'(m,m) = \frac{1}{2^m} \tag{6}$$

3. Corollaries 1–8

Corollary 1 (Conjugate Measurement Process Decomposition). *The conjugate decomposition process involves five core parts, namely, the linear input variables $2m$, the conjugate logic state space 2^{2m}, and the full conjugate logic function space $2^{2^{m+1}}$, the narrow projection measure space 2^{2m}, and the general projection measure space $(m+1)^2$ expressed by the dual-binomial.*

Corollary 2 (Conjugate Complementary Measurement Process Decomposition). *The conjugate + complementary decomposition process involves five core parts, which are the linear input variable $2(m+1)$, the conjugate logic state space $2^{2(m+1)}$, the full conjugate logic function space $2^{2^{m+1}+2}$, the narrow projection measure space $2^{2(m+1)}$, and the general projection measure space $4(m+1)^2$ represented by the dual-binomial.*

Corollary 3 (Number of Elements on Narrow Distribution). *In a complete narrow distribution, each entry contains one element without any degeneration.*

Corollary 4 (Number of Elements on General Distribution). *In a complete general distribution, each i,j position is linked to a coefficient of $\binom{m}{i}\binom{m}{j}$ degenerate states and the total numbers of degenerate states are 2^{2m} (one block) and $2^{2(m+1)}$ (four blocks).*

Corollary 5 (Four-Block Narrow Information Entropy Function). *For $m+1$ variables, $2^{2(m+1)}$ measures, $(s, s' : \tilde{s}, \tilde{s}')$ extended conjugate function system, the narrow information entropy function satisfies*

$$E_a((s, s' : \tilde{s}, \tilde{s}')(m)) = 2(m+1) \tag{7}$$

Corollary 6 (Two-Block General Entropy Function). *For the $m+1$ variables, choose any two 2^{2m} measures $(s, s' : \tilde{s}, \tilde{s}')$ conjugate function, and the general entropy function satisfies*

$$E_b((s, \tilde{s})(m)) \leq 1 + 2log_2(m+1) \tag{8}$$

Corollary 7 (Four-Block General Entropy Function). *For the $m+1$ variables, select the four-block 2^{2m} regions $(s, s' : \tilde{s}, \tilde{s}')$ conjugate function, and the general entropy function satisfies*

$$E_b((s, \tilde{s})(m)) \leq 2 + 2log_2(m+1) \tag{9}$$

Corollary 8 (Four-Block 0-Lines Combined General Entropy Function). *When $\{(s, \tilde{s}), (s', \tilde{s}), (s, \tilde{s}'), (s', \tilde{s}')\}$ are selected, if the four blocks are combined, then the difference of the general entropy function satisfies*

$$E(m, m) - E'(m, m) > \frac{1}{2^{m-1}} \tag{10}$$

Appendix 4

Lemmas, Theorems, and Corollaries in Chapter 4

Lemma 1, Theorems 1–13, Corollaries 1–12:

1. Lemma 1

Lemma 1 (Conjugate Feature Group + Multiplicative Identity Element). *K4 group* $K = \{e, u, v, w\}$ *contains four elements,* $e = \begin{pmatrix} 1 & 0 \\ 0 & 1 \end{pmatrix}$ *is the multiplication unit, adding the extended conjugate structure* $\tilde{K} = \{\tilde{e}, \tilde{u}, \tilde{v}, \tilde{w}\}$, $\{K, \tilde{K}\}$ *form* D_4 *group including eight elements* $\{K, \tilde{K}\}$.

2. Theorems 1–13

Theorem 1 (Eigenvector Length Constraint). *For arbitrary states* $B = (a, b)$, $K(B)$ *and* $\tilde{K}(B)$ *The distinguishable eigenvector length of* ≤ 4.

Theorem 2 (Additive Group + Inverse/Zero Combination). *For arbitrary states* $B = (a, b)$, $K(B)$ *and* $\tilde{K}(B)$, $\{K, \tilde{K}\}$ *The combined state of the components each contains three sets of zero elements of the zero matrix corresponding to the additive group, and the four elements are two pairs of eigenvalues, which are inverse features of each other value. Reverse paired eigenvalues ensure that after the eigenvalue clusters are accumulated as a whole, the sum is expressed as a zero element* $= (0, 0)$.

Theorem 3 (Combination Eigenvalues). *According to* **Theorem 2** *and* **Corollary 2**, *for any state* $B = (a, b)$, $K^1(B)$ *or* $\tilde{K}^1(B)$, $2^4 = 16$ *combined eigenvalues of, respectively, including four zeros and four pairs of identical features value, the length of distinguishable eigenvectors after merging is* ≤ 9.

Theorem 4 (Combined Lattice Eigenvalues Sum Zero). *All* $K^1(B) \bigoplus \tilde{K}^1(B)$ *elements sum to zero after accumulation.*

$$\sum_{0 \leq i,j \leq 8} \left[K^1(B) \oplus \tilde{K}^1(B)\right]_{i,j} = \sum_{i=0}^{8} \sum_{j=0}^{8} \left[K_i^1(B) + \tilde{K}_j^1(B)\right] = (0, 0) \qquad (1)$$

Theorem 5 (Combination Lattice Equivalence). *Using $\{K(B), \tilde{K}(B)\}$ eigenvectors constructed $K^1(B) \bigoplus \tilde{K}^1(B)$, which is equivalent to the combined lattice structure formed by the eight element feature clustering contained in $D_4(B)$.*

$$K^1(B) \oplus \tilde{K}^1(B) = D_4^1(B) \tag{2}$$

Theorem 6 (Combinatorial Eigenvalue Structure Extension). *For any state $B = (a, b)$, if $T(B) = \{u, v, \cdots, x\}$ is a cluster containing m distinguishable eigenvalues, $T'(B) = \{u, v, \cdots, x, y\}$ is an extension of $T(B)$ containing $m+1$ distinguishable eigenvalues, then the combined cell $T^1(B)$ contains at most 2^m combined eigenvalues, and $T'^1(B)$ contains at most 2^{m+1} combined eigenvalues, all lattice points contained in $T^1(B)$ are completely contained in $T'^1(B)$. Additional features do not reduce the existing grid distribution.*

Theorem 7 (Combinatorial Lattice Partition Optimization). *For any state $B = (a, b)$, if $T(B) = \{u, v, \cdots, x\}$ is a cluster containing m distinguishable eigenvalues, $\exists l \in Z, 0 \leq l \leq m$ divides the combined lattice into two sub-lattices, including $\{l, m-l\}$ generator correspondences $\{2^l, 2^{m-l}\}$ lattice points. When $m = 0(mod\ 2), l = \frac{m}{2}$ or $m = 1(mod\ 2), l = \{\frac{m+1}{2}, \frac{m-1}{2}\}$, the combinatorial lattice has overall optimization characteristics.*

Theorem 8 (Conjugate Lattice Global Optimization). *For any state $B = (a, b)$, $T(B) = \{u, v, \cdots, x\}$ is a cluster containing m distinguishable eigenvalues, and the number of distinguishable eigenvalues of the conjugate combination lattice $\|[T^1(B) \oplus \tilde{T}^1(B)]\| \leq 2^{2m}$ balances the $\{T^1(B), \tilde{T}^1(B)\}$ feature clusters to meet the overall computational optimization characteristics.*

Theorem 9 (Conjugate Symmetric Permutation). *If the feature numbers for any m terms are $\{P(1), P(2), \cdots, P(i), \cdots, P(m)\}, 1 \leq i \leq m$, the corresponding m term conjugate number is $\{\tilde{P}(1), \tilde{P}(2), \cdots, \tilde{P}(j), \cdots, \tilde{P}(m)\}, 1 \leq j \leq m$, choose any permutation P and its conjugate permutation $\tilde{P}$ is the horizontal and vertical alignment coordinates because $P(i, j) = P(i) + \tilde{P}(j)$, then the formed $m \times m$ indicates that the matrix has conjugate symmetry.*

$$P(i, j) = (x, y) \& \tilde{P}(i, j) = P(j, i) = (y, x)$$

Theorem 10 (Antisymmetric Permutation). *If the feature numbers for any m terms are $\{P(1), P(2), \cdots, P(i), \cdots, P(m)\}, 1 \leq i \leq m$, the corresponding m item inverse item number is $\{P^{-1}(1), P^{-1}(2), \cdots, P^{-1}(j), \cdots, P^{-1}(m)\}, 1 \leq j \leq m$, choose any permutation P for the number and its inverse transformation P^{-1} are horizontal and vertical alignment coordinates $P(i, j) = P(i) + P^{-1}(j)$, then the formed m timesm indicates that the matrix is antisymmetric.*

$$P(i, j) = (x, y) \rightarrow P^{-1}(i, j) = P(j, i) = -(x, y)$$

Theorem 11 (Four Groups of Conjugate Combination Lattice Points). *For any order pair (a, b), the accompanying conjugate combination lattice has four distinguishable numbers:*

$$\{(0, 0), \{(1, 0), (1, 1)\}, \{(2, 1), (3, 1)\}, \ Other\}$$

Value	$(0, 0)$	$\{(1, 0), (1, 1)\}$	$\{(2, 1), (3, 1)\}$	*Other*
Cluster No.	1	25	73	81
Value condition	$(0, 0)$	$\{(1, 0), (1, 1)\}$	$\{(2, 1), (3, 1)\}$	*other*
Number of clusters	1	25	73	81

Theorem 12 (Conjugate Combination Lattice Mean Limit). *Choose a larger n, and the number of conjugate combination lattices at any position on the complex plane is 81.*

Theorem 13 (Local Entropy Measure). *For any order pair (a, b), the accompanying conjugate lattice has four distinguishable numbers:*
$\{(0,0), \{(1,0),(1,1)\}, \{(2,1),(3,1)\}, other\}$, *corresponding to four kinds of local entropy values, respectively.*

3. Corollaries 1–12

Corollary 1 (Conjugate Global Invariance). *For the D_4 group, the eight-element cluster under the action of the conjugate operator, $D_4 = \tilde{D}_4$ as a whole remains unchanged.*

Corollary 2 (Complementary Digestion Properties). *Choose any combination of three primitives from the four nonzero primitives of $K(B)$ or $\tilde{K}(B)$ addition, equivalent to selecting a single primitive.*

Corollary 3 (Partial Sum Zero). *Accumulate from nine combined primitives of $K^1(B)$ or $\tilde{K}^1(B)$. The sum is zero.*

$$\sum_{0 \leq i \leq 8} \left[K^1(B)\right]_i = \sum_{i=0}^{8} K_i^1(B) = (0,0) \tag{3}$$

$$\sum_{0 \leq j \leq 8} \left[\tilde{K}^1(B)\right]_j = \sum_{j=0}^{8} \tilde{K}_j^1(B) = (0,0) \tag{4}$$

Corollary 4 (Conjugates Satisfy Combinatorial Conditions). *For any choice state B, conjugates $K^1(B) \bigoplus \tilde{K}^1(B)$ satisfy all the basic conditions defined by abstract lattice theory.*

Corollary 5 (Maximum Number of Distinguishable Conjugated Combinatorial Lattice Points). *For any choice state B, conjugated combinatorial lattice $K^1(B) \bigoplus \tilde{K}^1(B)$. The feature distribution contains at most eighty-one distinguishable elements in the 9×9 table.*

Corollary 6 (Number of Non-Zero Conjugate Combinatorial Lattice Points). *Under general conditions, the conjugated combinatorial lattice $K^1(B) \bigoplus \tilde{K}^1(B)$ provides eighty nonzero eigenvalues to distinguish lattice distributions.*

Corollary 7 (Conjugate Combination Lattice Condition). *For any state $B = (a, b)$, if $T(B) = \{u, v, \cdots, x\}$ is a cluster containing m distinguishable eigenvalues, $\exists l \in Z, 0 \leq l \leq m$ can divide the combined lattice into two sublattices, including $\{l, m - l\}$generators corresponding to $\{2^l, 2^{m-l}\}$ lattice points.*

Corollary 8 (Conjugate Symmetry and Antisymmetric Association Conditions). *If $a = -b, (b, a) = \tilde{P}(i) = P^{-1}(i) = -(a, b)$ conjugate symmetric and antisymmetric matrices have the same value, then $P(i, i) = (0, 0)$.*

Corollary 9 (Number of Distinguishable Conjugate Combination Lattices). *Corresponding to the value of any order pair (a, b), there are six distinguishable state distributions.*

Corollary 10 (Regional Convergence Condition of Label Lattice). *For a given n and $n > 7$, the sum of the three states can be distinguished by the conjugate combination lattice topology and the number of lattice points in the given region. The ratio is a descending sequence.*

Corollary 11 (Labeled Area of Conjugate Combination Grid). *Select the area where the number of conjugated combination grid nodes is not equal to 81, gathered near the* $(0,0)$ *point, and its level* $(1,0)$*/vertical* $(0,1)$*, two diagonals* $(1,1)$*, paired bisections* $(1,2),(2,1)$*, paired three-point line* $(1,3),(3,1)$.

Value condition	$(0,0)$	$\{(1,0),(1,1)\}$	$\{(2,1),(3,1)\}$	other
Local entropy	$=0$	$\leq \log_2(25)$	$\leq \log_2(73)$	$\leq \log_2(81)$

Corollary 12 (Maximum Entropy of Conjugate Lattices). *On the complex plane, the maximum entropy of conjugated lattices at any position is measured as* $\log_2(81)$.

Appendix 5

Lemmas and Corollaries in Chapter 5

Lemmas 1–22, Corollaries 1–20:

1. Lemmas 1–22

Lemma 1 (Complex Geometry). *A complex number can be represented as a point $a + ib$ in the complex plane or as a directed radial vector from the origin to this point.*

Lemma 2 (Fundamental Theorem of Algebra). *Every nonzero, univariate polynomial of degree n with complex coefficients has n complex roots.*

Lemma 3 (Complex Exponential Representation). *Elegant and concise complex function representation, opening up the complex magic door for complex exponential function operations.*

Lemma 4 (Modern Representation of Complex Numbers). *The algebraic number-even theory of conjugate functions uses algebraic structures to establish ordinal pairs of complex variable functions.*

Lemma 5 (Complex Number Function Theory). *Cauchy–Riemann differential equations provide local differential optimization conditions for complex functions, and Cauchy definite integral equations provide global integral optimization conditions for complex functions from the integrated optimization level.*

Lemma 6 (Fourier Mathematical Physics Equations). *The Fourier Mathematical Physics equations strictly establish the representation of the step integral between the 2n eigenfunction and the same number of eigencoefficient values, exhibiting parsimony effects under orthogonal conditions.*

Lemma 7 (Automorphic Function). *As a generalization of many types of periodic functions, the autologous function exhibits complex transformation effects under orthogonal conditions for supporting complex periodic transformation modes.*

Lemma 8 (Hilbert Space). *Hilbert space is the product space L^2 of multivariable complex variable functions, which constitutes the core part of the foundation of modern mathematical physics, and promotes mathematics with modes such as inner product, orthogonal decomposition, and normal form in advanced physics.*

Lemma 9 (Hermite Complex Function 2 x 2 Matrix). *In complex function matrix representation 2×2 Hermite matrix conjugate transpose operation has core importance, and norm is multivariate vector converted to scalar. The inner product operation of the pseudo-orthogonal basis is accompanied by the conjugate operator in the selected n-element basis vector representation. When the transformation group is accompanied by the unit determinant* $\det\ U = 1$, *the corresponding transformation groups are $SU(n)$ and $SU(p, q)$.*

Lemma 10 (Real/Complex Linear Regular Transformation Matrix). *The real linear regular transformation LCT represents the special linear group $SP(2, R)$ of 2×2 real symmetric matrix, and the regular transformation constitutes a noncommutative group. The complex linear regular transformation CLCT forms a symplectic group $SP(2, C)$, which maps the real square-integrable function to the analytic function of the Hilbert space.*

Lemma 11 (Klein Quaternary Matrix). *The Klein four group is the smallest acyclic group, which can be represented as a 2×2 matrix with four distinguishable elements.*

Lemma 12 (Pauli Matrix). *The irreducible representation of a Pauli matrix is a set of three matrices. The transformation group contains a total of sixteen 2×2 matrices. Eight are real matrices, and the other eight are imaginary matrices. The first four of the real matrices are Klein quaternions.*

Lemma 13 (Dirac Equation). *The Dirac equation has three forms: simplified representation, natural unit representation, and strict representation. In strict representation, the four $4 \times 4\gamma$ matrices are three 2×2 Pauli. The matrix plus an identity matrix 4×4 represents the structure.*

Lemma 14 (Lattice Operator Specification). *The core part of lattice theory is $\{+, \times\}$ providing two sets of unit elements $\{0, 1\}$ and corresponding inverse operations and inverse operations element. Operator operations satisfy the associative law of addition, the commutative law of addition, the zero element of addition, and the negative element of addition, as well as the identity element of multiplication, the associative law of multiplication, and the distribution rate of multiplication.*

Lemma 15 (Complex Function Quantum State Representation). *The quantum state complex function representation expresses different rotation angles, superposition states, superposition state probabilities, and paired entangled state representations for the relevant state through the complex variable function formula.*

Lemma 16 (Two-Body Entangled State Discrimination). *The two-body entangled state is symmetric with time reversal, and the following properties can be used to determine: 1) if there are two entangled quantum states, they cannot be expressed as a direct product; 2) it can be written in the form of a direct product as entanglement-free, and the nonzero measured state satisfies the reciprocity condition; and 3) the entangled state time inversion operator converts any state into an orthogonal state.*

Lemma 17 (Symmetry + Supersymmetry). *Symmetry + Supersymmetry Criterion Poisson Operator Forming Extension Operator $[a, b]_+ = ab + ba$ Supersymmetry Operator $[]^s$ is represented as $[a, b]^s = ab \pm ba$.*

Lemma 18 (Tensor Product Expansion). *The tensor product operator forms an expansion vector representation, and n primitives are expanded into 2^n quantum state vector representations after tensor product expansion.*

Lemma 19 (Two-Quantum Entangled State). *The quantum entangled superposition state vector related to two qubits can be expressed as: {Bell, GHZ, NOON} and other pairwise description forms.*

Lemma 20 (Global Phase Difference). *In Bell quantum state representation, the sign feature of the global phase difference between quantum states does not need to be distinguished in classical quantum systems.*

Lemma 21 (Hadamard Matrix Tensor Product). *Hadamard matrix tensor product operator forms an extended matrix representation, and n tensor product operators become 2^{2n} element quantum state matrix representation after expansion.*

Lemma 22 (Modern Optical Measurement Conversion Equipment). *Modern optical equipment based on semitransparent and total reflection sheets, providing a variety of complex variable function transformation modes in the measurement path channel: Basic complex functions such as total reflection sheet and conjugate transflective sheet transform the measurement mode.*

2. Corollaries 1–20

Corollary 1 (Imaginary Representation). *Paired complex numbers have been around since the 16th century. The imaginary number is officially taken seriously and has gone through a long-term historical evolution process.*

Corollary 2 (Riemann Surface). *The Riemann face shows the highest state of the differential conjugate analytical local constraint. To satisfy the Dirichlet principle, this condition ensures that the complex variable function is a minimum value after being integrated through the whole space.*

Corollary 3 (Elliptic Function). *A series of studies on elliptic functions are carried out from the inverse functions of elliptic integrals and have the characteristics of a series of periodic functions.*

Corollary 4 (Green Mathematical Physics Equations). *Green mathematical physics equations show a typical mode of differential conjugate analytical local constraints, which partially satisfy the Dirichlet principle, ensuring that the Green function is a minimum value after spatial integration.*

Corollary 5 (Partial Differential Equations). *Navier–Stokes equations and Maxwell equations are both PDEs associated with typical frontier applications. Such representations of partial differential equations form a central part of modern mathematical physics applications.*

Corollary 6 (Lagrange Equation). *The Lagrange equation is a system of second-order differential equations with n elements $q_i(t)$.*

Corollary 7 (Hamilton–Jacobi Equation). *The Hamilton–Jacobi equation is a system of first-order differential equations with 2n metavariables p_i, q_i being a function of time t. The variational condition is that the typical differential optimization condition satisfies the Dirichlet principle, which ensures that the minimum value of the corresponding equation after the full-space integration is accompanied by the evaluation condition that each corresponding solution satisfies 2n constant coefficients.*

Corollary 8 (Jones Vector and Matrix). *In physical optics, various optical signal measurement metrics, in the complex form of the Jones vector and Jones matrix, systematically deal with complex optical system measurement metrics that can be represented by complex variable functions.*

Corollary 9 (Antiparticle Representation). *Majorana Fermion Antiparticle represents itself, with reflex invariance.*

Corollary 10 (Yang–Mills Equation). *The Yang–Mills equation consists of two parts: the first part involves the differential operator of the primitives, and the second part, the binary nonreciprocal condition via Poisson brackets, determines the difference feature under the bivariate exchange product.*

Corollary 11 (Complex System Stability). *A one-dimensional system with a small amplitude periodic oscillation of n elements determines the system stability from the batch eigenvalues. There are three kinds of system solutions: periodic, neutral equilibrium, and uncertainty.*

Corollary 12 (Identification of Topological Gravitational Variables). *In the research environment of topological gravitation, there is an obvious difference between the number of components of descriptive variables and observable variables, which needs to be identified in specific exploration.*

Corollary 13 (Hadamard Matrix Balance Positive and Negative Sign). *The positive and negative quantities formed by the Hadamard matrix tensor product are unbalanced in the single block representation, and the balanced state distribution of positive and negative quantities can be formed by four region expansion. The total number of elements involved is* $2^{2n+2} = 4 \times 2^{2n}$.

Corollary 14 (EPR Paradox Measurement). *The EPR paradox can be described under the mechanism of measuring the output using a complex function formal representation, distinguished from four aspects: full correlation, locality, reality, and completeness.*

Corollary 15 (Bell Inequality). *Bell's inequality provides multiple quantum state correlation combinations based on state probability-based quantitative measurement conditions.*

Corollary 16 (CHSH–Bell Inequality). *The CHSH–Bell inequality provides multiple quantum state correlation combinations based on state probability-based quantitative measurement conditions.*

Corollary 17 (Single Quantum Entanglement). *For special value conditions, single qubits can be used to represent entanglement.*

Corollary 18 (Noninequalities Express Multiquantum Entanglement). *For multiquantum state detection mechanisms, such as GHZ ternary entangled state detection, the noninequalities discriminant condition can be used to effectively detect the composite superposition feature of multientangled states.*

Corollary 19 (Hardy Theorem—Double Quantum Entanglement). *Hardy's theorem strictly describes the interaction of two-particle entanglement. Using modern optical equipment such as semitransparent sheets and fully transparent sheets, under the transformation of complex functions, the series may result in an impossible result analysis.*

Corollary 20 (Cluster State Quantum Entanglement). *For more complex entanglement state representations, the cluster state representation formula provides further extended modes of many types of quantum entanglement effects.*

Appendix 6

Lemmas, Assumptions, Predictions, Theorems, and Corollaries in Chapter 6

Lemmas 1–5, Assumptions 1–6, Predictions 1–6, Theorems 1–16, Corollaries 1–16:

1. Lemmas 1–5

Lemma 1 (n qubit phase space 2^n). *The discrete form partially satisfies the Wigner condition to be a classical quantum information phase space, and the combined state of n qubits is a 2^n phase space.*

Lemma 2 (Conjugate Klein Complex Group). *For any complex number $x = (p, q)$, the conjugate Klein complex group is expressed as two clusters $\{K(x), K(\tilde{x})\}$, each containing four elements, forming a rectangle spanning four quadrants in the complex plane.*

Lemma 3 (Four Real Eigenvalues of Pauli Matrices). *The eigenvalues of the four Pauli matrices $\{I, \sigma_1, \sigma_2, \sigma_3\}$ are $\{I : \{\lambda = ++\}, \sigma_1 : \{\lambda = +-\}, \sigma_2 = \{\lambda = -+\}, \sigma_3 : \{\lambda = +-\}\}$. All eigenvalues are $\{+1, -1\}$ combinations of unit real numbers.*

Lemma 4 (16 Matrix Product Expansions). *Using matrix product expansion, the 2×2 matrix group contains eight real matrices and eight imaginary matrices, for a total of sixteen matrices.*

Lemma 5 (Six Dirac Matrices). *Six Dirac matrices $\{I_4, -I_4, \gamma_0, \gamma_1, \gamma_2, \gamma_3\}$ have eigenvalues*

$$I_4 : \{\lambda = ++++\}, \gamma_0 : \{\lambda = ++--\}, \gamma_2 : \{\lambda = +--+\}$$

$$-I_4 : \{\lambda = ----\}, \gamma_1 : \{\lambda = --++\}, \gamma_3 : \{\lambda = -++-\}$$

2. Assumptions 1–6

Assumption 1 (Quantum Optics and Optical Quantum Measuring—Application Constraints). *There are correlation measurement, analysis model method and its system, belonging to the field of quantum optics and optical quantum science, involving constraints on the interaction characteristics of optical quantum, multiqubit feature description and analysis, entangled state interaction, quantum information, and comprehensive measurement result as well as distribution visualization scope.*

Assumption 2 (Geometric Optics+Cobweb Projection+Coincidence Counting—Measurement Visualization). *Based on light quanta detector and geometric optics transformation mechanism, the measurement satisfies geometric optics, cobweb projection, coincidence counting, and other modes, as discrete metric information interaction on visualization support of the quantitative feature.*

Assumption 3. *[Paired Conjugate Complex Function Eigenvalue—Conjugate Measurement] In the configuration space of the optical quantum information measurement system, the eigenvalue measure of the paired optical quantum is the conjugate complex pair* $\{x, \tilde{x}\}$, *the measure parameter is* $\{\theta, \tilde{\theta}\}$, *and the intrinsic vector radius r is the region invariant eigenvalue.*

Assumption 4 (Photon Quark Conjugate Discrete Cluster—Smallest Complete Set). *In the configuration space, paired light quanta conjugate complex eigenvalues, the four attribute regions represented by* $\{0, \pi/2, \pi, 3\pi/2\}$ *are the starting point to distribute in each region* $\pi/2$ *containing eight light quanta, associated with the region property code* $\{0231, 0132\}$ *conversion to determine the inverse of the interaction in the clock/clockwise rotations on the order of time. There are at most four groups of coherent light quanta pairs at the same time, forming the conjugate Klein group-quark discrete cluster.*

Assumption 5 (*n* Iterative Photons Form Single Energy Level Continuous Ring—Single Ring). *In the n iterative processing mode, four groups of coherent photon sequence pairs form complex entangled particles in the measurement. A closed loop is formed in the phase space, which satisfies the conjugate complex vector function, and the quantum eigenvalues after n iterations remain unchanged.*

Assumption 6 (*n* Iterative Quanta Form Multiple Energy Level Continuous Rings—Multiple Rings). *Under the strong interaction, the light quanta n iterative processing under the condition of Assumption 3, the complex entanglement composite states are distributed in the measurement phase space to form multiple closed loops, which satisfy the conjugate group qualifying conditions, and the light quanta grouping is constrained in the specific eigenvalue region.*

3. Predictions 1–6

Prediction 1 (Complete Quantum Phase Space). *The dimension of the complete phase space of n cobits is* 2^{2n}, *the imaginary eigenvalue quantifies the solution space of the system, and the statistical distribution of the phase space is transformed by the conjugate orthogonal transformation into a visualized image distribution.*

Prediction 2 (Conjugate Phase Space: Complete Phase Space 1:3). *Under the condition that the conjugate operator satisfies the time transformation order, the conjugate eigenvalue solution space covers one-third of the complete phase space, and two-thirds of the complete phase space transcend the conjugate systems.*

Prediction 3 (Determination of Dirac and Conjugate Eigenvalue Solutions Based on Rotational Symmetry). *Using the matrix diagonal and auxiliary eigenvalues of the complex plane, the rotational symmetry distribution accurately determines the Dirac equation of the system* *eigenvalue solution and common differences between solutions of eigenvalue systems. The solution space of the conjugate structure eigenvalue system has complete global rotational symmetry, but the solution space of the Dirac eigenvalue system does not provide rotational symmetry.*

Prediction 4 (Complex Plane Rhombus Link Rotational Symmetry). *Satisfying* $|j| + |k| = n, n > 0, j, k \in \mathbf{N}$ *on the complex plane, the rhombus link corresponds to the conjugate orthogonal normalized solution space of n quantum states, the dimension of the solution space is* $4n$*, and the independent measure of each projection space dimension* $n+1$ *is* $n-1$*. The four coordinates* $(n, 0), (0, n), (-n, 0), (0, -n)$ *are shared as the start and end of the link.*

Prediction 5 (Precise Conjugate Expansion Operators). *Four conjugate operators* $\{\sim, \sharp, \prime, *\}$ *paired with conjugate variables* $\{y, \tilde{y}\}, \theta, \tilde{\theta}$ *are the smallest cluster of quantum entanglement structures. The enigma of entangled states, which has plagued quantum optics and quantum information for a long time, can be naturally eliminated by accurately measuring the extended structure of complete conjugate vector state clusters.*

Prediction 6 (Strong Interaction Characteristic Cluster Distribution). *In addition to zero points, five nonzero distribution patterns in the quantum optics of a strong interaction environment can be rigorously tested by quantitative precision measurement mechanisms. Six sets of conjugated groups qualified a typical distribution.*

(0).	$p = q = 0$	*Single zero point on the* $(0, 0)$ *position*
(1).	$p = q$	*Diamond 25 points*
(2).	$(p, 0)\|(0, q)$	*Rectangular 25 points*
(3).	$p = 2q$	*Octagon 73 points*
(4).	$p = 3q$	*Octagon 73 points*
(5).	*Others*	*Octagon 81 points*

4. Theorems 1–16

Theorem 1 (Conjugate Hermite Matrix Quadruple on Primitive Eigenmatrix Representation). *Utilizing the* $\{0, \{1, i, -1, -i\}\}$ *primitive representation, the conjugate extension structure of the Hermite form is four groups of* $\{1, i, -1, -i\}$*. The feature matrix represents* $\{K(E), K(\overline{E}), K(\tilde{E}), K(\tilde{\overline{E}})\}$*, and each group of feature clusters contains two real matrices and two imaginary matrices.*

Theorem 2 (Complete Phase Space of Conjugated Structures). *The conjugated structure realizes the Wigner phase space based on discrete geometry, which ensures that the discrete quantum information architecture can provide complete support for four qubit interaction modes.*

Theorem 3 (Square Root Product of Conjugate Vectors in First Quadrant of Complex Plane). *The vector* $z = re^{i\theta} = (x, y), \theta = \angle z \in [0, \pi/2)$; $\tilde{z} = re^{i\tilde{\theta}} = (y, x), \tilde{\theta} = \angle\tilde{z} \in (0, \pi/2]$ *in the first quadrant of the complex plane corresponds to the square root vectors* $\{\sqrt{z}, \sqrt{\tilde{z}}\}$ *in the first quadrant of the complex plane. The product of vectors is* $\sqrt{z}\sqrt{\tilde{z}} = ir$.

Theorem 4 (Product of Squares of Conjugate Vectors in First Quadrant of Complex Plane). *The vector* $z = re^{i\theta}, \theta, 2\theta \in [0, \pi/2)$; $\tilde{z} = re^{i\tilde{\theta}}, \tilde{\theta}, 2\tilde{\theta} \in (0, \pi/2]$ *are in the first quadrant of* *the square complex plane, the corresponding vector* $\{z^2, \tilde{z}^2\}$. *The product of paired vectors is* $z^2\tilde{z}^2 = ir^4$.

Theorem 5 (Complex Plane First Quadrant Conjugate Vector on *n* Power Root Product). *Vector* $z = re^{i\theta}\tilde{z} = re^{i\tilde{\theta}}$ *in the first quadrant of the complex plane corresponds to the vector* $\{\sqrt[n]{z}, \sqrt[n]{\tilde{z}}\}$. *The product of paired vectors is* $\sqrt[n]{z}\sqrt[n]{\tilde{z}} = ir^{2/n}$.

Theorem 6 (Product of First Quadrant Conjugate Vector *n* Power of Complex Plane). *Vector* $z = re^{i\theta}, \theta \in [0, \pi/2)$; $\tilde{z} = re^{i\tilde{\theta}}, \tilde{\theta}, n\tilde{\theta} \in (0, \pi/2]$ *in the first quadrant of the n power complex plane, the corresponding vector* $\{z^n, \tilde{z}^n\}$. *The product of paired vectors is* $z^n\tilde{z}^n = ir^{2n}$.

Theorem 7 (Noncommutative Cross-level Operator). *Complex vector product operation with nonlinear projection, the operation involves power exponent function n power root | n power operation. The order of operation of operators is not commutative* $n > 1$, $\sqrt{z}\sqrt{\tilde{z}} = ir \neq i^{1/2}r = \sqrt{z\tilde{z}}$; $\sqrt[n]{z}\sqrt[n]{\tilde{z}} = ir^{2/n} \neq i^{1/n}r^{2/n} = \sqrt[n]{z\tilde{z}}$; $z^2\tilde{z}^2 = ir^4 \neq i^2r^4 = (z\tilde{z})^2$; $z^n\tilde{z}^n = ir^{2n} \neq i^n r^{2n} = (z\tilde{z})^n$. *After opening the n power root | n power transformation, the vector product of the n power root | n power is not equal to the cross-level multiplication of conjugate vectors in the original space and then n power root | n power. In the results of a power operation, the order of operations follows the transformation rules of the same level to avoid cross-level operations.*

Theorem 8 (Cobweb Projection Operator Eight-direction Projection Invariance). *The cobweb projection operator can keep the projection direction invariant in eight directions. They are* $\{0, \pi/4, \pi/2, 3\pi/4, \pi, 5\pi/4, 3\pi/2, 7\pi/4\}$, *respectively.*

Theorem 9 (Cobweb Projection Operator Quantitative Measurement Real Function Output Mode). *In the complex plane* $[-N, N]^2$ *region, it involves the quantitative measurement value carried by the complex primitive coordinate point, through the correlation functions represented as real measurements at z coordinates perpendicular to the complex plane.*

Theorem 10 (Number of Equipotential Loops in $[-N, N]^2$ Region of Complex Plane). *The number of points involved in the* $[-N, N]^2$ *region of the complex plane. The number of cycles more than 2 is* $2N - 1$.

Theorem 11 (Three-dimensional Rhomboid Equipotential Torus in Complex Plane $[-N, N]^2$). *From* $(0, 0)$ *position in* $[-N, N]^2$ *complex plane, it starts to reach* $(N, 0), (0, N)$ *link diamond area before stepping up and expanding outward, reaching* $(N, 0), (0, N)$ *link diamond area then step and shrink until* (N, N) *contains a step number of* $2N + 1$ *to form a rhomboid three-dimensional structure.*

Theorem 12 (Paired Conjugate Vector Detection ≡ Paired Quantum Entangled State Detection). *Eight vectors of a conjugate quark cluster are located in the four attribute regions; in the same region, paired conjugated states provide measurable characteristic information carried in the representation of paired entangled states in the description of quantum information interaction.*

Theorem 13 (Single Loop Detection). *Select a single diamond-shaped loop on the complex plane and use the spider web transformation of the square root mode to determine the density matrix of each coordinate point on the complex plane of the accompanying projected density values, forming a circular feature visualization distribution.*

Theorem 14 (Maximum Value of Complete Distribution Located on Diagonal of Four Attribute Regions). *On the complex plane satisfying the complete distribution condition, the accompanying density matrix has maximum values near the four diagonals passing through the zero point.*

Theorem 15 (Maximum Value of Product Distribution Located on Four Coordinates). *The distribution is formed by the product of conjugate vectors in the complex plane. The matrix has a maximum value of the accompanying density on the four coordinate lines of the complex plane.*

Theorem 16 (Strong Interaction Detection). *The density matrix formed on the coordinates of the complex primitives in the complex plane for the conjugate group qualification, through* $n = 1$ *cobweb transformation, shows from diamond to rectangular distribution features, fine distribution details are determined by the real and imaginary parts of the input vector quantization scale and the number of iterations of the interaction.*

5. Corollaries 1–16

Corollary 1 (Conjugated Hermite Matrix Eigenvalues). *For a complex number* $x = (p, q)$ *Hermite* 2×2 *matrix representation,* $K(x)$ *contains the four matrices corresponding to the complex cluster. These matrices provide real eigenvalue solutions. The* $K(\tilde{x})$ *conjugate complex cluster of four matrices provides the imaginary eigenvalue solutions.*

Corollary 2 (Four Groups of Imaginary Matrix Eigenvalues). *The eigenvalues of the four* $K(\tilde{E})$ *matrices* $K(\tilde{E}) \equiv \left\{\left\{\begin{pmatrix} i & 0 \\ 0 & i \end{pmatrix}, \begin{pmatrix} -i & 0 \\ 0 & -i \end{pmatrix}\right\}, \left\{\begin{pmatrix} 0 & -1 \\ 1 & 0 \end{pmatrix}, \begin{pmatrix} 0 & 1 \\ -1 & 0 \end{pmatrix}\right\}\right\}$ *The main diagonal is an imaginary number, and the eigenvalues of the four matrices,* $\{\{\lambda_1 = +i, \lambda_2 = +i\}, \{\lambda_1 = -i, \lambda_2 = -i\}, \{\lambda_1 = +i, \lambda_2 = -i\}, \{\lambda_1 = -i, \lambda_2 = +i\}\}$ *All are unit imaginary numbers.*

Corollary 3 (Dirac Matrix Product Produces 64 Matrices). *Using matrix product expansion, the* 4×4 *matrix group contains corresponding* $4 \times 8 = 32$ *real matrices and the same number of imaginary matrices, including a total of sixty-four matrices.*

Corollary 4 (Quantum Mechanics and Quantum Information/Conjugated Structure Distinguishable Eigenvectors). *Compared with a set of quaternary unit eigenvectors in classical quantum mechanics and quantum information, the conjugate complex vector function provides sixteen sets of quaternary distinguishable output eigenvectors, and the dimension ratio of the extended structure is* $1/16$.

Corollary 5 (Complete Solution Space of Pauli Matrices). *The complete solution space of the Pauli matrices needs to expand the auxiliary diagonal elements to include the adjoint solution space* $\{+i, -i\}$. *The dimension of the complete phase space is* $2^2 \times 2^2 = 2^4 = 4^2 = 16$ *to completely cover the* $\{K(E), K(\tilde{E})\}$ *cluster. Each set of matrices is guaranteed to provide a pair of eigenvalue solutions.*

Corollary 6 (Complete Solution Space of Dirac Matrices). *The complete solution space of the Dirac matrices needs to expand the auxiliary diagonal elements to include the quaternion solution structure represented by the imaginary number of the adjoint solution space* $\{+i, -i\}$, *and the dimension of the phase space is* $2^4 \times 2^4 = 2^8 = 256$.

Corollary 7 (Product of Pairs of Conjugate Vectors in Other Quadrants). *The product of conjugate vectors* $\{\sqrt[n]{z}, \sqrt[n]{\bar{z}}\}$ *vectors is* $\sqrt[n]{z}\sqrt[n]{\bar{z}} = \pm ir^{2/n}$.

Corollary 8. *Start from any* $(\sqrt[n]{m}, 0), m \geq n$ *position to* $(0, \sqrt[n]{m})$ *end, there are* $m+1$ *vectors* $\{(\sqrt[n]{j}, \sqrt[n]{k})\}_{j+k=m}, j, k \in [0, m]$ *associated with the ring chain.*

Corollary 9 (Classical Logic, Quantum Logic, and Conjugate Logic). *Classical logic is based on a single primitive* $\{0,1\}^n$ *linear structure, quantum logic is based on two primitive vectors based on* $\{|0\rangle, |1\rangle\}^n$ *planar structure, and conjugate logic is based on four primitive vectors* $\{(1,0),(0,1),(-1,0),(0,-1)\}^n$ *complex plane structure.*

Classical Phase Space :	$\{0,1\}$/bit	→	n bits	→	2^n States Linear Combination Solutions
Quantum Phase Space :	$\{|0\rangle, |1\rangle\}$/qubit	→	n qubits	→	2^n States $\{1,-1\}$ Real Number Eigensolutions
Conjugate Phase Space :	$\{0,1,\tilde{0},\tilde{1}\}$/cobit	→	n cobits	→	2^{2n} States $\{1,i,-1,-i\}$ Complex Eigensolutions

Corollary 10 (Number of Classical, Quantum, and Conjugate Phase Space Eigensolutions). *From the perspective of primitive distribution, classical logic is a linear structure, quantum logic is generalized to be two vectors, and conjugate logic is generalized to be four meta-vectors on balanced distribution structure, the numbers of phase space states involved in the n meta-variable are* $\{2^n, 2^n, 2^{2n}\}$.

Quantum Transformation Group :	$xx^* = 1, U_T^{-1} = U_T^*$ $tr(U_T U_T^*) = tr(U_T^* U_T) = I$	→	$SU(n)$ Transformation Group Lie Algebra Real Eigensolutions
Conjugate Transformation Group :	$y\tilde{y} = i, Y^{-1} = \overleftarrow{\tilde{Y}}$ $tr(Y\tilde{Y}) = tr(\tilde{Y}Y) = iI$	→	$CTS(n)$ Transformation Group Conjugate Algebra Complex Eigensolutions

Corollary 11 (Quantum and Conjugate Transformation Groups). $SU(n)$ *is the core of the classical quantum transformation group, and the inner product sum of pairs of complex conjugate vectors is the unit vector on the global transformation structure supported by Lie algebras.* $CTS(n)$ *is the core of the conjugate transformation group, and the product of pairwise conjugate vectors is an imaginary unit vector.*

$x(1:1) \rightarrow$ [$\{*\}(2)$ Vector Group, Ring 2:4] $\rightarrow$ [$\{-1\}(2)$ Vector Field 4:16]
$\rightarrow$ [$\{-\}(2)$ Vector Reflection 8:64] $\rightarrow$ [$\{\leftarrow\}(4)$ Rotation 32:1024]
$\rightarrow$ [$\{[,]_d, \{,\}_d\}(2)$ Symmetry/Antisymmetry 64:4096]
$\rightarrow \{X(64), H(4096)\}$

$y(1:1) \rightarrow$ [$\{\sim, \sharp, \prime, *\}(8)$ Conjugate Vector Group, Ring 8:64]
$\rightarrow$ [$\{-1\}(2)$ Vector Field 16:256] $\rightarrow$ [$\{-\}(2)$ Vector Reflection 32:512]
$\rightarrow$ [$\{\leftarrow\}(4)$ Rotation $128:2^{14}$]
$\rightarrow$ [$\{[,]^u_d, \{,\}^u_d\}(4)$ Symmetry/Antisymmetry $512:2^{18}$] $\rightarrow \{Y(512), C(2^{18})\}$

Corollary 12 (Lie and Conjugate Algebras). *From basic variables to symmetric operators, both Lie algebra and conjugate algebra need to go through six levels of extended structure, where on* $\{1,3,4,5\}$ *layers, the content of the level expansion is the same, and the essential difference lies in the second and the sixth levels. In the second layer, the classical quantum involves pairs of complex conjugate vectors* $\{x, x^*\}$ *and pairs of complex conjugate angles* $\{\theta, -\theta\}$*, while conjugate structures involve four conjugate extension operators* $\{\sim, \sharp, \prime, *\}$*, paired conjugate angles* $\{\theta, \tilde{\theta}\}$*, and eight conjugate vectors with an expansion factor of 1:4. In the sixth layer, the symmetric and antisymmetric operators are expanded from the main diagonal to the main (real number solution)/auxiliary (imaginary number solution) double diagonals, indicating that the expansion factor is 1:2. The composite structure expansion factors are 1/8 and 1/64. A unit cobit different from a qubit requires 64 bits of additional information to support.*

Corollary 13 (Classical Quantum Structure Is Simplified Representation of Conjugate Structure). *From Lie algebra to conjugate algebra representation, it can be observed that the classical quantum structure is a conjugate algebra that simplifies the eight-variable vector cluster into two argument vectors on complex conjugate pair representation. The complete quantum solution space structure needs to be equipped with principal and auxiliary double diagonal complex eigenvalues, and the n cobit phase space representation is extended from* 2^n *to* 2^{2n}*.*

Type A Measurement Description
Input: Photon flow
Output: Pairwise conjugate features
Processing: Two-stage half-lens separation of spin and complex pairs

Input: $\rightarrow$ [I. BS1: $\{x, \tilde{x}\}$ $\rightarrow$ II. BS2: $\{x = (p,q), \tilde{x} = (q,p)\}$] $\rightarrow$ Output

Type B Measurement Description
Input: Photon flow
Output: Single/multiple concentric rings
Processing: Multilevel interference, polarization, filtering, projection

Input: $\rightarrow$ [I. DM Interference $\rightarrow$ II. Filters $\rightarrow$ III. PBS Splitter] $\rightarrow$ Output

Type C	Measurement Description
Input:	Photon flow
Output:	Diamond/rectangle and other point-like feature clusters, point group feature distribution
Processing:	Multilevel interference, polarization, filtering, strong interaction, diffractive projection

Input: → [I. Interference] → [II. Interaction Filters] → [III. Grating] → Output

Corollary 14. *The Type A measurement pipeline is for detecting and separating pairs of conjugate eigenvectors* $\{z, \tilde{z}\}$ *satisfying Theorem 9.*

Corollary 15 (Multiple Loop Detection). *Select multiple diamond-shaped loops on the complex plane. Through the square root mode spider web transformation, the density matrix of each coordinate point on the complex plane determines the accompanying projected density values to form multiple circular features on visualization distributions.*

Corollary 16 (Distribution of Maximum Value in Eight Directions). *Both complete distribution and product distribution are superimposed to form a distribution of maximum value in eight directions, located at* $\{0, \pi/4, \pi/2, 3\pi/4, \pi, 5\pi/4, 3\pi/2, 7\pi/4\}$, *decomposed into two sets of conjugate cross structures.*

Appendix 7

Lemmas, Assumptions, Predictions, Theorems, and Corollaries in Chapter 7

Lemmas 1–10, Theorems 1–12, Corollaries 1–11, Predictions 1–3:

6. Lemmas 1–10

Lemma 1 (Simple Paradoxes). *Paradox representations such as a white horse is not a horse, substantial implication, and groups can use the local variables involved to form regional judgment conditions. However, to systematically describe Russell's paradox, it is necessary to establish the overall decision rules related to clusters.*

Lemma 2 (Visual Representation of Godel's Paradox). *Based on the optical reflection effects of complex lenses, it is possible to show the limits of what is possible with objects through recursive transformations. This provides a visual paradox in the reflected world to contain an* $A = \neg A$ *property with self-contradiction.*

Lemma 3 (Mobius Ring and Klein Bottle). *The Mobius ring forms a 3D single leaf surface with twisted 2D strips; the Klein bottle uses the twisted 3D embedding hypersurface connections to show four-dimensional Mobius structures. On both constructions, there are intrinsic logic paradoxes on* $A = \neg A$ *naturally.*

Lemma 4 (Riemann Surface). *The Riemann surface based on the multivariate complex function formula from the overall distribution projection connection can observe the inherent Mobius ring paradox characteristics for the* $A = \neg A$ *property in the self-contradiction.*

Lemma 5 (Escher's Visual Paradox). *Based on the complex diagrams displayed by Escher's visual art, from the overall distribution projection connection and other levels, the inherent characteristics of the Mobius ring paradox can be observed for the* $A = \neg A$ *property in the self-contradiction.*

Lemma 6 (Classical Tai Chi and Visual Paradox). *Based on the complex diagrams displayed by the classic Tai Chi diagram, the inherent characteristics of the Mobius ring paradox can be observed from the overall distribution projection connection and other levels. Escher's visual paradox is equivalent to the abstract dynamic change characteristic of traditional oriental Tai Chi visual expressions. Since chaotic states are located in the central position of the Tai Chi diagrams, they all contain the logic paradoxes* $A = \neg A$ *in dynamic activities of multiple integrated levels in their phase spaces.*

Lemma 7 (Modern Progress in Entangled Measurements of Quantum Optics). *The frontier quantum theory system and its corresponding various representations of the quantum entangled state are reviewed, and their typical complex geometric invariants satisfying the characteristics of global transformation are described.*

Lemma 8 (Non-orthogonal Measurements in Quantum Entanglements).
Wootters has listed various measurement schemes to indicate entanglement vectors without orthogonal properties in quantitative measuring conditions.

Lemma 9 (Real Eigenvalues of Quantum Optics and Quantum Information). *Quantum optics and quantum information under Dirac brackets provide a necessary and sufficient condition to support {Real Eigenvalue} solutions of* $\{+1, -1\} \leftrightarrow \{+\theta, -\theta\}$ *and* $\{x, x^*\}$ *a pair of measures, complex vectors, and complex conjugate operator* $*$.

Lemma 10 (Recursive Structures of Vortex Math). *Vortex mathematics expresses a series of repetitive loops as generated sequences to be four classes: (1) single convergent points; (2) fixed length anticlockwise loops; (3) fixed length clockwise loops; and (4) injective convergent points.*

7. Theorems 1–12

Theorem 1 (Group Paradox). *Group paradox provides the smallest and compact case of a state space as a pair of two states to satisfy the logic constraint* $A = \neg A$.

Theorem 2 (Conjugate Ring Resolve Logic Paradoxes on Mobius Ring Types). *Since the double-layer torus can form a set of close Mobius strip structures, after one rotation of the ring, the left torus staggers into the right torus and ensures that the values taken are different to satisfy* $A \neq \neg A$. *Under the action of* $\neg$ *logic operators, the system resolves the inherent paradoxes of the Mobius ring types using the pairing mode of the left and right exchanges.*

Theorem 3 (Fundamental Theorem of Algebra Support Real Coefficients Only). *The fundamental theorem of algebra is a necessary and sufficient condition to correspond the solution space on the condition of {real coefficients}* $\leftrightarrow \{+\theta, -\theta\}$ *and* $\{x, x^*\}$ *a pair of measures, complex vectors, and complex conjugate operator* $*$.

Theorem 4 (Conjugate Complex Vector Systems). *Four conjugate operators* $\{\sim, \sharp, \prime, *\}$ *are defined associated with a pair of conjugate angles* $\{\theta, \tilde{\theta}\}$ *and the four attribute regions in a complex plane on the synchronous parallel processing and recursive organization conditions. Eight feature vectors are composed of two crosses, and the whole construction provides six dynamic transformation modes associated with the basic recursive operations.*

Theorem 5 (Output from Conjugate Transformation Structure). *The conjugate transformation structure processes the initial 0–1 vector of N elements to be 2n feature classes and transforms feature vectors into* 2^{2n} *configurations. To output a pair of conjugate classes, eight elements are conjugate complex number vectors.*

Theorem 6 (Two Projects of Clock Reference Signals). *Conjugate measurements of clock reference signals provide the conjugate complementary measurement space, the narrow projection on the conjugate configuration, and the general projection on the complex number plane.*

Theorem 7 (Conjugate Group Combinatorial Lattice and Abel Groups). *Three extension steps are required from a single vector to generate a conjugate group combinatorial lattice. Step 1. The conjugate Klein group expands one vector to eight vectors of two conjugate crosses in a conjugate cluster; Step 2. Four vectors of each cross take sixteen combinations to generate nine nontrivial vectors; Step 3. Two sets of eighteen nontrivel vectors create an integrated Abel group under complex addition to generate a conjugate group combinatorial lattice.*

Theorem 8 (Numbers of Logic, Quantum, and Conjugate Eigensolutions). *From the primitive distributions, classical logic is a linear structure, quantum logic is generalized to be two vectors, and conjugate logic is generalized to be four meta-vectors on balanced distributions. The numbers of state spaces involved in the n meta-variables are* $\{2^n, 2^n, 2^{2n}\}$.

Theorem 9 (Paired Conjugate Vectors ≡ Paired Quantum Entangled States). *Eight vectors of a conjugate cluster are located in the four attribute regions; in the same region, paired conjugated states provide measurable characteristic information carried in the representation of paired entangled states in the description of quantum information interaction.*

Theorem 10 (Single Loop Detection). *Select a single diamond-shaped loop on the complex plane and use the spider web transformation of the square root mode to determine the density matrix of each coordinate point on the complex plane of the accompanying projected density values, forming a circular feature visualization distribution.*

Theorem 11 (Strong Interaction Detection). *Through* $n = 1$ *cobweb transformation, the density matrix in the complex plane shows from diamond to rectangular feature distributions, fine eigenvalues are determined by the quantization scales of real and imaginary parts of the input vector, and the number of iterations of the interactions.*

Theorem 12 (Variant Construction). *Variant construction is composed of variant logic, variant measurement, variant map, and hierarchical knowledge model, including description + process knowledge model, local + global analysis structure, and classification + retrieval mechanism. According to the relevant local characteristics of the system, variant construction forms a foundation to provide integrated holistic structural support on the multiple levels of a hierarchy on complex system analysis, design, implementation, and other engineering activities.*

8. Corollaries 1–11

Corollary 1 (Gongsun Long Paradox). *The paradox "A white horse is not a horse" can be formally expressed as a logical statement.*

Corollary 2 (Paradox in Gauss's Theorem). *Gauss's fundamental theorem of algebra is incomplete in covering the solution spaces under the coefficients on imaginary numbers to naturally contain the Mobius ring and intrinsic logic paradoxes* $A = \neg A$ *in the phase space.*

Corollary 3 (Paradox Free on Complete Solution Spaces). *Based on the four distribution regions (four colors of the conjugate ring) and a pair of conjugate vectors (opposite rotations), and other comprehensive approaches, this extends the fundamental theorem of algebra to cover {complex coefficients} in a complete solution space. The conjugate complex vector systems guarantee without logic paradoxes on the vector phase space in synchronous dynamic environments.*

Corollary 4 (Restriction of Poisson, Lagrange, Lie, Dirac Brackets, and Lie Algebra). *Poisson, Lagrange, Lie, and Dirac Brackets support multiple real eigenvalue solutions* $\{-1, +1\}$ *of a dynamic system to identify symmetric/antisymmetric properties of the principal diagonal eigenvalues on the Hermite matrices, and those ignore multiple imaginary solutions of the auxiliary diagonal eigenvalues completely.*

Corollary 5 (Conjugate Brackets and Conjugate Algebra). *Using conjugate algebra, conjugate brackets provide the complex eigenvalue solutions* $\{+1, +i, -1, -i\}$, $\{+1, -i, -1, +i\}$ *beyond Poisson, Lagrange, Lie, and Dirac brackets to identify vector operators* $\{-1, \leftarrow\}$ *of symmetric/antisymmetric properties under {Group, Ring, Field} domains. The time order satisfies vector and tensor operational environments under a single circle of the eigenvalue matrix condition.*

Corollary 6 (Narrow and General Projections on Complex Plane). *A narrow projection provides* 2^{2m} *positions or* 4×2^{2m} *positions in a configuration space of one or four 2D uniform distributions without a metric order, and a general projection provides* $(m+1)^2$ *positions or* $4(m+1)^2$ *positions in 2D double-binormal distributions corresponding to a complex number plane on one quadrant or four quadrants.*

Corollary 7 (Lattice and Strong Interaction). *Normally entangled interactions appear in the eight vectors of a conjugate cluster for the conjugate Klein group in Step 1, and extended entanglements are restricted in the same eigenvalue on a cycle with a constant radius* r. *However, the extensions in Steps 2 and 3 generate multiple cycles under combinatorial and Abel group operators to indicate that there are strong interactions involved beyond a single cycle condition.*

Corollary 8 (Reflective Mirrors ≡ Basic Complex Operators).
In modern quantum optics theory and experimental devices, various reflection mirrors play a key role in implementing the basic complex operators in quantum optical and photonic environments.

Corollary 9 (Paradox in Real Eigenvalue Solutions).
The general solution of the Dirac bracket and Lie algebra schemes is a reverse version of the Gaussian fundamental theorem of algebra. It is incomplete to cover the complex solution spaces under the parameters of any imaginary numbers containing the Mobius ring and intrinsic logic paradoxes $A = \neg A$ *in the phase space.*

Corollary 10 (Classical Quantum Formula ≡ Simplified Conjugate Structure). *From Lie algebra to conjugate algebra, the classical quantum formula simplifies the eight-variable vectors into two vectors on a pair of complex conjugate representations. The complete quantum solution needs to be equipped with principal and auxiliary double diagonal complex eigenvalues, and the* n *cobit phase space representation expands from* 2^n *to* 2^{2n}.

Corollary 11 (Maximum Distributions in Eight Directions). *Both complete distribution and product distribution are superimposed to be a distribution with maximum values in eight directions* $\{0, \pi/4, \pi/2, 3\pi/4, \pi, 5\pi/4, 3\pi/2, 7\pi/4\}$ *that can be decomposed into two conjugate crosses.*

9. Predictions 1–3

Prediction 1 (Conjugate Complex Vector Solutions Useful in Vortex Mechanisms). *Conjugate complex vector functions provide powerful synchronous functions to recursively generate dynamic effects of increasing and decreasing radii in the complex plane to provide a foundation to support future exploration and applications on complicated vortex mechanisms.*

Prediction 2 (Coming General Computers). *Modern computers are restricted in Type B* $n = 0, State = 2$ *mode on* $\{0, 1\}$ *Turing machine in 1D real manifolds. The coming general computers are based on Type* $\{A, B\}$ $n = 1, State = 4$ *mode on full* $\{(1, 0), (0, 1), (-1, 0), (0, -1)\}$ *eigenvalue solutions on conjugate machines to support 1D complex manifolds for the complete quantum computations.*

Prediction 3 (Extensive Applications on Supercomputation and Visualizations). *Resolved Mobius paradoxes, it will be useful to implement higher performance supercomputing and visualization targets on conjugate complex vector systems in hypersynchronous network environments for knowledge and intelligent content management for subsequent generations.*

Index

D

F

G

H

N

O

P

Q

R

S